P9-CEG-152

High-Performance Communication Networks

The Morgan Kaufmann Series in Networking
Series Editor, David Clark

High-Performance Communication Networks
Jean Walrand and Pravin Varaiya

Computer Networks: A Systems Approach
Larry Peterson and Bruce Davie

Forthcoming:

Frame Relay Applications: Business and Technical Aspects
James P. Cavanagh

Cable Television Technology
Walter Ciciora and David Large

Multicasting
Stephen Deering, Deborah Estrin, and Lixia Zhang

Optical Networks
Rajiv Ramaswami and Kumar N. Sivarajan

High-Performance Communication Networks

Jean Walrand
Pravin Varaiya

University of California,
Berkeley

Morgan Kaufmann Publishers, Inc.
San Francisco, California

Sponsoring Editor Jennifer Mann
Production Manager Yonie Overton
Production Editor Cheri Palmer
Editorial Assistant Jane Elliott
Cover Design Ross Carron Design
Text Design, Composition, and Illustrations Windfall Software
(Paul C. Anagnostopoulos, Joe Snowden)
Copyeditor John Hammett
Proofreader Jennifer McClain
Indexer Ty Koontz
Printer Courier Corporation

Morgan Kaufmann Publishers, Inc.
Editorial and Sales Office
340 Pine Street, Sixth Floor
San Francisco, CA 94104-3205
USA
Telephone 415 / 392-2665
Facsimile 415 / 982-2665
E-mail mkp@mkp.com
Web site http://www.mkp.com

Printed in the United States of America

00 99 98 97 96 5 4 3 2 1

Library of Congress Cataloging-in-Publication Data

Walrand, Jean.
High-performance communication networks / Jean Walrand, Pravin Varaiya.
p. cm.
Includes bibliographical references and index.
ISBN 1-55860-341-7 (hardcover)
1. Computer networks. 2. Multimedia systems. 3. High performance computing. 4. Integrated services digital networks. I. Varaiya, P. P. (Pravin Pratap) II. Title.
TK5105.5.W353 1996
384–dc20 96-21937
CIP

We dedicate this book

to *Annie* and *Isabelle* and *Julie,*

and to *Ruth*

Contents

Preface xv

1 Overview 1

1.1 History of Communication Networks 4
1.1.1 Telephone Networks 5
1.1.2 Computer Networks 10
1.1.3 Cable Television Networks 16

1.2 Networking Principles 20
1.2.1 Digitization 20
1.2.2 Economies of Scale 23
1.2.3 Network Externalities 24
1.2.4 Service Integration 25

1.3 Future Networks 26
1.3.1 The Internet 26
1.3.2 Pure ATM Network 28
1.3.3 Video Dial Tone 29
1.3.4 And the Winner Is . . . 29

1.4 Summary 32

1.5 Notes 32

1.6 Problems 33

2 Network Services and Layered Architectures 37

2.1 **Applications** 39
- 2.1.1 Constant Bit Rate 41
- 2.1.2 Variable Bit Rate 41
- 2.1.3 Messages 41
- 2.1.4 Examples 42
- 2.1.5 Other Requirements 43

2.2 **Network Services** 45
- 2.2.1 Connection-Oriented Service 45
- 2.2.2 Connectionless Service 46

2.3 **High-Performance Networks** 47

2.4 **Network Elements** 48
- 2.4.1 Principal Network Elements 48
- 2.4.2 Network Elements and Service Characteristics 50
- 2.4.3 Examples 51

2.5 **Basic Network Mechanisms** 54
- 2.5.1 Multiplexing 56
- 2.5.2 Switching 60
- 2.5.3 Error Control 65
- 2.5.4 Flow Control 74
- 2.5.5 Resource Allocation 75

2.6 **Layered Architecture** 77
- 2.6.1 Layers 77
- 2.6.2 Implementation of Layers 78

2.7 **Open Data Network Model** 82

2.8 **Summary** 86

2.9 **Notes** 86

2.10 **Problems** 86

3 Packet-Switched Networks 91

3.1 **OSI Reference Model** 92
- 3.1.1 Layer 1: Physical Layer 92
- 3.1.2 Layer 2: Data Link Layer 92
- 3.1.3 Sublayer 2a: Media Access Control 94
- 3.1.4 Sublayer 2b: Logical Link Control 94
- 3.1.5 Layer 3: Network Layer 96
- 3.1.6 Layer 4: Transport Layer 98
- 3.1.7 Layer 5: Session Layer 99

3.1.8 Layer 6: Presentation Layer 99
3.1.9 Layer 7: Application Layer 100
3.1.10 Summary 100

3.2 **Ethernet (IEEE 802.3)** 101
3.2.1 Physical Layer 102
3.2.2 MAC 104
3.2.3 LLC 107

3.3 **Token Ring (IEEE 802.5)** 109
3.3.1 Physical Layer 110
3.3.2 MAC 110
3.3.3 LLC 112

3.4 **FDDI** 112

3.5 **DQDB** 117

3.6 **Frame Relay** 120

3.7 **SMDS** 123
3.7.1 Internetworking with SMDS 126

3.8 **Summary of Packet-Switched Networks** 129

3.9 **Internet** 130
3.9.1 IPv4, Multicast IP, Mobile IP, IPv6, RSVP 131
3.9.2 TCP and UDP 143
3.9.3 SMTP, rlogin, and TFTP 144
3.9.4 Real-Time Protocols 145
3.9.5 Faster Transport Protocols 146
3.9.6 Internet Success and Limitation 146

3.10 **Summary** 149

3.11 **Notes** 150

3.12 **Problems** 150

4 Circuit-Switched Networks 155

4.1 **Performance of Circuit-Switched Networks** 157

4.2 **SONET** 161
4.2.1 SONET Frame Structure 165

4.3 **Fiber to the Home** 172
4.3.1 The AT&T Subscriber Loop System 172
4.3.2 The British Telecom TPON System 174
4.3.3 Passive Photonic Loop 176
4.3.4 Hybrid Scheme 177

4.4 **ISDN** 178

4.5 **Intelligent Networks** 181

4.5.1 Service Examples 181
4.5.2 Intelligent Network Architecture 183
4.5.3 Functional Components 185

4.6 **Video Dial Tone** 186

4.6.1 Layout 187
4.6.2 Control and Video Networks 188
4.6.3 MPEG 191

4.7 **Summary** 193

4.8 **Notes** 193

4.9 **Problems** 194

5 Asynchronous Transfer Mode 197

5.1 **Main Features of ATM** 198

5.1.1 Connection-Oriented Service 199
5.1.2 Fixed Cell Size 203
5.1.3 Statistical Multiplexing 208
5.1.4 Allocating Resources 209

5.2 **ATM Header Structure** 210

5.2.1 VCI and VPI 211
5.2.2 Other Fields 213
5.2.3 Reserved VCI/VPI 214

5.3 **ATM Adaptation Layer** 214

5.3.1 Type 1 215
5.3.2 Type 2 216
5.3.3 Type 3 216
5.3.4 Type 4 217
5.3.5 Type 5 217

5.4 **Management and Control** 217

5.4.1 Fault Management 219
5.4.2 Traffic and Congestion Control 223
5.4.3 Network Status Monitoring and Configuration 223
5.4.4 User/Network Signaling 224

5.5 **BISDN** 225

5.6 **Internetworking with ATM** 226

5.6.1 Multiprotocol Encapsulation over AAL5 227
5.6.2 LAN Emulation with ATM 227
5.6.3 IP over ATM 228

5.6.4 Multiprotocol over ATM (MPOA) 231
5.6.5 FR and SMDS over ATM 232

5.7 Summary 232

5.8 Notes 233

5.9 Problems 234

6 Control of Networks 237

6.1 Objectives and Methods of Control 239
6.1.1 Overview 239
6.1.2 Control Methods 239
6.1.3 Time Scales 241
6.1.4 Examples 242
6.1.5 Quality of Service 244

6.2 Circuit-Switched Networks 246
6.2.1 Blocking 246
6.2.2 Routing Optimization 248

6.3 Datagram Networks 252
6.3.1 Queuing Model 253
6.3.2 Key Queuing Result 254
6.3.3 Routing Optimization 255
6.3.4 Flow Control 261

6.4 ATM Networks 264
6.4.1 Control Problems 264
6.4.2 Deterministic Approaches 266
6.4.3 Statistical Procedures 276
6.4.4 Deterministic or Statistical? 293

6.5 Summary 296

6.6 Notes 297

6.7 Problems 297

7 Control of Networks: Mathematical Background 301

7.1 Markov Chains 301
7.1.1 Overview 301
7.1.2 Discrete Time 302
7.1.3 Continuous Time 308

7.2 **Circuit-Switched Networks** 313
7.2.1 Single Switch 313
7.2.2 Network 316
7.3 **Datagram Networks** 320
7.3.1 M/M/1 Queue 320
7.3.2 Discrete-Time Queue 322
7.3.3 Jackson Network 326
7.3.4 Buffer Occupancy for an MMF Source 328
7.3.5 Insensitivity of Blocking Probability 331
7.4 **ATM Networks** 335
7.4.1 Deterministic Approaches 335
7.4.2 Large Deviations of iid Random Variables 339
7.4.3 Straight-Line Large Deviations 343
7.4.4 Large Deviation of a Queue 344
7.4.5 Large Deviation of a Multiclass Queue 349
7.4.6 Bahadur-Rao Theorem 355
7.5 **Summary** 356
7.6 **Notes** 357
7.7 **Problems** 357

8 Economics 361
8.1 **Network Charges: Theory and Practice** 363
8.1.1 Economic Principles 363
8.1.2 Charges in Practice 366
8.1.3 Vulnerability of the Internet 367
8.2 **A Billing System for Internet Connections** 368
8.2.1 Stage 1 369
8.2.2 Stage 2 372
8.2.3 Stage 3 373
8.2.4 Functions of Internet Charges 374
8.3 **Internet Traffic Measurements** 377
8.3.1 Connection Statistics 378
8.3.2 Diversity of Usage 378
8.3.3 Potential for Congestion Pricing and Traffic Shaping 379
8.4 **Pricing a Single Resource** 380
8.4.1 Usage-Based Prices 381
8.4.2 Congestion Prices 385
8.4.3 Cost Recovery and Optimum Link Capacity 387

8.5 **Pricing for ATM Services** 389
8.5.1 A Model of ATM Resources and Services 390
8.5.2 Revenue Maximization 395
8.6 **Summary** 397
8.7 **Notes** 397
8.8 **Problems** 398

9 Optical Links 401

9.1 **Optical Link** 402
9.2 **Fiber** 404
9.2.1 Attenuation 404
9.2.2 Dispersion 409
9.3 **Sources and Detectors** 415
9.3.1 Light-Emitting Diodes 416
9.3.2 Laser Diodes 419
9.3.3 Detectors 421
9.4 **Nondirect Modulation** 425
9.4.1 Coherent Detection 425
9.4.2 Subcarrier Multiplexing 426
9.4.3 Wave-Division Multiplexing 427
9.5 **Quantum Limits on Detection** 429
9.6 **Other Components** 432
9.7 **Summary** 433
9.8 **Notes** 434
9.9 **Problems** 434

10 Switching 437

10.1 **Switch Performance Measures** 438
10.2 **Time- and Space-Division Switching** 442
10.3 **Modular Switch Designs** 444
10.4 **Fast Packet Switching** 450
10.5 **Distributed Buffer** 453
10.5.1 Impact of Hot Spots 457
10.5.2 Input Buffers 458
10.5.3 Combating Hot Spots 462

10.6 **Shared Buffer** 466
10.6.1 Queuing Analysis 467
10.7 **Output Buffer** 468
10.7.1 Knockout 469
10.8 **Input Buffer** 470
10.8.1 HOL Blocking 471
10.8.2 Overcoming HOL Blocking 473
10.9 **Summary** 475
10.10 **Notes** 476
10.11 **Problems** 476

11 Towards a Global Multimedia Network 479

11.1 **Attributes of the Global Network** 480
11.2 **Technology Areas** 482
11.2.1 Architecture 483
11.2.2 Networking 484
11.2.3 Signal Processing 485
11.2.4 Applications 485
11.3 **Challenges** 486
11.3.1 Architecture 486
11.3.2 Quality of Service 487
11.3.3 Mobility 491
11.3.4 Heterogeneity 494
11.3.5 Scalability and Configurability 496
11.3.6 Extensibility and Complexity Management 498
11.3.7 Security 499

Bibliography 501

Index 507

Related Titles from Morgan Kaufmann 537

Preface

A high-performance network is built with high-speed communication links and switches and designed to support a large variety of user applications. The network is scalable, accommodating a growing number of users without degrading performance. As more sophisticated applications are supported, the demand for network services deepens; with increases in network scale, this demand broadens.

By accommodating growth in depth and breadth, high-performance networks are revolutionizing our lives—how we interact, study, acquire information, entertain, and conduct business and politics—in ways that we cannot yet fully perceive. But the core principles and technological basis underlying this revolution are visible, and they can be described in the language of the communications engineer, the computer scientist, and the economist. This book attempts to give that description.

We have conducted research in networking for more than fifteen years. For the past six years we have taught an introductory graduate course on networking at the University of California-Berkeley and offered a series of short courses to practitioners in the United States, Europe, and India: engineers and computer scientists who specialize in networking, technical managers, and persons responsible for acquiring and managing networking resources for their organizations.

During those six years our understanding has changed, and that change is reflected in our teaching and in this book. We appreciate better the student's needs, the perspectives of the network designer and the user, the concerns of those who must invest in new network technologies while making sure that the new investment can be integrated with legacy systems, and the economist's interest in questions of cost recovery and pricing.

Intended Audience and Approach

If you are a student, this book will provide you with a firm grounding in networking and a background for the critical appreciation of current research. If you are a practicing network engineer or computer scientist who has little time to keep up with the scholarly literature, you can selectively read this book to address the questions outside your expertise. If you are an engineer or scientist with an interest in networking, bewildered by the thicket of acronyms—ATM, ISDN, SONET—this book can serve as an accessible reference. If you are interested in questions of networking economics or policy, this book will help you understand how these questions relate to networking technology.

We have tried to address the different concerns of our intended audience. The student's needs are met by a systematic development of the subject matter, emphasis on fundamentals, problem sets that test understanding, and pointers to the literature. The practitioner will find a reasonably self-contained treatment of each major network development, including the technology, the potential application, and the place in the market. If you manage a networking group or if you are responsible for networking in your organization, you will find an analysis of trends and a description of future evolutions that can help clarify the implications of some of the decisions you must make.

In general, we have attempted to present these several perspectives from the basis of first principles, rather than engaging in discussions that demand advanced concepts from communications engineering, computer science, operations research, or economics. It is our hope that you can then gain a greater appreciation of these multiple views and a deeper understanding of how networks are built, how they are used, and who will pay for them. In particular, we have minimized the use of advanced mathematics by presenting in an intuitive manner the results of mathematical argument separately from the argument itself, which may be skipped without loss of continuity.

How to Use This Book

This book can be used for study at different levels; the table below lists three possibilities. We ourselves have used this material in two ways. At Berkeley, we have taught a one-semester, 45-hour introductory graduate course to students from electrical engineering, computer science, and operations research. Students need no prior exposure to communication networks—the emphasis is on descriptive breadth that conveys the excitement of the technological advances and the challenges posed by speed, distance, and demanding applications. Three or four times each year we have taught a short course to practitioners, between 8 and 20 hours long. The aim there is to provide an overview of recent developments, to decipher trends, and to speculate about opportunities.

Audience	Focus	Chapters
Practicing engineers	Current networks and trends	1, 2, 3, 4, 11
	ATM	5, 6
	Optical links and ATM switching	9, 10
Undergraduate students	Introduction to networks	1, 2, 3, 4
Graduate students	Current networks and trends	1, 2, 3, 4, 11
	ATM	5, 6
	Control of networks	7
	Optical links and ATM switching	9, 10
	Pricing	8

Throughout this book we treat each major topic in a self-contained manner so it can be tailored to meet the needs of a course or study of a particular development.

In Chapter 1 we give a quick historical account of networking, discuss the principles that drive the demand for network services, and take a glimpse at how networks might evolve in the future.

In Chapter 2 we explain network layered architectures and introduce the basic network functions used to implement network services. Chapters 3 and 4, respectively, describe the innovations in packet-switched and circuit-switched networks. In Chapter 5 we describe Asynchronous Transfer Mode (ATM).

Chapters 6 and 7 are devoted to the control of networks: the first chapter is descriptive, the second is mathematical. In Chapter 8 we take up economic questions: how should network services be priced, and how can we implement such pricing schemes?

Chapters 9 and 10, respectively, deal with the technologies of optical links and fast packet switching that make high-performance networks affordable. Lastly, in Chapter 11 we discuss the networking issues that will be raised by the challenging applications of the future.

Support Materials

In both types of courses we have used a multimedia presentation consisting of lecture slides and animations, now available from the publisher. See the Web page for our book at http://www.mkp.com for details.

Each chapter of the book ends in problems that test understanding of the material and challenge the reader to use that understanding in situations that may arise in practice. A solutions manual is also available from the publisher.

Acknowledgments

This book is an attempt to synthesize the different viewpoints of networking offered by specialists who know more about each view than we do. Inevitably, errors of fact and judgment and balance of treatment have crept into the book. We would be very grateful to our readers for bringing those errors to our attention and for providing us with feedback about their experiences in learning or teaching from this book. We can be reached via e-mail at {wlr, varaiya}@eecs.berkeley.edu. We will post corrections and comments at the Web site http://www.mkp.com.

We have benefited enormously from the comments of our students, both at Berkeley and in our short courses. We wish particularly to acknowledge the help of Richard Edell, Farokh Eskafi, and Karl Petty.

Various versions of this book and particular chapters were reviewed by Costas Courcoubetis, University of Crete, Greece; Gunner Danneels, Intel Corporation; Gary Delp, IBM Corporation; Gustavo de Veciana and Takis Konstantopoulos, University of Texas-Austin; Anthony Ephremides, University of Maryland; William Halverson, Pacific Bell; Ivy Hsu, Telesis Technologies Laboratory, Pacific Telesis; Kamal Jabbour, Syracuse University; Chuck Kalmanek and S. Keshav, AT&T Research; Jeffrey Kipnis, Ameritech Advanced Data Services; Jean-Yves Le Boudec and Stephan Robert, EPFL, Switzerland; Nick McKeown, Stanford University; Mart L. Molle, University of Toronto; Ramesh Rao, University of California-San Diego; Galen Sasaki, University of Hawaii–Manoa; Anujan Varma, University of California-Santa Cruz; Sergio Verdu,

Princeton University; and Alfred Weaver, University of Virginia. A version of the lecture slides and animations were reviewed by Gustavo de Veciana and Robert Thomas, Cornell University; S. Keshav, AT&T; Ramesh Rao, University of California-San Diego; and Anujan Varma, University of California-Santa Cruz.

We are immensely grateful to all of these reviewers for their patience and their suggestions, most of which are incorporated in the book. Parts of this book are based on research supported by the National Science Foundation (NSF), Pacific Bell, and the California Micro Program, for which we are very appreciative. The first author is grateful to the Swiss Federal Insitute of Technology of Lausanne (EPFL) for the quiet hospitality when he was finishing work on the manuscript.

This book would never have been completed without the sympathetic and persistent encouragement of our editor, Jennifer Mann. She made bearable the seemingly interminable effort that went into the various revisions that were necessary before the book could make its way into print. Her assistant, Jane Elliott, was always available to help. Finally, our thanks to Cheri Palmer, who has been a thoroughly professional production editor keeping to an impossibly tight schedule.

1

CHAPTER

Overview

All activities of production and consumption, whether in business or government, in work or play, involve the process of material transformation—the storage, manipulation, and transport of physical matter—and the process of information transformation—the storage, processing, and communication of information. The impact of changes in the technologies of material and information transformation manifests itself over time in changes in the nature, composition, and importance of the production and consumption activities to which we devote much of our physical and mental effort.

Technological advances of the last hundred years have steadily reduced the effort that society devotes to material transformation. This reduction is seen, for instance, in the fact that a hundred years ago in the United States nearly all workers were engaged in manual labor in agriculture, mining, and manufacturing. Today those sectors employ less than one-quarter of the workforce.

More than three-quarters of workers who today are engaged in the processes of information transformation could be called "information workers." We are now in the midst of another great wave of change—the "information revolution." This revolution consists of a core of advances in information technologies and their impact on ever-widening circles of social activity as depicted in Figure 1.1. The first circle of impact is seen in the computer and communications industries that provide the hardware and software that embody the new technologies. In the next circle are the organizations that incorporate this equipment into their own processes of production and consumption. Effective incorporation requires restructuring those processes by

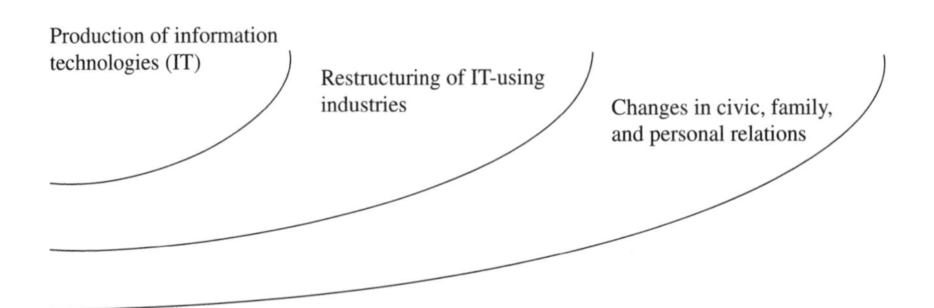

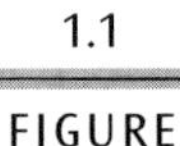
1.1 FIGURE

Initiated by the production of information technologies (IT), the information revolution restructures IT-using industry and eventually alters civic, family, and personal relations.

promoting changes in corporations and government, in education and entertainment. The outermost circles of impact will alter relations within society itself, at the civic, family, and personal levels.

These outer circles of the information revolution are still obscure, but some changes are visible. One widely observed trend is "globalization." Corporations are becoming global as communication networks permit the collection of data and the coordination of corporate decisions around the world. The foundations of the "global village" may be seen in the homogeneous culture resulting from the worldwide broadcast of the same TV entertainment and news programs. Government may become more intrusive as databases of information about citizens, hitherto isolated by agency or distance, now become interconnected, centralized, and accessible to all agencies of government.

However, there are also trends that run counter to globalization and centralization. Because access to information is much cheaper than before, citizens have better means to counter the propaganda from powerful organizations. By communicating over the Internet, geographically isolated, like-minded individuals can come together in small groups to pursue their common interests and to form new cultural islands outside mass culture. A movement such as this would have been impossible a decade ago. Further in the future lies the possibility that education will become a truly lifelong process, enabling a much fuller development of our potential.

Peering successfully into the future and the outer reaches of the information revolution requires the imagination of the poet and the luck of the seer. Yet the technologies at the core of the revolution are now visible and can be described using the language and concepts of the communications engineer, the computer scientist, and the economist. These technologies comprise advances in computers, communications, signal processing, and their applications in diverse domains. A subset of these advances relates to networks.

Advances in networking form the substance of this book. We begin with a formulation of the services that networks offer and the important models of network architecture. We then survey the major network designs that implement those architectures, concluding with one of the most important designs, Asynchronous Transfer Mode, or ATM networks. An extended discussion follows in which we describe how high-performance networks can be managed and controlled to provide services efficiently. We conclude with a description of the two network elements, optical links and high-speed switches, that have made high-performance networks affordable. A final chapter attempts to peer into the more distant future to discern the outlines of global multimedia networks.

After reading this book, you will have a system-level understanding of the networking technologies and the actual networks that lie near the center of the information revolution. You will learn that while there are many details, only a few principles are sufficient to grasp the field of networking. You will then know what questions to ask in order to compare different networks, and in many instances you will know how those questions should be answered. If you work in an organization, you will have the basis to judge how well different networking solutions will meet the needs of the organization, now and in the future. If you are a student, you will be able to read critically many of the recent research contributions to networking, and you will gain the sense of what directions of research are likely to be fruitful.

Advances in networking technology feed on, and are constrained by, the current state of networking. Within those constraints, the advances are guided by wider technological and economic forces. This chapter presents a highly abbreviated history of the key innovations in telephone, data, and cable TV networks and the principles that can serve as a compass for judging which directions of technological advance are likely to be more successful. By the end of the chapter you will be acquainted with the main contenders for the role of "network of the future," and you will appreciate that no single network technology will be the winner: the future network will integrate all of the major networking solutions.

In section 1.1 we review the history of networking technology by dividing it into three parts: the telephone network, computer or data networks, and cable television or CATV. In the past, these three types of networks used different technologies that were well suited to the information services they provided.

In section 1.2 we explain the four principles that underlie the forces driving the industry towards convergence and leading to the interpenetration of telephone, computer, and CATV networks.

In section 1.3 we discuss some plausible scenarios for networks of the future. These future networks will offer services ranging from telephone to interactive video to high-speed file transfers.

1.1 HISTORY OF COMMUNICATION NETWORKS

Communication networks enable users to transfer information in the form of voice, video, electronic mail or e-mail, and computer files. Users request the communication service they need through simple procedures using a telephone handset or set-top TV box or through applications running on a host computer such as a PC or workstation.

The communications industry is the United States' fastest-growing industry, with annual revenues of $400 billion in 1994. Table 1.1 provides a breakdown of revenues from this industry by sector.

This table hides a wide diversity in industry structure. The telephone industry consists of a small number of large companies and a large number of small companies. It provides 250 million subscriber lines; 90% of its revenue goes to service providers and 10% to equipment manufacturers. The broadcast industry comprises 1,000 commercial TV stations and 9,000 radio stations. The print news industry produces 1,600 daily newspapers and 11,000 magazines; there are some very large newspaper companies. The computer communications industry is competitive, with hardware revenues to $75 bil-

Industry segment	$B/year
Telephone	200
Broadcasting	50
Newspaper	60
Computers	80
Books	15

1.1 TABLE

Communications industry revenue, by sector, in billions of dollars.

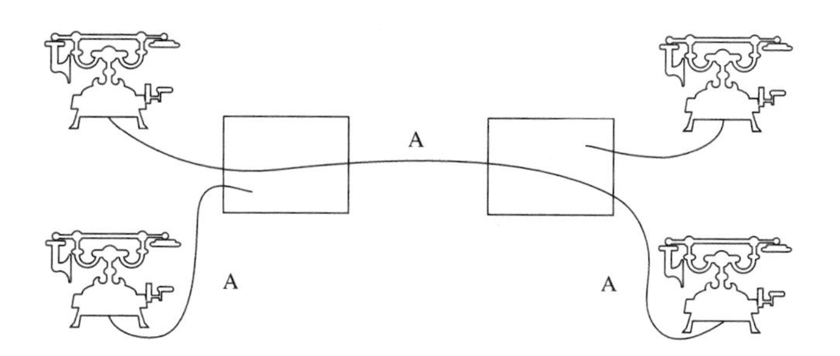

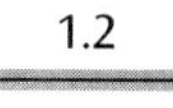

1.2 FIGURE Telephone network around 1890. The transmissions are analog and the switches are manually operated.

lion and software revenues to $5 billion. Table 1.1 should perhaps include the movie and the music-recording companies alongside the news industries.

The industry is in a stage of upheaval, intense competition, reconfiguration, and consolidation, all spurred by innovations that have dissolved the technological barriers that once separated the industry into distinct sectors. In the following sections, we identify the steps in the evolution of communication networks that, in retrospect, marked major advances in those networks. Of course, there were many other innovations, but the steps emphasized here proved to be decisive.

1.1.1 Telephone Networks

From our viewpoint, the key innovations in telephony are circuit switching, digitization, separation of call control from voice transfer, optical links, and service integration.

In 1876, Alexander Graham Bell invented a pair of telephones. Around 1890, simple networks connected telephones by manually operated switches. In this network, as shown in Figure 1.2, the signal is analog, as indicated by the letter A on the links. To call another telephone, a customer first rings the operator and provides the phone number of the other party. The operator then determines the line that goes either directly to the other party or to another operator along a path to the other party. In the latter case, the operators talk to one another, decide how to handle the call, and the procedure of constructing the path to the other party continues, possibly involving other operators. Eventually, one operator rings the destination, and, if the telephone is picked up, the two parties are connected. The parties remain

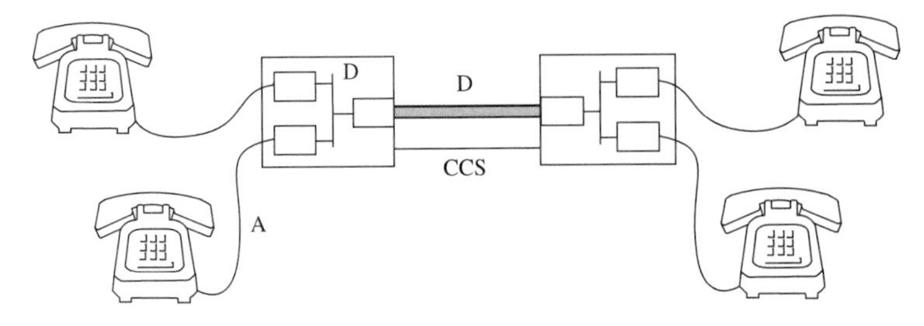

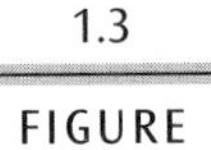
FIGURE 1.3 Telephone network around 1988. The transmissions are analog (A) or digital (D). The switches are electronic and exchange control information by using a data network called common channel signaling (CCS).

connected for the duration of the conversation and are disconnected by the operators at the end of the conversation.

Note how the transmission lines are allocated to the phone conversation. This is accomplished by means of *circuit switching,* where "circuit" refers to the capability of transmitting one telephone conversation along one link. To set up a call, a set of circuits has to be connected, joining the two telephone sets. By modifying the connections, the operators can switch the circuits. Circuit switching occurs at the beginning of a new telephone call. Operators were later replaced by mechanical switches and, eventually, by electronic switches.

Figure 1.3 illustrates the telephone network around 1988. One major development at this stage is that the transmission of the voice signals between switches is digital, as indicated by the letter D, instead of analog.

An electronic interface in the switch converts the analog signal traveling on the link from the telephone set to the switch into a digital signal, called a *bit stream.* The same interface converts the digital signal that travels between the switches into an analog signal before sending it from the switch to the telephone.

The switches themselves are computers, which makes them very flexible. This flexibility allows the telephone company to modify connections by sending specific instructions to the computer. Figure 1.3 also shows another major development—*common channel signaling* (CCS). CCS is a data communication network that the switches use to exchange control information among themselves. This "conversation" between switches serves the same function as the conversation that took place between operators in the manual network. Thus CCS separates the functions of call control from the transfer of voice. Combined with the flexible computerized switches, this separation of function facilitates new services such as call waiting, call forwarding, and call back.

Medium	Signal	No. of voice circuits	Rate in Mbps		
			North America	Japan	Europe
T-1 paired cable	DS-1	24	1.5	1.5	2.0
T-1C paired cable	DS-1C	48	3.1		
T-2 paired cable	DS-2	96	6.3	6.3	8.4
T-3 coax, radio, fiber	DS-3	672	45.0	34.0	32.0
Coax, waveguide, radio, fiber	DS-4	4032	274.0		

1.2 TABLE

Digital carrier systems. This is the hierarchy of digital signals that the telephone network uses. Note that the bit rate of a DS-1 signal is greater than 24 times the rate of a voice signal (64 Kbps) because of the additional framing bits required.

In current telephone networks, the bit streams in the trunks (lines connecting switches) and access links (lines connecting subscriber telephones to the switch) are organized in the digital signal (DS) hierarchy. The links themselves—the "hardware"—are called *digital carrier systems.* Trunk capacity is divided into a hierarchy of logical channels. In North America these channels, listed in Table 1.2, are called DS-1, . . . , DS-4 and have rates ranging from 1.544 to 274.176 Mbps (megabits per second). The basic unit is set by the DS-0 channel, which carries 64 Kbps (kilobits per second) and accommodates one voice circuit. Larger-capacity channels multiplex several voice channels. The rates in Japan and Europe are different.

Observe in the table that the rates are not multiples of each other: the DS-1 signal carries 24 DS-0 channels, but its rate is more than 24 times 64 Kbps. The additional bits are used to accommodate DS-0 channels with rates that deviate from the nominal 64 Kbps because the signals are generated using clocks that are not perfectly synchronized.

Since the 1980s the transmission links of the telephone network have been changing to the SONET, or Synchronous Optical Network, standard. SONET rates are arranged in the STS (Synchronous Transfer Signal) hierarchy shown in Table 1.3. In North America and Japan the basic SONET signal, STS-1, has a rate of 51.840 Mbps. (In Europe the basic signal is STS-3 and has a rate of 155.52 Mbps. The hierarchy is called Synchronous Digital Hierarchy, or SDH.)

Carrier	Signal	Rate in Mbps
OC-1	STS-1	51.840
OC-3	STS-3	155.520
OC-9	STS-9	466.560
OC-12	STS-12	622.080
OC-18	STS-18	933.120
OC-24	STS-24	1244.160
OC-36	STS-36	1866.240
OC-48	STS-48	2488.320

1.3 TABLE

SONET rates. The rates of multiplexed STS-1 signals are exact multiples; no additional framing bits are used.

Two differences are immediately apparent when comparing the STS and DS hierarchies. First, SONET signals have much higher bit rates, thanks to the much higher rates that optical links can support compared with the copper links of the current network. Second, the STS-n rate is exactly n times the STS-1 rate. Because all clocks in a SONET network are synchronized to the same master clock, it is possible to compose an STS-n signal by multiplexing exactly n STS-1 signals. As a result, multiplexing and demultiplexing equipment for STS signals is less complex than for DS signals.

The last major innovation in telephony is the integration of voice and data signals through the introduction of the *Integrated Services Digital Network* (ISDN), illustrated in Figure 1.4. The ISDN basic access offered to a customer consists of two B channels and one D channel (both B and D channels are digital). Each B channel is a bidirectional, or full-duplex, channel at 64 Kbps. One B channel can carry either a circuit-switched connection, a packet-switched transmission service (described below), or a permanent digital connection. The D channel carries a 16-Kbps packet-switched service. ISDN makes available to subscribers the digital transmission facilities that were previously used between the switches of the network, thus extending the digital transmission all the way to the users. Applications of the ISDN services include computer communication, high-speed facsimiles, remote monitoring of buildings, videotex, and low bit rate videophones. With ISDN, the telephone system is transformed into a network that can transfer information in many forms, if at modest speeds.

Figure 1.5 shows one possible future stage in the evolution of the telephone network, called *Broadband Integrated Services Digital Network* (BISDN). In such a network, basic access might consist of three incoming channels and

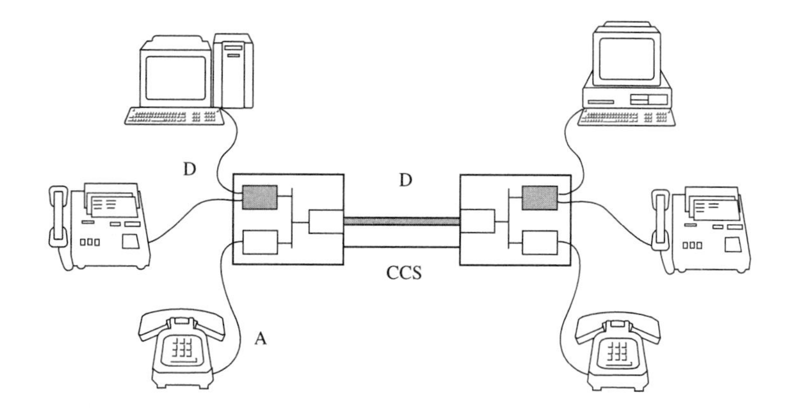

1.4 FIGURE

Integrated Services Digital Network. The basic access provides two bidirectional 64-Kbps links and one 16-Kbps link. These links can be used to transmit voice or data.

one outgoing channel at 155 Mbps. Applications that need such high-speed basic access include distribution of high-definition television (HDTV) and digital radio, interconnections of high-performance workstations for working at home, teleconferencing to replace business trips, remote learning, and interactive virtual reality entertainment. While the BISDN network represents one plausible evolution of the telephone network, other developments are at least as likely.

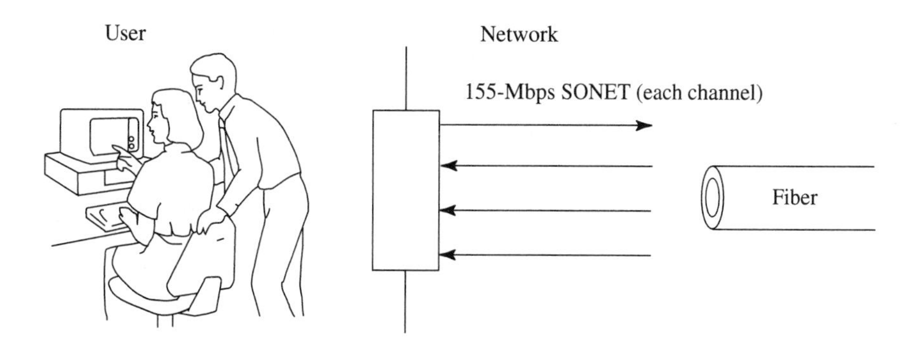

1.5
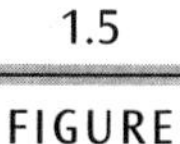
FIGURE

Broadband Integrated Services Digital Network. The basic access of this hypothetical network could consist of three incoming and one outgoing 155-Mbps channels.

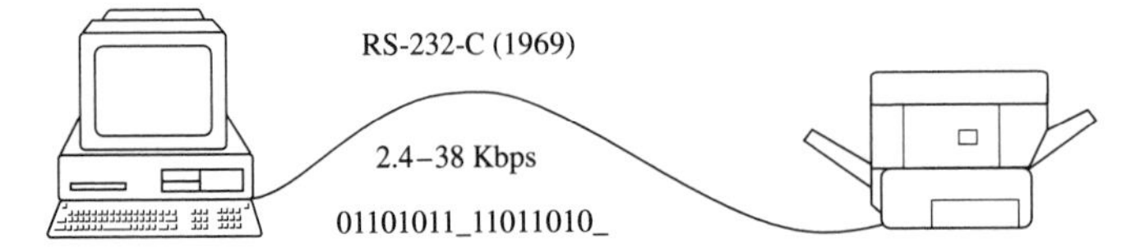

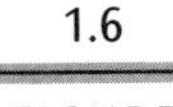

1.6 FIGURE The RS-232-C standard for the serial line specifies the transfer of one 8-bit character at a time, separated by time intervals. The speed and distance of the serial line are limited.

1.1.2 Computer Networks

This section discusses the following key innovations in computer or data networks: organization of data in packets, packet switching, the Internet Protocol hierarchy, multiple access methods, and service integration.

We begin our historical sketch with the publication in 1969 of the RS-232-C standard for the *serial port* of computer devices, illustrated in Figure 1.6 This standard is for low bit rate transmissions (up to 38 Kbps) over short distances (less than 30 m). Serial transmission proceeds one character at a time. The computer devices encode each character into seven bits, to which they can add a parity bit for error detection, and successive characters are separated by some time interval. When the receiver detects the beginning of a new character, it starts a clock that times the subsequent bits. Both bit rate and distance must be kept small, because transmissions take place over untwisted wires, which can introduce errors due to crosstalk. Crosstalk becomes more severe as the rate and the distance increase.

A serial link is often used to attach a computer to a *modem*. A modem, or modulator-demodulator, transmits data by converting bits into tones that can be transported by the telephone network as if they were voice signals. The receiving modem then converts these tones back into bits, thus enabling two computers with compatible modems to communicate over the telephone network as if they were directly connected by a serial link. In 1995 common modem speeds ranged from 300 bps to 28,800 bps.

Figure 1.7 illustrates the *synchronous transmission* standards introduced in the 1970s to increase the transmission rate and the usable length of transmission links. These standards are known as SDLC (Synchronous Data Link Control). A number of standards are based on SDLC, including HDLC (High-Level Data Link Control), LAPB (Link Access Procedure B), LAPD, and LAPM. The main idea of SDLC is to avoid the time wasted by RS-232-C caused by gaps between successive characters. To eliminate that lost time, SDLC groups many

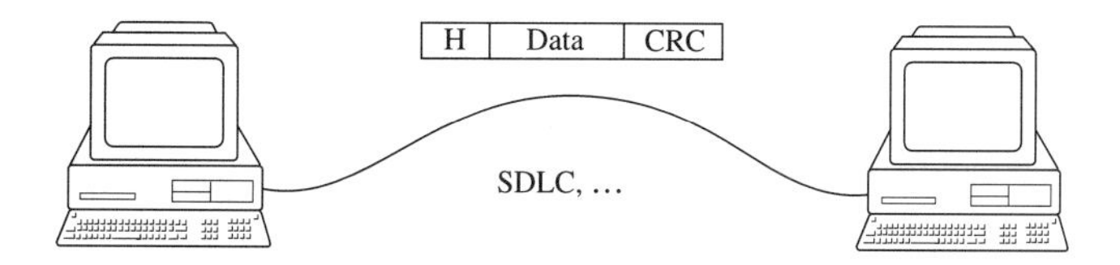

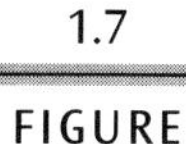

1.7 FIGURE

The Synchronous Data Link Control and related standards transmit long packets of bits. The header (H) contains the preamble that starts the receiver clock, which is kept in phase by the self-synchronizing encoding of the bits. The receiver uses the cyclic redundancy check (CRC) bits to verify that the packet is correctly received.

data bits into *packets.* A packet is a sequence of bits preceded by a special bit pattern called the *header* and followed by another special bit pattern called the *trailer.* The number of bits in a packet may be fixed or variable.

The receiver is synchronized by a preamble contained in the header (H) of the packet and by a self-synchronizing code that contains the timing information in addition to the data. Moreover, SDLC uses an error-detection code called the *cyclic redundancy check,* or CRC, that is more efficient and more powerful than the single parity bit of RS-232-C. Two computers, then, can exchange information over a transmission link using either RS-232-C or SDLC. But what if many computers are to be interconnected? In the early 1960s, communication engineers proposed the *store-and-forward packet-switching* method illustrated in Figure 1.8.

This figure shows computers connected by point-to-point links. To send a packet to computer E, computer A puts the source address A and the destination address E into the packet header and sends the packet to computer B. When B gets the packet from A, it reads the destination address and determines that it must forward the packet to D. When D gets the packet, it reads the destination address and forwards the packet to E. In this scheme, when a node receives a packet, it must first store it, then forward it to another node (if necessary). Hence the name store-and-forward given to this switching method.

When computers use store-and-forward packet switching, they use a given link only when they send a packet. As a result, the same links can be used efficiently by a large number of intermittent transmissions. This method for sharing a link among transmissions is called *statistical multiplexing.* Statistical multiplexing contrasts with circuit switching, which reserves circuits for the duration of the conversation even though the parties connected by the circuit may not transmit continuously.

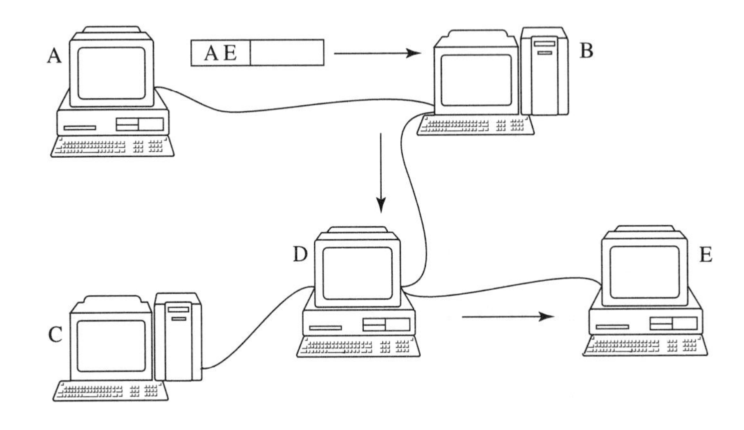

1.8 FIGURE

Store-and-forward transmissions proceed by sending the packet successively along links from the source to the destination. The packet header specifies the source and destination addresses (A and E, for example) of the packet. When it receives a packet, a computer checks a routing table to find out on which link it should next send the packet.

Starting in the late 1960s, the U.S. Department of Defense Advanced Research Projects Agency (ARPA) began promoting the development of packet-switched networks. The resulting network, ARPANET, began operations in 1969 by connecting four computers. The rules of operations, or protocols, ARPANET used were published in the open literature. By implementing these protocols, engineers in many research and educational institutions attached their computers to the ARPANET.

Through the ARPANET protocols, engineers agreed on a single packet format standard and a common addressing scheme. This allowed networks that conformed to this packet format to be easily interconnected. The benefits of interconnectivity soon became obvious, and the ARPANET evolved into the Internet, which today is used to interconnect a large number of computers and local area networks throughout the world. Until 1983, only 500 "host computers" had Internet access. In 1995, according to the Internet Society, there were 4.5 million such computers being used by 30 million people. In the 18 months ending in July 1995, users published three million pages of information, entertainment, and advertisement on the Internet, mostly on the World Wide Web. (We should note that these figures about the Internet are guesses and are not based on a survey.)

The single packet format of the Internet Protocol offered two advantages. On the one hand, the format could be supported by a variety of physical networks, including local area networks (LANs) such as Ethernet and token ring, as well as by point-to-point links. On the other hand, engineers and computer scientists could develop communications applications assuming that data would be transported in packets of a standardized format. The ARPANET implicitly implemented a three-layered architecture consisting of (1) the physical network that transfers bits, (2) groups of data encapsulated into packets with a common format and addressing scheme, and (3) applications that assume transfer of packets with no regard to the underlying physical network. This implicit layered architecture was subsequently elaborated and formalized in the Open Systems Interconnection, or OSI, model.

In the late 1960s and early 1970s, engineers proposed a new method for connecting computers. This method is called *multiple access.* It dramatically reduced the cost of interconnecting nearby computers in a LAN as well as the cost of access to a wide area network (WAN).

Figure 1.9 illustrates a popular implementation of multiple access called *Ethernet.* In an Ethernet network, computers are attached to a common coaxial cable via an interface that today consists of a small chip set mounted on the main board. When computer A wants to send a packet to computer E, it puts the source address A and the destination address E into the packet header and transmits the packet on the cable. All the computers read the packet, but only the computer with the destination address indicated on the packet copies it. The original Ethernet transmission rate was 10 Mbps; 100-Mbps Ethernet is now available.

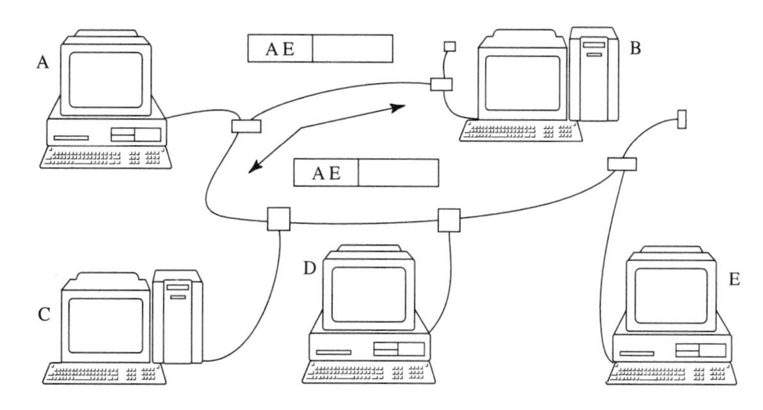

1.9 FIGURE Ethernet. In this network, computers are attached to a common coaxial cable. The computers read every transmitted packet and discard those not addressed to them.

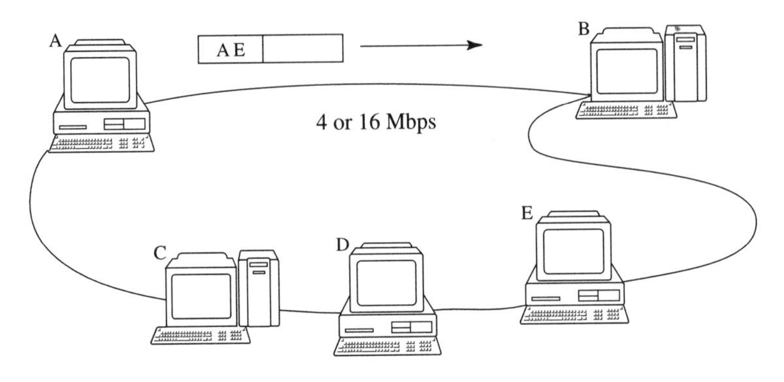

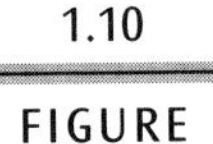
1.10 FIGURE

Token ring. The computers share a ring. Access is regulated by a token-passing protocol.

In the early 1980s, IBM developed another multiple access method, called *token ring,* illustrated in Figure 1.10. When networked as a token ring, computers are attached by point-to-point links in a unidirectional ring configuration using token ring interface boards. When the computers have no information to transmit, the interfaces pass a *token* around the ring. The interface boards between the computers and the network are configured so that they put back on the ring whatever information they receive with a delay equal to a few bit transmission times. This enables the token to circulate very fast around the ring.

Suppose computer A wants to send a packet to computer E. Computer A puts the source address A and the destination address E into the packet header and gives the packet to its interface, which then waits for the token. As soon as it gets the token, the interface of computer A transmits its packet, instead of forwarding the token. The other computers keep forwarding the packet they receive while making a copy for themselves. In particular, the interface of computer E copies the packet destined to it. The other interfaces discard their copy of the packet when they find out that it is not for them. When A receives the last bit of its own packet, after the packet has traveled around the ring, it puts the token back on the ring. What is important is that the computers get to transmit in turn, when they get the token. Hardware is available for token ring networks at 4 Mbps and at 16 Mbps.

The maximum time a computer waits before it gets to transmit in a token ring or Ethernet network is small enough for many applications but too large for interactive audio or video applications. Also, the transmission rate of token ring (4 or 16 Mbps) or 10-Mbps Ethernet networks is too slow for

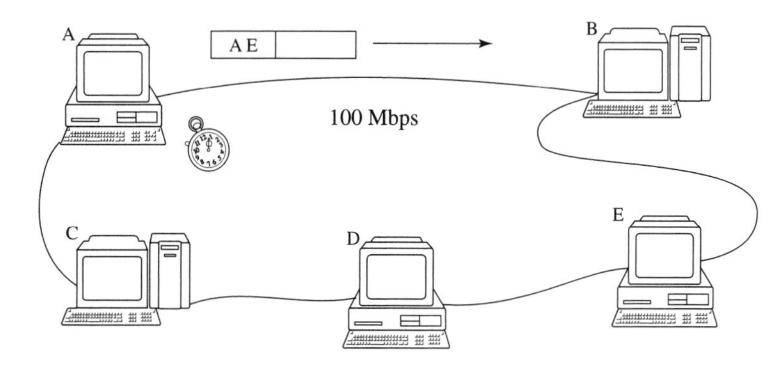

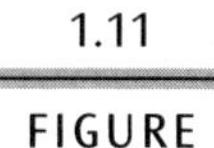

1.11 FIGURE

Fiber Distributed Data Interface (FDDI). A token-passing protocol is used to share the ring. The computers time their holding of the token. This network guarantees that every computer gets to transmit within an agreed-on time.

some multimedia applications. These two limitations led engineers in the late 1980s to develop a new network called *Fiber Distributed Data Interface* (FDDI), illustrated in Figure 1.11. FDDI networks use optical fibers to transmit at 100 Mbps; access to the channel is regulated by a timed-token mechanism. This mechanism is similar to the access control of a token ring network except that with FDDI the arrivals of tokens are timed to assure that they are retransmitted within a fixed time. FDDI networks can use twisted wire pairs to connect nodes over distances up to 100 m.

The high speed of FDDI makes it suitable for networking workstations with instruction rates of a few hundred Mips (millions of instruction per second). Because it can guarantee a token rotation time, FDDI can offer integrated services for applications that combine audio and video with data.

Engineers are currently developing faster networks than FDDI using *Asynchronous Transfer Mode* (ATM). With ATM, a computer transmits information at a typical rate of 25, 45, 100, 155, or 622 Mbps in packets of 53 bytes (1 byte = 8 bits). These fixed-size packets, called *cells,* can be switched rapidly by ATM switches. Figure 1.12 illustrates an ATM network with one switch.

With the appropriate control software, network engineers can connect many ATM switches together to build large networks. Moreover, the links between ATM switches can be long optical fibers. Using this technology, then, the telephone companies could build a worldwide network. In an ATM network data is transferred from source to destination over a fixed route, just like in a telephone connection. Unlike telephone networks, however, an ATM connection is not allocated a fixed bandwidth. The ATM network determines how

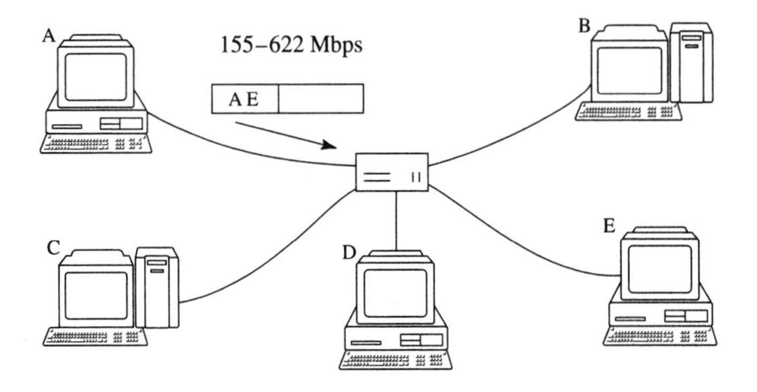

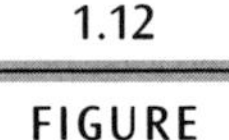

1.12 FIGURE

Asynchronous Transfer Mode (ATM) network. The network transports information in 53-byte cells. Total throughput of this network is much larger than that of FDDI or of a 100-Mbps Ethernet.

much bandwidth to allocate so that information is transported with very low loss rate or delay, as required by the application. Thus this technology is well suited for building large integrated services networks.

The next step in the evolution of computer networks could be the Broadband Integrated Services Digital Network or BISDN, introduced earlier in Figure 1.5. BISDN could be implemented as an ATM network that uses SONET transmission for transferring ATM cells. We can thus imagine that telephone and data networks may fuse by the end of the decade to provide BISDN services.

Figure 1.13 summarizes the dramatic increase in speed of data networks. Over the 25 years from 1970 to 1995, speed increased by five orders of magnitude, from 10 Kbps to 2.4 Gbps (gigabits per second).

1.1.3 Cable Television Networks

The key innovations in cable TV discussed in this section are optical feeder links, digital compression techniques, and service integration.

The cable television systems deliver television signals to more than 50% of U.S. households and could serve close to 90% with relatively minor additional infrastructure. Changes in CATV networks are depicted in Figure 1.14.

Today, CATV uses frequency-division multiplexing to transmit up to 69 analog TV channels, each 4.5-MHz wide. Transmission is over coaxial cables arranged as a unidirectional tree, with wideband amplifiers used to compensate for the attenuation of the cable signal (top panel in Figure 1.14). The

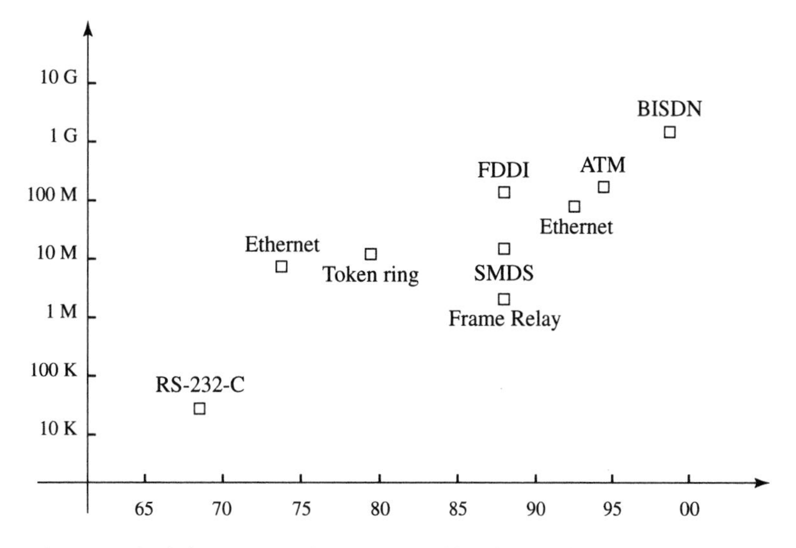

1.13 FIGURE

The speed of data networks increased by five orders of magnitude between 1970 and 1995.

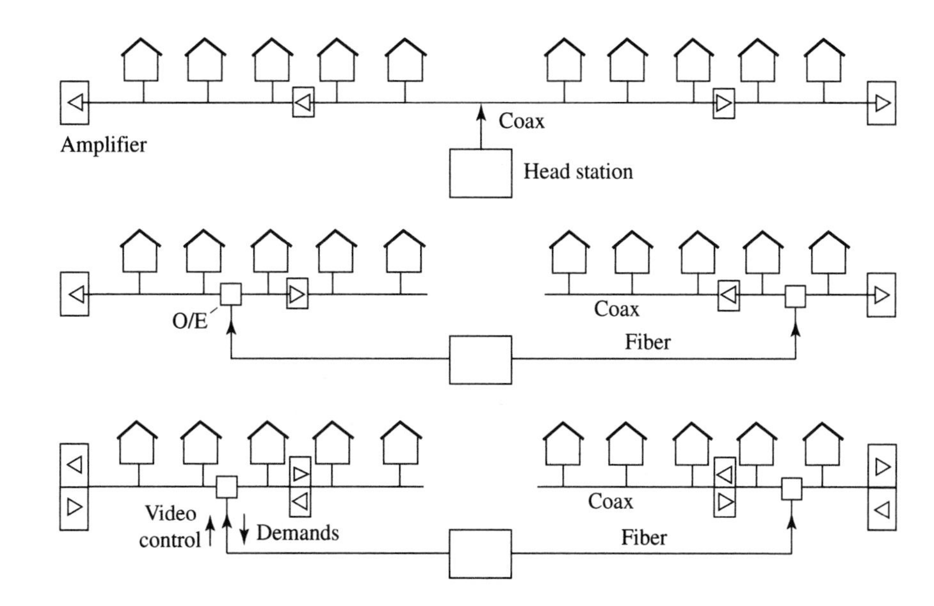

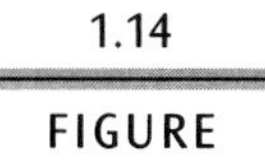

1.14 FIGURE

CATV networks have improved in two steps. In the first step the coax distribution system is replaced by fiber. In the second step channels are provided from the user to the head station.

number of TV channels is limited by the bandwidth of coaxial cables. The span of a CATV network is limited by the noise power, which increases as more amplifiers are added to compensate for the signal power loss during propagation.

The next major technological step in CATV is to utilize optical fibers to transmit the TV signals over longer distances (middle panel in Figure 1.14). Fibers have a much lower attenuation than coaxial cables, so they can transmit signals over longer distances before it becomes necessary to use an amplifier. In this implementation, the transmission over the fiber is still analog. The signal is fed into the coaxial cable network at various points, where the optical signal is converted into electrical signals (indicated by the box labeled O/E). The cost of each optical transmission line is spread over a few hundred users. Moreover, existing coaxial cables can be reused. This hybrid fiber/coaxial cable distribution system has a longer span and a higher signal quality than a coaxial cable network. The network is now a tree whose first level is a fiber network and whose bottom levels are coaxial cable. (In the top panel of Figure 1.14, both levels of the tree are coax.) This network is called a *fiber-to-the-curb* network, where "curb" designates a location in some neighborhood where the fiber is connected to the local coaxial distribution network.

To increase the number of TV channels, the CATV industry is now migrating to a digital transmission technology. Before transmitting the TV signals, the CATV company uses a TV codec (coder-decoder) that converts each signal into a bit stream that represents the video frames. Using compression algorithms that have been standardized by the *Motion Pictures Expert Group* (MPEG), the codec compresses the bit stream to reduce its rate. This is accomplished by eliminating redundant information as well as information that does not contribute significantly to the image quality, as perceived by the viewers. The bit streams are transmitted over fibers to the curb and are then distributed by the neighborhood coaxial network. The compression gain now allows the network to transmit about 500 TV channels. Using the first version of the MPEG standard, MPEG1, a moderate-quality TV signal is encoded as a 1.5-Mbps bit stream, which can be modulated in a signal that has a bandwidth of about 600 kHz. (By comparison, the analog NTSC TV signal has a bandwidth of 4.5 MHz, about seven times larger. NTSC is the North American standard for commercial broadcast color TV.) The decompression is performed by set-top boxes at the user residence. This CATV network is still unidirectional.

To provide new services, such as video on demand and interactive TV, the CATV industry is designing bidirectional networks. Such a network, depicted in the bottom panel of Figure 1.14, connects video servers to users by means of control messages. The user chooses these messages to select the video program, and the video program is sent over the network to the user. In one proposed standard, the network uses fiber to the curb and a local coaxial network. The bandwidth of the coaxial cable is divided into 500 digital video channels from the curb to the users and a multiple access network for control messages. The equipment at the curb relays the control messages between the head station and the coaxial network.

This arrangement could provide 500 homes with 500 simultaneous different programs or video clips that they requested over the control network. Users would interact with the network by using remote controls that communicate with set-top control boxes attached to the coaxial network. The advantage of this strategy for deploying interactive TV services is that it requires only an incremental investment over the fiber-to-the-curb network that the CATV industry is building now. The economics of video servers, however, may limit the flexibility of interactive services by restricting the number of available movies and their start times.

A cost-effective alternative to a coaxial or fiber/coaxial network is offered by a technology that can transmit relatively high-speed data over untwisted or twisted pair cables for distances up to 4,000 m. The technology can use existing digital telephone subscriber lines. Although the telephone voice channel is limited to a bandwidth of 3 kHz, the twisted pair cable connecting to the central office has a much wider bandwidth, limited by signal attenuation and noise. The High Bit-Rate Digital Subscriber Line (HDSL) offers bidirectional transmission at 1.5 Mbps with a transmission bandwidth of 200 kHz. Asymmetric Digital Subscriber Line (ADSL) can transmit four one-way 1.5-Mbps video signals, in addition to a full-duplex 384-Kbps data signal, a 16-Kbps control signal, and analog telephone service. ADSL has a transmission bandwidth of 1.1 MHz. It appears that 66% of subscriber loops in the United States can support the HDSL/ADSL technology. The main impairment of the other loops are the loading coils (inductors) present in loops longer than 4,000 m.

There are two other important network technologies that we will not discuss further. The first is wireless transmission by radio (and sometimes by infrared light). The second is microwave satellite transmission. Wireless transmission permits mobility among users. Because the bandwidth is shared by users, users are grouped into small cells. Users in each cell communicate with a single "base station," and base stations are linked together by a wired network. Transmission power is kept low so that spillover into adjacent cells is minimized, permitting the same frequency band to be reused in different cells. Satellite transmission also facilitates mobility. Because the transmission area covered by a satellite is very large, it is well suited for broadcasting video or audio programs.

Telephone networks	Computer networks	Cable TV
Circuit switching, CCS, and separation of call control from voice transfer	Packets, packet-switched networks, multiple-access networks	Digitization and compression using signal processing techniques
ISDN and service integration	Layered architecture, ARPANET	Fiber-to-the-curb network
Optical links, SONET	Internet, OSI model	Two-way, interactive TV
SONET to BISDN	Integrated services, ATM to BISDN	Service integration

1.4 TABLE Key innovations in telephone, computer, and CATV networks.

The key innovations in telephone, computer, and CATV networks are summarized in Table 1.4. The differences among these three types of networks are still great. However, the table reveals that each type of network is now able to provide services that were formerly the exclusive province of other networks. We can discern in this the tendency towards "convergence." In the next section we study the forces underlying this tendency.

1.2 NETWORKING PRINCIPLES

Worldwide demand for and supply of communication network services are growing exponentially. This is not a new phenomenon. In affluent countries, the telephone network has grown so that for many years it has been accessible to virtually every business and home. In developing countries where penetration of the telephone network is significantly below 100%, there is a large unfilled demand, leading to a waiting time for telephone access of several months to years. The growth of data services, of which the Internet offers an impressive example, is more recent. In this section we describe the four principles that underlie the growth of communication network services: digitization, economies of scale, network externalities, and economies of scope or service integration.

1.2.1 Digitization

There are two aspects to digitization. First, any information-bearing signal can be represented by a binary string with an arbitrarily high degree of accuracy (explained later in this section). The second aspect is that it is much cheaper to store, manipulate, and transmit a digital signal than an analog signal, because advances in electronics have made digital circuits much more robust and cheaper than analog circuits. Because of these two aspects, the overwhelming majority of today's communication systems are digital.

If a signal is in the form of a sequence of discrete symbols, it is obvious that this signal can also be represented by a binary string. For example, since there are fewer than $2^7 = 128$ characters on the computer keyboard, we can represent each character as a unique 7-bit sequence. Thus, any character string can be represented as a binary string seven times as long. Conversely, by decoding the binary string, 7 bits at a time, we can recover the character string.

Binary represention of an analog signal, such as voice, requires two steps, illustrated in Figure 1.15. The first step, called *sampling,* consists of measuring periodically the value of the analog signal $V(t)$, a real-valued function of continuous time. These values, called *samples,* are represented by the small dark circles in the figure. The second step, *quantization,* consists of representing

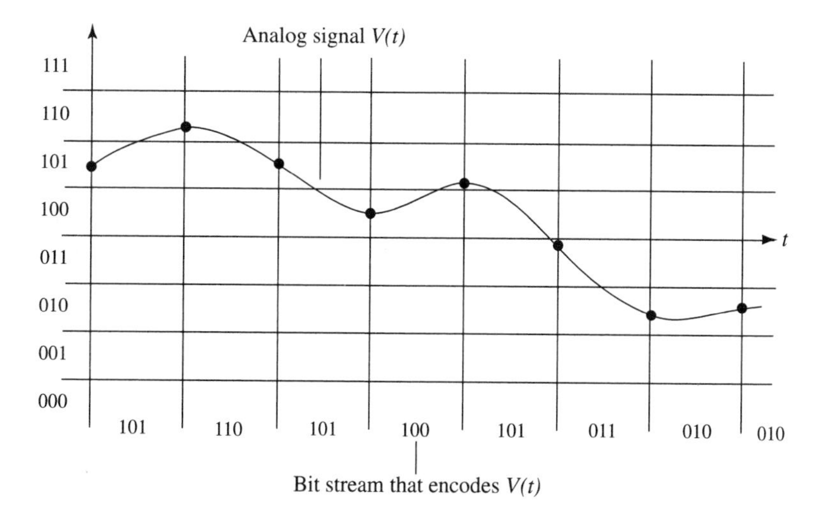

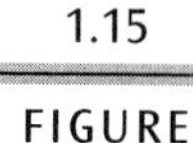

1.15

FIGURE

Digitization of an analog signal $V(t)$. The signal is sampled at a rate $2 \times f_{max}$ where f_{max} is the maximum frequency in the signal. The samples are quantized with N bits per sample. The resulting bit stream has rate $2N \times f_{max}$, and the signal-to-noise ratio is $6N$ dB.

the samples by a fixed number of bits. To quantize the samples, the digitization hardware decomposes the range of possible sample values into a finite set of intervals, called *quantization intervals,* and associates a different binary number with each interval. The hardware then represents a given sample by the binary number associated with the quantization interval of the sample. (If there are 2^n quantization intervals, each interval is represented by an n-bit word. In the figure, the eight intervals are represented by 3-bit words.) Using these two steps, the hardware represents the analog signal $V(t)$ by the bit stream composed of successive binary numbers associated with the quantization intervals of the samples.

The theoretical basis for sampling is *Nyquist's theorem.* That theorem states that no information is lost by the sampling provided that the sampling rate is at least twice the maximum frequency, or *bandwidth,* of the analog signal. This theorem makes precise the fact that the faster a signal changes—that is, the larger its bandwidth—the more frequently it must be measured to observe its variations.

The quantization of a sample approximates its value by one representative value in the quantization interval. Consequently, quantization introduces errors. The errors are small if the quantization intervals are small, which is the case if the range of values of the signal is decomposed into a large number of quantization intervals. Equivalently, we can view these errors as the addition of some analog noise to the original signal. That equivalent noise is called the *quantization noise.* We measure the importance of the quantization errors by the ratio of the power of the signal over that of the quantization noise. This signal-to-noise ratio is usually measured in *decibels* (dB). If the ratio is equal to R, then its value expressed in decibels is $10 \log R$, where log designates the logarithm in base 10. For example, 3 dB means a signal power that is twice as large as the noise power, since $10 \log 2 \approx 3$. In telephone transmission, a signal-to-noise ratio of about 48 dB is acceptable. In high-fidelity audio applications, a signal-to-noise ratio of 65 dB or better is desired. A low-quality cassette deck, for example, has a signal-to-noise ratio of about 55 dB, where the noise is due mostly to the granularity of the magnetic material on the tape. A high-quality cassette deck has a signal-to-noise ratio of about 68 dB, where the magnetic noise is attenuated by some signal processing such as Dolby C.

The signal-to-noise ratio due to quantization is approximately equal to $6 \times N$ dB, where N is the number of bits used to represent each sample. (Thus in Figure 1.15 the ratio is 18 dB.) This result can be explained as follows. If N bits are used to number the quantization intervals, then there are 2^N such intervals, and a typical error has a magnitude proportional to 2^{-N}. Since the power of the quantization noise is proportional to the square of the magnitude of that noise, we conclude that the noise power is proportional to 2^{-2N}, yielding a signal-to-noise ratio in decibels of about $10\log(2^{2N}) = 2N \times 10\log 2 \approx 6N$.

If the digitization hardware uses N bits per sample and samples the signal f_S times per second, then it produces a bit stream with rate $N \times f_S$ bps. For example, the telephone network transmits frequencies in the voice signal up to 4 kHz and achieves a signal-to-noise ratio approximately equal to 48 dB. To meet these objectives, the sampling rate must be 2×4 kHz $= 8$ kHz, and the number of bits per sample must be equal to $48/6 = 8$. Consequently, the digitized voice signal has a rate of 8 kHz $\times\ 8 = 64$ Kbps.

In a compact disc, or CD, the target specifications are a maximum frequency of 20 kHz and a signal-to-noise ratio of 96 dB. These specifications require a sampling rate of at least 40 kHz and at least 16 bits per sample. A stereo signal, then, corresponds to a bit stream rate of at least 1.3 Mbps. Consequently, a 70-minute CD must store at least $70 \times 60 \times 1.3 \times 10^6/8 =$ 682.5 MB. This large storage capacity of CDs explains why CD-ROMs are used to distribute digital information.

As a final example, consider a television signal. The NTSC TV signal has a maximum frequency of 4.5 MHz. If the signal-to-noise ratio must exceed 48 dB, then the bit stream must have a rate of about 72 Mbps. Note that these examples ignore the savings in bit rate that can be accomplished by data compression techniques.

1.2.2 Economies of Scale

Communication networks exhibit scale economies. That is, the average cost per user of the network declines as the network increases in size, measured by a factor such as number of users, subscribers, or host computers. There are several reasons for this declining average cost. First, owing to advances in communications technology, primarily optical communications, the cost of a transmission link grows at a slower rate than does its capacity or speed. Hence, if the bit streams generated by n users can be made to occupy the same link with a capacity of nB bits per second at an average cost of $C(nB)$, then each user may continue to generate a bit stream at rate B, but the per user cost $C(nB)/n$ will decline as n grows. Note that to take advantage of cheaper high-speed transmission, the low-speed bit streams of individual users must be combined into a high-speed bit stream. This is possible to do with the techniques of multiplexing and switching explained in section 2.5.

Second, a network has certain fixed costs of operations, administration, and maintenance. Because these costs are not sensitive to network size, the per user share of these fixed costs declines with the number of users. Third, in networks dedicated to distribution services, such as cable TV, where the same bit stream is delivered to many users, the sharing of transmission facilities allows the per user cost to decline with the number of users.

Scale economies leading to declining average cost are present in many industries, including telephone, CATV, electric power transmission, and water distribution. In these industries, a large company enjoys a lower average cost than a smaller company. The large company can lower its price and drive its smaller competitors out of business. (This happened in the United States in the early years of the electric power and telephone industries.) Thus, these industries have a tendency to create one large monopoly, resulting in government regulation of these industries in most countries. In the United States the communications industry is regulated by the Federal Communications Commission (FCC) at the national level and by public utility commissions at the state level. The objective of this regulation is to prevent large companies from reaping monopoly profits or unfairly competing with small companies.

1.2.3 Network Externalities

A network service is said to have positive externalities if its value to a user increases with the number of users. The clearest example of externalities is the case of telephone service: the value to a telephone subscriber increases as more people subscribe, since the subscriber can then talk to more people. A less obvious example is the formation of special interest groups of users of a data network. These groups provide services to their members, such as operating an electronic bulletin board dedicated to a special topic of interest. The presence of network externalities can be inferred from the fact that the group must become sufficiently large before it is viable.

Externalities provide a powerful incentive for internetworking. When two independent networks are interconnected, the value to the users of both networks increases. The extra resources needed to implement the interconnection are few if the two networks follow compatible standards. The combination of network scale economies, which reduces per user cost as the network grows, and network externalities, which increases per user benefits as the network grows, creates positive feedback that can lead to an exponential growth in demand and supply of network services. Examples of this growth occurred in the demand for facsimile machines and Internet access.

This growth phenomenon is illustrated in Figure 1.16. If the number of users is below a critical size, then the per user cost exceeds the benefit, and users would not be willing to pay for that service. Once the critical size is

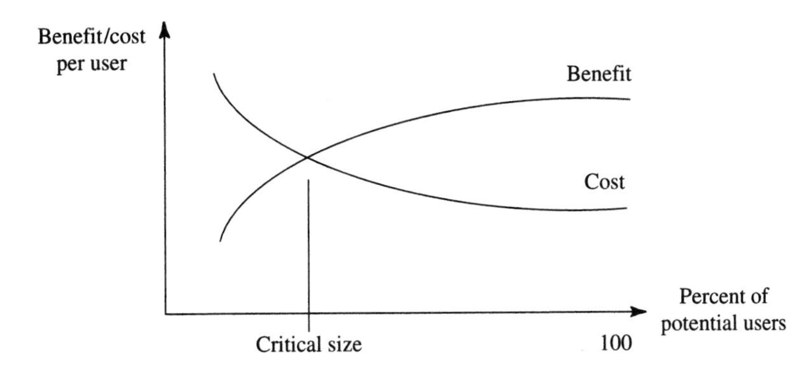

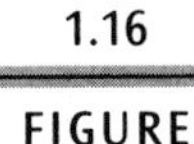
1.16 FIGURE As more users subscribe to a network, the cost per user decreases, and the benefit per user increases. This positive feedback further fuels the growth of the network.

exceeded, users are increasingly willing to bear the cost. For example, the French telephone company gave free terminals to its subscribers for use in accessing a variety of private on-line services such as news, train schedules, and restaurant menus on the Minitel network. The minimal device for accessing the network is a dumb terminal with a monochrome display and dual mode 75/1200-baud modem (PCs can also be used). Within a short time a very large set of services was offered, and the telephone company quickly recovered the cost of the terminal through greater use of the telephone network.

A counterexample is provided by AT&T, who in the 1960s developed a video telephone called Picturephone. The product was a commercial failure because the number of subscribers remained below the critical size.

Figure 1.16 makes it clear that to initiate a new network service there must be an initial period in which users are subsidized. The subsidy, which may come from the government, lowers the cost that users face, and hence the critical size. For example, the ARPANET computer network was paid for entirely by the U.S. government. It later developed into the Internet, which now enjoys no subsidy. The initial subsidy is often provided by users of existing services, through service integration.

1.2.4 Service Integration

Economies of scope, or service integration, refers to the fact that a network that currently provides one set of services may be expanded to provide new services at an additional cost that is much less than if a separate network were built to provide those new services. Economies of service integration are possible because communications engineers now design services in a modular and standardized way so that new services can be introduced using existing hardware and software modules.

The widespread deployment of ATM, described in Chapter 5, will facilitate service integration to such an extent that one can imagine a single network that will provide all of the services that today are provided by separate networks. These services include telephone, data, broadcast TV and radio, and CATV. Information carried by newspapers, magazines, books, and other forms of print media could also be provided over this network. The potential economies and profits are enormous. This potential explains the current struggle among telephone, computer, TV, and entertainment companies to form coalitions that will own and operate this "universal" network.

1.3 FUTURE NETWORKS

In this section we compare the capability of the Internet, ATM, and cable TV networks to provide the core technologies that can be used to develop the information superhighway. We use this popular term to designate a high-performance, flexible communication network that can provide all of the services now provided by the three types of networks.

1.3.1 The Internet

The Internet today comprises tens of thousands of local area networks (LANs) worldwide, interconnected by a backbone wide area network (WAN). LANs typically operate at rates of 10 to 100 Mbps. Links of the WAN support much lower bit rates of 0.1 to 1 Mbps, though they are now increasing to 45 and 155 Mbps. The low bandwidth of WAN links compared with LAN speeds is appropriate because LANs support the high bit rate traffic between workstations and file servers within a single organization. The WAN, on the other hand, supports electronic mail and infrequent file transfers that can be accomplished with low-speed connections. For example, in 1995 40,000 students, faculty, and staff at the University of California, Berkeley, had access to the Internet using 20,000 workstations and PCs. Within the university campus these computers were interconnected by 10-Mbps Ethernets and 100-Mbps FDDI rings. The Internet traffic between the campus and the rest of the world was handled by two 1.5-Mbps links, with an average utililization of 30%. By comparison, the telephone links between the campus and the rest of the world have a capacity of 200 Mbps.

Users access the Internet in one of two ways. Within a large company, government agency, or university, the user's PC or workstation is attached to a LAN that is part of the Internet. Users at home and in small companies subscribe to a network access provider, or NAP. Subscribers use low-speed modems to connect their PCs to NAP hosts, which, in turn, have Internet access. A growing proportion of users access the Internet in this second way, using one of hundreds of NAPs in the United States. Many of the smaller NAPs are simply retailers of Internet access that they purchase from the large NAPs.

We consider here some of the many factors that have contributed to the spectacular success of the Internet, which by 1995, at 20 years of age, connected 30 million users worldwide. For users connected to LANs the incremental cost of Internet access is a small fraction of the cost of the LANs. This insignificant incremental cost, combined with the network externalities of services like e-mail and formation of special interest groups, has led to an exponential growth. Second, more and more people own PCs, and newer PCs come equipped with networking hardware, including a built-in modem, and software. For these users, the cost of Internet access is only the charges of their NAP provider. By 1995 competition among NAPs had driven charges down to an affordable $20 for monthly access and a few hours daily of free connect time. The combination of positive networking externalities and the steady reduction in the cost of computers accounts for some of the Internet's exponential growth.

Additional growth is fueled by innovative, low bit rate, delay-insensitive applications such as the World Wide Web (WWW), with icon-driven interfaces such as Mosaic and Netscape that make browsing easy. Internet applications such as e-mail and file transfer can be provided at a cost that no alternative network can match. Lastly, designers of these new applications often distribute them freely. (They do so because the Internet has been developed by, and in turn has helped to sustain, a remarkable cadre of experts who strongly support keeping the Internet a free and open network. The successful introduction of commercial software such as WWW browsers may also require an initially subsidized distribution of the software to overcome the critical size depicted in Figure 1.16.)

One future development of the Internet, then, is more growth of the same kind: more users and more low bit rate, delay-insensitive applications for which the Internet has an overwhelming cost advantage.

Another possibility is that the Internet will develop into the information superhighway by supporting real-time, high bit rate, delay-sensitive applications such as interactive voice and video applications. To support those applications, the Internet will need to change in three ways. The backbone links will have to be upgraded to 155 Mbps or 622 Mbps, and the network switches for those links must be replaced by switches with very large throughput and low delays. At that point network designers may replace the IPv4 (Internet Protocol version 4) network layer, which cannot guarantee the delay, bandwidth, and loss bounds that real-time applications need. IPv4 could be replaced with a newer version, IPv6, or other protocols that meet some of those needs, or with ATM. (The Internet Protocol is discussed in Chapter 3.) Many commercial routers developed in the 1990s support some rudimentary resource reservation needed for real-time applications. The decision to migrate to ATM for the high-speed, low-delay links of the Internet may be forced by the advantages of ATM over IP.

These latter considerations point to an Internet growth path built on ATM technology and high-capacity links. However, that growth path will require an extremely costly upgrading of the Internet. It is unlikely that hundreds of network access providers will be able to raise the capital needed to deploy such a high-speed transmission infrastructure. Moreover, the real-time high bit rate applications such as interactive video will also be offered by telephone or CATV companies that have either the high-capacity links already in place or the revenues needed to support large investments. Thus, it is more likely that these applications will be supported on networks owned by the telephone or CATV companies.

In summary, two nonexclusive development paths for the Internet seem plausible. The Internet will continue to grow as now, and network access providers will offer very cheap, low bit rate, delay-insensitive services, taking advantage of its unique network externalities; or Internet protocols, backbone network links, and switches will be changed to permit much higher speeds. This latter change will likely be controlled by the telephone companies.

1.3.2 Pure ATM Network

ATM technology is capable of carrying a wide range of information transfers, from e-mail to videoconferences, with a range of quality of service that can match user needs. Thus, ATM could provide the full range of services contemplated as offerings on the information superhighway.

To implement such a network, the telephone companies would install high-speed ATM access lines to users over optical-fiber subscriber loops. The ATM cells would be switched by high-speed ATM switches in the telephone network. The result would be a "universal" network, built on sound theoretical principles.

Several obstacles must be overcome before this scenario is realized. First, the cost of installing dedicated optical subscriber loops will remain high for a number of years. Second, we lack the large, high-speed ATM switches that would allow the network to provide the wide range of services that are theoretically possible. Third, user equipment, from TV to PCs, with ATM interfaces does not yet exist or is too costly. Finally, the main direction of high bit rate traffic is expected to be from the network to the user in the form of entertainment video, controlled by a low bit rate stream originating at the user. As we will see next, that pattern of demand can more economically be met by CATV.

1.3.3 Video Dial Tone

As explained in section 1.1.3, the CATV industry is developing a bidirectional network that can deliver video programs controlled by users. The user interface rate of such a network is orders of magnitude larger than an ISDN network. The increased rate comes from using fiber to the curb plus local coaxial networks instead of using twisted pair subscriber loops.

By using frequency-division multiplexing, this CATV network can also provide data, telephone, and compressed videophone services. Connecting this CATV network to the telephone network would allow a wide area multimedia network to be implemented. (Government regulation has prevented this from happening in the United States. That situation has changed with the passage of the 1996 Telecommunications Bill in the U.S. Congress. In the United Kingdom CATV companies can now offer telephone service; telephone companies will be able to offer video programs after five years.) On such a network, information could be transported locally by the CATV network and, over long distances, by ATM over the SONET network of the telephone company. This fusion of the CATV and telephone networks will be facilitated by a collaboration of two companies, as has already occurred between US West and Time-Warner, Bell Atlantic and Tele-Communications, and Southwestern Bell and Cox. A similar outcome will result if the phone company upgrades its low-speed copper subscriber loop network with a high-speed fiber-to-the-curb plus local coaxial network. Such a plan is being implemented by some phone companies.

1.3.4 And the Winner Is . . .

We can be sure that no single network technology will emerge as the undisputed winner. The reason is in part technological, in part economic. The technological reason is that all three different technologies (ATM, Internet, CATV) are converging to provide a widely overlapping set of services. Thus, to a limited extent each technology can substitute for the others. The economic reason is that the large investments being made in all three types of networks mean that they will all be deployed for a long time. The 1996 Telecommunications Bill eliminates many ownership and regulatory barriers in long-distance, local telephone, and cable TV markets. The bill will spur mergers of companies and intensify competition, especially between local and long-distance telephone companies. Thus the information superhighway will be characterized by a collection of heterogeneous networks, offering a variety of services.

In the year 2000 we are likely to see the wide deployment of a large number of video services, such as movies, video on demand, teleshopping, using frequency-division multiplexing, fiber-to-the-curb, and local coaxial cable networks. Those networks will be deployed by telephone companies, sometimes in collaboration with CATV companies. Once this network is in place, the communication cost of distributing video programs to a large number of users will be reduced considerably, and profits will accrue to those companies that can provide video services or other content that people are willing to purchase.

The Internet will grow using IP with faster links and routers and new protocols that better control the quality of service. Growth will come from the interconnection of more and more LANs and from the introduction of new services. Internet protocols are evolving to include resource reservation and call admission and may well eventually incorporate a pricing component to regulate the reservation of resources. Thus, the Internet will continue to provide the best-effort services, such as e-mail, file transfer, and WWW, for which it was initially designed. The potential of such applications is enormous, as the example of the French network Minitel suggests. Minitel connects a larger proportion of households and businesses than any other data network. Thousands of businesses and consumer services such as telephone directories, travel agencies, restaurant reservations, stock market quotes, "chat" groups, and database searches can be accessed on-line. Minitel services in many cases provide greater convenience than is provided today by the 800 number phone service. Internet applications, by contrast, are just entering the commercial arena. In addition, the Internet will increasingly attempt to capture the potentially large market of real-time traffic. The capability of the Internet to carry real-time traffic will increase as high-speed ATM networks become extensive in the Internet backbone network.

The demand for ATM will first be expressed through ATM LANs, as organizations find it worth providing multimedia communication, with its stringent requirements on service quality. With the introduction of ATM LANs we will see the accommodation of data, telephone, and video services in the same network. The widespread deployment of ATM LANs may promote the intro-

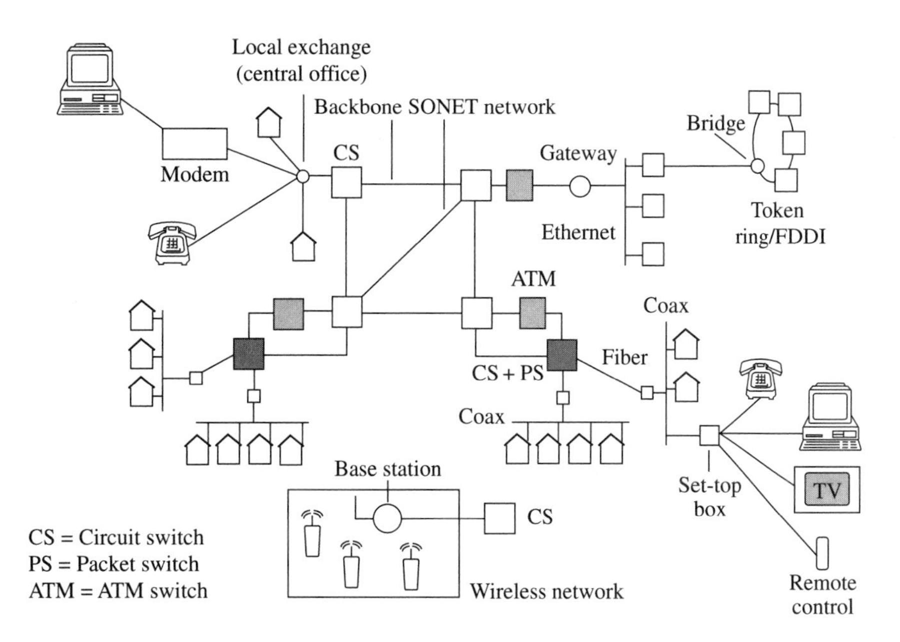

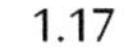
FIGURE 1.17 The future network will be heterogeneous, scalable, and flexible.

duction of ATM in part of the Internet WAN. The vision of the pure ATM "universal" network will then begin to be realized.

Finally, there will be a large increase in mobile and wireless access to networks. Terminals resembling today's notebook computers, connected to base stations by wireless radio or infrared links, will become common.

Figure 1.17 sketches a version of the information superhighway with these features. It includes a backbone circuit-switched SONET network with links at speeds of gigabits per second. The SONET network provides transport for ATM services, as well as circuit-switched connections for carrying video programs to local CATV head stations. The latter distribute those programs over a fiber-coaxial distribution network. Control messages from users are sent over a packet-switched network. ATM services are used to provide wide area transport for Internet traffic generated over FDDI or Ethernet local area networks. Lastly, wireless terminals access base stations connected to the wired networks.

The challenge is to interconnect these networks in ways that accommodate this heterogeneity, that are extensible, and that provide the range of quality of service needed to support a large variety of information services. In the rest of this book we will be concerned with precisely defining these challenges and offering plausible responses.

1.4 SUMMARY

Communications services today are provided by telephone, CATV, and data networks. These three types of networks serve different markets using different technologies. However, technological advances are dissolving those differences, and the three networks are converging in their ability to provide integrated services. The drive towards convergence is spurred by digitization, economies of scale, networking externalities, and economies of scope. ATM represents one major technological integration of these developments. The future information superhighway will comprise a collection of heterogeneous networks: circuit-switched SONET networks that carry telephone and video traffic and provide for the transfer of ATM cells, fiber-coaxial cable distribution networks for video, and a ramified Internet that provides the most economical best-effort packet transport and modified protocols to control the quality of service. The challenge to network engineers is to make it possible to interoperate these heterogeneous networks in ways that are extensible and secure and that provide the necessary range of service quality.

1.5 NOTES

A brief history of the Bell System with an extensive account of the economic and political factors that led to the deregulation of the long-distance telephone industry in the United States is given in [TG87]. A description of the ARPANET philosophy is available in [Cl88]. The current cable TV technology is described in [C90]. The applicability of ADSL for video dial tone is studied in [CW94]. Technical discussions of high-speed digital subscriber loop technology, including HDSL and ADSL, are presented in [JSAC95]. There is a large volume of literature on digital signal processing; [GG92] is devoted to compression techniques. There was a flurry of papers in the 1970s on the economics of networking externalities of which a typical example is [AA73]. The example of Minitel is reviewed and compared with Internet in [LLKS95]. Many popular books on the Internet have appeared recently. There are no reliable figures on its size or growth (measured by number of hosts, subnets, users, activity), how it is used, its cost, the extent of congestion, and so forth. A summary of a survey of Internet users is available on the Web [CN95]. ATM is described in Chapter 5. A 1995 issue of *IEEE Network* is devoted to video dial-tone networks [IN95].

1.6 PROBLEMS

1. Using modems to exchange information can be inefficient. Suppose you use a 9.6-Kpbs modem to type an e-mail message from your home computer to a friend's computer. During this transmission, a telephone connection is established. If you are a fast typist, you can type 80 words (400 characters) per minute. Show that you are using less than 0.5% of capacity of the modem connection. By first typing your mail into a file and then transmitting it, you can send it at the full line rate.
2. In a packet-switched network, the transmission line is shared by many users. How many users' e-mail needs can be accommodated by a shared 9.6-Kbps line? Assume that the average e-mail message consists of text composed by typing and copying an existing file in the proportion of 1 to 5.
3. We develop a model to estimate the feasible length of a CATV distribution network. The network consists of a series of sections. Each section comprises a length of cable, which attenuates the signal, followed by an amplifier, which boosts the signal strength but also adds noise. If the input power to the cable is x, the output power is $y = \alpha x$, where $0 < \alpha < 1$ (α depends on the length of the cable). If the input power to the amplifier is y, its output power is $z = \gamma y + n$ where $\gamma > 1$ is the gain and n is noise (refer to Figure 1.18). Suppose the power of the CATV signal at the beginning of the distribution network is S, and suppose the signal goes through a network that is K sections long. At the end of this network the signal Z is a sum of two terms, $Z = X + N$, where X is the CATV signal power and N is the noise power. For proper reception, we need (1) minimum CATV signal

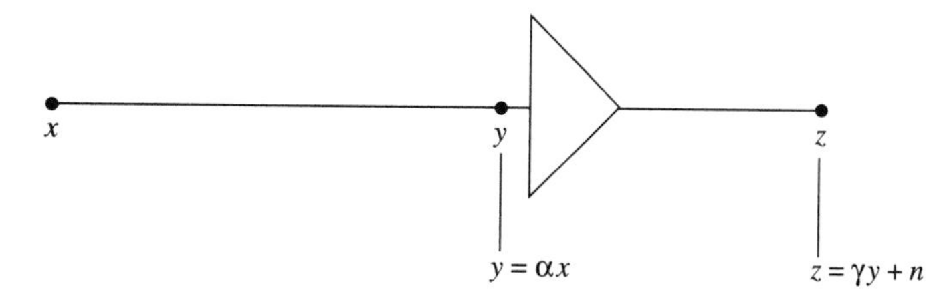

FIGURE 1.18 A section of a CATV distribution network consists of a length of cable, which attenuates the signal, followed by an amplifier, which boosts the input signal and adds noise.

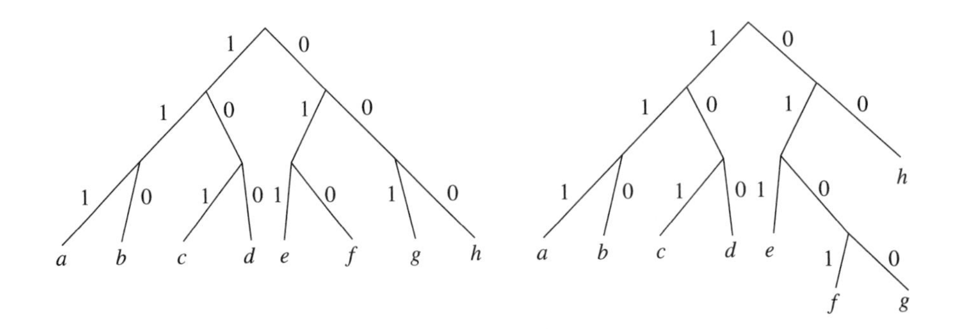

1.19

FIGURE

A finite alphabet is encoded into binary words by associating each letter with a leaf of a binary tree. The tree on the left gives a fixed-length encoding; the tree on the right gives a variable-length encoding.

strength $X \geq X_0$ and (2) a minimum signal-to-noise ratio $X/N \geq R_0$. How long can the network be? We can increase the network length by increasing the input signal power S. We can also replace the cable by an optical fiber whose α is much larger than that of cable, but for which the input signal level S is much smaller. Discuss the trade-off between increasing S and increasing α.

4. Suppose sequences of letters from a finite alphabet are to be transmitted over a binary communication channel. Assume that there are 2^n letters in the alphabet, so that each letter can be encoded into an n-bit word, as in the left panel of Figure 1.19. An m-letter sequence is thus encoded into a binary sequence of length mn. A variable-length encoding is obtained, as in the right panel, by associating each letter with a leaf of the tree. Thus the letters f and g are encoded into 4-bit words, h is encoded into a 2-bit word, and the rest into 3-bit words. A variable-length encoding is preferred if some of the letters are used more frequently than others. Suppose, in the example above, the letter h is used with probability 0.5, and the remaining seven letters are used with probability 1/14. Show that the average length of the binary encoding on the right is smaller than the average length on the left. If there are N letters in the alphabet and they occur with probabilities $p(1), \ldots, p(N)$, then the minimal length encoding requires H bits per letter, where H is the entropy:

$$H = -\sum_k p(k) \log_2 p(k).$$

Show that if all letters are equally likely, the minimal length encoding requires $\log_2 N$ bits per letter.

5. A customized CD company might operate as follows. All its music would be available by means of central servers. Customers would come into any branch store and select the songs they want. Bit streams representing those songs would be transported over the network from the servers and written onto a CD. The company would save money since it would have no inventory. Is the scheme feasible? As we have seen, a 70-minute CD stores 66 MB. How large a communication bandwidth should be provided between server and store so that a customer would not need to wait for more than 10 minutes? If a 45-Mbps link has a monthly rent of $20,000, and a CD sells for $20, and if a 100% markup is needed for other costs, is this a profitable idea? How much should the rent be to make the scheme profitable? Can you think of other business opportunities where communication substitutes for inventory?

6. Sketch the data network in your campus or company. How many hosts are there, and how large is the user population? What is the speed of the access link to the Internet? How do you gain access to the Internet? How much does home access to the Internet cost?

7. Sketch how your telephone is connected to the central office of your telephone company. How large is the switch located in that central office? How can you gain access to different long-distance phone companies using the same local telephone company?

8. Sketch the CATV network in your city. It is likely that the city has granted this network a franchise, meaning that it has the exclusive right to operate in your city. What arguments can you muster for and against such a franchise?

9. Besides CATV and broadcast TV, what other means do you have to receive video programs? How would you characterize the differences between these alternative sources of supply from the customer's point of view?

10. Is there a newspaper in your area that you can read on-line, for example on the World Wide Web? If there is, compare that service with the printed version of the same newspaper.

11. Recent purchases of CATV companies suggest that they are valued at $3,000 per customer, so that if a CATV company has one million customers, its market value is about three billion dollars. To obtain such a market value, the CATV company must make profits of about $300 per

customer per year. How does the company make such profits? Look at the annual reports of some CATV companies in your library.

12. What is the revenue per residential and business customer of your local phone company?

13. Compare two proposals to distribute movies. Assume that in both schemes 100 video channels are available to distribute movies to 300 households. In the first scheme, which uses the existing CATV network, the CATV operator finds out the 25 most popular movies and plays them continuously starting every 15 minutes. In the second scheme, households order the movies they want by using a control channel. This scheme requires costly modification of the existing CATV network to create the control channel and additional terminal equipment that users must employ to register their demand.

    Households must pay for watching a movie. Build a model for household demand that can be used to predict which scheme will make the greater profit. How would you go about collecting data to estimate or validate your model? What is your guess as to which scheme is more likely to be profitable and why?

14. One common approach to public regulation of a monopolistic telephone company is called *rate of return regulation,* in which a company's profit is limited to $\rho \times V$, where ρ is a "fair" rate of return (say, 15%) and V is the value of the company's assests. (V can be measured as the depreciated value of the company's past investments.) It has been suggested that rate of return regulation encourages the telephone companies to invest in more capital equipment and hire fewer workers than is optimum from society's viewpoint. Can you argue in favor of or in opposition to this suggestion?

2

CHAPTER

Network Services and Layered Architectures

Many networks provide transportation services. The postal system, whose services include the transfer of letters and parcels, is a familiar example. The postal system's services are differentiated by quality: there is registered mail, overnight delivery, surface mail, third-class mail. These services are built from basic transportation services, such as truck or rail or air transport. The postal system uses these basic services to create the more sophisticated services that its customers purchase. If you mail a letter, the system selects a route over which to send it; puts the letter with others going on the same route in a larger container; ships the container using air transport, say; transfers the container to a post office near the destination using truck transport; and finally brings the letter to the destination using yet another service, namely hand delivery by the letter carrier.

Characteristics of the basic services and system performance are summarized and measured by a few parameters: the volume of letters that can be handled per day or per hour, the speed of delivery, the fraction of letters that are lost. To a considerable extent these characteristics are determined by the capabilities of the postal system "hardware": the number of trucks, train cars, and airplanes the postal system has; their speeds; the routes they can use; and so on. These hardware-determined capabilities are then managed and controlled by the postal system's "intelligence"—embodied in hardware or software systems or postal system employees—to produce the more sophisticated services offered to customers. Naturally, the characteristics of the underlying basic services limit the range and quality of the sophisticated services that can be provided. For example, if the postal service used only railway and truck transport, it would not be able to offer overnight delivery. Over time, postal services of different countries have been *interconnected*. Lastly, through service integration, the postal system uses its existing resources to offer new services: you can "wire" money and telegrams, you can mail a letter and send a fax, you can open a post office box and a postal savings account, etc.

We will find that the key concepts introduced in this simple description of the postal system carry over with appropriate reinterpretation to the case of communication networks. Those concepts are user services and service quality, basic or bearer services and their characteristics, the underlying "hardware," and management "intelligence." We have already encountered the concepts of network interconnection and service integration in section 1.2.4.

The services provided by communication networks enable users to exchange information. There is a wide range of services: users can talk to each other over a telephone network, transfer data over a computer network, and watch a video program over a cable TV network. Since users interact with the network through some terminal device (telephone, computer, TV controller), it is sometimes more appropriate to say that the network services are used (consumed) by user applications (processes running on the terminal device).

Engineers build a network by interconnecting two types of "hardware" or network elements: transmission links and switches. Links transfer strings of bits from one location to another. Switches are computers that store, route, and manipulate those bit strings. This hardware supports the network's *bearer services:* the transfer of bit strings, in a few standard formats, from one source or user to one or more network destinations. Bearer service performance characteristics are summarized in a few parameters: the acceptable formats; the connectivity and selection of routes from source to destination; and the speed, delay, errors, etc., of the bit string.

A network can effectively support a particular user application only if its bearer services have the requisite characteristics. To support voice conversations, for example, the end-to-end delay should not be more than 200 milliseconds (ms), say. To support data transfer the error rate should not be more than 10^{-4}, say, and so on. The most demanding applications, such as the transfer of X rays with a fidelity and speed that make them acceptable for use by radiologists for diagnosis and interactive videoconferencing, require a high-performance network.

More sophisticated services are built from less sophisticated ones in a layered hierarchy or architecture. Each layer adds functionality. Each layer comprises specific rules of control and management, implemented in software or hardware. These rules and the software that implements them are called *protocols*. The protocols reside in the switches and host computers. Some architectures have become standardized, and interconnection of networks is facilitated if they conform to the same standard.

This chapter introduces the concepts of how networks function and the logical layers used to divide networks into smaller subsets of functionality. By the end of this chapter you will be able to calculate the bit rate needed for various high-performance applications and understand the reduction in the bit rate needed for video signals using MPEG in comparison with uncompressed NTSC signals. You will understand the mechanisms used to implement network functions, including error-control schemes such as the Alternating Bit and Go Back N protocols.

We start by discussing the communication traffic generated by user applications in section 2.1. We comment on the information that different types of user applications exchange. In particular, we explain why different applications require network services with different characteristics.

In section 2.2 we describe network services and discuss their characteristics. These characteristics match those required by user applications.

In section 2.3 we identify high-performance networks in terms of their scalability, their ability to support demanding and diverse user applications, and the ease with which they can be connected to other networks.

In section 2.4 we discuss network elements and explain how the properties of these elements affect the characteristics of the services that they implement.

We explain the mechanisms that the network elements use to implement services in section 2.5.

In section 2.6 we introduce the layered organization of network operations. We explain that the main advantages of a layered organization are modularity and standardization.

In section 2.7 we present the Open Data Network (ODN) model. This model provides a useful framework for describing how networks are organized. The Open Systems Interconnection or OSI reference model, used in data networks, is presented in Chapter 3.

2.1 APPLICATIONS

Communication networks enable users to exchange information. Users exchange information through applications (automated processes) implemented in computers or in communication devices. This information can be in many forms: text, voice, audio, data, graphics, pictures, animations, and videos.

Information form	Traffic type	Size of bit stream
Voice	CBR	64 Kbps
Video	CBR	64 Kbps, 1.5 Mbps
	VBR	Mean 6 Mbps, peak 24 Mbps
Text	ASCII	2 KB/page
	Fax	50 KB/page
Picture	600 dots/in, 256 colors, 8.5 × 11 in	33.5 MB
	70 dots/in, b/w, 8.5 × 11 in	0.5 MB

2.1 TABLE Characteristics of bit streams for some common forms of information.

Moreover, the information transfer may be one-way, two-way, broadcast, or multipoint.

The information exchanged can be analog or digital. CATV networks, for example, deliver analog video signals to television sets. The telephone network transmits analog or digital voice signals between telephones. Computer networks transfer bit files or bit streams representing text, data, still images, and audio or video signals. Most networks transmit analog signals by first converting them into bit streams, as we explained in section 1.2.1. In this chapter we limit discussion to digital transmission of information, so user applications eventually require the communication network to transmit strings of bits. We call these bit strings the *traffic* generated by the application. In order to support a user application, the network must be able to transport in a satisfactory manner the traffic that the application generates. Table 2.1 presents some characteristics about the bit stream generated by common forms of information.

Notice that the bit streams generated by a video signal can vary greatly depending on the compression scheme used. When a page of text is encoded as a string of ASCII characters, it produces only a 2-Kilobyte (KB) string; when that page is digitized into pixels and compressed as in facsimile, it produces a 50-KB string. The LaTeX file for this book, including figures, compresses into a 1-megabyte (MB) file. A high-quality digitization of a color picture (similar quality to a good color laser printer) generates a 33.5-MB string; a low-quality digitization of a black-and-white picture generates only a 0.5-MB string.

We classify all traffic into three types. A user application can generate a constant bit rate (CBR) stream, a variable bit rate (VBR) stream, or a sequence of messages with different temporal characteristics. We briefly describe each type of traffic, and then consider some examples.

2.1.1 Constant Bit Rate

To transmit a voice signal, the telephone network equipment first converts it into a stream of bits with a constant rate of 64 Kbps (see section 1.2.1). Some video-compression standards convert a video signal into a bit stream with a constant bit rate (CBR). For instance, MPEG1 converts a video signal into a 1.5-Mbps bit stream.

For the voice or video application to be of an acceptable quality, the network must transmit the bit stream with a short delay and corrupt at most a small fraction of the bits. (This fraction is called the *bit error rate* or *BER*.)

The end-to-end delay should be less than 200 ms for real-time video and voice conversations, since people find larger delay uncomfortable. That delay can be a few seconds for non-real-time interactive applications such as interactive video and information on demand. The delay is not critical for noninteractive applications such as distribution of video or audio programs.

The maximum acceptable BER is about 10^{-4} for audio and video transmission, in the absence of compression. When an audio and video signal is compressed, however, an error in the compressed signal will cause a sequence of errors in the uncompressed signal. Therefore, the tolerable error rate for transmission of compressed signals is much less than 10^{-4}.

2.1.2 Variable Bit Rate

Some signal-compression techniques convert a signal into a bit stream that has a variable bit rate (VBR). For instance, MPEG2 is a family of standards for such variable bit rate compression of video signals. The bit rate is larger when the scenes of the compressed movie are fast moving than when they are slow moving.

To specify the characteristics of a VBR stream, the network engineer specifies the average bit rate and a description of the fluctuations of that bit rate. We study such descriptions in Chapter 6.

The acceptable delay and BER of these applications are similar to those of CBR applications.

2.1.3 Messages

Many user applications on a network are implemented by processes that exchange messages. (For present purposes, a message is a variable-length bit string.) For instance, to consult a remote database, a user sends queries to the database server, which replies by sending the requested records to the user. As another example, a distributed computing application generates remote procedure calls for remote machines, which then return the results of the execution of the procedures.

The message traffic generated by various user applications can have a wide range of characteristics. Some applications, such as e-mail, generate isolated messages. Other applications, such as a distributed computation, generate long streams of messages. The rate of messages can vary greatly across applications and devices.

To describe the amount of traffic generated by an application that produces a stream of messages, the network engineer may specify the average traffic rate and some measure of the fluctuations of that rate, in a way similar to the case of a VBR specification.

The network must transfer the messages with an acceptable delay, and it can corrupt only a small fraction of the messages. Typical acceptable values of delays are 200 ms for real-time applications, a few seconds for interactive services, and many seconds for noninteractive services such as e-mail. The acceptable fraction of messages that can be corrupted ranges from 10^{-8} for data transmissions to much larger values for noncritical applications such as junk mail distribution.

Among the applications that exchange sequences of messages, we can distinguish those applications that expect the messages to reach the destination in the correct order and those that do not care about the order.

2.1.4 Examples

Figure 2.1 indicates ranges of peak and average values of bit rates generated by different applications.

Uncompressed HDTV generates a bit stream at a constant rate (equal to a fixed number of frames per second multiplied by a fixed number of bits per frame) so its peak and average rate coincide at several hundred Mbps. By contrast, compressed HDTV generates a bit stream at a variable rate with a peak to average rate ratio of about 10. (The figure shows a range of bit rates needed by compressed and uncompressed HDTV.) As we mentioned earlier, MPEG1 is a standard for compressing NTSC TV into a 1.5-Mbps constant bit rate stream. Adaptive compression schemes such as MPEG2 lead to randomly varying bit rates, requiring statistical descriptions that cannot be simply summarized by peak and average values. We study such descriptions in Chapter 6. A file transfer involves reading from a disk at a rate determined by the network (and the characteristics of the file server). Typically, this rate is constant, and applications do not require speeds exceeding a few hundred Kbps. Users

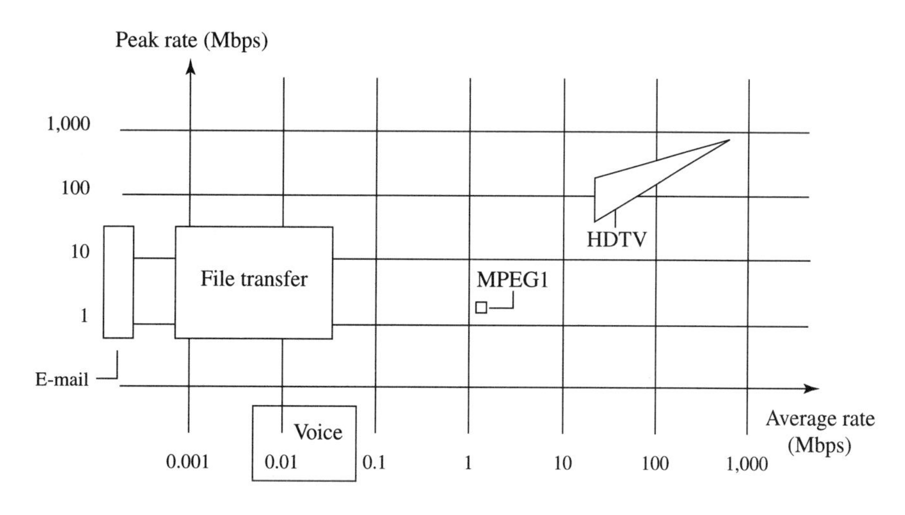

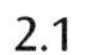
2.1 FIGURE Peak rate and average rate of a few representative applications.

are accustomed to e-mail being delivered with a delay of several minutes or hours, and so this application does not require speeds exceeding a few Kbps.

Figure 2.2 indicates acceptable ranges of peak rates and delays. A remote control application (as in the case of controlling an electric power plant) may require a very small delay, while TV distribution, e-mail, and file transfers may tolerate larger delays.

Figure 2.3 compares acceptable error rates for some applications (note that the horizontal axis is in bits/Mb). For voice and TV, bit error rates on the order of 10^{-4} are acceptable because people do not perceive them. File transfer applications demand much lower error rates because an error in a file transfer will usually trigger retransmission of a part of the file, so that too many errors will cause an intolerably large number of retransmissions and waste of transmission bandwidth. Control applications may similarly be intolerant of large error rates, because the control system may perform poorly in the presence of delays introduced by retransmission.

2.1.5 Other Requirements

In this book we will consider only the delay and loss requirements that applications impose on the network. We should remember, however, that other requirements may be important. We mention here reliability and security.

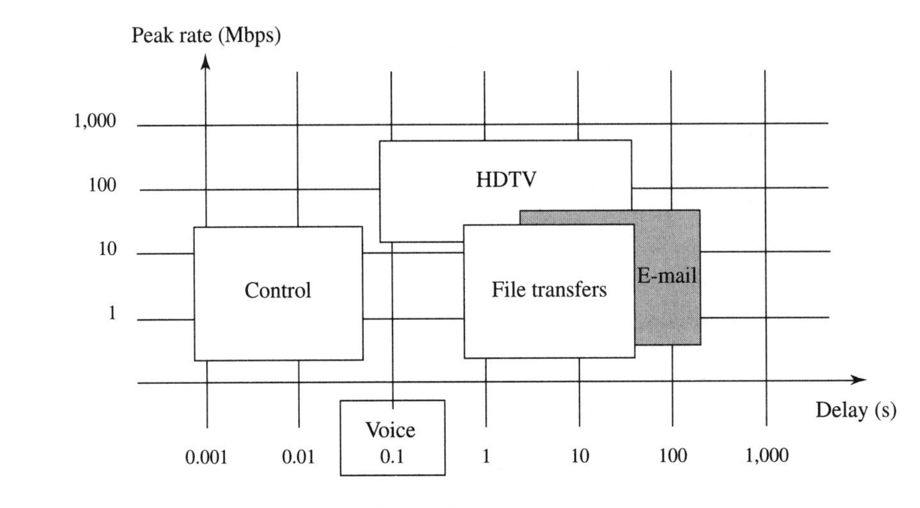

2.2 FIGURE Peak rate and acceptable delays of applications.

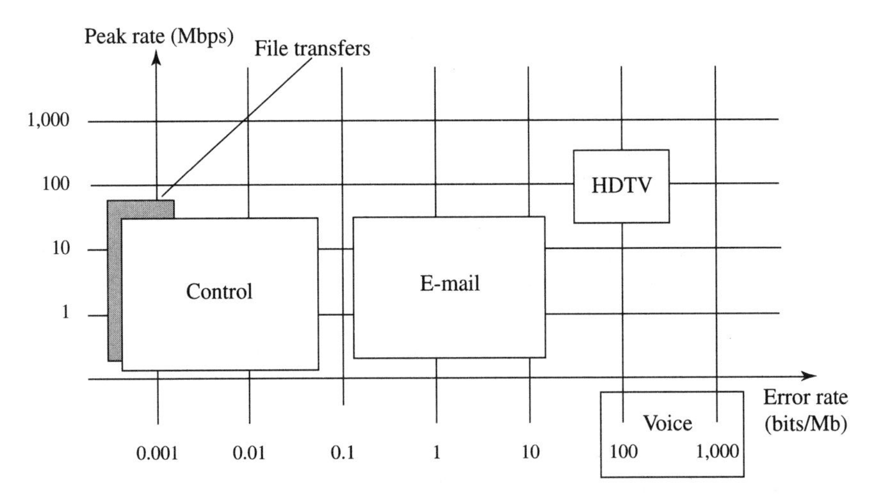

2.3 FIGURE Peak rate and acceptable error rates of applications.

When one or more links or switches fail, the network may be unable to provide a connection between source and destination until those failures are repaired. *Reliability* refers to the frequency and duration of such failures. Some applications (e.g., control of electric power plants, hospital life support systems, critical banking operations) demand extremely reliable network operation. Typically, we want to be able to provide higher reliability between a few designated source-destination pairs. Higher reliability is achieved by providing multiple disjoint routes between the designated node pairs.

Recall that in a multiple-access network such as Ethernet, every computer "hears" every packet that is transmitted. The case of wireless phone transmission is similar. In these networks, to guarantee privacy of transmissions, it will be necessary to encrypt those transmissions. More generally, *security* is concerned with measures that can be taken to prevent unauthorized access to data or information transfer. As money and other assets take on an electronic form that can be transmitted over a network, issues of security will become more pressing.

2.2 NETWORK SERVICES

We have just seen that user applications exchange bit streams or messages with widely different traffic characteristics. These applications expect the network to deliver the bit streams or messages within a specific delay and to corrupt only a small fraction of the bits or messages.

Network engineers distinguish two types of information transfer services: connectionless and connection-oriented. We explain that important distinction next.

2.2.1 Connection-Oriented Service

When a network implements a connection-oriented service, it delivers messages from the source to the destination in the correct order. Thus, the data transfer in a connection-oriented service appears to take place over a dedicated transmission line, except for the variability in the transmission delay of different packets. A connection-oriented service is required by user applications that expect reliable and ordered transmissions of messages. A CBR or VBR bit stream is delivered by a connection-oriented service.

A connection-oriented service involves three phases: a connection setup phase, a data transfer phase, and a connection teardown phase.

The quality of service (QoS) in some connection-oriented network services specifies whether the transmission is error free and may assign some priority level to packets with the understanding that the network will attempt to transmit high-priority packets before low-priority packets. Thus the delays are likely to be smaller in a high-priority connection-oriented service than in a low-priority connection-oriented service.

Some networks permit a more detailed QoS specification. That specification includes the delay, delay jitter, and packet error rate. It also includes a description of the amount of traffic that the service can transport. These more detailed specifications are required by real-time and interactive applications, as we explained in section 2.1.

When requesting a connection-oriented service in these networks, at the connection setup time the user specifies the QoS that is required by the user application. The network can then determine whether it has sufficient resources available to handle that connection with the requested QoS, and the network can then set aside these resources for the connection. In order to undertake these tasks, the network retains "state" information about existing connections. How these tasks can be carried out is discussed in Chapters 6 and 7.

2.2.2 Connectionless Service

When it implements a connectionless service, the network transfers each packet of data to the destination one at a time, independently of the other packets. Unlike the case with connection-oriented services, the network has no state information to determine whether a packet is part of a stream of other packets. In particular, the network has no knowledge of the amount of traffic that will be sent by the user. Consequently, the network cannot set aside resources that would be needed to achieve a specific quality of service.

Because of this limited information, only a restricted range of service quality can be offered. Typical QoS parameters include a bound on the maximum packet size and service priority: a higher-priority packet is transmitted before a lower-priority packet.

Connectionless service is characterized by the average or typical delay and a specification of the way the service handles errors. Some connectionless services do not indicate when they fail to deliver a message. Other services acknowledge the correct delivery of messages.

2.3 HIGH-PERFORMANCE NETWORKS

We define a high-performance network (HPN) as a communication network that supports a large variety of user applications and that is scalable. In order to support many applications, the network must be able to transfer user traffic at high speed and with low delay. It must be able to allocate resources in ways that match the application requirements. Network organization and management must be flexible so that new applications can be supported as the need arises.

A scalable network can accommodate growing numbers of users without degradation in performance. Growth is usually accommodated by interconnecting distinct networks. The network must be able to provide connectivity over a growing span. The span may be expressed in distance, number of links, or number of subnetworks.

An HPN that supports a wide range of current and future applications and that can accommodate growth is built and managed differently from networks that are designed for a specific application or user population.

In Figure 2.4 we differentiate HPNs from other networks in terms of speed and distance. At one extreme, phone networks have many nodes and operate over very large distances, but users can transfer data only at low

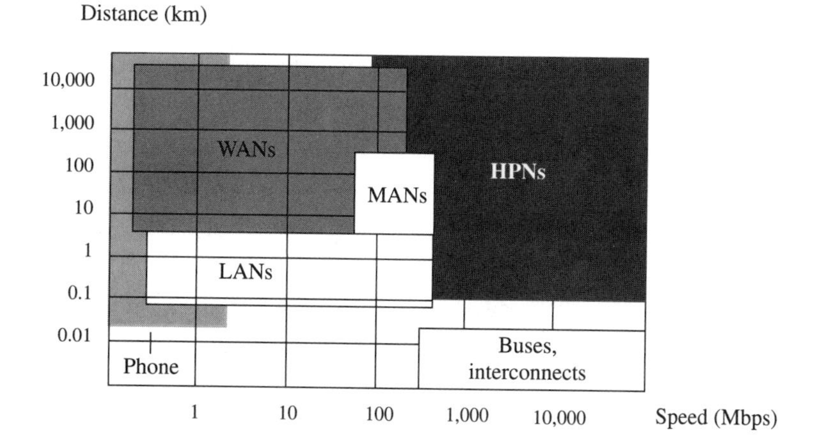

2.4 FIGURE

High-performance networks are defined by the characteristics of the services they provide.

speeds. (Thus phone networks are scalable, but support limited applications.) At the other extreme, a computer backplane bus operates at high speeds but connects only a small number of devices very close to each other. Local area networks or LANs (e.g., Ethernet) can transfer data between tens of nodes at moderate speeds (10 Mbps) over moderate distances (1 km). More recent LANs support speeds of 100 Mbps. Wide area networks or WANs, such as X.25 networks and Internet, connect hundreds of nodes over hundreds of kilometers, but they operate at limited speeds of a few Mbps or less. Metropolitan area networks (MANs—e.g., DQDB, FDDI, SMDS) have higher speed and connect users separated by about 100 km. Frame Relay networks are streamlined versions of X.25 networks that can operate at high speed. The "backbone" telephone network is an HPN: it comprises the switches and links or trunks connecting them, but excludes the low-speed links connecting user telephones to the switches.

The precise values of the user transfer rate, the acceptable delay, the network span, and the number of users that characterize an HPN are somewhat arbitrary. We have in mind a user rate that exceeds 100 Mbps, delays on the order of 100 ms, a span of at least 100 m, and a number of users that can exceed 100. What is essential for a network to qualify as an HPN is for it to be able to support demanding services such as interactive MPEG video and LAN interconnections among many users.

2.4 NETWORK ELEMENTS

A communication network is a collection of network elements interconnected and managed to support the transfer of information from a user at one network location or node to a user at another node. In this section we discuss the two principal network elements and we examine how the properties of these elements affect the characteristics of the services that they implement.

2.4.1 Principal Network Elements

The principal network elements are (transmission) links and switches.

A *link* transfers a stream of bits from one end to the other at a certain rate with a given bit error rate and a fixed propagation time. Links are unidirectional. The most important links are optical fiber, copper coaxial cable, and microwave or radio "wireless" links. Optical fiber and copper links are

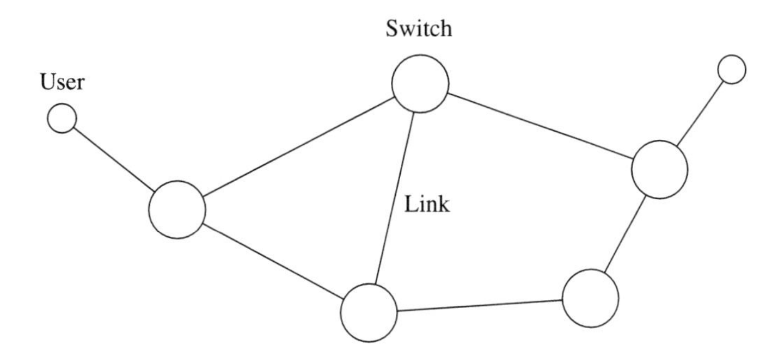

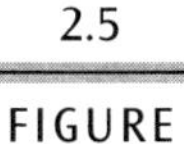

2.5 FIGURE

The principal network elements are transmission links and switches. Transmission links connect user nodes and switches.

usually point-to-point links, whereas radio links are usually broadcast links. We study links in Chapter 9, focusing on optical links, which are essential for high-speed networks.

Several incoming and outgoing links terminate at a *switch,* which is a device that transfers bits from its incoming links to its outgoing links. Whenever the rate of incoming bits exceeds that of outgoing bits, the excess bits are buffered at the switch. We study switches in Chapter 10, focusing on switches that can handle high-speed traffic.

When we view a network as an interconnection of network elements, we may represent the network as in the graph of Figure 2.5. In the graph, edges denote links, large circles denote switches (including buffers), and small circles denote user nodes where bits are generated or consumed. We will use the names *user, source, destination, station,* and *node* as near synonyms.

Another very important view of an interconnection of network elements is provided by the queuing network model. Consider a switch with n incoming links and m outgoing links, as in the left of Figure 2.6. The diagram on the right is the queuing model. Each incoming link's receiver writes into its input buffer, and each outgoing link's transmitter reads from its output buffer. The switch transfers bits or packets from input buffers to the appropriate output buffers. This queuing network model of switches and links is used to describe and evaluate network performance. For example, the queuing delay encountered by a packet in an output buffer is proportional to the number of packets in the buffer in front of it. If packets arrive into a full buffer, there will be packet loss. It is difficult to calculate queuing delay and packet loss. We describe several approaches in Chapter 6.

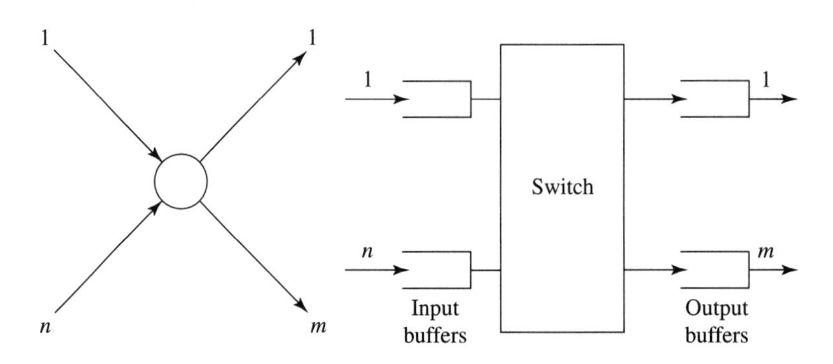

2.6 FIGURE The queuing network model is used for performance analysis. Each switch has a buffer corresponding to each incoming and outgoing link, which is serviced at the link rate.

2.4.2 Network Elements and Service Characteristics

A packet generated by a source travels over one link, gets buffered at a switch, is then routed to another link, and so on, until it arrives at its destination.

The delay packets experience through a network depends on the elements that constitute the network, the traffic that goes through these elements, and the way the network is operated.

The detailed analysis of the delay through a network is rather involved. For now, note that we can decompose the total delay into four components:

$$\text{total delay} = \text{TRANS} + \text{PROP} + \text{QD} + \text{PROC}. \tag{2.1}$$

In (2.1) TRANS is the time required to transmit a packet, so

$$\text{TRANS} = (\text{packet size})/(\text{transmission speed}). \tag{2.2}$$

For example, for a 10,000-bit packet and a transmission speed of 1 Mbps, TRANS is 10 ms. PROP is the signal propagation time, so

$$\begin{aligned}\text{PROP} = {} & (\text{distance from source to destination})/ \\ & (\text{speed of electrical or optical signal}).\end{aligned} \tag{2.3}$$

Propagation time for an electrical or optical signal is between 3.3 and 5 μs/km. QD is the queuing delay in the switch. It occurs whenever the bit rate of the traffic coming in to the switch exceeds the capacity of the outgoing link. The excess bits are queued in the switch buffers. In contrast with the other two sources of delay, queuing delay is significantly affected by the network control policy. Finally, PROC is the processing time required by the network switches. We will assume that this processing time is negligible.

Suppose as a rough rule of thumb that the network is controlled so that each packet coming into the switch has to wait on average for four previous packets to be transmitted. Then the average queuing delay is four times the transmission delay, so

$$\text{total delay} = 5 \times \text{TRANS} + \text{PROP}. \tag{2.4}$$

By combining the expressions (2.2)–(2.4), we see how the delay depends on the transmission rate and length of the links and on the length or size of messages.

2.4.3 Examples

We illustrate how we can use the preceding analysis in three examples.

We consider a potential telecommuting application in which a company employee responds to customer billing inquiries from a terminal at home. When a customer calls, an automatic call-transfer service transfers the call to the employee's home. The employee then requests the customer's record from the company's database computer. The record is transferred by the network and displayed on the terminal. The employee then answers the customer's questions and updates the record as necessary. For the response to be satisfactory to the customer, the retrieval of the record and its display should not take more than one second, say. A screen full of text takes about 16,000 bits, so the network response will be satisfactory provided the transmission speed of B bps is such that

$$\text{total delay} = 5 \times 16{,}000 \times B^{-1} + \text{PROP} < 1.$$

In this example, we may neglect PROP, which is a few μs. We see then that this application needs a transmission speed of 80 Kbps. This rate is supported by (narrowband) ISDN services described in Chapter 4. An ISDN connection costs much less than $200 per month (1996 prices), including the cost of ISDN user interface equipment, spread over the equipment lifetime. Comparing this cost with the savings to the employee and employer resulting from fewer work commute trips, less office and parking space, and more flexible work schedules, we can imagine a large potential demand for communication services needed to support telecommuting. (It is estimated that a telecommuter working at home 1–2 days per week can save most companies \$6,000–\$12,000 a year because of increased productivity, lower staff turnover, and reduced office space. The annual cost per telecommuter is estimated to be up to \$10,000 for equipment, space at home, and phone services.)

As another telecommuting example, consider a design engineer or architect using a workstation at home. Most of the material needed for work is stored in the workstation's local disk. However, from time to time, the engineer needs to retrieve or replace a large file (e.g., several bit maps) stored in a file server at the workplace. Such a file is about 3 MB long and if the workstation is connected by a 1-Mbps link, the retrieval would take about 24 seconds. This may be adequate if such transmissions are infrequent. However, if 10 transmissions every minute are needed, this may require a 4-Mbps link, which at current costs may make this telecommuting application uneconomical. (At the workplace, the engineer's workstation and file server are connected by a relatively inexpensive 10-Mbps local area network such as Ethernet.)

The third example provides a more general illustration of the requirements that applications place on the delay and speed of bearer services. Figure 2.7 compares three bearer services that transport data across the United States through 10 switches at various combinations of speed and delay.

The first service is implemented by a circuit-switched network using wired (optical or copper coaxial) links. The end-to-end delay equals the propagation delay, PROP, independent of the speed. This delay is about 25 ms (5,000 km $\times$ 5 μs/km). This service is summarized by the horizontal line labeled "Wired circuit-switched."

The second and third services are implemented by a packet-switched network using 1-kilobit (Kb) and 100-Kb sized packets, respectively. If at each switch, a packet encounters between 0 and 100 packets waiting in queue, then the queuing delay (through the 10 switches) is between 0 and 1,000 packet transmission times. The transmission time, TRANS, is 1,000/bit rate for the smaller and 10^5/bit rate for the larger packet. So for the second service the total delay is

$$\text{PROP} < \text{total delay} < \text{PROP} + 10^6/\text{bit rate},$$

which is the lower shaded region in the figure. For the third service the total delay D is

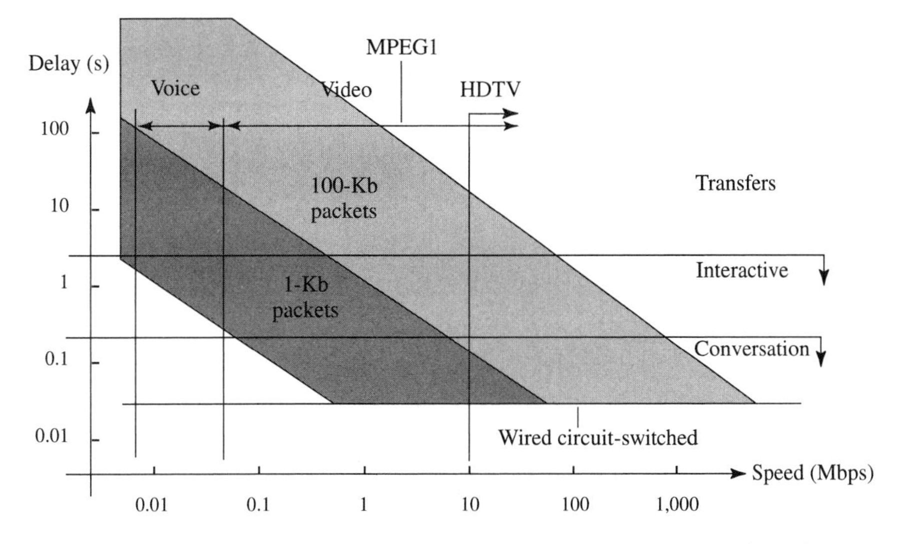

2.7 FIGURE

Delay as a function of transmission rate for a U.S.-wide network with ten nodes and a queue at each node of 1 to 100 packets. Delay for the circuit-switched connection equals the propagation time.

$$\text{PROP} < \text{total delay} < \text{PROP} + 10^8/\text{bit rate},$$

which is the upper shaded region.

We now consider three applications: a conversation, an interactive exchange, and a transfer of a voice or video file. The conversation can tolerate a delay of 250 ms, the interactive exchange can accept a few seconds of delay, and the file transfer can tolerate a large delay. These bounds are shown in the figure. Finally, depending on the compression technique used, voice will generate traffic at a rate between 8 and 64 Kbps, video between 64 Kbps and 10 Mbps, and HDTV will generate higher-speed traffic.

When we compare the delay and speed requirements of the application with the characteristics of the three services, we see that at speeds above 10 Mbps there is little difference between circuit-switched connections and a 1-Kb packet-switched service. With 100-Kb packets, higher speed is needed. For example, at T3 speed of 45 Mbps the delay of a 100-Kb packet service is indistinguishable from a circuit-switched connection, provided queues are shorter than 100 packets long. For interactive and transfer applications, the advantage of a circuit-switched connection over the packet-switched services similarly disappears at higher speeds.

2.5 BASIC NETWORK MECHANISMS

A network's bearer services comprise the end-to-end transport of bit streams, in specific formats, over a set of routes. These services are differentiated by quality: speed, delay, errors. They are produced using five basic mechanisms: multiplexing, switching, error control, flow control, and resource allocation. We discuss those mechanisms in this section.

As we noted in section 1.2.2, there are economies of scale in transmission. That is, the cost of a transmission link connecting two nodes does not increase in proportion to its capacity. Individual users located at those two points, however, typically need a small transmission bandwidth for short durations. *Multiplexing* combines data streams of many such users into one large bandwidth stream for long durations. Users thereby can share in the scale economies of transmission. However, users are not concentrated in a few locations; they are geographically dispersed. *Switching* allows us to bring together the data streams of these dispersed users. (In essence, the communication network industry makes profits by installing large transmission capacity and by "renting" this capacity to users in smaller amounts using multiplexing and switching.)

In the left side of Figure 2.8 we see a network with no multiplexing or switching. Each pair of phones is connected by a dedicated link of capacity 64 Kbps needed to carry one voice conversation. In this network, the average number of links per phone, and hence the average cost, increases with the number of nodes. Since a telephone is engaged in at most one conversation at any time, link utilization, i.e., the fraction of time the link is busy, decreases as the number of nodes increases. Thus the cost per phone grows with the size of the network. This network wastes resources.

The network on the right employs multiplexing and switching. Each phone is connected by a dedicated access link to one of two local switches, which are connected by a single link called a trunk. (In telephone networks, a switch is located in a *central office;* a link between two switches is called a *trunk;* a link between a subscriber telephone and a switch is called an *access line* or *subscriber loop*.) The access line capacity is 64 Kbps, and the trunk capacity is a multiple of 64 Kbps. (In this example it may be 2×64 Kbps.)

Depending on its capacity, a trunk can carry several voice conversations simultaneously by multiplexing. A conversation between two phones will occupy only two access lines—if both phones are connected to the same local switch—or, in addition, it will occupy a fraction of the trunk capacity. It is the

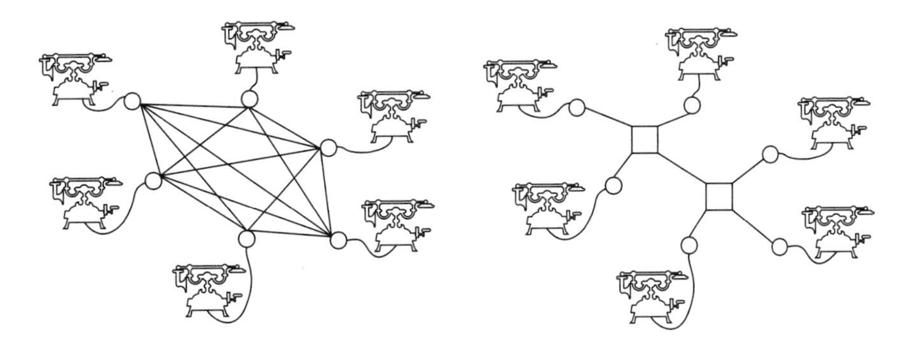

2.8
FIGURE

The left-hand part of the figure shows a fully connected network. The right-hand part of the figure shows a network where some links are shared. The sharing of links is made possible by multiplexing and switching.

task of the switch to determine whether a call originating from a local phone is destined for another local phone or for a phone connected to the remote switch. In this network, there is one access line per phone, but the switch and trunk capacities grow much less rapidly than the number of phones. As a result, the average cost of the network per user decreases with the number of phones. This decreasing cost structure of communication networks is made possible by multiplexing and switching.

Another important mechanism is *error control*. All transmission links occasionally corrupt the messages they transmit. Although carefully designed and maintained links have a very small bit error rate (e.g., 10^{-12}), even the rare errors in the transmissions may not be acceptable. Moreover, in addition to transmission errors, a message may fail to reach its destination because it arrived at a switch whose buffer was full and so the message was discarded. It is therefore important for the network to control such errors. We explain two methods that networks use to control errors later in this section.

The end-to-end delay is the sum of a fixed propagation delay and a variable queuing delay. In order to keep this delay within an acceptable range, the rate at which packets enter the network must be controlled. *Flow control* is the generic name for a set of mechanisms designed to limit the rate or number of packets introduced into the network by a source or a switch. If the flow-control mechanism does not function properly, an excessive number of packets may accumulate in the switch buffers causing unacceptable delay or loss.

We have seen that some applications require the network to provide a minimum bandwidth to ensure acceptable performance. For example, a voice conversation requires 64 Kbps and MPEG1 requires 1.5 Mbps. A variable bit rate application will require a guaranteed combination of minimum bandwidth and buffers. Because network resources—link bandwidth and switch buffers—are shared by many applications at the same time, *resource allocation* mechanisms must be designed to ensure that each application receives the necessary resources to maintain its quality of service.

Multiplexing is carried out in switches or specialized hardware. Switching and resource allocation are implemented in switches. Error control and flow control are implemented in switches and host computers.

2.5.1 Multiplexing

We now explain three important techniques for multiplexing N incoming channels onto one outgoing channel. These are time-division multiplexing, statistical multiplexing, and frequency-division multiplexing. For our present purposes a *channel* is a communication link of fixed capacity measured, say, in bits per second (bps). That is, each second of time on the channel is divided into a number of intervals called *bit times*, the number being equal to the channel capacity. For example, in a 1-Mbps channel, a bit time is 1 μs long. Each bit time may be occupied by an information or data bit, or it may be empty. We also say, accordingly, that the channel is busy or idle at that time. The average data rate divided by channel capacity, or the fraction of time the channel is busy, is the channel *utilization*. Utilization is equal to 100% only if the instantaneous data rate is always equal to the channel capacity.

Multiplexing is the process by which information bits from N incoming channels are transferred into bit times on one outgoing channel. *Demultiplexing* is the reverse process: the information bits on one incoming multiplexed channel are separated and transferred onto N outgoing channels. The multiplexed outgoing channel contains some extra bits (in addition to the incoming channels' data bits) that the demultiplexer uses to determine which data bits belong to which incoming channel.

Time-division multiplexing or TDM is illustrated in Figure 2.9. In TDM, the capacity of the outgoing channel is divided into N logical channels, and data in each of N incoming channels is placed in a designated outgoing logical channel. This is achieved as follows. Time on the outgoing channel is divided into fixed-length intervals called *frames*. Frames are delimited by a special bit sequence called a framing pattern, not shown in the figure. Time in each frame is further subdivided into N fixed-length intervals called *slots* (this name is not always used). Thus each frame consists of a sequence of slots: slot 1, slot 2, . . . , slot N. (A slot is usually 1 bit or 1 byte wide.) A logical

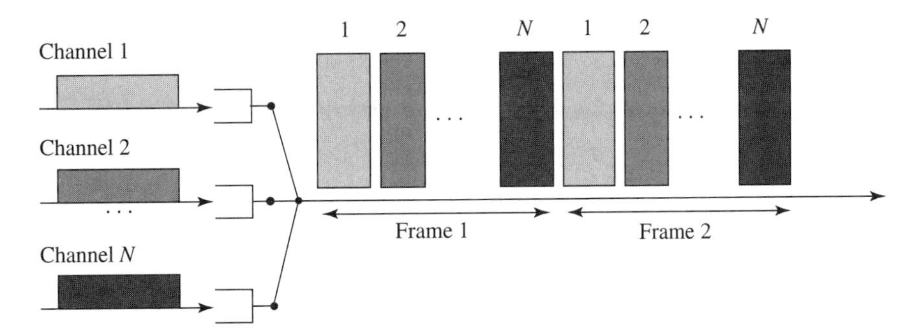

2.9

FIGURE

When a communication link is shared by time-division multiplexing, time is divided into frames. Each frame is divided into time slots that are allocated in a fixed order to the different incoming channels.

channel occupies every Nth slot. There are thus N logical channels. The first logical channel occupies slots $1, N+1, 2N+1, \ldots$; the second occupies slots $2, N+2, 2N+2, \ldots$; and so on.

The multiplexer operates as follows. The data bits in each incoming channel are read into a separate FIFO (first in, first out) buffer. The multiplexer reads this buffer in sequence for an amount of time equal to the corresponding slot time: buffer 1 is read into slot 1, buffer 2 is read into slot 2, etc. (If there are not enough bits in a buffer, the corresponding slot remains partially empty.) The bit stream of the outgoing channel is easily demultiplexed: the demultiplexer detects the framing pattern from which it determines the beginning of each frame, and then each slot.

TDM is easy to implement. The overhead due to extra framing bits is small. However, the utilization of the outgoing channel may vary a great deal depending on the burstiness of the incoming data streams. To see this, observe that the capacity of the outgoing channel must be as large as the sum of the capacities of the N incoming channels. As a result, the utilization of the outgoing channel will be low or high accordingly as the utilization of the incoming channels is low or high. TDM leads to high utilization if the incoming data is not bursty. Thus TDM is ideal for constant bit rate traffic.

Statistical multiplexing or SM, illustrated in Figure 2.10, is most effective in the case of bursty input data. As in TDM, the data bits in each incoming channel are read into separate FIFOs. The multiplexer reads each buffer in turn until the buffer empties. (It is customary to call the data read in one turn a data *packet*.) In TDM each FIFO is read for a fixed amount of time—one slot—and so each incoming channel is allocated a fixed fraction of the

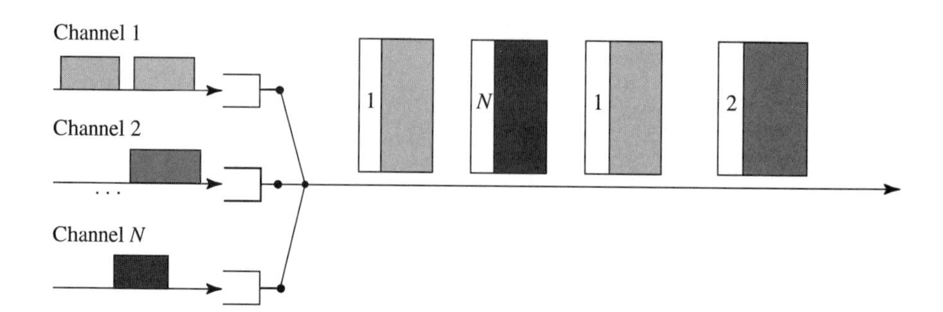

2.10 FIGURE

In statistical multiplexing, the multiplexer visits the incoming channel buffers in some order. The multiplexer empties a buffer before moving to the next one. The buffer contents are tagged to indicate their incoming channel. An idle channel does not waste transmission time.

outgoing channel capacity, independent of the data rate on that channel. By contrast, in SM, the capacity allocated to each incoming channel varies with time, depending on its instantaneous data rate: the higher the rate, the larger the capacity allocated to it at that time. As a result, the capacity of the outgoing channel needs to be only as large as the sum of the average data rates of the incoming channel, which, for bursty traffic, may be much smaller than the sum of the peak data rates. Hence the capacity of the outgoing channel may be smaller than the sum of the incoming channel capacities. We call the ratio of the total incoming capacity to the total outgoing capacity the *multiplexing gain*. This gain is unity for TDM but can be much larger for SM. Figure 2.11 illustrates the possible multiplexing gains for SM relative to TDM and FDM for various applications.

As we have seen, in SM the size of packets read from each FIFO can vary across channels and over time within each channel. Therefore, the demultiplexer cannot sort the packets belonging to different channels merely from their position within a frame. Consequently, additional bits, which delimit each packet and identify the corresponding incoming channel or source, must be added to each packet. The resulting overhead is significantly larger than under TDM. It also becomes more difficult to implement the multiplexer (which must now add the packet delimiter and channel or source identifier) and the demultiplexer (which must locate and decode those bit patterns). These increases in complexity and overhead must be balanced against high utilization in the face of bursty data to determine whether SM or TDM is more efficient. In general, constant bit rate traffic, such as voice, fixed-rate video, and control and sensor signals, are better handled by TDM, whereas mes-

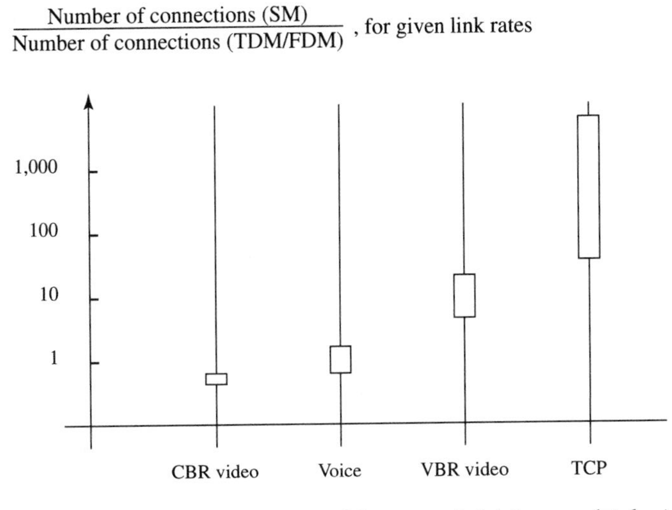

FIGURE 2.11 Statistical multiplexing can achieve much higher multiplexing gain relative to TDM and FDM, especially for bursty traffic.

sage traffic such as database transactions and variable bit rate video are better handled by SM. Not surprisingly, telephone networks use TDM, whereas computer communication networks use SM.

Frequency-division multiplexing or FDM is illustrated in Figure 2.12. The frequency band of the outgoing channel is divided into distinct fixed bands, one for each incoming channel. The signal in each incoming channel is modulated to fit into its assigned band. The signal on the outgoing channel is simply the sum of these modulated signals. Thus the bandwidth of the outgoing channel must be greater than or equal to the sum of the bandwidths of the incoming channel. (In this sense, FDM is similar to TDM.) For demultiplexing, the FDM signal is passed through an appropriate filter. By demodulating the filter output, we obtain the appropriate input signal. FDM is used in AM and FM radio and TV broadcast as well as in CATV distribution. It is also used in cellular radio. FDM is more flexible than the other two multiplexing schemes. Indeed, one incoming channel may contain an analog signal, and another may be digital.

FDM has two disadvantages. First, it is wasteful of bandwidth since the frequency bands assigned to an incoming channel must be separated by a "guard band" from the other channels. Second, if the transmission link exhibits significant nonlinearities, there will be "cross-talk" among the different signals, leading to errors.

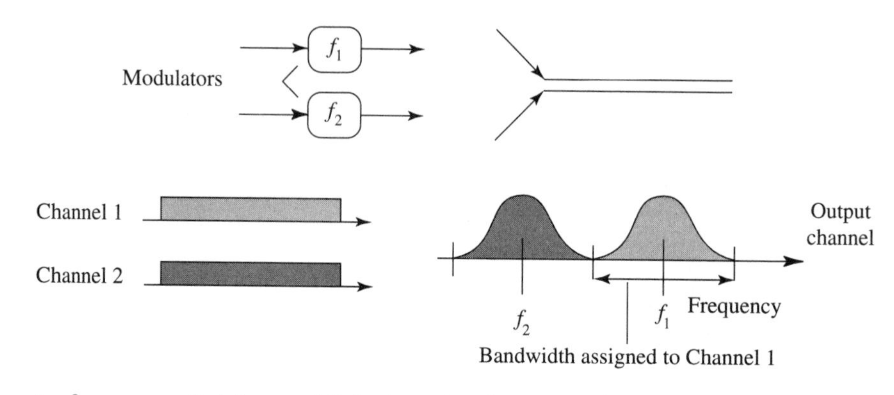

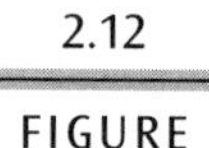

2.12 FIGURE In frequency-division multiplexing, the frequency band is divided into distinct fixed bands, one for each incoming channel. The signal in each incoming channel is modulated to fit into its assigned band.

We have used bandwidth (or size of a link's frequency band in Hz) and speed or bit rate of a link (in bps) interchangeably, but that is not strictly correct. A (digital) bit stream is converted into an analog signal by a modulation scheme like phase-shift keying (PSK). The frequency spectrum of the resulting analog signal must lie within the link's frequency band. As a result the bit rate is proportional to the bandwidth. The ratio bit rate/bandwidth measured in bps/Hz is called the *spectral efficiency,* which typically ranges between 0.5 and 8.0. The maximum value of the achievable spectral efficiency is given by Shannon's theorem as a function of the noise in the communication channel. When bandwidth is scarce, as in wireless radio and satellite channels, communication engineers spend a lot of effort designing modulation schemes to increase spectral efficiency.

2.5.2 Switching

We now describe the two most important switching techniques: circuit switching and packet switching. When a network is to transfer a stream of data from a source to a destination (e.g., from A to D or C to B in the following figures), it must assign to the stream a *route,* that is, a sequence of links or channels connecting the source to the destination, and then allocate to the stream a portion of the capacity or bandwidth in each channel along the route to be used to transfer the stream. Those decisions are implemented in switches. (Route selection and bandwidth allocation are examples of resource allocation.) The

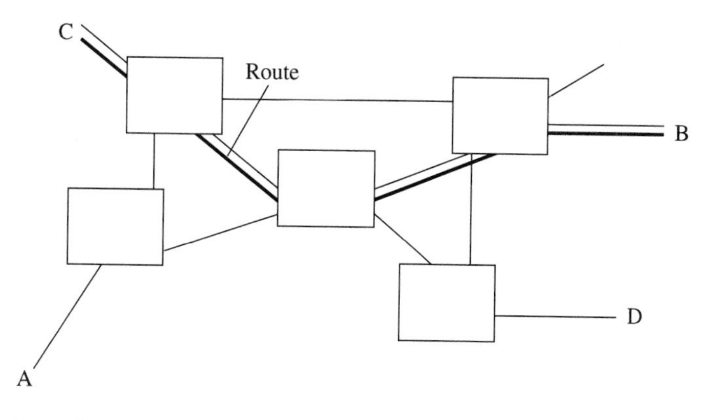

2.13 FIGURE In order to transmit information, a circuit-switched network finds a route along which it has free circuits. The network connects the circuits together and reserves them for the transmission.

name *switch* is used in telephony; in computer communications, the device that performs routing is also called a *router.* We shall use *switch* and *router* interchangeably.

In *circuit switching,* illustrated in Figure 2.13, the route and bandwidth allocated to the stream remain constant over the lifetime of the stream. Further, the capacity of each channel is divided into a number of fixed-rate logical channels, called *circuits*. The division is usually accomplished by TDM. Thus circuit switching involves assigning to the bit stream a route and one circuit in each link along the route. This assignment is made before any bits are transferred in a phase known as *call* or *connection setup*. At the end of this phase, data transfer is carried out. When the transfer is complete, the route and the circuits are deallocated. That phase is called *connection teardown*.

Circuit switching thus involves three phases: (1) the source makes a *connection* or *call request* to the network, the network assigns a route and one idle circuit from each link along the route, and the call is then said to be admitted (if the network is unable to make this assignment, the call is rejected); (2) data transfer now occurs—the duration of the transfer is called the call *holding time;* (3) the call is then torn down. The switch computers maintain information about which circuits are busy (i.e., currently allocated to calls in progress) and the routing tables. The computers also execute algorithms implementing call admission policies and routing strategies. They also record call duration and other statistics needed for purposes of administration, billing, and maintenance.

Circuit switching is easy to implement relative to other schemes, but because a stream is assigned a fixed-rate circuit, capacity utilization may be low if the data stream is bursty. It is therefore used in voice networks but not in networks designed for data transfer. Since circuit switching assigns a fixed bandwidth to a call, there is no queuing delay, and we can guarantee ahead of time the maximum end-to-end delay from source to destination experienced by the data stream. This guaranteed-delay feature may be essential in control, videoconferencing, and other real-time applications. From (2.1), this delay equals the transmission plus propagation times.

Figures 2.14–2.16 illustrate *packet switching*. The data stream originating at the source is divided into packets of fixed or variable size. The time interval between consecutive packets may vary, depending on the burstiness of the stream. As the bits in a packet arrive at a switch or router, they are read into a buffer. When the entire packet is stored, the switch routes the packet over one of its outgoing links. The packet remains queued in its buffer until the outgoing link becomes idle. This *store-and-forward* technique thereby introduces a random queuing delay at each link; the delay depends on the other traffic sharing the same link. (Packets from different sources sharing the same link are statistically multiplexed.) The total delay at a switch is the queuing delay plus the time taken to transmit the entire packet.

The routing decision is determined in one of two ways: datagram and virtual circuit. In *datagram* packet networks, each packet within a stream

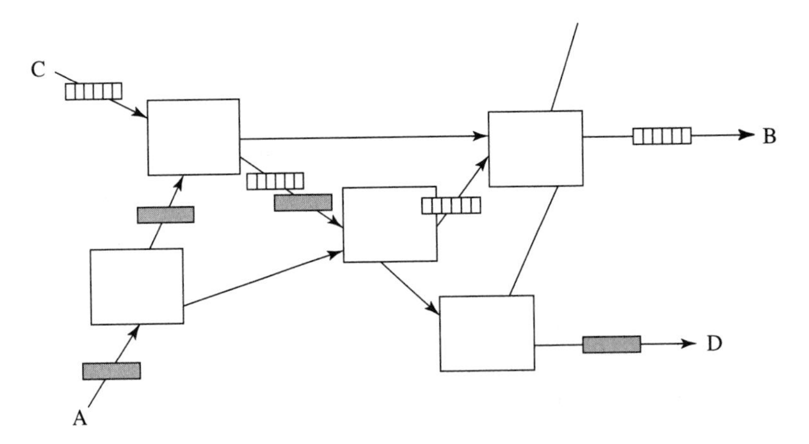

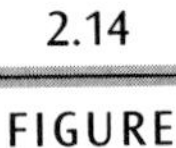

2.14 FIGURE A packet-switched network first divides the information it has to transmit into packets. The packets are then sent along links that are multiplexed statistically.

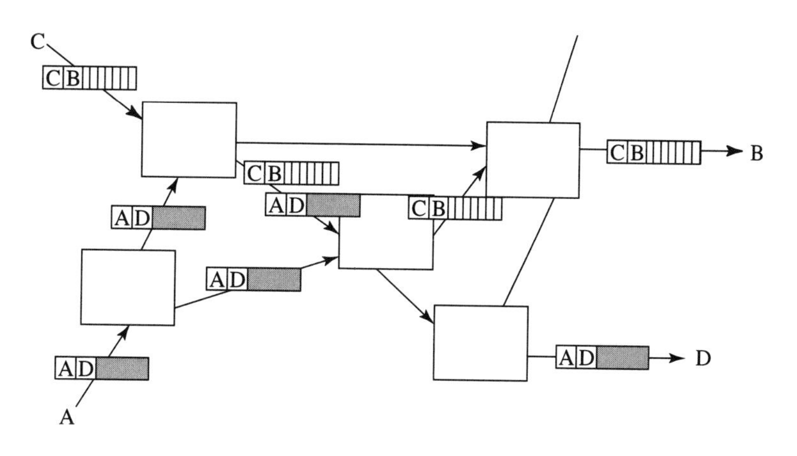

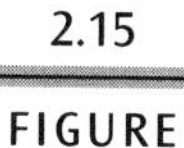

2.15 FIGURE

A datagram network transports packets individually. Each packet contains its source and destination addresses.

is independently routed, as illustrated in Figure 2.15. A routing table stored in the router (switch) specifies the outgoing link for each destination. The table may be static, or it may be periodically updated. In the latter case, the outgoing link assigned to a destination depends on the router's estimate of the shortest path to the destination. Since the estimate may change with time, consecutive packets may be routed over different links. (Observe in the figure that packets from A to D travel over two routes.) Therefore each packet must contain bits denoting the address of the source and destination. This may be a significant overhead (since the addresses are often quite long) if the average packet size is small; the overhead may be negligible if the packet size is long, but then the packet transmission time and the queuing delay are also long.

In *virtual circuit* packet networks, a fixed route is selected before any data is transmitted in a call setup phase similar to circuit-switched networks. (See Figure 2.16.) However, there is no notion of a fixed-rate circuit or logical channel. All packets belonging to the same data stream follow this fixed route, called a virtual circuit. Packets must now contain a virtual circuit identifier; this bit string is usually shorter than the source and destination address identifiers needed for datagrams. However, the call setup phase takes time and creates a delay not present in datagram packet networks.

Datagram switching achieves higher link utilization than circuit switching, especially when traffic is bursty. It is used in data networks. Datagram switching, however, also has potential disadvantages in comparison with circuit switching. First, the end-to-end delay may be so large or so random as

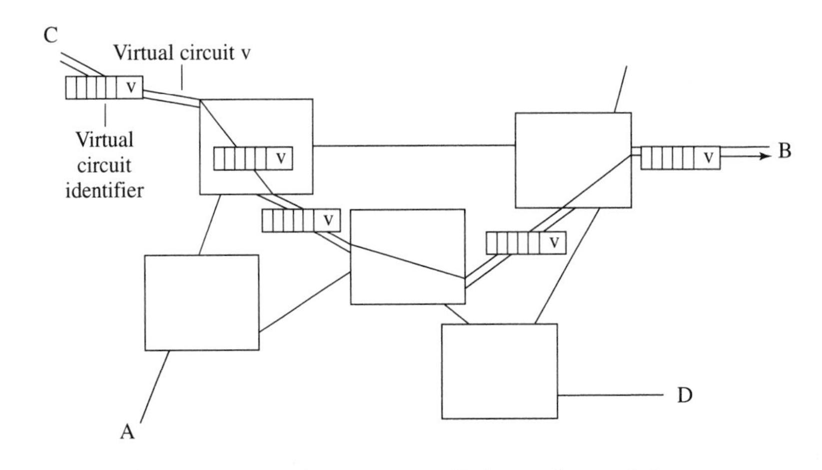

2.16
FIGURE

A virtual circuit network transports all the packets of the same connection along the same path, called a virtual circuit. Capacity along the virtual circuit is not reserved for a connection.

to preclude applications that demand guaranteed delay. Second, the overhead due to source and destination identifiers and bits needed to delimit packets may waste a significant fraction of the transmission capacity if the packets are very short. Since consecutive packets may travel over different routes, packets may not arrive at the destination in the same sequence in which they were sent. The host at the destination may then have to buffer several packets and resequence them, adding to the end-to-end delay. Lastly, a datagram switch does not have the state information to recognize if a packet belongs to a particular application. Hence the switch cannot allocate resources (bandwidth and buffers) that the application may require. Thus it is difficult to implement sophisticated resource allocation schemes in a datagram network.

Virtual circuit switching is a compromise between datagram and circuit switching. The overhead is comparable to circuit switching. Since packets arrive at the destination in order of transmission, no resequencing is needed. Statistical multiplexing of packets at the router or switch can achieve better utilization than in circuit switching. But the utilization may be lower than under datagram switching, since routing decisions in the latter may be changed at the time scale of individual packet durations. Under virtual circuit switching those decisions can be changed only at the time scale of call durations. Lastly, since packets contain their virtual circuit identifiers or VCIs, the switch can allocate resources depending on the VCI. During the connection setup phase the switches may be notified that a particular VCI should be given extra resources.

In circuit switching, bit streams are allocated fixed-capacity circuits, and incoming bit streams are time-division multiplexed into outgoing streams. Only a tiny amount of buffering is needed at the circuit switch—enough to compensate for short-term fluctuations in the frequency and phase of the incoming and outgoing bit streams. In packet switching, the rate of the incoming stream may exceed the capacity of the outgoing link for significant time intervals. Thus packet switches must be able to buffer the excess incoming traffic. Consequently, buffer management is another task performed at the switch in addition to routing and multiplexing.

2.5.3 Error Control

Figure 2.17 illustrates the basic ideas behind error control. To control errors, a transmission link can use two methods: *error detection* and *error correction*.

A simple error-detection method is the parity bit used, for example, in the RS-232 serial line discussed in section 1.1.2. When the link uses that method, the transmitter appends a 1 or a 0 to the packet it transmits so that the resulting string (packet + parity bit) contains a number of 1s that has an agreed parity, say odd. For instance, if the original packet is 0100100, then the transmitter appends a 1 and transmits the resulting string 01001001, which contains an odd number (three) of 1s. If transmission errors modify the string into one whose number of 1s is even, then the receiver detects an error. For instance, if the receiver gets the string 01101001, then it knows that the transmission modified at least one bit. This parity bit method does not detect errors that modify an even number of bits. Thus, this method is not very reliable for long packets: it is not unlikely that transmission errors modify 2 or 4 bits in long packets. Communication engineers developed error-detection methods that detect all but very unlikely errors by adding more than one error-detection bit to the packets. Two such methods are the *cyclic redundancy check* (CRC) and the *checksum code* (CKS). We discuss the CRC method in the next two paragraphs. The discussion is highly technical, and the uninterested reader may skip it without loss of continuity.

The CRC method works as follows. A binary word G forms the basis of the CRC calculation. Standards specify the word G that should be used for different networks. Denote by $r + 1$ the number of bits of G. Assume that we want to transmit a packet that consists of the finite string of bits P. We denote by $P2^r$ the binary word obtained by appending r 0s to the right of the packet P. By performing a long division, an electronic circuit in the transmitter calculates

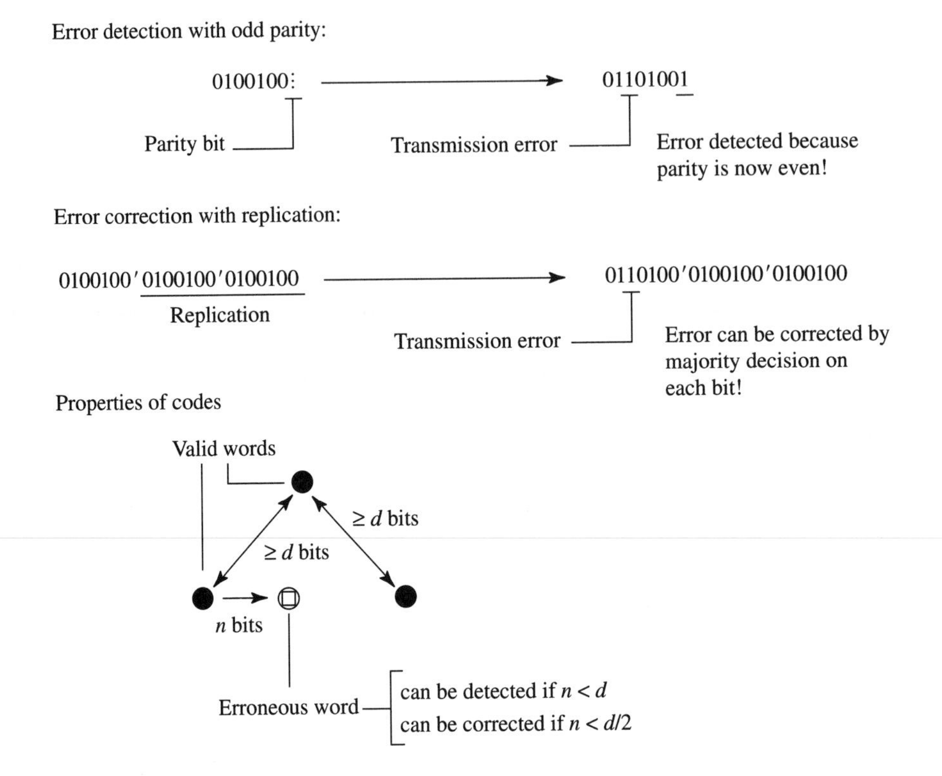

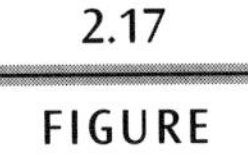

2.17 FIGURE

The figure illustrates error detection and error correction. The bottom panel shows that the error-control properties of the codes depend on the minimum distance between valid codewords.

the remainder R of the division of $P2^r$ by G. This long division is calculated by performing the additions of the binary words bit by bit modulo 2 and without carry. With these rules, 10011 + 01011 = 11000. By definition, the remainder R has r bits and is such that $P2^r = A.G + R$. The binary word R is the CRC code that the transmitter appends to the packet. That is, the transmitter sends the binary word $P2^r + R$. Note that, because of the rules used to perform the operations, $P2^r + R = A.G + R + R = A.G$. Thus, every transmitted packet, together with its CRC bits, is an exact multiple of G. If the receiver gets a packet that is not a multiple of G, then it knows that transmission errors corrupted the packet.

We can represent transmission errors by a binary word E whose 1s indicate which bits the transmission corrupts. With this representation the packet that the receiver gets is $A.G + E$. This received packet is a multiple of G, and the receiver does not detect the errors if and only if E is a multiple of G. Thus, to be a robust error-detection code, the CRC should use a binary word G that is not likely to divide an error word E. For instance, we can show that if $G = 1'0001'0000'0010'0001$, then E cannot be a multiple of G if E has fewer that 32,768 bits and has fewer than four 1s. Thus, the 16-bit CRC that uses this specific G detects all errors that corrupt up to 3 bits in a packet of up to 32,768 bits.

The middle panel of Figure 2.17 illustrates a simple error-correction method that transmits every packet three times. By performing a majority vote on every bit, the receiver can correct transmission errors as long as they affect each packet bit in at most one of the three transmissions. Thus, if the third bit in the packet is a 0 and if one of the transmissions modifies it into a 1, then the receiver can decide that the transmitter sent a 0 and thereby correct the error.

This replication method is wasteful and is not very robust. Communication engineers have designed better methods, including the *Bose-Chaudhury-Hocquenghem* (BCH) and the *Reed-Solomon* (RS) codes.

In general terms, an error-control code adds R bits to a packet of M bits. The code computes the R bits from the original M bits so that the transmitter can send only 2^M different strings of $M + R$ bits, called *codewords*. Assume that any two codewords differ by at least d bits. (See bottom panel of Figure 2.17.) Then, the receiver can detect transmission errors that modify fewer than d bits. Indeed, such errors cannot modify a codeword into another codeword, and the receiver detects the error when it finds that the received string is not a codeword. The receiver can correct errors that modify fewer than $d/2$ bits because the transmitted codeword is the codeword that differs the least from the received string.

Computer networks today use error detection. The transmitter retransmits packets that do not arrive intact at the receiver. The transmitter and receiver follow a specific set of rules, a *protocol,* for making sure that the receiver eventually gets every packet correctly and that it discards corrupted packets and duplicated correct copies. These protocols use timers and acknowledgments.

Figure 2.18 sketches a typical implementation of such a protocol. Before sending a packet, the transmitter makes a copy that it keeps. The transmitter then sets a count-down timer to a specific value and sends the packet. We say that the packet *times out* when the count-down timer reaches the value 0. If the receiver gets the intact packet, then it sends back an acknowledgment to the transmitter. If the transmitter gets the acknowledgment of the packet

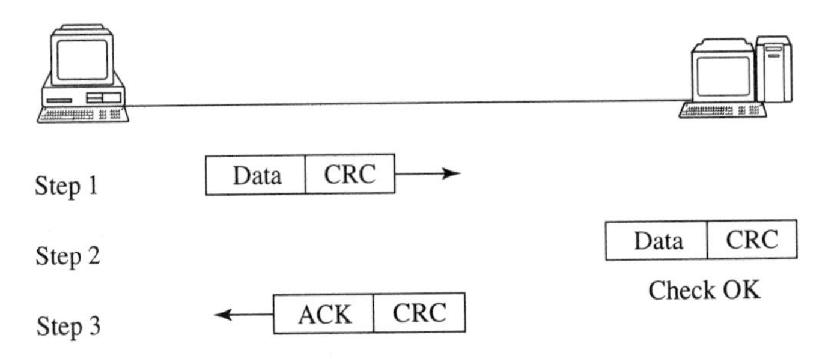

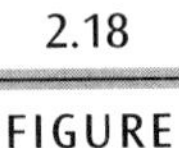

2.18 FIGURE

The figure shows an error-free transmission. The transmitter sends the packet (step 1). When it receives the packet, the receiver verifies that it is correct (step 2). The receiver sends an acknowledgment of the packet to the transmitter (step 3).

before the packet times out, then the transmitter knows that the receiver got the packet. In that case, the transmitter discards its copy of the packet and repeats the procedure with the next packet.

If the receiver gets a corrupted packet (see Figure 2.19), then it does not send an acknowledgment. When the packet times out, the transmitter assumes that something went wrong during the packet transmission, and it repeats the procedure with a copy of the packet. The packet also times out when its acknowledgment is corrupted during its transmission.

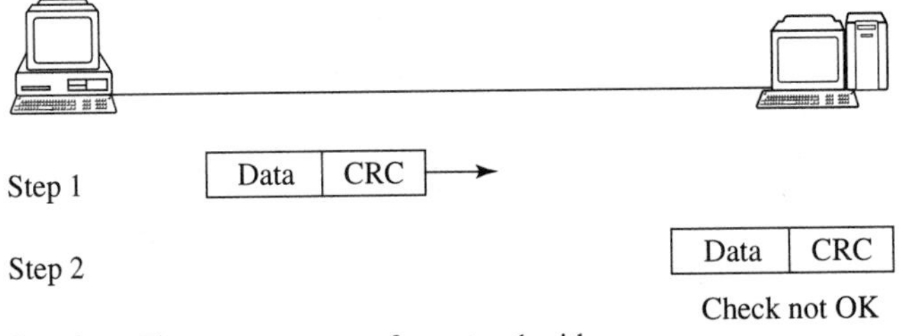

2.19 FIGURE

The figure shows a transmission corrupted by errors. The transmitter sends the packet (step 1). When it receives the packet, the receiver finds out that it is incorrect by checking the error-detection bits in the CRC (step 2). The receiver does not send an acknowledgment of the packet to the sender. When its packet times out, the sender transmits a copy of the packet (step 3).

If there is no permanent hardware problem and if the initial value of the timer is large enough, the transmitter eventually gets an acknowledgment before it times out and every packet eventually reaches the receiver.

We explain the retransmission protocols that networks use below. For real-time applications and storage, the network should use error correction instead of error detection.

Alternating Bit Protocol

The simplest retransmission protocol is the *Alternating Bit protocol* (ABP). The sender numbers the packets alternately 0 and 1. The receiver acknowledges every correct packet that it receives with the same number as the received packet. The sender waits for the receiver's acknowledgment. Figure 2.20 shows the sequence of events when there is no transmission error.

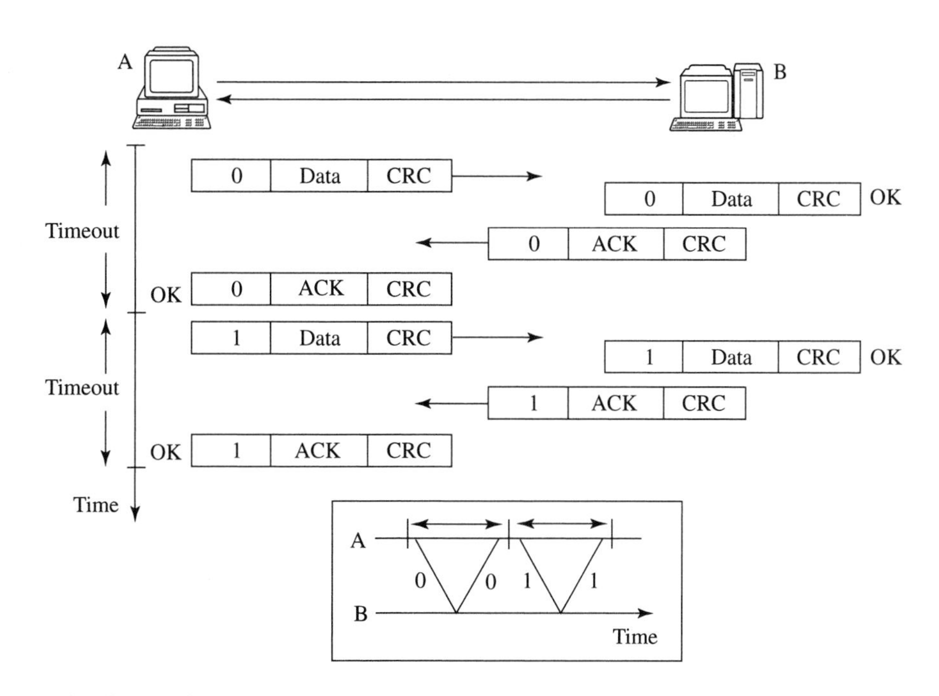

2.20 FIGURE

The figure shows packet 0 arriving correctly at the receiver, which then sends an acknowledgment 0. The acknowledgment arrives before the timeout value. The transmissions then repeat with sequence number 1. This sequence of transmissions is summarized in the timing diagram at the bottom of the figure.

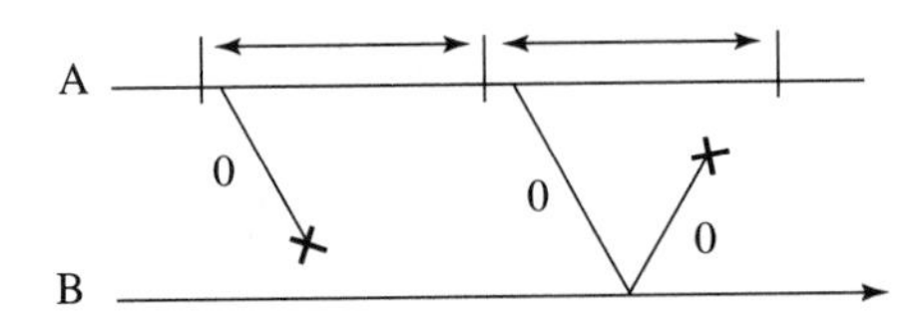

2.21

FIGURE

A cross on the downward line marked 0 represents a transmission error that corrupts packet 0. The receiver does not acknowledge the reception of that incorrect packet. After the timeout value, the transmitter retransmits the packet, again numbered 0, which arrives correctly. Transmission errors corrupt the acknowledgment.

If the acknowledgment does not reach the sender before the packet times out, then the sender retransmits the packet (with the same number), as illustrated in Figure 2.21. The receiver gets the packets in order and needs only to keep track of the number (0 or 1) of the last correctly received packet to identify duplicates that may arrive when acknowledgments are lost or late. Figure 2.22 shows that the transmitter must number packets. In the situation of the figure, if the packets were not numbered, the receiver could not tell whether the second packet it receives is a copy of the first or a new packet.

Figure 2.23 shows why the receiver must number the acknowledgments. The figure assumes that the acknowledgments are not numbered and shows a sequence of events that lead the protocol to fail to retransmit a packet. The figure shows that the transmitter can be led to confuse the acknowledgment of a packet 0 with the acknowledgment of a packet 1. To prevent such confusion, it is necessary for the receiver to number the acknowledgments.

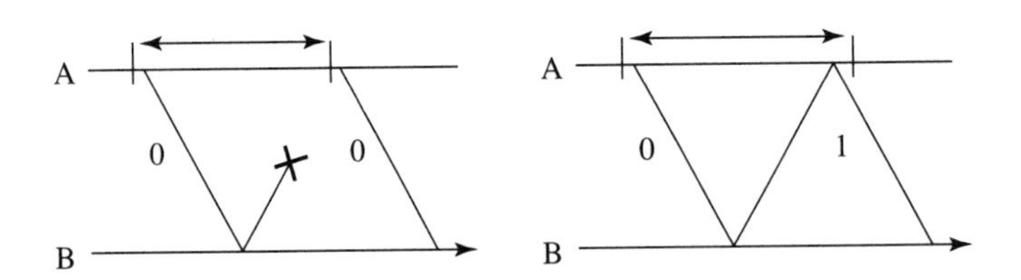

2.22

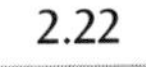

FIGURE

In the left-hand part of the figure, a packet is correctly transmitted but its acknowledgment is not. Consequently, the transmitter sends a copy of the packet. In the right-hand part of the figure, the packet and its acknowledgment are correctly received so that the transmitter sends a new packet. The receiver could not distinguish these two cases if the packets were not numbered.

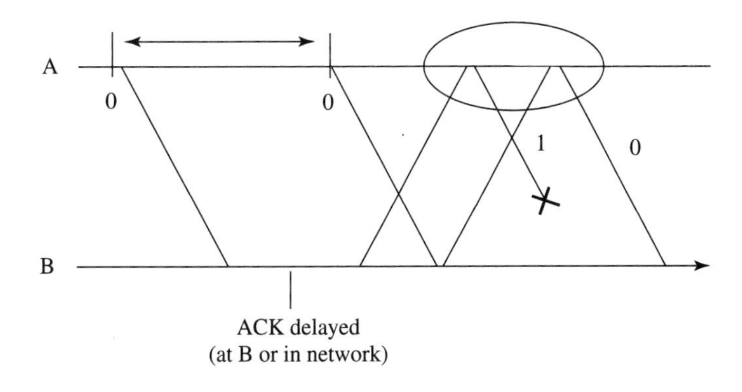

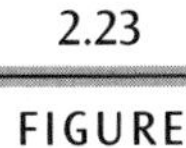

2.23
FIGURE

In the sequence of events shown here, the protocol fails to transmit a packet if the acknowledgments are not numbered.

The ABP protocol is inefficient if the propagation time is long compared to the transmission time, because the sender waits for the acknowledgment before sending the next packet. The efficiency of the Alternating Bit protocol is defined as the fraction of time that the transmitter transmits new packets. When the bit error rate is negligible, the efficiency is equal to TRANS/(TRANS + 2PROP + ACK). (See Figure 2.24.) In this expression, TRANS denotes the time to transmit a packet, ACK the time to transmit an acknowledgment, and PROP denotes the time taken for the signal to travel from

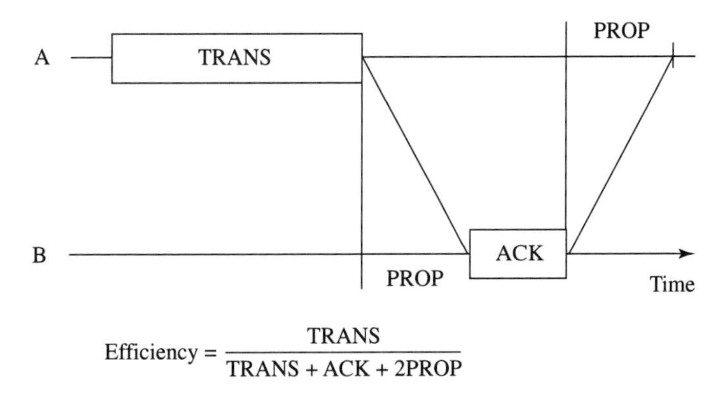

2.24
FIGURE

The figure shows the time that the transmitter takes to transmit a packet and get its acknowledgment.

the sender to the receiver and from the receiver to the sender. (We assume these two times to be equal.)

Go Back N

A protocol that is more efficient than ABP for long propagation times is Go Back N. The network designer or user selects a window size N. Typically, N is just large enough so that the pipe is full: the sender gets the acknowledgment of the first packet when it finishes transmitting packet number N. The sender numbers the packets sequentially modulo $N+1$, i.e., $1, 2, \ldots, N, 1, 2, \ldots,$ N, etc.

The receiver acknowledges every correct packet that it receives with an acknowledgment that it numbers the same as the packet. Before transmitting packet number $n+N$ (modulo $N+1$), the sender waits until it receives the acknowledgment of packet number n. Whenever the sender sends a packet, it starts a timer with a timeout T selected by the network designer. When the sender fails to receive the acknowledgment of a packet before the packet times out, the sender retransmits that packet and all the packets that it has transmitted since it last transmitted that packet. Therefore, the receiver need not store any packet since it eventually gets them correctly in sequence. The sender must be able to store up to N packets. Note that with this protocol the sender may have to retransmit packets that were correctly received by the receiver. Figure 2.25 illustrates the sequence of events in Go Back N with $N = 4$.

In Figure 2.26 we illustrate the calculation of the efficiency of Go Back N where there is no transmission error. The efficiency is given by

$$\text{Efficiency} = \min\left\{\frac{N\ \text{TRANS}}{\text{TRANS} + \text{ACK} + 2\text{PROP}}, 1\right\}.$$

The efficiency increases linearly with N up to 100%. To appreciate the magnitudes that may be involved in practice, we consider a numerical example. Suppose the packet size is 2,000 bits, the ACK is 80 bits, transmission is over a 30-km fiber link at 155 Mbps. Then

$$\begin{aligned}\text{Efficiency} &= \min\left\{\frac{N\ \text{TRANS}}{\text{TRANS} + \text{ACK} + 2\text{PROP}}, 1\right\}\\ &= \min\left\{\frac{N \times 2000}{2000 + 80 + 46{,}500}, 1\right\}.\end{aligned}$$

So the efficiency is 4.2% when $N = 1$, and 100% for $N \geq 24$. The network engineer should propose a window size equal to 24 in this example. Note

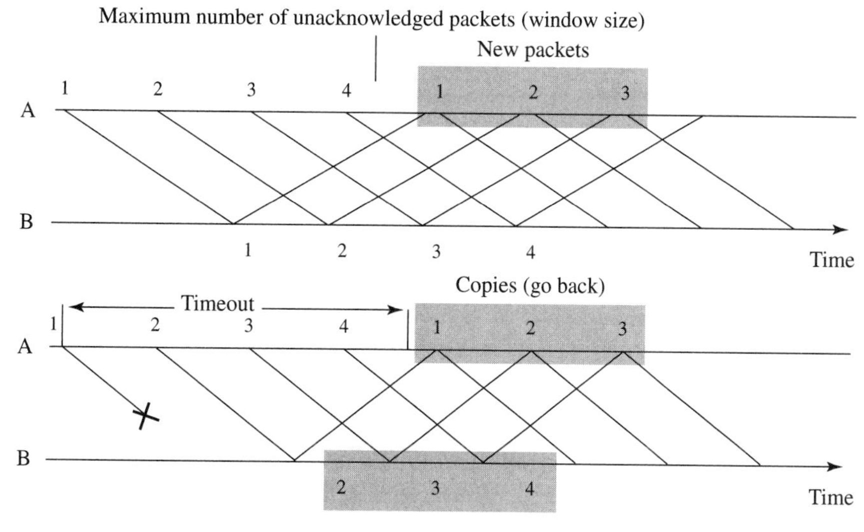

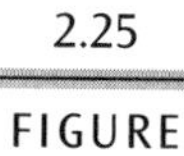

2.25

FIGURE

The top panel shows the sequence of transmissions using Go Back N with a window size of four with no transmission errors. The bottom panel shows the "go back" being triggered when the transmitter fails to get an acknowledgment of packet 1 before the timeout.

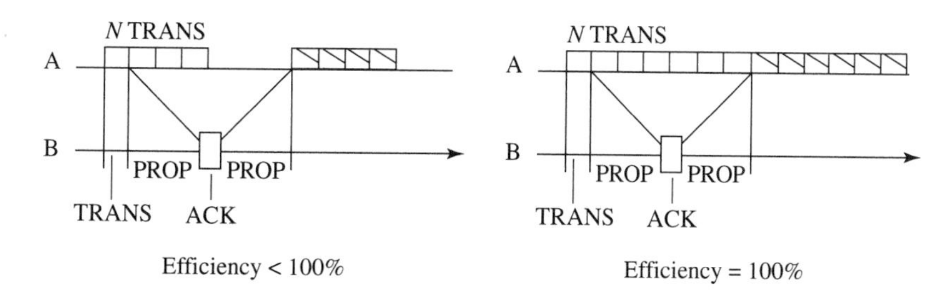

2.26

FIGURE

The transmitter can send N packets before it gets the acknowledgment of the first one. Consequently, when no errors corrupt the transmissions, the packets are transmitted in cycles of N transmissions. That observation enables us to calculate the efficiency in the text.

that Go Back N with $N = 1$ is exactly ABP. Transmission errors reduce the efficiency of Go Back N. The analysis of the efficiency when errors occur is more complicated.

Selective Repeat Protocol

The final protocol that we explain is the *Selective Repeat Protocol* (SRP). SRP is similar to Go Back N but retransmits only packets that were not correctly received. A window size $2N$ is agreed on, and the packets and acknowledgments are numbered modulo $2N$. Once again, the sender waits for the acknowledgment of packet n before sending packet $n + N$ (modulo $2N$). The sender retransmits packets that have not been acknowledged within a timeout after their transmission, and the receiver acknowledges all the correct packets. The sender and the receiver must both be able to store up to N packets. Figure 2.27 shows typical sequences of events when the nodes use SRP (with $N = 2$).

2.5.4 Flow Control

The bit streams of many sources get multiplexed and read into a switch buffer. If the sources are unregulated, occasionally the buffer will get filled up, causing long delays and, possibly, buffer overflow and resulting packet loss. Flow-control mechanisms can cause the sources or switches to reduce the number of packets they inject into the network when conditions of congestion are detected.

If the sources use a window mechanism as in Go Back N or SRP, then this mechanism can be adapted to serve the purposes of flow control. Notice that the sources can detect congestion or packet loss from the fact that their

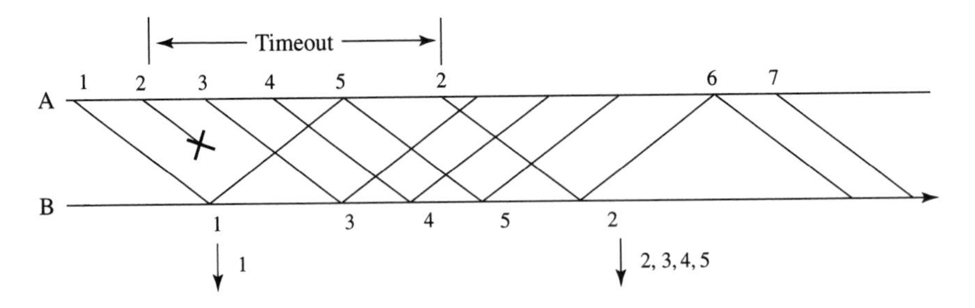

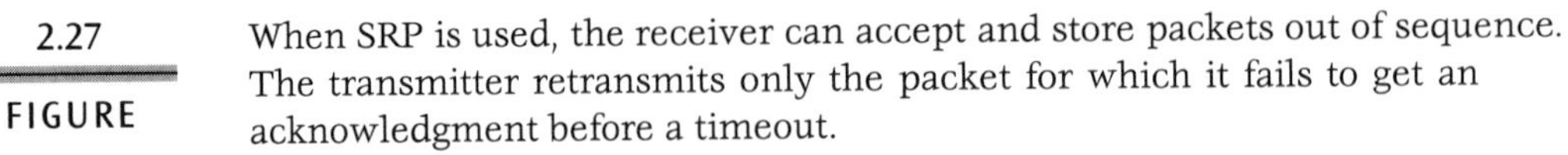

2.27 FIGURE

When SRP is used, the receiver can accept and store packets out of sequence. The transmitter retransmits only the packet for which it fails to get an acknowledgment before a timeout.

packets time out. Thus as soon as several timeouts occur in quick succession, the source should reduce the window size. This will automatically reduce the number of outstanding packets, and hence network congestion.

There is a conflict between keeping the window size small to reduce congestion and keeping it large enough to "fill the pipe" to maintain high efficiency. Suppose the delay-bandwidth product is large and the window size is large. Then most of the outstanding packets are propagating through the links, rather than waiting in the buffers. As a result, once congestion is detected and the window size is reduced, the large number of packets in the links will be unaffected by the reduced window size. Those packets will arrive into the buffers, increasing congestion and packet loss. Thus window flow control is not effective when individual sources have large rates and the end-to-end propagation delay is large. In that case, "open loop" flow control in the form of so-called rate control is more effective. We study these techniques in detail in Chapters 6 and 7.

2.5.5 Resource Allocation

Different applications require bearer services of different qualities (delay, error rate, etc.). A network can guarantee an application a particular service quality only if it can dedicate resources (bandwidth, buffers) to that application.

A circuit-switched network dedicates a fixed bandwidth to each connection, so it can guarantee minimum delay to an application whose peak rate is less than the fixed bandwidth.

A datagram network is connectionless, and the network switches do not have the state information needed to distinguish packets from different applications. As a result, the network cannot dedicate resources that are specific to individual applications. Some crude resource allocation is possible, however. For example, as will be seen in Chapter 3, the TCP/IP protocol permits expedited packets to be treated with a higher priority than nonexpedited packets.

In virtual circuit-switched networks, each packet carries its virtual circuit identifier or VCI. This allows switches to treat packets differently based on their VCI. Thus, at the time its virtual circuit is set up, an application can negotiate with the network for certain service quality. The network switches can then reserve bandwidth and buffers for that virtual circuit so as to provide that quality. Guaranteed service quality is thus a possibility. In Chapters 6 and 7 we consider resource allocation mechanisms that provide this guarantee.

Table 2.2 summarizes the main techniques used to implement the basic network operations.

Function	Technique	Characteristics	Reference
Multiplexing	Time division (TDM)	Fixed bit rate channels, flexible, minimum overhead to identify channels	
	Frequency division (FDM)	Fixed bandwidth channels, very flexible, suited for analog signals	
	Statistical (SM)	Channel bandwidth on demand, overhead for channel identifier	
Switching	Circuit (CS)	Good for CBR traffic, no queuing delay, low utilization for bursty traffic	Ch. 4
	Packet (PS)	Good for connectionless message traffic, variable queuing delay, high utilization for bursty traffic, robust against link failures	Ch. 3
	Virtual circuit (VC)	Good for connection-oriented VBR traffic, variable queuing delay, good utilization	Ch. 5, 6
Error control	Detection	CRC codes can detect most errors	
	Retransmission	Alternating Bit, Go Back *N*, etc., used to control errors when reverse path is available	
	Error correction	Used when reverse path is not available, e.g., storage or high-delay satellite links	
Flow control	Window flow control	Combined with end-to-end error control protocol, effective in low delay-bandwidth connections	Ch. 6, 7
	Rate control	Effective in high delay-bandwidth connections, used in ATM services	Ch. 6, 7
Resource allocation	Routing, bandwidth, buffer allocation; admission control	Used in virtual circuit networks to guarantee service quality	Ch. 6, 7

2.2 TABLE

Techniques and characteristics of basic network operations.

2.6 LAYERED ARCHITECTURE

An *architecture* is a specific way of organizing the many functions performed by a computer network when it provides services such as file transfer, e-mail, directory services, and terminal emulation.

In this section we first introduce the *layered* architecture of network functions. We then comment on the implementation of layers.

2.6.1 Layers

In most networks, the functions are organized into *layers*. As shown in Figure 2.28, in a layered decomposition, services of layer n are implemented by protocol entities (processes) at layer n using the services of layer $n - 1$. Examples of such a layered decomposition of communication functions abound in everyday situations. For instance, two executives can exchange messages by using the services of their secretaries. The secretaries themselves exchange messages by using facsimile machines. The machines transmit facsimiles by using the services of the telephone network. (See Figure 2.29.) We covered a more technical example when we explained error control: by using some

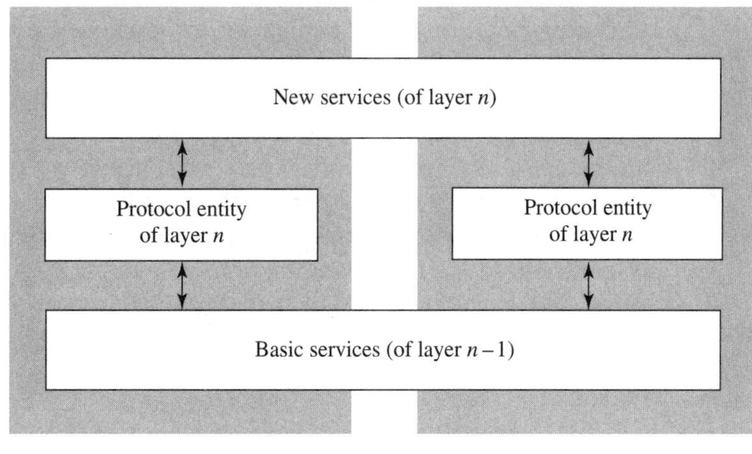

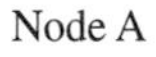

FIGURE 2.28 In a layered architecture, protocol entities of layer n implement services by using the services implemented by layer $n - 1$.

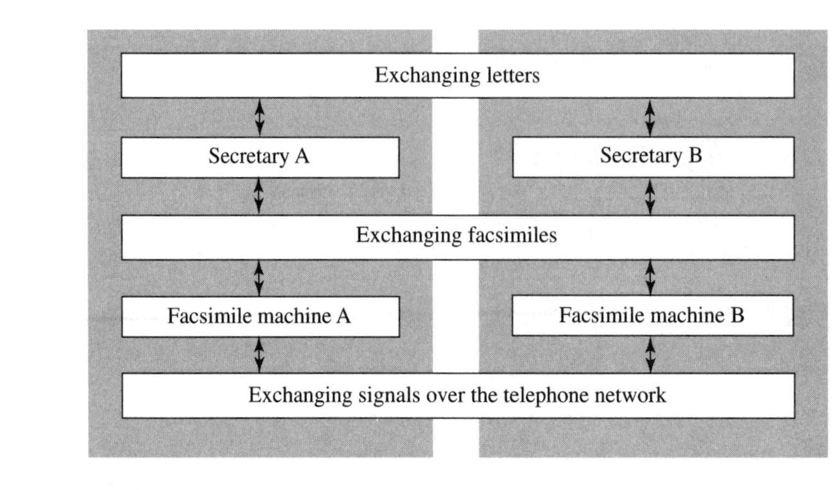

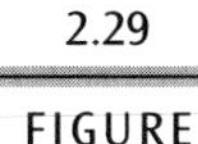

2.29 FIGURE
This figure illustrates that we can view the exchange of letters by secretaries using facsimile machines as a multilayered process.

suitable protocol, a transmitter and a receiver can implement reliable packet transmissions over an unreliable transmission link.

By decomposing network functions into layers, the network engineers partition a complex design problem into a number of more manageable subproblems—those of designing the different layers. This decomposition simplifies the design and its verification. Moreover, the decomposition permits standardization, makes possible the competitive implementations of different layers, and facilitates network interconnection. For the layered decomposition to provide these benefits, the network engineers must specify without ambiguity the services provided by different layers and the interfaces between layers. Standardization bodies such as the ITU (International Telecommunication Union), ISO (International Organization for Standardization), IEEE (Institute of Electrical and Electronics Engineers), and ANSI (American National Standards Institute) organize working groups that develop and publish these specifications. The resulting specifications, called *standards,* are necessarily detailed and lengthy.

2.6.2 Implementation of Layers

When protocols are arranged into layers, the protocol entities of adjacent layers exchange messages. Since the protocol entities are computer processes, this implies that a number of processes communicate. Once it gets a message, a protocol entity performs some operations before it transmits the message to the next protocol entity.

To make the interprocess communication and the operations of a protocol entity more concrete, we examine how these functions are implemented. Our discussion does not cover all possible implementations nor all possible types of operations. However, it captures some of the main features of implementations.

Our objective is to explain the steps shown in Figure 2.30. The figure shows protocol entities in two computers. The protocol entity P_{n+1} at layer $n+1$ in computer 1 sends a message M to the peer entity P'_{n+1} in computer 2. To send that message, P_{n+1} sends a control message C to the protocol entity P_n, asking it to transmit M to P'_n. The entity P_n adds a header H to the message M before it sends it to P'_n. This header may contain addresses, sequence numbers, control fields, and error-detection bits that P_n and P'_n need to supervise the transmission of M.

We first examine how P_{n+1} passes the message C to P_n. This message passing can be implemented by using a shared memory or by using a queue.

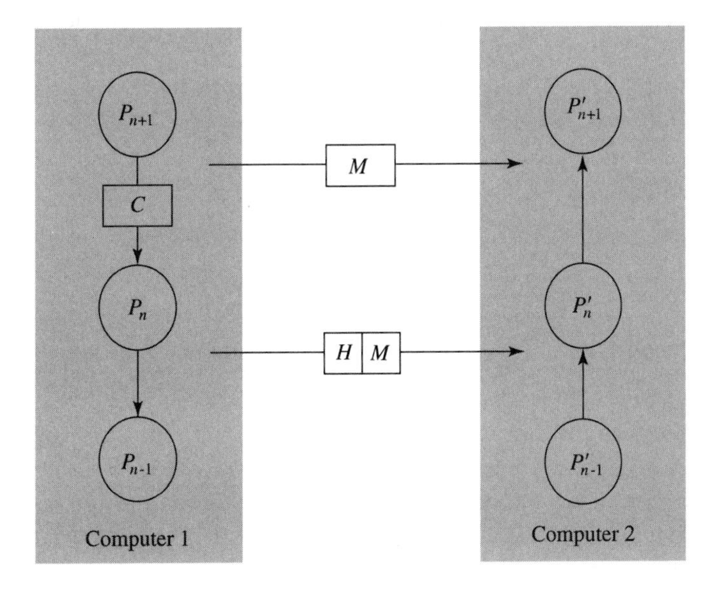

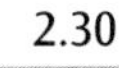

2.30 FIGURE

The protocol entity P_{n+1} asks layer n to transmit a message M to P'_{n+1} by sending a control message C to P_n. The entities P_n and P'_n execute the protocol of layer n. The text describes implementations of the message passing between P_{n+1} and P_n and the execution of the protocol by P_n and P'_n.

When using the shared memory method, the processes P_{n+1} and P_n have access to a common memory segment that is divided into N locations that store data and to a common variable X, called a *semaphore,* that can take the values $0, 1, \ldots, N$. The variable X represents the number of locations that are available to be written. Process P_{n+1} can write as long as $X > 0$, and it decrements X by 1 whenever it has written into a location. Process P_n can read whenever $X < N$, and it increments X after it has read a location. Semaphores can also be used by multiple writer and reader processes. When this possibility is implemented, special care must be taken to avoid conflicting manipulations of the semaphore value and also for handling situations when a process aborts before resetting the semaphore.

When the processes use queues to communicate, process P_{n+1} writes C into a queue that is read by process P_n. A queue is organized as a first-in, first-out array of data. The queue has some reserved capacity, and it can be implemented by the operating system as a linked list with a pointer to the head of the queue and another to the tail of the queue. Process P_{n+1} writes into the queue (at the tail), and process P_n reads from the queue (at the head). The operating system checks that there is data to be read when P_n wants to read and that there is space available when P_{n+1} wants to write. Typically, the operating system can handle a large number of queues between various processes by sharing a large memory among these queues. The capacities of the different queues can be adjusted dynamically by creating a new linked list whenever a new interprocess queue is needed. The different queues can be used to pass messages that should be handled differently. For instance, one queue may contain high-priority messages and another low-priority messages. A number of processes may write into the same queue or read from the same queue if, for instance, each message in the queue contains an identification number that specifies the process for which it is intended.

Typically, P_{n+1} does not transmit the message M to P_n. Instead, it transmits a pointer to that message that indicates where M is in memory and its length. The actual transfer of M must occur when the message must move across different computer boards. In many systems, the network interface board has its own memory that stores the packets ready to be transmitted. In such an implementation, it may be that P_n is implemented by the network interface board while P_{n+1} is implemented by the main CPU. In that case, P_{n+1} actually copies M to P_n across the computer bus to which the interface board is attached. Obviously, the message must also be copied by the bottom layer, which implements the actual transmission between computers.

Let us now turn to the implementation of the functions that P_n performs. The entities P_n and P'_n implement the protocol of layer n. This protocol specifies that P_n must compute the header H, start some timer, and update some counters. When a timer expires, P_n typically initiates a new call to P_{n-1} to retransmit the message. The entity P'_n must read the header H and perform a set of operations such as verifying that the packet is correct, send an acknowledgment, and indicate to P'_{n+1} that a packet has arrived.

In some protocols, the header H contains an error-detection field whose value depends on the message M. In that case, to calculate H, the process P_n must read the message M, which, together with the calculation of H, requires a large number of instructions. In other protocols, the header H contains an error-detection field that is independent of M and that protects only the header itself. The execution of such a protocol is typically much faster.

This discussion points to the implications of both the design and the implementation of protocols for the achievable rates of execution of such protocols. Fast protocols are designed to limit the need for protocol entities to read full messages. Protocol implementations are faster when they minimize the number of actual message transfers by passing pointer values instead of copying the messages. Finally, protocol executions can be speeded up by implementing protocol entities on dedicated hardware that frees up the main CPU. Ideally, the execution of the protocols should impose a minimum burden on the main CPU, and it should be fast enough to keep up with the communication link and with the source of the data to be transferred.

We make this discussion more concrete with the help of Figure 2.31. The panel on the left shows a basic host computer architecture. The CPU and the main memory communicate over the CPU bus; there also is an I/O bus for communication with the network. A dedicated Network Interface Card (NIC) implements many of the functions dealing with the physical transmission and reception of packets. The panel on the right shows that four copies of a file across the CPU bus are needed when the file is transferred by FTP (File Transfer Protocol), studied in Chapter 3. The file, initially in user space in memory, is first copied into system space, where it is handled by FTP (1). Then FTP copies the file over to the next layer protocol TCP (Transmission Control Protocol) (2). The TCP protocol entity fragments the file into IP (Internet Protocol) packets. The CPU now computes the CRC bits of each packet, which requires a third copy (3). Finally, each completed IP packet is forwarded to the NIC (4). The NIC transmits the packet over the network. If the CPU bus has a throughput of 320 megabytes per second (MBps) (80-MHz clock and 4-byte-wide bus), the four copies have reduced this to 80 MBps.

We have introduced the concept of layered architectures. We now describe a very useful architecture, the Open Data Network or ODN model.

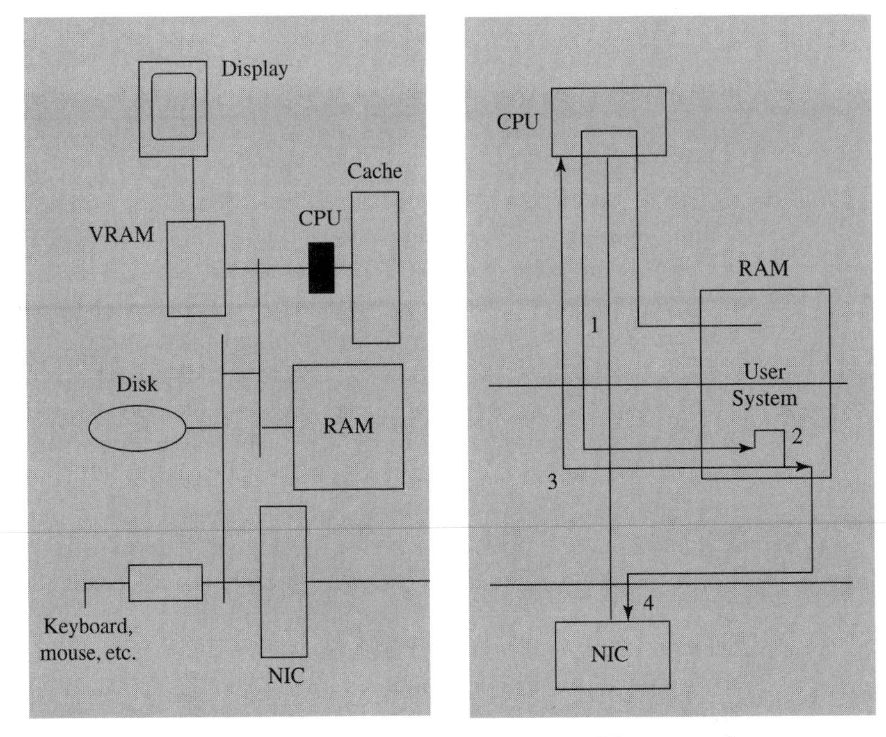

2.31 FIGURE

The left panel gives a simple architecture of a host computer and its connection to the network. The right panel shows that four copies may be involved across the CPU bus to run an application, reducing the host throughput.

2.7 OPEN DATA NETWORK MODEL

The Open Data Network or ODN model was recently proposed by a panel of network engineers as a framework within which the telephone, computer, and CATV networks can be located and compared. The purpose of the ODN model is not to develop a standard like the OSI model studied in Chapter 3, but to help understand how it may be possible to interconnect these three types of networks, despite the differences in their technologies, services, and markets.

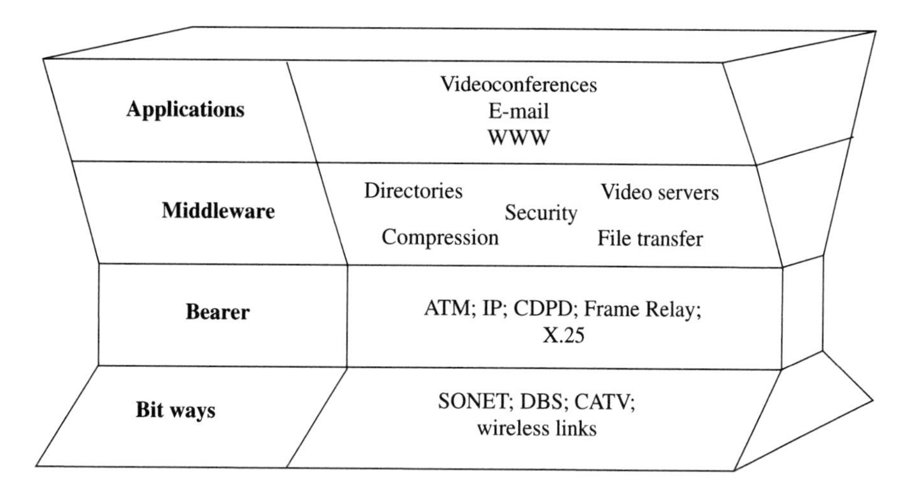

2.32 FIGURE The Open Data Network model has four layers.

The ODN model is displayed in Figure 2.32. It has four layers, called *bit ways, bearer, middleware,* and *applications,* as shown on the left side of the figure. On the right side are listed a sample of implementations of those layers in specific networks.

The service provided by a *bit way* is the transport of bit streams over a link. The bit way may be implemented by a SONET link of the telephone network, by a direct broadcast satellite or DBS link, by a CATV link from the head station to a user, by a cellular radio channel, or some other wireless connection. The bit way provided by a specific link technology can be characterized by its speed, delay, and error rate.

A *bearer* service is the end-to-end transport of bit streams in specific formats. For example, in the ATM bearer service, 53-byte cells are transported end-to-end over virtual circuits. The format of the cells is fixed. In the network or IP layer of the Internet, the bearer service transports variable-sized datagrams in a specified format from source to destination over a packet-switched network. As we will see, the bearer services layer is the most important layer from the viewpoint of network interconnectivity.

The *middleware* services are generic services that are used by a large number of applications. Examples of such middleware services are file transfer, directories, and video servers. These services could be provided by individual user computers. This is indicated in Figure 2.33, which shows middleware services being provided by the operating system. In a campus environment,

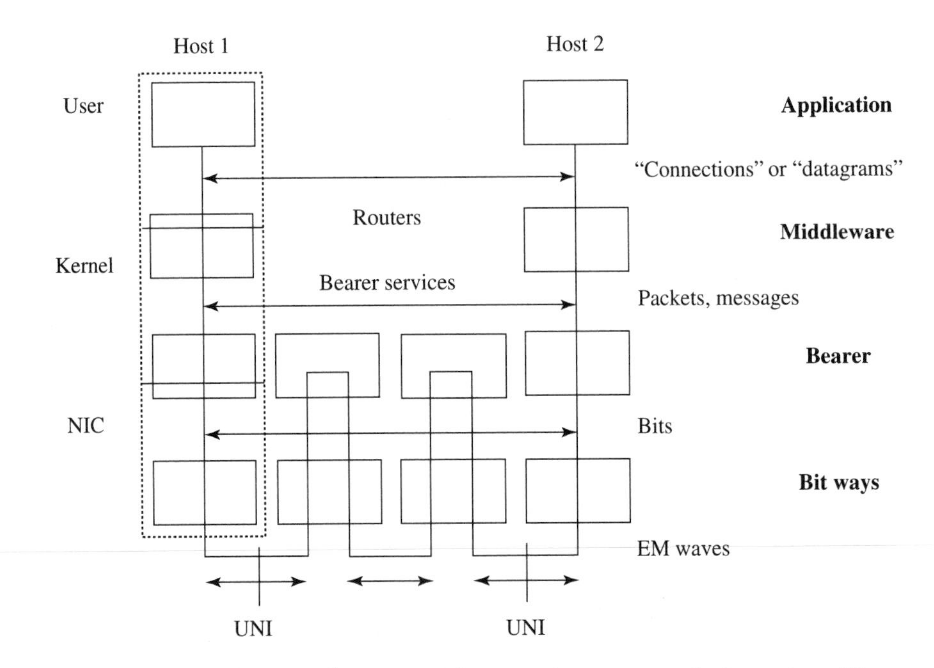

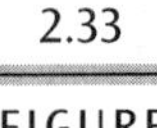

2.33 FIGURE

Networks with the same bearer services can be connected via routers. The simpler the bearer services, the easier is network interconnection.

however, economies of scale motivate the installation of dedicated servers for directories and other databases, software "warehouse," etc.

For some middleware services, the economies of scale are even greater, and this may justify installing special hardware and software in network nodes to provide those services. For example, cellular phone standards like GSM and IS-54 compress a voice signal into a 16-Kbps bit stream. The telephone company switches contain hardware that decompresses those bit streams to recover the original voice signal. (If this were not done, only telephone sets equipped with decompression hardware could receive the compressed signal.) Similarly, CATV networks may transport a compressed video signal to the curbside where it is decompressed before distribution to individual users.

Finally, the *application* layer provides the services that users want. Examples include e-mail, WWW (World Wide Web), video on demand, and videoconferencing. As indicated in Figure 2.33, the application layer is usually implemented in the host computer. This often involves proprietary software, like Eudora for e-mail and Intuit's Quicken for home banking. Sometimes specialized hardware is needed to run the application. Thus, "pay TV" requires a "set-top box" that unscrambles or decompresses a TV signal. The set-top box helps create a record for billing purposes. Because applications often involve proprietary and expensive "progam content" (e.g., a movie) as well as transport services, provision must be made for billing for this program content.

Much profit will be made by hardware and software manufacturers whose set-top boxes become a *de facto* standard. If a manufacturer's equipment is widely adopted by users, program content providers will have an incentive to conform to that equipment, and the equipment manufacturer can then extract a monopoly rent from the content providers for use of that equipment. In 1995 the contenders for this set-top "prize" included the large software companies, such as Microsoft and Sybase, and the telephone companies. On the other side, companies such as Sun Microsystems, which have not yet developed competitive products, are lobbying for government regulation that will maintain an "open" application layer.

We can now see how different layers cooperate to produce sophisticated user services. An application may make use of middleware services such as file transfer. File transfer service in turn is implemented using a bearer service such as IP datagram transport. And datagram transport is implemented using a bit way provided by a local area network such as Ethernet.

Finally, we study network interconnectivity. Observe that the ODN model of Figure 2.32 has a narrow waist at the bearer service layer. This is intended to suggest two things. First, a very small set of bearer services can be provided by a large variety of bit way implementations. Second, the small set of bearer services is sufficiently versatile to support a large variety of applications.

The small set of bearer services greatly promotes internetworking, as can be seen with the help of Figure 2.33. The figure shows two hosts, Host 1 and Host 2, belonging to two separate networks. These two networks can be interconnected by a router or switch, provided both networks support the same bearer service. (The router is attached to the bit ways of both networks.) The application running on Host 1 produces a bit stream that is recovered by the router using the bearer services of the first network. The router then forwards that bit stream to Host 2 using the bearer services of the second network.

We thus see that the simpler the set of bearer services, the easier it will be to interconnect networks. The most important example of this is the Internet Protocol (IP) and ATM, both of which specify a single bearer service. On the other hand, the set of bearer services should be versatile enough to support a wide range of applications. We will see that in this respect ATM is better than IP.

2.8 SUMMARY

Networks provide communication services needed to support user applications. Those services are provided from more elementary services in a layered architecture, like the Open Data Network model. Some of the layers are implemented in the network switches, others are implemented in host computers. The network itself provides bearer services, i.e., the end-to-end transport of bit streams. The bearer services are implemented by network links and switches using mechanisms of multiplexing, switching, error control, flow control, and resource allocation.

Applications generate constant or variable bit rate traffic or message exchanges. The applications impose certain requirements on the bearer services that transport this traffic. Those requirements are expressed in terms of bandwidth, delay, and error rates. Circuit switching meets the most stringent requirements in terms of delay and bandwidth but may lead to such poor utilization that it becomes uneconomical. Datagram switching is the most efficient, but it may not be able to provide guaranteed delay or bandwidth. Virtual circuit switching can combine high utilization with the ability to meet guarantees in delay and bandwidth.

In Chapter 3 we study packet switching, in Chapter 4 we study circuit switching, and in Chapter 5 we study ATM networks, the most important type of virtual circuit switching network.

2.9 NOTES

Shannon's theorems, which form the basis of information theory, are discussed in many texts; see [CT91].

The notion of layered architecture, modularity, and hierarchy shows up in various parts of computer science as well as in communication networks; see [T88, W91]. For details on implementation of protocols in UNIX, see [P93]. The Open Data Network model appears in [Kle94].

2.10 PROBLEMS

1. It is very expensive to store and archive X rays for medical diagnosis. The current system uses large photographs. An electronic image of comparable quality would require a display of 1,000 × 1,000 pixels, with a 16- or

24-bit grayscale, and with a zoom-in capability. How expensive is a monitor capable of displaying so much information? (How many pixels does your monitor display, and how many grayscale levels does it permit?) If a radiologist is retrieving the X ray from an archive, the acceptable delay is 10 s, say. What should be the bit rate of the links connecting the archive to the radiologist's office? If the X rays that the radiologist is going to view are known, say 10 min in advance, one could retrieve the X rays early and buffer them locally. This permits a reduction in the bit rate at the cost of increasing local storage. What is the bit rate/buffer trade-off? Suppose the cost of increasing the bit rate by 1 Mbps is r times the cost of 1 Mbps of disk. For what values of r is it worth reducing the link rate and increasing local disk storage?

2. As seen in Figure 2.5, we may represent the interconnection of network elements as a network graph whose edges are the links and whose vertices are the switches and user nodes. Formally, we represent the network as a graph $G = (V, E)$ where $V = \{1, 2, \ldots, N\}$ is the set of vertices and $E \subset V \times V$ is the set of edges. The interpretation is that $(i, j) \in E$ if there is a one-way transmission link from switch i to switch j. This representation is useful to specify and verify many network algorithms.

   Shortest-path algorithm. Suppose there is a cost $C_{ij} > 0$ associated with each link $(i, j) \in E$ representing delay or dollar cost of transmitting one packet over that link. We assume that the knowledge of C_{ij} is *local,* i.e., router i knows only the costs C_{ij} of links that originate at i.

   The problem is to build a shortest-path routing table at each router. The table at i has the following form:

Destination router	Outgoing link	Cost to destination
1	j_1	D_{i1}
2	j_2	D_{i2}
⋮	⋮	⋮
N	j_N	D_{iN}

   The first row in the table is interpreted like this. If a packet for destination #1 arrives at router i, it should be forwarded over link (i, j_1). The minimum cost incurred by this packet from router i onwards is D_{i1}. The other rows are interpreted similarly.

   In order that each router be able to construct its own table, it will need to exchange some information with its neighbors.

(a) Construct a distributed algorithm in which each router iterates over a two-phase cycle. In the first phase it updates its estimate of the table; in the second phase it communicates its table with its minimum cost estimate to its immediate upstream neighbors. How is the table initialized? Show that your algorithm converges after a finite number of iterations.

(b) Obtain an upper bound on the number of iterations needed for convergence.

(c) Is there any formal way in which you can say that the information that is exchanged between the routers is the *minimum* amount of information that must be exchanged?

3. What is the uncompressed bit rate for the NTSC TV signal? What compression ratio is achieved by MPEG1?

4. What is the propagation delay of a link from an earth station to a geostationary satellite? What would be the end-to-end delay of a voice conversation that is relayed via such a satellite?

5. A very common way to compress an audio signal is the following: Sample the audio signal. Denote the samples by $x_0, x_1, \ldots$. Transmit x_0. Then transmit $x_1 - x_0, x_2 - x_1, \ldots$. The dynamic range of the sample differences, $x_i - x_{i-1}$ is much smaller than the samples themselves, so the sample differences can be coded into fewer bits to achieve the same quantization noise. Develop a mathematical model that shows this.

6. Consider a video source that produces a periodic VBR stream with the following "on-off" structure. The source is "on" for 1 s with a rate of 20 Mbps; it is then "off" for 2 s with a bit rate of 1 Mbps.

   (a) What are the peak and average rates of this source? Suppose this source is served at a constant bandwidth of c Mbps, where c is larger than the average rate. Calculate the buffer size $b = b(c)$ in MB needed to prevent any loss as a function of c. What is the queuing delay as a function of c?

   (b) If this traffic is produced by a videoconferencing application, which permits a maximum delay of 200 ms, what should c be?

   (c) If this traffic is produced by a video server that is downloading a 1 hour-long video program, how much disk storage do you need? Suppose you want to play the program, but your disk access rate is only 10 Mbps. How many parallel disks would you need, and how would you store the video program on the disk?

(d) In the description above, the period of the source is 3 s, and there is a duty cycle of 1/3. Consider another source with the same duty cycle but with a smaller period. Would you say the second source is more or less bursty? Why?

7. The analog phone access line has a bandwidth of 4 kHz. The line can be used to transmit digital voice at 64 Kbps or, using a modem, to transmit data at 9.6 or 11.4 Kbps. What is the spectral efficiency in each case? What kind of modulation scheme would you use to increase spectral efficiency?

8. Give examples of applications that can lead to multiplexing gain ranges displayed in Figure 2.11.

9. Explain why error-correction schemes are used (instead of error detection followed by retransmission) in data storage applications (such as audio CDs and magnetic disks) and in real-time applications (e.g., controlling a satellite).

10. The ABP protocol is used with packets of size n. The transmission link has a BER of p. (Assume the acknowledgments are received error free.) What is the average number of packet transmissions per correctly received packet?

11. Consider a 10,000-km round-trip route with a transmission rate of 100 Mbps. Suppose a propagation time of 5 μs/km. Consider a packet size of 1,000 bits. How many packets are needed to fill up the links along the route? How large should be the minimum window size in the Go Back N protocol to achieve 100% efficiency?

12. Consider the Go Back N protocol. Suppose that the packet error probability is p. (Errors in different packets are independent.) How would you calculate the efficiency, assuming that N is chosen so that the pipe is just full? How will efficiency change as N increases?

13. Consider the retransmission protocol described in Figure 2.19. Suppose that the acknowledgment following a correct transmission arrives after a random delay d. (The randomness is due to random queuing delay in the network.) Let F be the cumulative probability distribution of d, i.e., $F(t) = \text{Prob}\{d < t\}$. Thus the probability of receiving an acknowledgment before the timeout is $F(T)$, and the probability of not receiving it before the timeout is $1 - F(T)$.

(a) Show that the expected number of timeouts that the same packet is sent is

$$N(T) = \sum_{n=1}^{\infty} nF(T)[1 - F(T)]^{n-1}.$$

Show that $N(T)$ decreases as T increases.

(b) Show that the expected time $\tau(T)$ before an acknowledgment is received before a timeout is between $T[N(T) - 1]$ and $TN(T)$. What is the exact value of $\tau(T)$?

(c) The timeout T is a design parameter. How would you choose it?

3

CHAPTER

Packet-Switched Networks

As we saw in Chapter 2, more sophisticated services demanded by user applications are built from basic services in a layered architecture. We also discussed the ODN architecture for communication networks. In this chapter we study the seven-layer Open Systems Interconnection (OSI) architecture for data networks. The OSI model can be regarded as a more detailed specification of the ODN model. (However, the OSI model was developed long before the ODN model.) Layer 3 of the OSI model is called the *network layer.* It corresponds to the *bearer service layer* of the ODN model.

This chapter presents the OSI and Internet Protocol (IP) models for logical layering of network functions. By the end of the chapter, you will become familiar with the major local and metropolitan area network solutions from Ethernet and token ring to FDDI and DQDB (Distributed Queue Dual Bus). You will be able to estimate the limitations of each solution in terms of speed, delay, and versatility and judge which solution best meets an organization's needs. You will also understand how services provided by Frame Relay, Switched Multimegabit Data Service (SMDS), and the Internet can be used to interconnect networks. We discuss the IP protocol suite as it exists in 1996 and the major proposals for upgrading IP so that it can support high-performance applications requiring guaranteed bandwidth and delay.

Section 3.1 describes the OSI model. Section 3.2 is devoted to Ethernet and section 3.3 to the token ring network; sections 3.4–3.7 describe the FDDI, DQDB, Frame Relay, and SMDS networks, respectively. Section 3.9 considers the Internet Protocol and related developments. The reader may skip particular sections without loss in continuity.

3.1 OSI REFERENCE MODEL

In this section we explain the Open Systems Interconnection reference model. It is a seven-layer decomposition of network functions published by the ISO. We explain the main functions performed by these layers. (See Figure 3.9 for a summary.) Many networks do not strictly follow the OSI model. In some cases, the networks were developed before the OSI model was published, and these networks established standards. However, despite these differences, the OSI model helps in understanding the design of packet-switched network architectures.

3.1.1 Layer 1: Physical Layer

The bottommost layer is the *physical layer.* It implements an unreliable bit link. A link consists of a transmitter, a receiver, and a medium over which signals are propagated. These signals are modulated electromagnetic waves that propagate either guided (by a copper cable, wire pair, or an optical fiber) or unguided in free space (as in radio). The transmitter converts the bits into signals, and the physical layer in the receiver converts the signals back into bits. The receiver must be synchronized to be able to recover the successive bits. To assist the synchronization, the transmitter inserts a specific bit pattern, called a *preamble*, at the beginning of the packet, indicated by sync in Figure 3.1. The bit link is unreliable because synchronization errors and noise can corrupt a packet.

3.1.2 Layer 2: Data Link Layer

Figure 3.2 illustrates the *data link layer* for a point-to-point link between two computers. As we explained in our discussion of error detection (see section 2.5.3), the transmitter and receiver can execute a specific protocol to retransmit the corrupted packets. In the figure, the protocol entities are represented by the boxes labeled 2. These entities are programs that are usually executed by dedicated electronic circuits (in the NIC or network interface card, shown in Figure 2.31) because of the high speed.

The transmitter numbers the packets and appends error-detection bits. The figure shows the fields that are appended to the packet that the data link layer transmits; first, the data link layer in the transmitter adds a sequence number to the packet and the error-detection bits (CRC). Then the physical layer adds a synchronization preamble (sync). In the receiver, the physical layer strips the preamble and gives the rest of the packet to the data link layer,

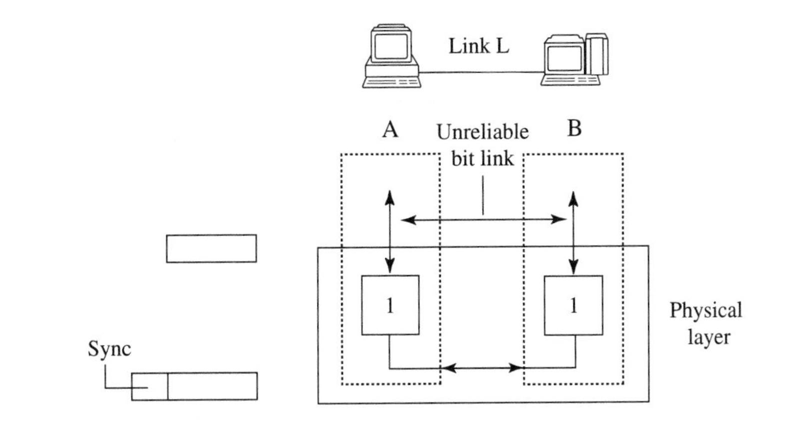

3.1
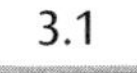
FIGURE

The physical layer transmits bits by converting them into electrical or optical signals. In many implementations, the physical layer uses synchronization bits (sync) to synchronize the receiver.

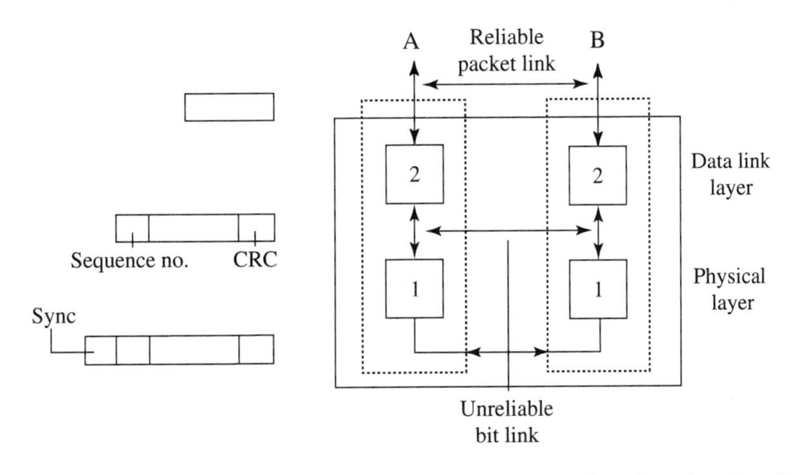

3.2

FIGURE

The data link layer supervises the transmission of packets by the physical layer. In a typical implementation, the data link layer adds a sequence number and error-detection bits (CRC). As we explain in the text, networks with reliable links control the errors from source to destination instead of controlling them on every link.

which uses the error-detection bits to verify that the packet is correct and the sequence number to check that the packet is one that it expected. The data link layer then strips the error-detection bits and the sequence number.

The fields that the protocol entities add to the packet contain control information that the protocols in the different layers use to monitor the transmissions. The appending of control fields to a packet is called *encapsulation*. The reverse process, stripping the control fields, is *decapsulation*. Observe that encapsulation and decapsulation may often be performed without examination of the packet, which simplifies the hardware and reduces packet processing time.

The errors can be controlled end to end instead or over every link. End-to-end control is preferable to link-level control when the network uses links with a small bit error rate. Indeed, in such a situation, most packets reach their destination without errors, and it is wasteful to verify them at every link. End-to-end error control is implemented at the transport layer (layer 4) with the same mechanism used by the data link layer.

3.1.3 Sublayer 2a: Media Access Control

Figure 3.3 shows computers attached to a common link. These computers must regulate the access to that shared link. This function, called access control, is performed by a sublayer called the *media access control* (MAC) sublayer. Thus, the MAC sublayers in the computers follow a set of rules—a protocol—to regulate access to the shared link. Because the link is shared, the MAC must append the physical address of the destination, the specific computer to which the packet is destined. The physical address identifies uniquely a device attached to a shared link. Such a physical address is not needed for a point-to-point link.

The MAC sublayer implements the unreliable transmission of packets between computers attached to the common link.

3.1.4 Sublayer 2b: Logical Link Control

Figure 3.4 shows the *logical link control* (LLC) sublayer. It uses the unreliable transmission of packets implemented by the MAC sublayer to implement reliable packet transmission between computers attached to a shared link. The functions of the LLC are those the data link layer executes for a point-to-point link.

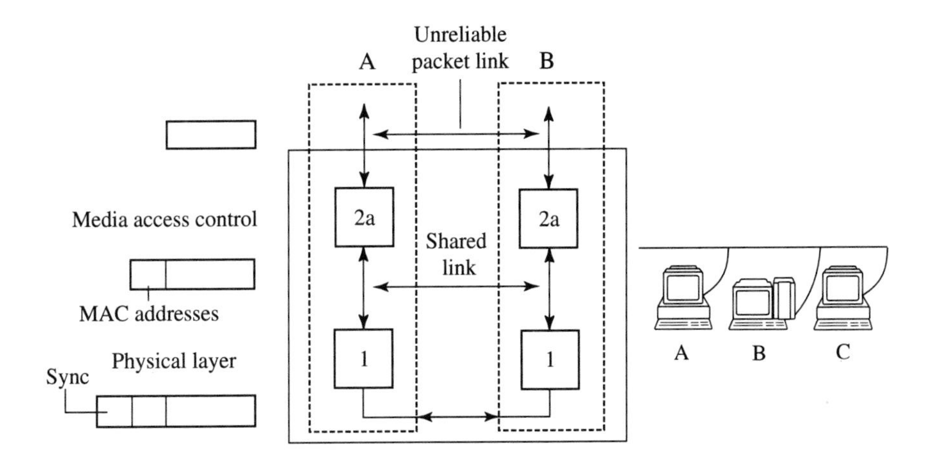

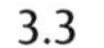

3.3
FIGURE

The figure shows three computers that share a common link. Access to such a common link is regulated by the media access control (MAC) sublayer. That sublayer adds the addresses of the source and destination to the packet before giving it to the physical layer for transmission.

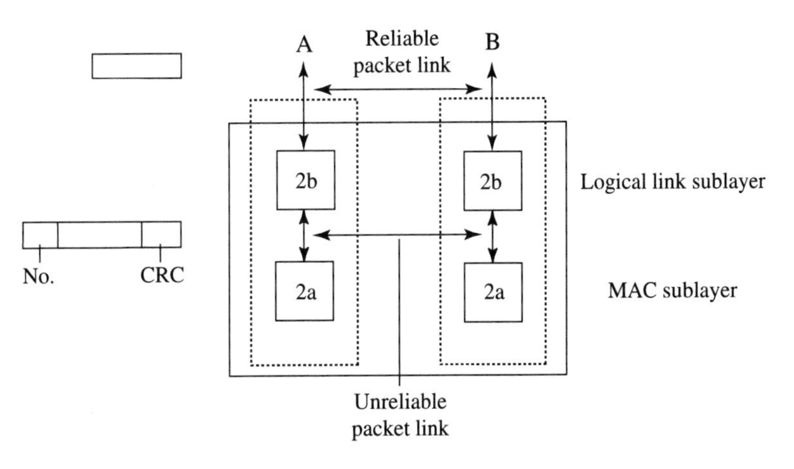

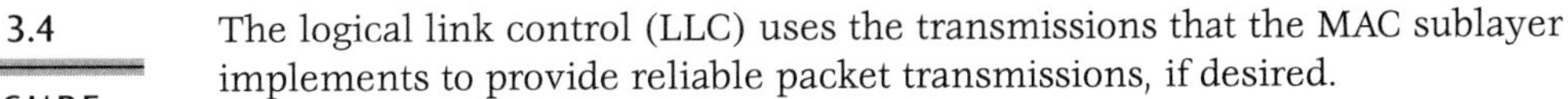

3.4
FIGURE

The logical link control (LLC) uses the transmissions that the MAC sublayer implements to provide reliable packet transmissions, if desired.

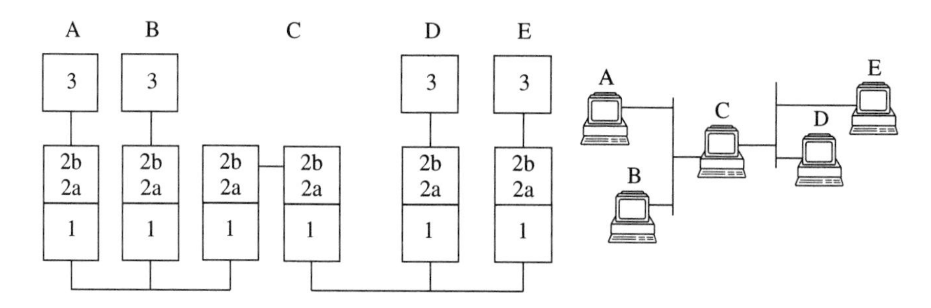

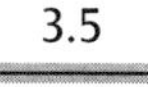

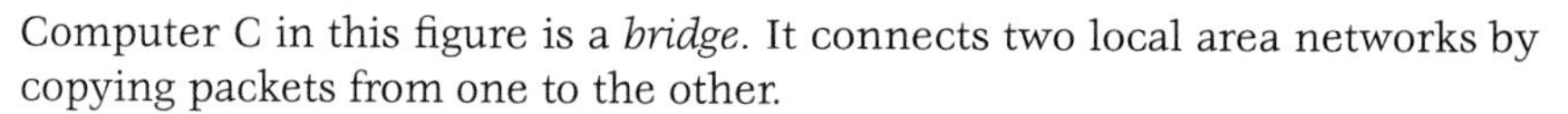

FIGURE 3.5 Computer C in this figure is a *bridge*. It connects two local area networks by copying packets from one to the other.

The MAC and LLC together constitute the data link layer for multiple access links: they use the unreliable bit link of the physical layer to implement a reliable packet-transmission service between computers attached to a common link.

Figure 3.5 shows a *bridge* (computer C) between two Ethernet networks. Computers A, B, and C are attached to one Ethernet. The right part of the figure shows these three computers attached to the same link. The MAC sublayer in the three computers implements unreliable packet transmissions that are made reliable by the LLC sublayer. The situation is similar for computers C, D, and E attached to the other Ethernet.

Consider a packet sent by computer A and destined for computer E. Computer C must store that packet and retransmit it on the second Ethernet. The routing decision of C is limited to whether to retransmit the packet or not. A bridge is a computer equipped with the hardware and software to perform such decisions and capable of retransmitting packets at a high rate. The bridge allows the computers on the two Ethernets to function as if they are on the same Ethernet (see section 3.2.3).

3.1.5 Layer 3: Network Layer

The data link layer implements a reliable packet link between computers attached to a common link. As we explained in our discussion of store-and-forward packet switching (see section 2.5.2), when computers are connected by a collection of point-to-point links, they must figure out where to send the packets that they receive: whether to send them out over another link and, if so, which one. (See Figure 3.6.) This function—finding the path the packets

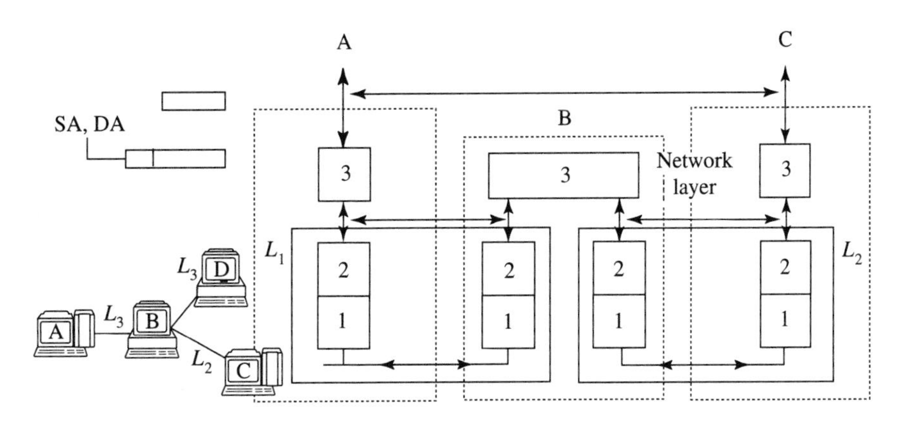

3.6

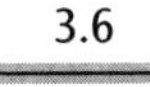

FIGURE

The network layer delivers packets between any two computers attached to the same network. That layer implements store-and-forward transmissions along successive links from the source to the destination.

must follow—is called *routing*. Routing is one of the main functions of the *network layer*.

Thus, the network layer uses the reliable transmission over point-to-point links provided by the data link layers of those links to implement the reliable transmission of packets between any two computers attached in a network.

Figure 3.7 shows a *router* attached to a number of links. When the router receives a packet, it must decide along which link it should retransmit the packet. This routing function is implemented by the network layer.

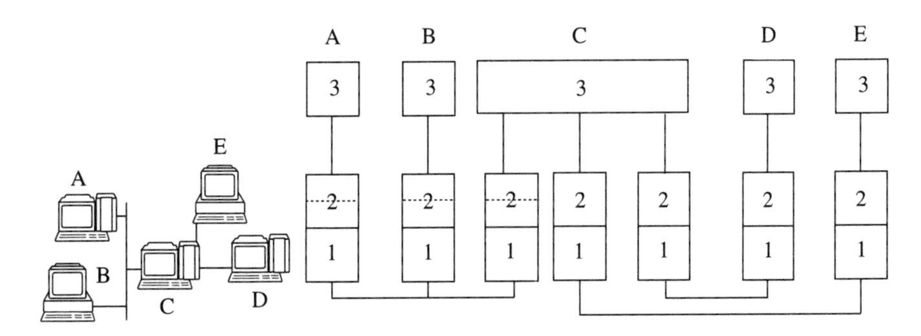

3.7

FIGURE

Computer C in this figure is a *router*. It is designed to relay packets at a high rate to the proper link and with a low delay.

3.1.6 Layer 4: Transport Layer

The *transport layer* delivers *messages* between *transport service access points* (TSAPs or *ports*) in different computers. The TSAPs differentiate the information streams. The transport layer may multiplex several low-rate transmissions with different TSAPs onto one virtual circuit or divide a high-rate connection into parallel virtual circuits.

Some frequently used processes such as e-mail, the Transmission Control Protocol (TCP), and the User Datagram Protocol (UDP) are allocated fixed TSAPs (also called *well-known ports*). To connect to a process with unknown TSAP, a remote process first connects to a process server attached to a fixed TSAP. The server then indicates the TSAP of the desired process.

The transport layer delivery of messages is either *connection-oriented* or *connectionless*. A connection-oriented transport layer delivers error-free messages in the correct order. Such a transport layer provides the following services: CONNECT, DATA, EXP_DATA, and DISCONNECT. CONNECT sets up a connection between TSAPs. DATA delivers a sequence of messages in the correct order and without errors. EXP_DATA delivers urgent messages by making them jump ahead of the nonurgent messages in the two end nodes. DISCONNECT releases the connection. A connectionless transport layer delivers messages one by one, possibly with errors, and with no guarantee on the order of the messages. The service of a connectionless transport layer is UNIT_DATA, the connectionless delivery of a single message.

The transport layer fragments messages into packets and reassembles packets into messages, possibly after resequencing them. To perform this fragmentation/reassembly function, the transport layer numbers the packets belonging to the same message. The layer also controls the flow of packets to prevent the source from sending packets faster than the destination can handle them. Moreover, the transport layer requests retransmissions of corrupted packets (see Figure 3.8).

To be able to reassemble messages, the transport layer numbers packets. When the source computer fails, it may lose track of the sequence number it had reached. Resuming the numbering arbitrarily might lead to delayed packets in the network having the same sequence numbers as the packets being transmitted; this would confuse the destination. One solution to this problem is to attach a "time to live," L, to each packet and to decrement that time to live every time the packet goes through a network node. Packets with a zero time to live are discarded. Thus, L is the maximum number of hops for each packet, and it corresponds to a maximum lifetime of T seconds inside the network, because a packet will remain in each node for some

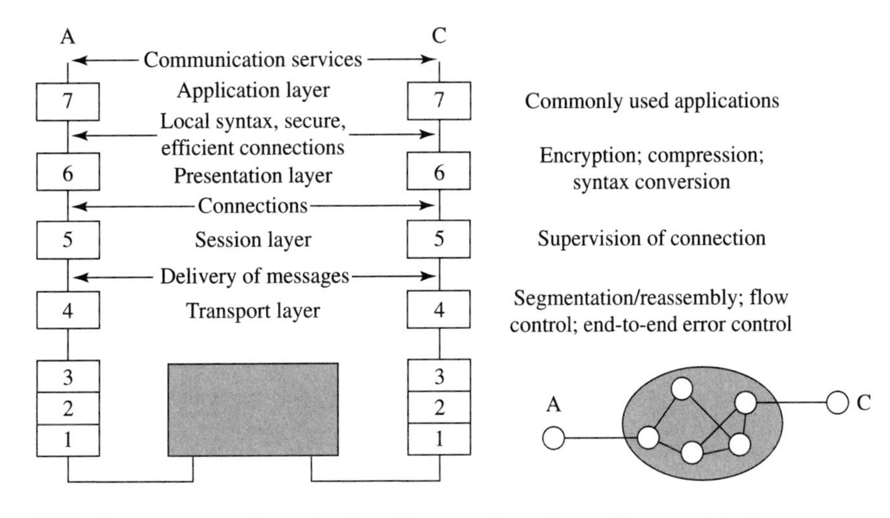

FIGURE 3.8 This figure summarizes the functions and services of layers 4 to 7.

bounded time. When a source computer recovers from a crash, it can wait for T seconds to make sure that all the delayed packets to the destination were discarded, thus avoiding any numbering ambiguity. The transport layer protocols in the two nodes agree on an initial sequence number by using a three-way handshake.

Figure 3.8 summarizes the functions of layers 4 to 7.

3.1.7 Layer 5: Session Layer

The *session layer* supervises the dialog between two computers. It can set up a connection prior to an exchange of information between the machines. The session layer can partition a transfer of a large number of messages by inserting *synchronization points*. These synchronization points are specific packets that divide the sequence of messages into groups. In case of computer malfunction, the transmission can restart from the last synchronization point.

3.1.8 Layer 6: Presentation Layer

Different application programs and computer devices use different conventions to represent information by binary numbers. For instance, some computers represent 16-bit words by placing the most significant byte before the least significant byte, whereas other computers use the opposite convention. As another example, different terminals use different control characters to specify backspace, line feed, and carriage return. Also, different application programs may follow different rules to encode data structures such as matrices and complex numbers. Computer scientists refer to a set of rules for representing information as a *syntax*. Thus, different computers use different syntaxes. The syntax used by a computer is its *local syntax*.

Say that computers using N different local syntaxes want to communicate. One possible method is to have every computer perform the $N - 1$ syntax conversions needed to communicate with the other computers. Another method is to adopt a common transfer syntax and have every computer perform the conversion between its local syntax and the transfer syntax. The second method is obviously more convenient because it does not require a computer to be aware of all the possible other local syntaxes. Communication networks use the second method, and the presentation layer is responsible for the conversion between local syntax and transfer syntax. As shown in Figure 3.8, one service provided by the presentation layer is the exchange of information between computers, each using its own local syntax.

An additional task of the presentation layer is to encrypt transmissions that must be secure. In abstract terms, encryption is a one-to-one transformation of a message into an encrypted version.

Another important task of the presentation layer is data compression. Such data compression eliminates some of the redundancy in the information to be transmitted, thereby reducing the number of bits to be transferred.

3.1.9 Layer 7: Application Layer

The *application layer* provides frequently needed communication services such as file transfer, terminal emulation, remote login, directory service, and remote job execution. These services are used by the user applications. For instance, an e-mail program uses a file transfer service that delivers a file to a list of addresses and informs the sender if some addresses cannot be reached. User applications, such as e-mail or WWW, are run on top of the application layer.

3.1.10 Summary

The OSI model is a detailed architecture that specifies how complex services such as file transfer and remote login are to be constructed out of the most basic services of unreliable bit transfer in a seven-layer hierarchy. Each layer

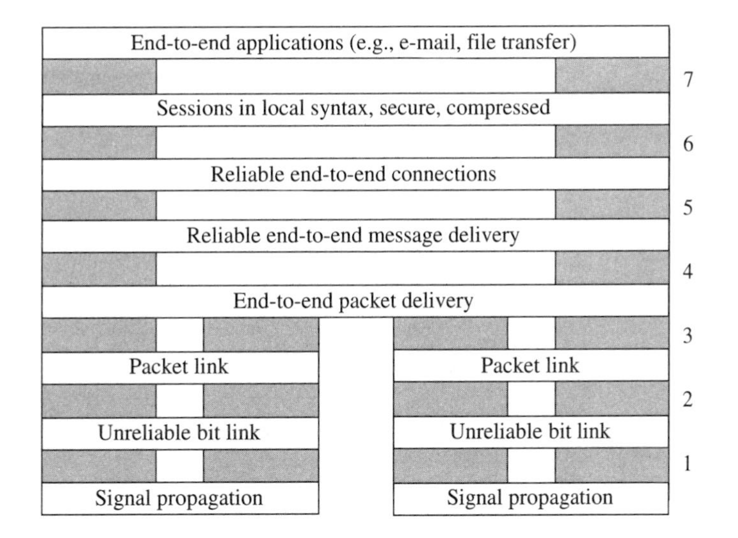

3.9 FIGURE The figure shows the services implemented by the seven layers of the OSI reference model.

adds further functionality to the services of the layer immediately below it. The OSI model is summarized in Figure 3.9.

We can place the OSI model in the context of the Open Data Network model of section 2.7 by identifying layers 1 and 2 with bit ways, layers 3 and 4 with bearer services, layers 4, 5, and 6 with middleware, and layer 7 with applications.

The OSI model is abstract like the ODN model. We now study the most important network implementations. We study Ethernet in considerable detail, mainly to illustrate the considerations that are involved in setting standards. The reader familiar with these standards and the uninterested reader may skip this material without loss of continuity.

3.2 ETHERNET (IEEE 802.3)

The IEEE 802.3 standards specify layers 1 and 2 of a family of local area networks or LANs, including 10BASE5 (thick Ethernet), 10BASE2 (thin Ethernet), 1BASE5 (StarLAN), 10BROAD36 (Broadband Ethernet), and 10BASE-T (Ethernet over twisted pairs). The Ethernet networks are the most popular local area networks. They are inexpensive and provide a relatively high throughput and low delays that can support many applications. Most importantly, Ethernet provides inexpensive, relatively high-speed network access to individual users. (The token ring network, in which users share some of the transmission costs, also provides inexpensive access.)

3.2.1 Physical Layer

The physical layer of IEEE 802.3 networks specifies the electrical and mechanical characteristics of the wiring and of the encoding of the bits. We describe the physical layers of the 10BASE5, 10BASE2, and 10BASE-T networks.

10BASE5

Figure 3.10 shows the physical layout of a 10BASE5 network. The wiring consists of segments. Each segment is a length of up to 500 m of coaxial cable with a diameter of 10 mm and a characteristic impedance of 50 ohms. Segments are connected by repeaters that can be up to 1,000 m apart. No two computers on the network can be more than 2,500 m apart. The transmission rate is 10 Mbps.

10BASE2

The 10BASE2 network is similar to 10BASE5 except that it uses a thin (5 mm) coaxial cable instead of a 10 mm coaxial cable. Also, the coaxial cable is attached directly to the interface card by a T connector (BNC). The segment length is limited to 185 m. The transmission rate is also 10 Mbps.

10BASE-T

The 10BASE-T wiring, shown in Figure 3.11, uses the same unshielded twisted pairs (UTPs) that telephone companies use to wire buildings for telephone service. Consequently, if an office building or residence has enough spare UTPs, these can be used to install a 10BASE-T network. (This explains the popularity of this form of Ethernet. Economies of scale have led to a steady reduction in the cost of Ethernet chips and boards. Today, the cost of wiring is likely to be more than the cost of the Ethernet hardware.)

In a 10BASE-T network, each computer is equipped with a network interface card that is attached to a medium access unit (MAU) with two twisted pairs. Two unshielded twisted pairs attach an MAU to a 10BASE-T hub. Hubs can be attached together with two UTPs to build larger networks.

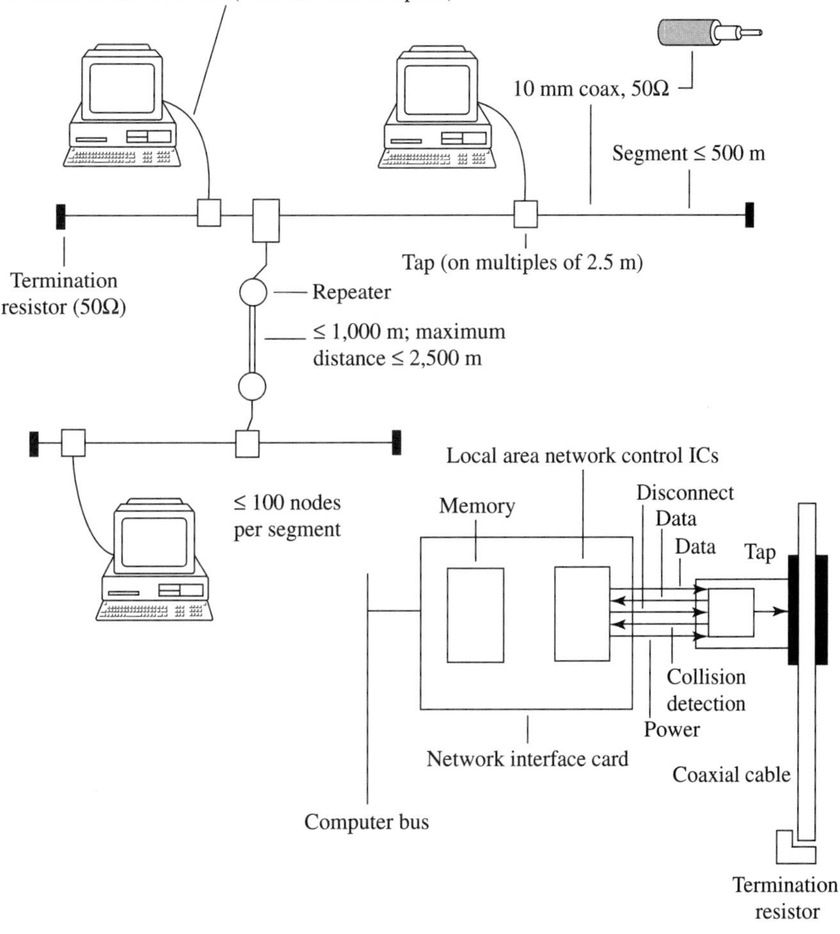

3.10 FIGURE Physical layout of a 10BASE5 (thick Ethernet) network. This network uses 10-mm coaxial cable. The nodes transmit at 10 Mbps.

The 10BASE-T network transmits data at 10 Mbps. Using *fiber extenders*, the network engineer can replace UTP segments with longer fiber segments.

Wireless Ethernet

A number of commercial products are available to set up wireless Ethernet networks. In such a wireless Ethernet network, all the stations share a radio

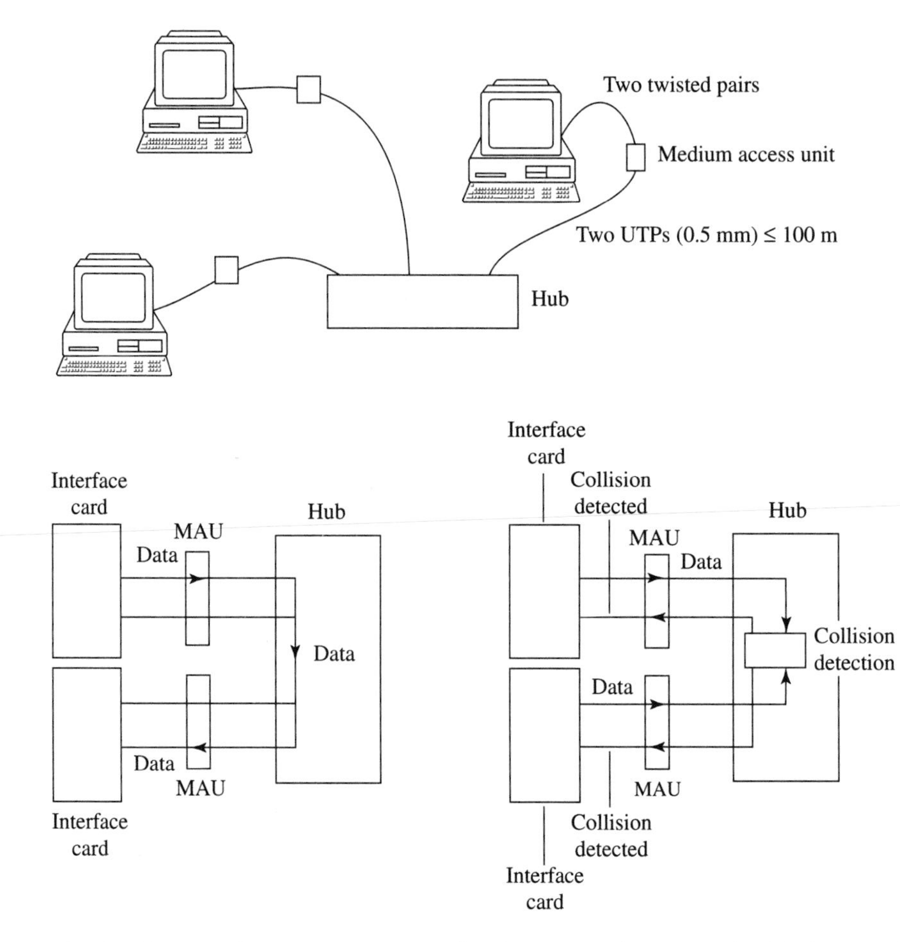

3.11 FIGURE

Physical layout of a 10BASE-T network. This network uses unshielded twisted pairs and can be wired with spare telephone pairs already in place in the building. The transmission rate is 10 Mbps. One can attach hubs together to build larger networks.

channel. Some products transmit using diffuse infrared light signals instead of radio waves.

3.2.2 MAC

The media access control sublayer of Ethernet specifies the MAC addresses of network interfaces, the frame format, and the MAC protocol for sharing the cable.

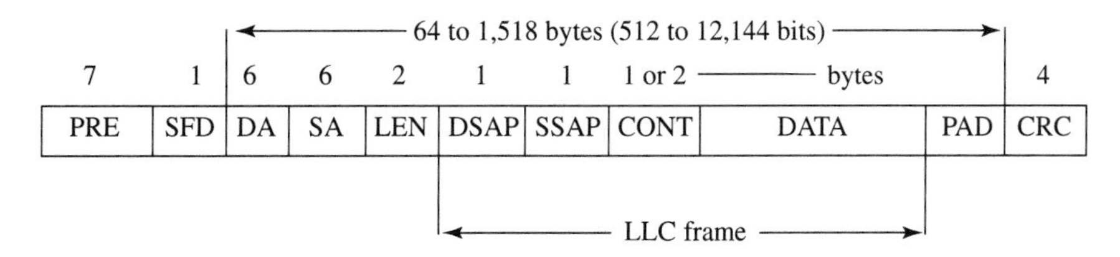

3.12 FIGURE

Format of Ethernet packets. The various fields that make up the frame are explained in the text.

The MAC address of an Ethernet interface card is a 48-bit string set by the manufacturer that is unique to that card.

The frame format of Ethernet packets is shown in Figure 3.12. The preamble (PRE) synchronizes the receiver. The start of frame delimiter (SFD) indicates the start of the frame. The destination (DA) and source (SA) MAC addresses are 48-bit-long strings unique to each interface card. The length indicator (LEN) eliminates the need for an end of frame delimiter and permits the use of a padding field (PAD) to make sure that the frames have at least 64 bytes. The cyclic redundancy check enables the receiver to detect most transmission errors, as we explained in section 2.5.3. The logical link control (LLC) frame specifies the destination (DSAP) and source (SSAP) service access points.

The Ethernet MAC protocol is CSMA/CD, Carrier Sense Multiple Access with Collision Detection. When using CSMA/CD, a node that has a packet to transmit waits until the channel is silent before transmitting. Also, the node aborts the transmission as soon as it realizes that another node is transmitting. After aborting a transmission, the node waits for a random time and then repeats these steps.

Figure 3.13 shows two nodes A and B that start transmitting. Node B starts transmitting shortly before the signal sent by node A reaches it. Note that node A detects that node B is also transmitting after about one round-trip propagation time of a signal between nodes A and B. This time is wasted because the two nodes must abort their transmission and restart after some random delay. This wasted time is proportional to the propagation time between the nodes. Figure 3.14 illustrates the sequence of events that takes place when a node transmits a packet. The figure shows that the node starts transmitting then detects that another node is also transmitting so that it must abort its transmission, and it then waits for some random time before trying again. Eventually, the node succeeds in transmitting its packet.

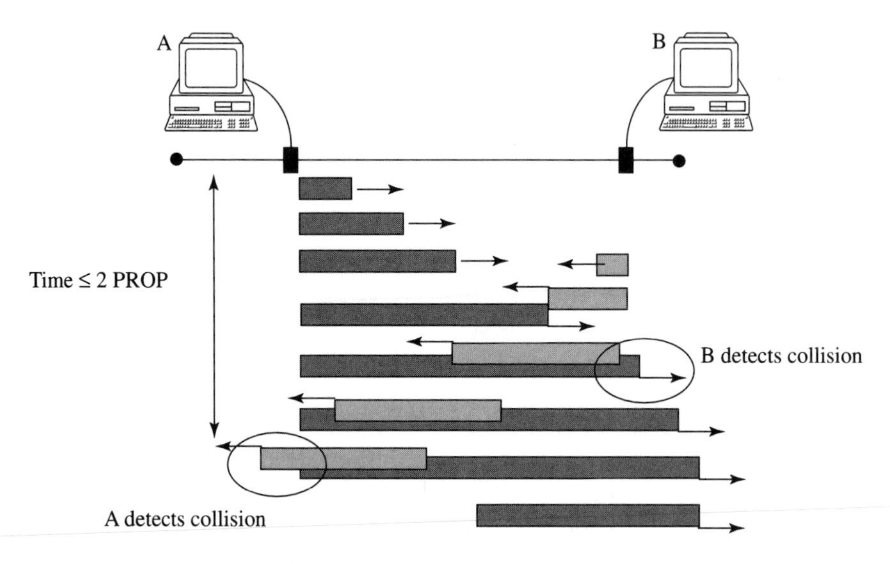

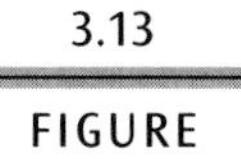
3.13 FIGURE

The maximum time until a node detects a collision is twice the propagation time of a signal between the nodes that are farthest apart.

It can be shown that when many nodes attempt to transmit, they waste, on the average, an amount of time approximately equal to 5 × PROP per successful transmission. Here, PROP designates the propagation time of a signal from one end of the cable to the other. Accordingly, the fraction of time that the nodes use the transmission channel to transmit packets successfully is approximately equal to $1/(1 + 5a)$ where a denotes the ratio of a propagation time (PROP) to a packet transmission time (TRANS). This fraction of useful time when many nodes want to transmit is called the *efficiency* of the MAC.

For instance, if a is very small, then the nodes learn very quickly about conflicting transmissions. Consequently, the nodes waste very little time be-

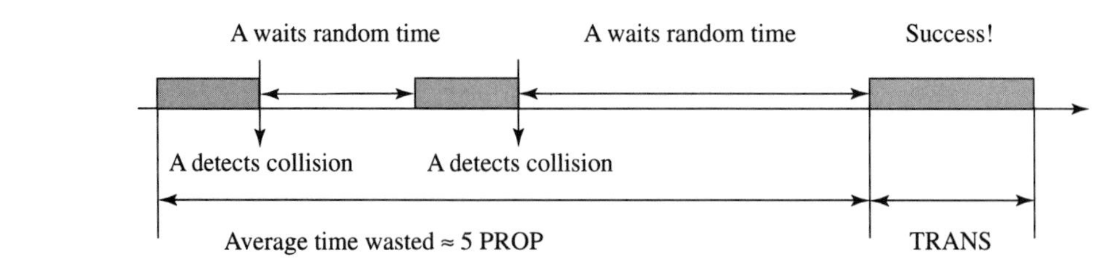

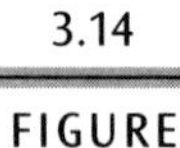
3.14 FIGURE

Sequence of events on an Ethernet network. The transmitters waste five propagation times per successful transmission, on the average.

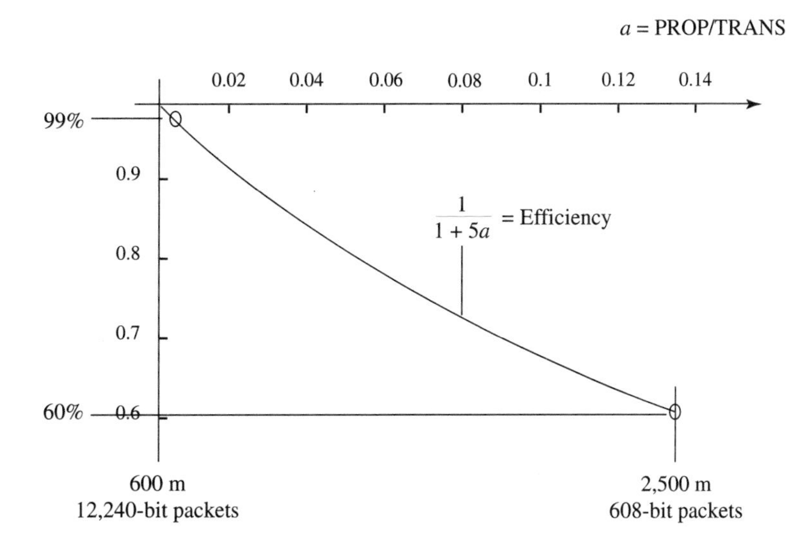

3.15 FIGURE

Two representative values of the efficiency of Ethernet. Typically, the efficiency is about 80%.

cause of such conflicts and the efficiency of the MAC is close to unity. At the opposite extreme, if the packet transmission times are comparable to a propagation time, then a node learns of simultaneous transmissions only after a long time (when the signal from another node reaches it), so that a significant fraction of time is wasted by transmissions that are aborted, and the efficiency is small.

Figure 3.15 shows two representative values of the efficiency of Ethernet. The first value corresponds to a relatively short Ethernet (600 m) with packets that have the maximum length admissible by the Ethernet standards. In that case, the calculations show the efficiency to be about 99%. The other case corresponds to an Ethernet with the maximum admissible separation between nodes (2,500 m) and with the smallest possible Ethernet packets. In this least-efficient case, the efficiency is about 60%. In a typical situation, the efficiency can be expected to be about 80%. This means that out of the raw bit rate of 10 Mbps, 80%, or 8 Mbps, are used to transmit successful packets.

3.2.3 LLC

The logical link control sublayer provides connection-oriented or connectionless (acknowledged or not) services. The LLC can also multiplex different transmissions that are differentiated by the service access point field (see

Figure 3.12). Finally, the LLC implements the *transparent routing* of packets between Ethernets attached together with bridges.

When it provides a connection-oriented or an acknowledged connectionless service, the LLC uses the CRC field to detect errors. To implement a connection-oriented service, the LLC uses the Go Back N protocol (see section 2.5.3) to arrange for the transmitter to retransmit packets that do not arrive error free.

We use Figure 3.16 to explain how a packet with a given MAC destination address finds its way in a network of Ethernets connected by bridges. The bridges use a simple procedure called *transparent routing.* Transparent routing requires the bridges (such as node C in the figure) to maintain tables of MAC addresses of the nodes on their Ethernets. To maintain such a table, a bridge reads the source addresses of the packets broadcast on the Ethernets. Say that a packet on an Ethernet is *local* if it is destined for another node of the same Ethernet. By maintaining address tables, a bridge learns which packets are local. When bridge C receives a packet that it does not think is local, bridge C retransmits the packet on the other Ethernet. If the packet was in fact local, no damage is done since all the nodes on the other Ethernet will disregard the packet.

Transparent routing is a very convenient procedure. However, it can make packets loop forever between bridges unless some care is taken to avoid that

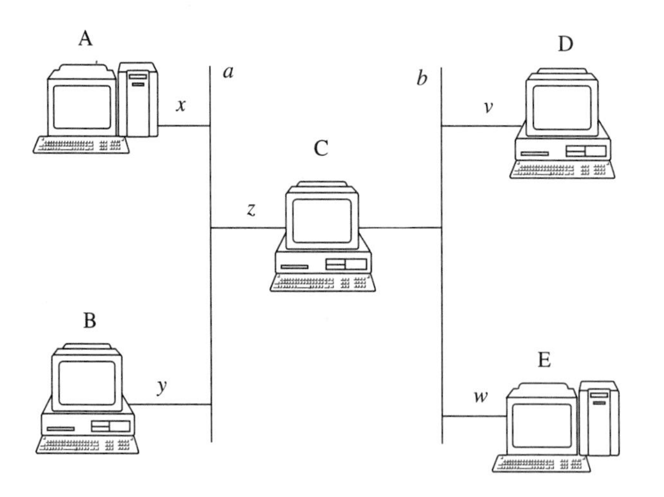

3.16 FIGURE

Ethernet attached by bridges. Transparent routing enables a computer to transmit a packet to another computer on one of these Ethernets as if it were on the same Ethernet.

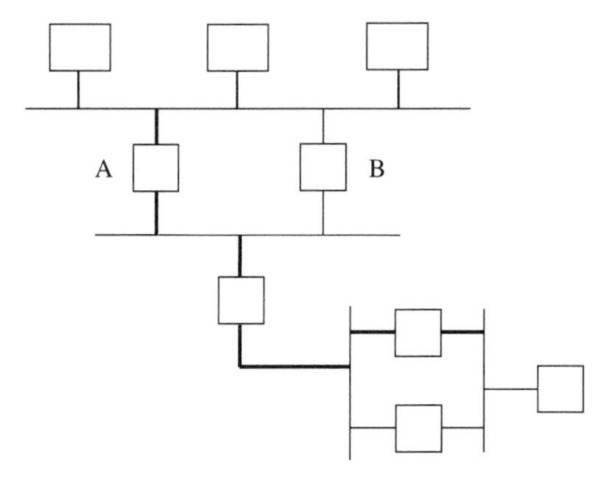

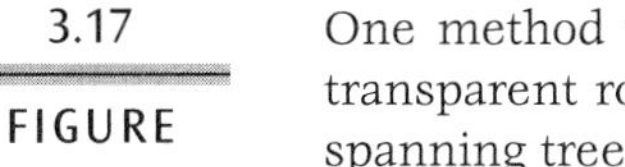

3.17 FIGURE One method that bridges use to avoid making multiple copies with transparent routing is spanning tree routing. Only the bridges along the spanning tree copy the packet.

behavior. Consider Figure 3.17. Assume that the top-left node in the figure sends a packet destined to the rightmost node. The packet is seen by two bridges, say A and B. Bridge A retransmits the packet on the second Ethernet because the packet is not local. When bridge B sees that packet on the second Ethernet, it retransmits it on the top Ethernet because the packet is not local to the second Ethernet. Thus bridge A gets the packet a second time, and the bridges A and B repeat this sequence of retransmissions indefinitely. To prevent such an infinite loop, the bridges use *spanning tree routing.* A spanning tree in a graph is a subgraph that is a tree (i.e., loop-free) and that spans all the nodes (but not necessarily all the bridges). If the bridges know a tree that spans all the network nodes, then they can avoid loops by agreeing that the packets will be retransmitted only by the bridges on the tree. A spanning tree is shown in thicker lines in Figure 3.17.

The bridges construct a spanning tree by using a distributed algorithm that finds the shortest path to destinations. The shortest path to any destination cannot contain any loop. Consequently, the set of shortest paths from all the bridges to a particular bridge must be a spanning tree.

3.3 TOKEN RING (IEEE 802.5)

The IEEE 802.5 standards specify layers 1 and 2 of a family of *token ring* networks. These networks transmit at 4 Mbps or 16 Mbps. These networks have

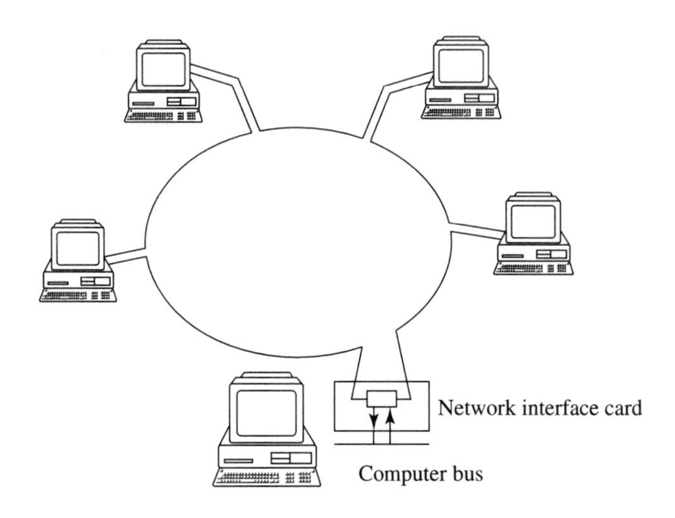

3.18 FIGURE

Layout of a token ring network. The computers are attached by unidirectional point-to-point links around a ring.

the advantage that, unlike in Ethernet networks, each node is guaranteed to get to transmit before a specific time. Also, the token ring networks are more efficient than Ethernet networks under high load.

3.3.1 Physical Layer

In a token ring network, the nodes are connected into a ring by point-to-point links. See Figure 3.18. A network interface has two possible configurations: repeater and open. In the repeater configuration, the interface repeats the incoming signal on the outgoing link with a delay of a few bit transmission times. At the same time, the interface copies the signal for the computer. In the open configuration, the interface transmits on the outgoing link and listens on the incoming link. The transmission rate is 4 Mbps or 16 Mbps, as already mentioned.

3.3.2 MAC

The frame format is similar to that of Ethernet packets (Figure 3.12), except that it uses an ending delimiter instead of a length indication. The token is a 3-byte frame that consists of a start of frame, an access control, and an ending delimiter, each 1 byte long. The access control field indicates that the 3-byte frame is a token that any station may grab and not a packet.

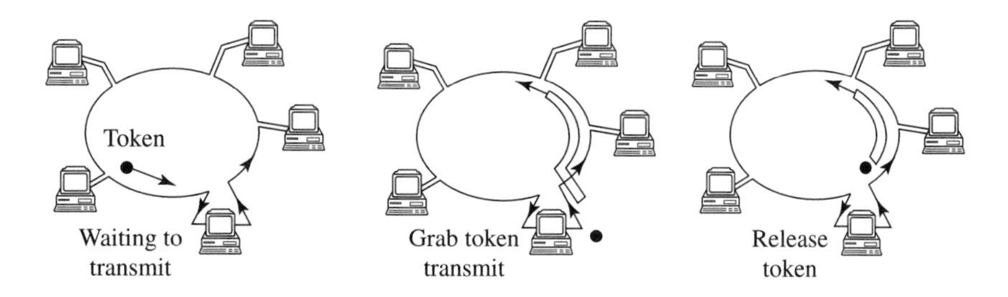

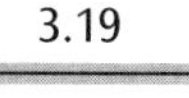

3.19 FIGURE Steps in the transmission of a packet when the computers use the release after transmission token-passing protocol.

The transmissions proceed in one direction along the ring. Figure 3.19 shows the sequence of events when a node wants to transmit a packet: the node waits for the token, then transmits some of its packets and releases the token. We call this version of the MAC protocol, where a node releases the token right after it finishes transmitting its packets, *release after transmission.* The 16-Mbps token ring networks use this protocol. In a 4-Mbps token ring network, a node that transmits waits until it has completely received its last packet before releasing the token. We call this version *release after reception.* The standard specifies that a node can hold on to the token and transmit for up to some time, called the *token holding time* (THT), before releasing the token. A typical value of THT is 10 ms.

We use Figure 3.20 to analyze the efficiency of the release after reception protocol. Assume that there are N nodes on a token-passing ring. We define T_n to be the time during which node n transmits when it gets the token, before it releases the token. Thus, T_n can range from 0 to THT. We assume that all the nodes want to transmit, so that $T_n > 0$ for $n = 1, \ldots, N$. At time 0, the first node starts transmitting a packet. At time T_1 the first node has transmitted its packets. The last packet has completely returned to the first node PROP seconds later, where PROP is the propagation time of a signal around the ring. Therefore, at time $T_1 + \text{PROP}$ the first node starts transmitting the token. A short time later, the first node finishes transmitting the token, which reaches the second node after a propagation time designated by $\text{PROP}_{1\to 2}$. Node 2 then goes through the same sequence of steps node 1 did, and so do the other nodes, one after another. Eventually, the token comes back to node 1. The efficiency of the token ring is the fraction of time that the nodes transmit packets. The figure shows that the efficiency is approximately equal to $1/(1 + a)$ where $a = \text{PROP}/E(T_n)$. In this expression, $E(T_n)$ is the average duration of a node transmission. The figure also shows representative values of the

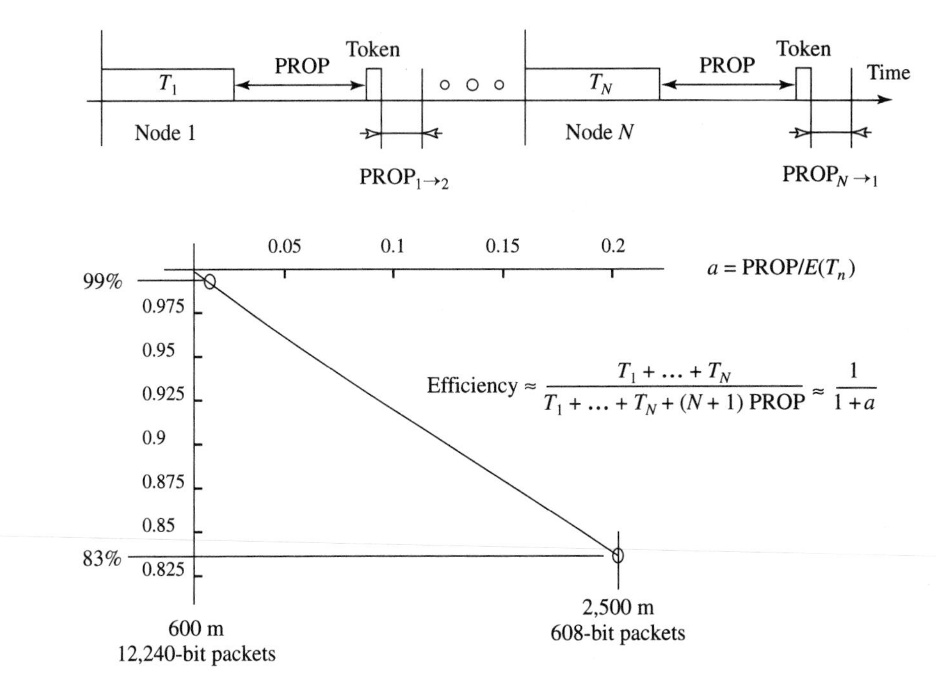

3.20 FIGURE

Timing diagram for the release after reception token-passing protocol when all the computers have packets to transmit.

efficiency, assuming that the nodes transmit a single packet of a fixed size. As we can see, the efficiency of a typical token ring network is more than 90%.

One can modify Figure 3.20 to study the release after transmission protocol used by the 16-Mbps token ring network and, with a similar analysis, show that its efficiency is approximately $(1 + a/N)^{-1}$, which approaches 100%.

3.3.3 LLC

The logical link control sublayer of IEEE 802.5 networks is the same as in IEEE 802.3 networks.

3.4 FDDI

The Fiber Distributed Data Interface (FDDI) (see Figure 3.21) is an ANSI (American National Standards Institute) standard for a 100-Mbps network. FDDI connects up to 500 nodes with optical fibers, in a dual ring topology.

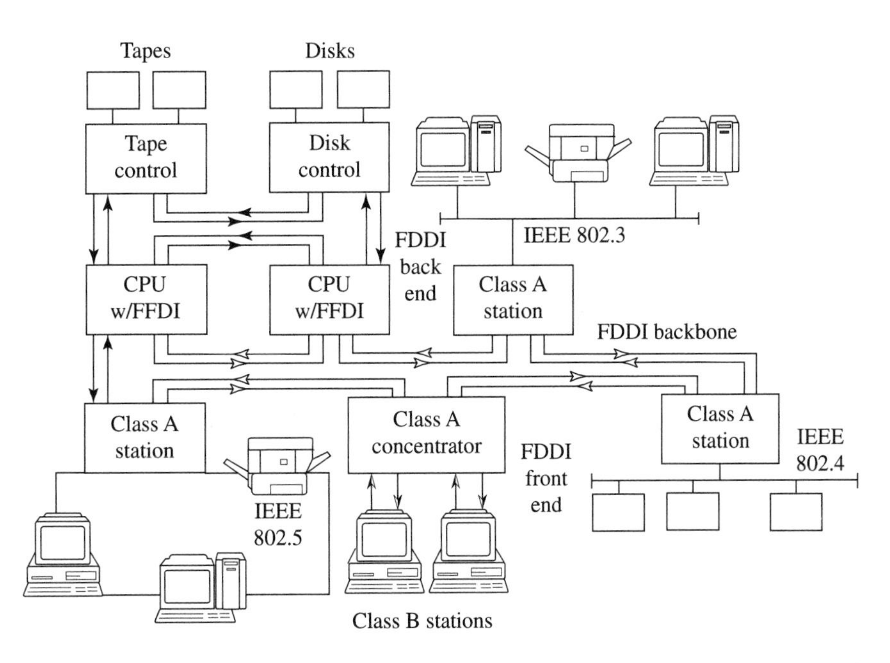

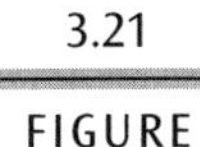

3.21 FIGURE

An FDDI dual ring network can support 500 stations with a total distance of 200 km and up to 2 km between adjacent stations. Stations are connected by 100-Mbps optical fiber, and a timed-token MAC protocol is used.

The distance between adjacent nodes cannot exceed 2 km when multimode fibers and LEDs are used. Longer separation is possible with single-mode fibers and laser diodes. The maximum length of the fibers is 200 km. Because of this length, FDDI networks are used to interconnect computers within a campus. Many vendors supply FDDI hardware and software for workstations.

The figure shows three configurations of FDDI as back-end, backbone, and front-end networks. A back-end network connects workstations to file servers and printers. A backbone network connects workstations. A front-end network attaches terminals or terminal emulators to workstations.

As indicated in Figure 3.22, the FDDI standards specify the MAC sublayer and the physical layer. The physical layer itself is divided into two sublayers. The standards also specify the station management (SMT) protocols. The PMD (physical medium dependent) sublayer specifies the fiber to be used as well as the optical sources and detectors. The specifications of PMD are summarized in Figure 3.22. It should be noted that vendors make alternative PMD products available. For instance, twisted pairs can be used to connect stations separated by less than 100 m.

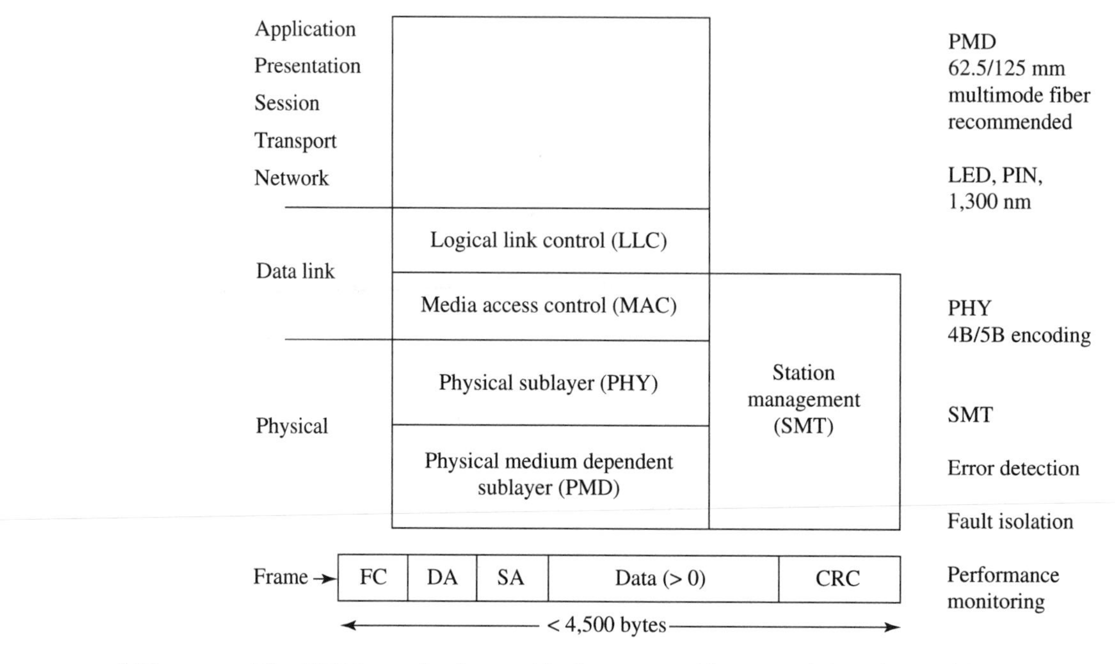

FIGURE 3.22 The FDDI standards specify the MAC sublayer and the physical layer of the protocol stack.

The PHY (physical) sublayer specifies that the stations must use the 4B/5B encoding. With this encoding, the transmitter groups the bits by 4 and converts each 4-bit word into a 5-bit word specified by the encoding table. The 16 words of 5 bits that the encoding uses were chosen so that the resulting optical signal contains enough transitions to keep the receiver synchronized. Note that with this encoding the 100-Mbps data rates result in a raw bit stream of 125 Mbps on the fibers. If the transmitters had used Manchester encoding, the optical signal would have transitions at 200 MHz, necessitating more expensive electronics.

The SMT must detect errors and isolate a fault on the ring, such as a failure of a station or link on the ring. Figure 3.23 illustrates how the dual ring is reconfigured as a single ring after the fault has been isolated. In addition, the SMT monitors the performance of the network. The MAC of FDDI specifies that the frames have a maximum length of 4,500 bytes. (The frame structure is illustrated in Figure 3.22.) The MAC uses a timed-token protocol. This protocol is similar to the token-passing mechanism of IEEE 802.5, except for the timing feature, as we explain next.

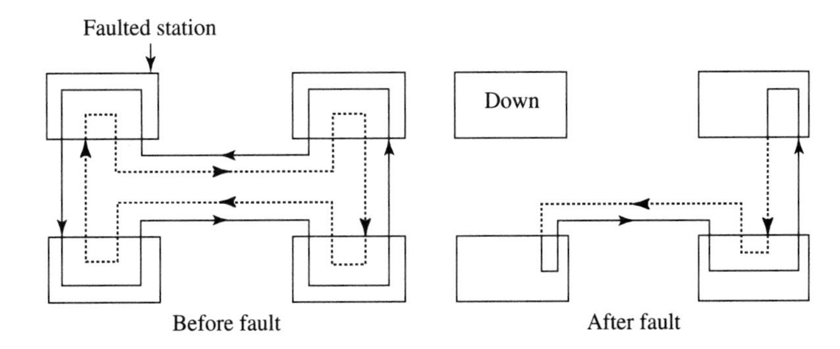

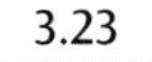
3.23

FIGURE

When a fault is detected, the rings are reconfigured to isolate the faulted station.

We first explain the MAC protocol when the stations transmit only asynchronous traffic with the help of Figure 3.24. Assume that the stations are initially idle, i.e., they have no packet to transmit. A token, which is a packet with a specific bit pattern, travels around the ring. Each station has two timers: TRT or token rotation time timer, which counts up, and THT or token holding time timer, which counts down. When a station has a packet to transmit, it waits until it gets the token. When the station gets the token, it does the following:

1. Grabs the token.
2. Sets THT = TTRT − TRT. (TTRT or target token rotation time is set by the network manager.)
3. Resets TRT = 0.
4. Transmits packets until THT = 0 or there is no packet left.
5. Releases the token.

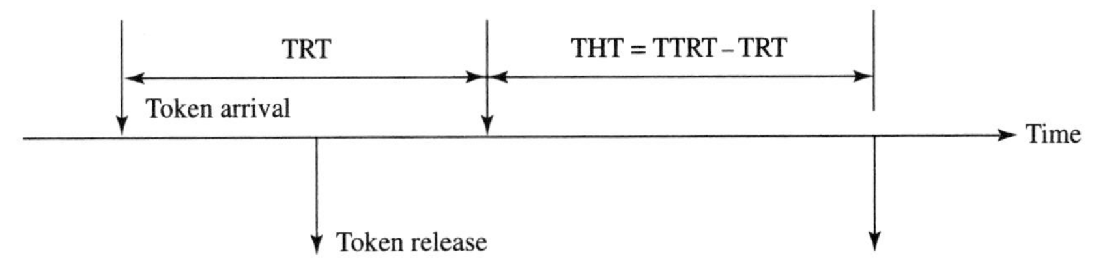

3.24

FIGURE

The timed-token protocol guarantees that each station will get a chance to transmit in less than TTRT. TTRT is the target token rotation time.

Figure 3.24 shows for a particular station two successive token arrival and release times. Suppose that the time, TRT, between successive arrivals is less than TTRT. The figure shows that in that case the time between successive token releases is also less than TTRT. But this is the time between successive arrivals for the next station. By repeating the argument, we conclude that every station will wait for time at most TTRT for a token arrival. If we assume that a station may complete transmitting its current packet even when THT = 0, then this argument must be slightly modified to conclude that each station waits at most TTRT + TRANS, where TRANS is the longest packet transmission time.

The transmitting station must remove its own packet from the ring: the station waits until it receives the packet that it transmitted, i.e., until it reads its own physical address as the source address of the packet, and it then removes the packet by transmitting "idle" symbols instead of repeating the packet.

Actually, the MAC protocol provides for two types of traffic: asynchronous and synchronous. As will be explained, the stations get to transmit their synchronous traffic at least every 2 TTRT seconds. For instance, if the stations agree on a value TTRT = 20 ms, then the stations that transmit synchronous traffic, say voice, get to transmit at least every 40 ms. If the voice is encoded into a 64-Kbps stream, then the stations need only be able to buffer $40 \times 10^{-3} \times 64 \times 10^{3} = 2{,}560$ bits of voice.

We now explain how the protocol accommodates synchronous traffic. The stations first request permission to transmit synchronous traffic. The network eventually decides which stations can transmit synchronous traffic, and it allocates a fraction of TTRT to each of those stations. The fractions add up to one. The protocol works as follows. When a station that can transmit synchronous traffic gets the token, it does so for up to the fraction of TTRT that it was allocated. It transmits asynchronous traffic as before, using the previously described protocol. When the stations use this protocol, they get the token at least once every 2 TTRT seconds.

As was explained in Figure 3.24, the MAC results in a bounded medium access time that is suitable for synchronous traffic. Note, however, that a station does not access the medium exactly at periodic times. Thus, the FDDI MAC does not implement an isochronous transmission facility. A later standard, FDDI.2, being formulated is intended to provide such a truly isochronous service. It can also be shown that the FDDI MAC provides a fair allocation of the bandwidth to the different stations for asynchronous traffic. (A fair allocation is one in which every station has the same probability of transmission access. The fairness of the FDDI MAC is not quite obvious from our description of the protocol.) Moreover, the FDDI MAC protocol is very efficient because the overhead that it imposes does not increase when the stations have many packets to send. Likely future applications of FDDI include multimedia connections where the stations exchange video, audio, text, and data. FDDI is being used to interconnect LANs. Back-end and front-end applications will also use FDDI.

In summary, by using a transmission rate of 100 Mbps instead of 10 Mbps for Ethernet or 4 or 16 Mbps for token ring, FDDI can achieve a higher throughput than these other LANs. Also, by using a timed-token mechanism instead of an untimed token-passing protocol or CSMA/CD, FDDI guarantees a bounded medium access time and is therefore suitable for synchronous transmission services in addition to asynchronous transmissions. Thus, the faster physical layer of FDDI increases the throughput, and the timed-token mechanism of its MAC enables the transport of constant bit rate traffic.

3.5 DQDB

The Distributed Queue Dual Bus (DQDB), illustrated in Figure 3.25, is the IEEE 802.6 standard for a MAN (metropolitan area network). The figure shows the topology of DQDB. Each station is attached to two unidirectional buses. The word *bus* is a misnomer, because the connections in each direction are implemented by a sequence of point-to-point links instead of a bus as in

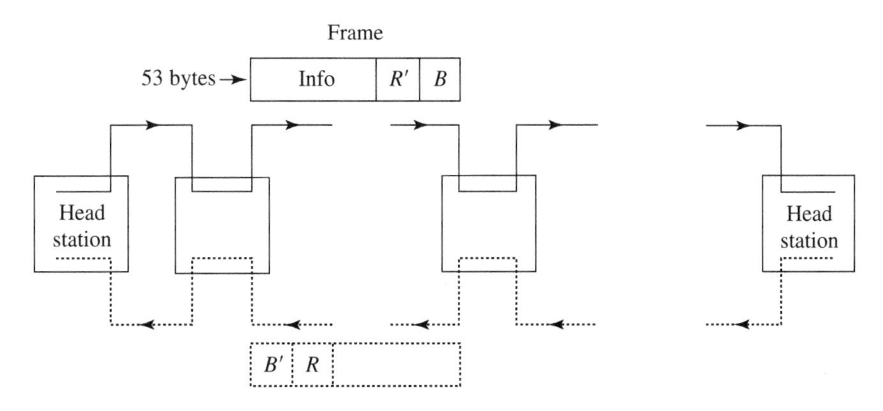

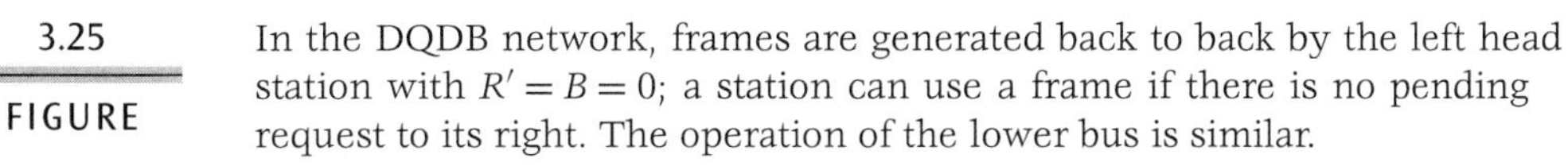

FIGURE 3.25 In the DQDB network, frames are generated back to back by the left head station with $R' = B = 0$; a station can use a frame if there is no pending request to its right. The operation of the lower bus is similar.

Ethernet or a token bus. (There are few DQDB vendors; given the popularity of FDDI, it seems unlikely that DQDB will be widely deployed. The DQDB protocol is used in the subscriber-network interface for SMDS.)

The DQDB MAC protocol is a clever way to regulate access to the medium as if all stations placed their packets in a single queue that is served on a first-come, first-served (FCFS) basis. Such a first-come, first-served protocol would be the fairest possible. However, it cannot be achieved perfectly because the queues are distributed in the different stations and no station knows exactly when the other stations got packets to transmit. We will see how the DQDB protocol approximates that FCFS behavior. A station wanting to transmit to another station situated to its right must use the upper bus, and it must use the lower bus to transmit to stations on its left. The operations of the two buses are identical, and we explain how the nodes transmit on the upper bus in Figure 3.26.

Fifty-three byte frames are generated back to back by the head stations. Each frame has two special control bits: the busy bit B and the request bit R. The head stations generate idle frames: the frames from the left head station have $R' = B = 0$, and frames from the right head station have $R = B' = 0$. When a station wants to transmit and when the protocol described below allows it to transmit, the station uses the next idle frame to transmit its own frame, setting the busy bit to 1. Each station copies each frame and retains copies addressed to itself. The final head station removes the frame from the bus.

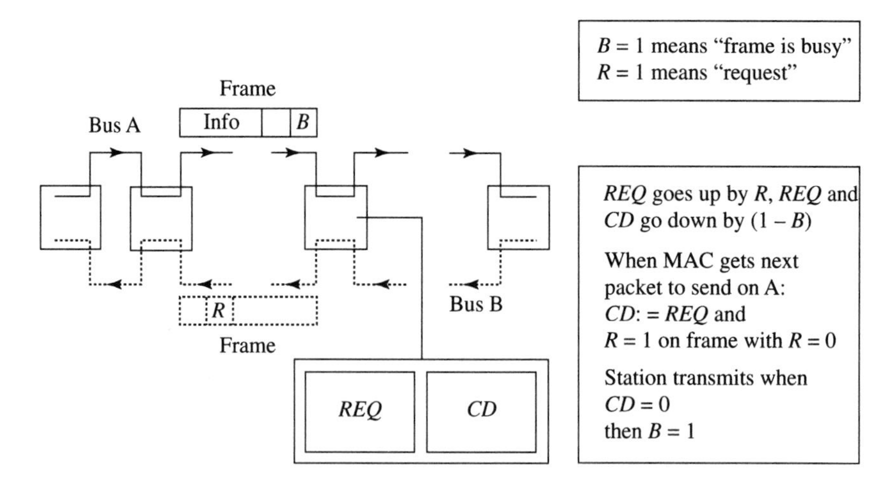

3.26 FIGURE The figure explains the operation of the DQDB protocol.

We now explain the MAC protocol. When a station S wants to transmit on the upper bus, it must first reserve a frame. To reserve a frame on the upper bus, S waits until it sees a frame on the lower bus with its request bit $R = 0$. S then sets bit $R = 1$. When that frame propagates, the stations to the *left* of S learn that one station to their *right* has a packet to transmit on the upper bus. By counting these requests, every station can keep track of the total number of packets that stations to its right want to transmit. More precisely, when station S gets a packet to transmit, it knows how many frames have been reserved by stations to its right. S stores that number in two counters: CD (count-down) and REQ (request). The DQDB protocol specifies that the stations must defer to their right. That is, station S cannot use an idle frame that comes by on its upper bus until all the CD reservations made by stations to its right have been serviced.

In order to implement this protocol, every time S sees an idle frame (indicated by $B = 0$) go by on its upper bus, it decrements REQ by one; and every time S sees a reservation (indicated by $R = 1$) on the lower bus, it increments REQ by one. So at each time, REQ is the number of outstanding requests from stations to the right of S. When S itself gets a packet to transmit, it loads the CD counter by the current value of REQ, $CD := REQ$. (This is the number of outstanding requests from stations to the right of S at the time it received a packet.) It then decrements CD by one each time an idle frame goes by on the upper bus. As soon as CD reaches zero, S knows that all the reservations to its right that were placed before it got its packet to transmit have been serviced. Station S is then allowed to use the next idle frame to transmit its own packet.

The DQDB MAC protocol is very efficient. Unlike Ethernet, there is no loss of capacity due to collision. Unlike token ring, an idle frame is continuously generated by the head station. If there always are stations with packets to transmit in both directions, utilization will be 100%. However, the protocol is not perfectly fair because its topology is not symmetric. For instance, the leftmost station must transmit all its packets on the upper bus, and it must defer to all the other stations to its right. By contrast, a station in the middle transmits half its packets on each bus and defers to only half the stations on each bus. To correct this unfairness, the standard specifies that each station be allocated an individual parameter F. F specifies the number of successive frames that the station can use to transmit. The network manager can select these parameters so that the resulting utilizations of the buses by the different stations are comparable. F is called the *bandwidth balancing* parameter. The IEEE 802.6 standard specifies only the MAC protocol of DQDB. The standard also provides for different traffic priorities. Priorities are implemented by having distinct B and R bits and distinct counters for different priorities.

The networks considered above are used as local area networks that connect nearby computers or campus networks that connect computers or LANs in nearby buildings. We now describe two wide area packet-switched networks, Frame Relay and SMDS. These networks are used to connect computers or LANs across a public switched network.

3.6 FRAME RELAY

Frame Relay is a connection-oriented data transport service for public switched networks. The Frame Relay protocols are a modification of the X.25 standards. Both X.25 and Frame Relay specify the lowest three OSI layers for virtual circuit networks.

X.25, introduced in 1974, was designed to operate with noisy transmission lines. Accordingly, the link level protocol of X.25 (called LAPB for Link Access Procedure B) performs error detection and recovery using the Go Back *N* protocol with a window size of 8 to 128. LAPB also provides for some link level flow control by enabling a receiver to stop the sender temporarily by sending it a control frame. The network layer of X.25 specifies that up to 4,096 virtual circuits can be set up on any given physical link. An end-to-end window flow control can be implemented along each virtual circuit independently of the link level flow control.

Frame Relay is simpler than X.25. It is designed to take advantage of links with a higher transmission rate and small bit error rate. (X.25 is intended to work with 64-Kbps links, Frame Relay with 1.5-Mbps and higher-speed links.) The main difference with X.25 is that Frame Relay does not control errors at the link level. Instead, error control and recovery are done by higher layers. Consequently, the packet or frame processing time at each node is smaller than for X.25. Moreover, the transmissions on a link are not slowed down as they would be by the Go Back *N* protocol of X.25 when its window has been transmitted and the sender waits for the acknowledgments to come back before resuming the transmissions.

We will explain why it is advantageous to replace link level error control by end-to-end control when the bit error rate is small. We will then show why Go Back *N* slows down transmissions when the bandwidth-delay product of the link exceeds the window size. These two observations justify the superiority of Frame Relay over X.25 for higher-speed, low-error links. Since Frame Relay is simpler than X.25, most vendors of X.25 equipment provide software to run Frame Relay on their switches. Frame Relay is a popular means to interconnect networks.

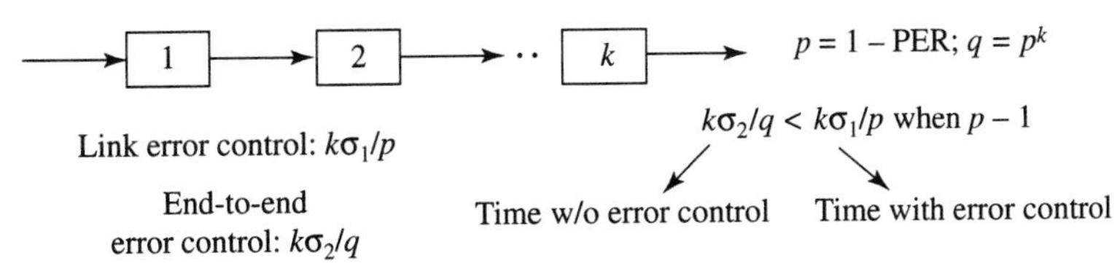

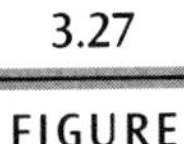

FIGURE 3.27 Frame Relay provides faster processing of packets because it does no link error control.

For the first observation, consider the virtual circuit connection model of Figure 3.27. The connection goes through k nodes. Each node takes a fixed time σ_1 to process and transmit each frame. Assume that each transmission from one node to the next corrupts the frame with probability $1-p$. If errors are controlled by the link and each frame is retransmitted until it is successfully received, then each frame must be transmitted $1/p$ times on average before being successfully received. Consequently, the average transmission time of the frame by the k nodes is equal to $k\sigma_1/p$. If no link level error control is performed, then the node processing and transmission time is assumed to be $\sigma_2 < \sigma_1$. (The difference in practice is large: $\sigma_1 \sim 50$ ms compared with $\sigma_2 \sim 3$ ms.) The total transmission time by the k nodes is now $k\sigma_2$. However, this end-to-end transmission time must take place $1/q$ times on average, where q is the probability that no transmission corrupts the packet, that is, $q = p^k$. Since $k\sigma_1/p > k\sigma_2/p^k$ whenever p is sufficiently close to 1 (i.e., the bit error rate is sufficiently small), the end-to-end error control becomes faster than the link level error control.

In section 2.5.3 we defined the efficiency of a transmission protocol as the fraction of time the transmitter is sending new packets. We showed that the efficiency of the Go Back N protocol is

$$\text{Efficiency} = \min\left\{\frac{N\ \text{TRANS}}{\text{TRANS} + \text{ACK} + 2\text{PROP}}, 1\right\},$$

where TRANS is the packet transmission time, ACK is the time to transmit the acknowledgment, and PROP is the propagation time between sender and receiver. Thus to achieve Efficiency = 1 we must have N TRANS $\geq$ TRANS + ACK + 2PROP. Neglecting ACK, this implies N should be at least 2PROP/TRANS, i.e., the window should be large enough to "fill up the pipe." For example, suppose the end-to-end distance is 5,000 km, so PROP equals $5 \times 5{,}000 = 25$ μs. For a 1,000-byte packet and a transmission speed of 50 Mbps, TRANS = 160 μs. This gives a window size of about 300 packets.

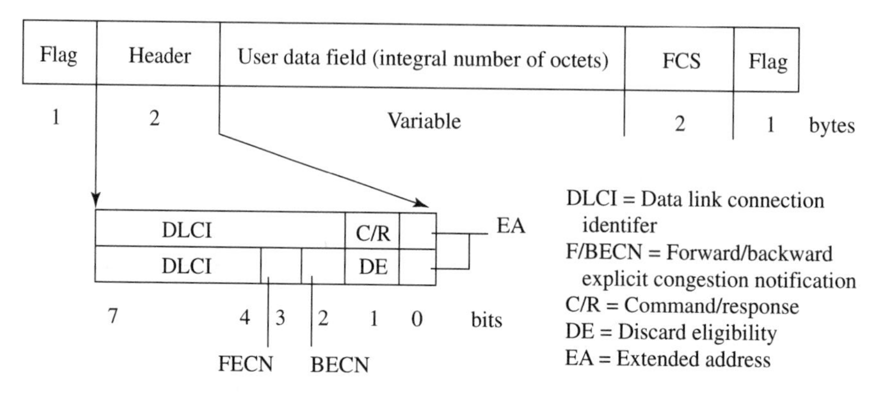

3.28 FIGURE Frame format of Frame Relay.

The frame format is shown in Figure 3.28. The 2-byte header contains address information (DLCI and EA, explained below) for routing, congestion-control information (F/BECN and DE, also explained below) for notification and enforcement, and the C/R bit whose usage is application-specific. The frame check sequence FCS is a 16-bit CRC for error detection: erroneous frames are discarded and are not retransmitted by the network. The standard specifies the use of Permanent Virtual Circuits (PVCs) for connections.

A PVC is a fixed route assigned between two users when they subscribe to a Frame Relay service. A PVC is identified at the network interface by a 12-bit data link connection identifier (DLCI). The DLCI field allows for 1,024 PVCs per access link. Of these, about 1,000 can be assigned to users, and the rest are reserved for control purposes. The header may be extended to 4 bytes to accommodate more DLCIs. The EA (extended address) bit is used for that purpose: EA = 0 indicates that the next byte is also an address byte; EA = 1 indicates the last address byte.

Frames are discarded at a node or switch when erroneous or when buffers overflow. To reduce buffer overflow, the switch can exercise flow control as follows. When a switch experiences some congestion, it notifies the sources and destinations of all the active PVCs passing through the node. This is done by setting the FECN (forward explicit congestion notification) bit in user frames going in the forward direction to inform the destination or the BECN (backward explicit congestion notification) bit in user frames going in the reverse direction to inform the source. FECN may be used by destination-controlled flow-control protocols, whereas BECN may be used by source-controlled flow-control protocols. The Frame Relay standard, however, does not define congestion, nor does it specify how users should respond to it.

The DE (discard eligibility) bit may be set by users to indicate low-priority frames such as some audio or imaging frames with less significant information. That bit may also be set by a network node. The network would preferentially discard frames with DE = 1 when necessary to alleviate congestion. (ATM packets also incorporate a 1-bit priority; see Chapter 5.) DLCI = 1,023 is a PVC reserved for communication between the user and the network. The user and the network periodically exchange "keep alive" messages on that PVC. A user could also poll the network on that PVC, at which point the network would report all active DLCIs on that access link and their traffic parameters such as CIR (explained below). That PVC can also be used for flow control, especially when there is too little traffic through a congested node in the reverse direction for a timely notification of congestion by the BECN.

At subscription time, each DLCI is assigned three parameters (T_c, B_c, B_e) for traffic shaping. These parameters are used as follows. Time is slotted into intervals of duration T_c. The network guarantees transport of B_c bytes of data in each interval. This guarantees a "committed information rate" $CIR = B_c/T_c$. If the user injects more than B_c bytes across the user-network interface in an interval, the network may admit the first B_e bytes of excess data with their *DE* bits set. Further frames in that interval may be discarded. The DLCI is guaranteed a long-term bandwidth of *CIR* and a maximum burst size of B_e. This traffic-shaping scheme regulates the input load to the Frame Relay network, thus reducing the likelihood of congestion. The scheme may be implemented by using a leaky bucket for each PVC at the network entrance. (The leaky-bucket scheme is described in section 3.7.) Traffic-shaping schemes are discussed in Chapters 6 and 7.

In summary, Frame Relay is an improvement over X.25 networks, taking advantage of better transmission links by streamlining the X.25 protocol. However, its switches lack the capability to reserve resources for individual connections, and so Frame Relay is unsuitable for applications that require guaranteed delay. It is interesting to note, nevertheless, that many of the developments used to differentiate service quality occur almost simultaneously in Frame Relay, SMDS, and ATM. Of these three designs, ATM will be the most successful in offering differentiated services.

3.7 SMDS

Switched Multimegabit Data Service (SMDS) is a public switched connectionless data transport service. Beginning in 1992, SMDS has been offered by regional Bell operating companies (BOCs) at DS-1 (T-1) access speed (1.54-Mbps line rate corresponding to 1.17-Mbps data rate). DS-3 (T-3) access speed

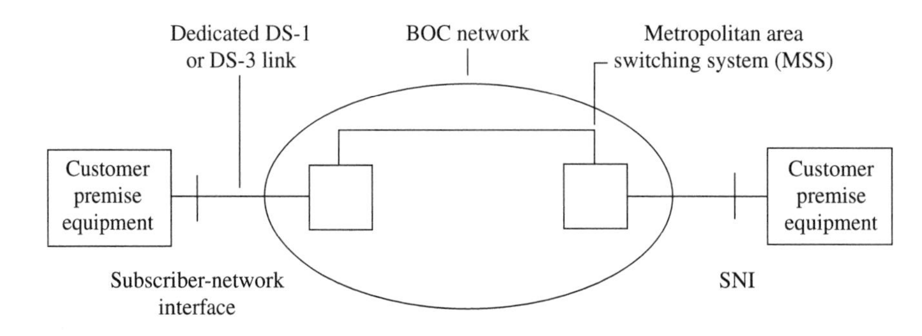

3.29 FIGURE The SMDS network and interfaces.

(45-Mbps line rate or 34-Mbps data rate) is also available. SMDS will later be offered at much higher rates over a SONET network. In some regions, SMDS is growing rapidly. Subscriber equipment is connected through the BOC network. Access to that network is via dedicated access lines. See Figure 3.29. The standard specifies the protocol at the subscriber-network interface.

The SMDS protocol roughly corresponds to the first three OSI layers. It is divided into three levels. Level 1 provides the physical interface to the digital network. Level 2 defines a cell structure similar to ATM cells and performs error detection. Level 3 handles addressing and routing. (The 53-byte cell structure will permit a migration path for SMDS to ATM, as we will see in Chapter 5.)

Figure 3.30 shows the formats of the protocol data units (PDUs) at levels 2 and 3. User data, up to 9,188 bytes, is encapsulated in an L3_PDU. The L3_PDU overhead contains the full source and destination addresses, the L3_PDU length, and, optionally, a CRC for detecting L3_PDU errors. Each address is specified by 15 BCD (binary coded decimal) digits. (The addressing scheme is identical to the North American telephone numbering system.) The total L3_PDU overhead may vary from 40 to 43 bytes. An L3_PDU is fragmented into a sequence of L2_PDUs. Lastly, the function of the Physical Layer Convergence Protocol or PLCP is to place one or more L2_PDUs into a frame of the physical link. As an example, the figure shows how 12 L2_PDUs are assembled into one 125 μs DS-3 frame.

Each L2_PDU is 53 bytes long and contains 44 bytes of L3_PDU payload. If the payload is less then 44 bytes, it is padded to make a 53-byte L2_PDU. The 2-byte trailer contains a payload length to indicate the size of the padding and a payload CRC to detect L2_PDU errors. The 7-byte header contains a MID (message identifier) and a sequence number for reassembly of L3_PDUs at the destination. Before any L3_PDU is transmitted, it is assigned a MID, which

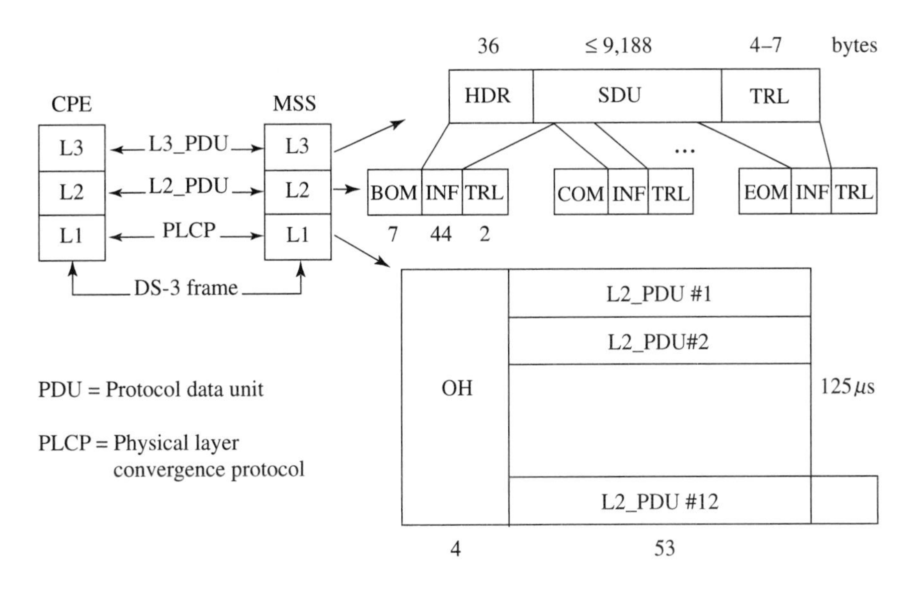

3.30 FIGURE The SMDS protocol stack and frame structure.

is borne by each of its L2_PDUs. The MIDs should be unique among all L3_PDUs being simultaneously transmitted. The sequence of L2_PDUs belonging to the same L3_PDU are ordered using the sequence number. The unique MID allows the destination to collect all L2_PDUs belonging to the same L3_PDU. The sequence number makes correct reassembly possible even when the L2_PDUs arrive out of order. (The standard specifies that the L2_PDUs must be delivered in order.)

Since an L2_PDU does not contain the full destination address, it cannot be routed individually. A simple connection-oriented implementation of the SMDS service is as follows. A virtual circuit is set up to transfer each L3_PDU, identified by its MID. All L2_PDUs then follow the same path using the MID as the virtual circuit identifier. They can be reassembled at the destination using the L3_PDU length field. This method ensures delivery of the packets in the correct order and hence does not require the sequence number. With this implementation, SMDS provides a datagram service for L3_PDUs using a connection-oriented L2_PDU delivery service.

Unlike previous data networks, SMDS offers several service-quality levels defined by service parameters, chosen at time of subscription. We discuss three parameters: address screening, limit on number of simultaneous packets, and information rate. Address screening means that packets may be received from and delivered to only a specified list of destination and source

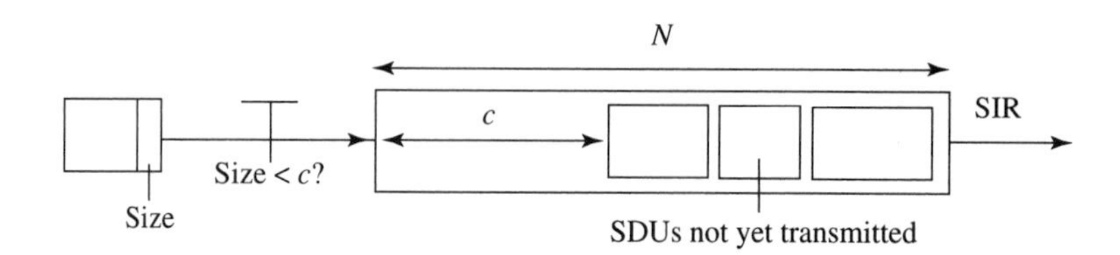

3.31 FIGURE The leaky-bucket scheme guarantees that a user's traffic will not have a sustained information rate higher than SIR or a burst of size larger than N.

addresses. (Address screening is a security device.) To understand the second parameter, observe that the network may interleave L2_PDUs belonging to different L3_PDUs and intended for the same destination. Before forwarding these L2_PDUs, the switch must deinterleave them and forward the L2_PDUs of one L3_PDU together. The switch must buffer these L2_PDUs to do this deinterleaving. A limit on the number of simultaneous packets will place a limit on the needed buffer size. The maximum information rate is specified by two parameters, the sustained information rate, SIR, and the maximum burst size, N.

The restriction on the information rate is implemented by the leaky-bucket scheme of Figure 3.31. A buffer or "bucket" of size N bits is served (read out) at a constant rate of SIR bps. A packet is accepted into the buffer only if its size is smaller than the size, c, of the free buffers; otherwise the packet is blocked. It can be seen that if the subscriber submits packets of size n_1 at time t_1, size n_2 at time t_2, and so on . . . , then no packets will be blocked provided that for every $i < j$,

$$\sum_{i}^{j} n_k \leq (t_j - t_i) \times \text{SIR} + N.$$

This formula makes precise the restriction that the subscriber's traffic on average cannot exceed SIR bps, and it cannot have a burst of more than N bits.

3.7.1 Internetworking with SMDS

SMDS and Frame Relay are used to interconnect local area networks. We conclude by describing the functions that must be carried out in order that users connected to different FDDI LANs can transparently interconnect over

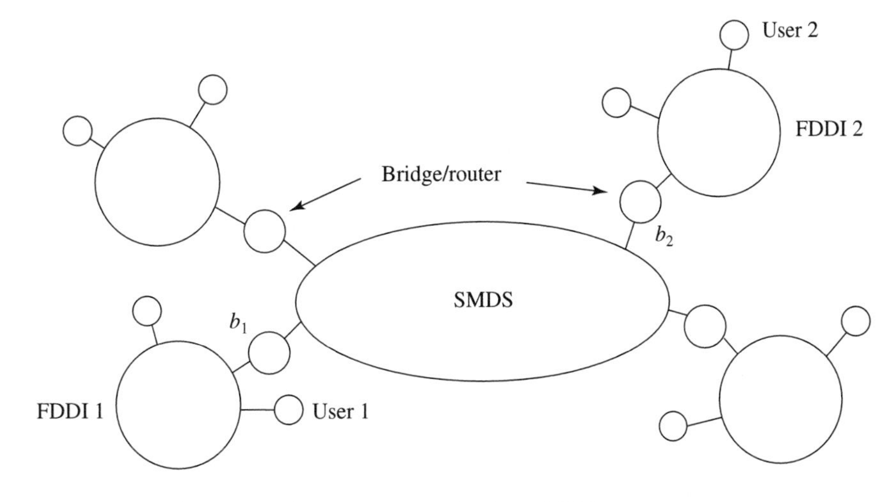

3.32 FIGURE FDDI networks can be interconnected using SMDS.

an SMDS network. (The functions are similar if Frame Relay is used.) Figure 3.32 shows user 1 on one FDDI ring who wishes to send a packet to user 2 on another ring.

Assume first that user 1 knows the MAC address of user 2. In that case user 1 places a frame on FDDI 1 with destination address that of user 2. Bridge b_1 on that ring must

1. copy that frame;
2. "convert" it into an L3_PDU, address it to station b_2 (using its SMDS address), and submit it to the SMDS network.

Note that bridge b_1 is a station on FDDI 1 and on the SMDS network. In order to "submit" the L3_PDU, it must go through the SMDS protocol stack, fragmenting the L3_PDU into L2_PDUs, and then organizing them for transmission into L1_PDUs as in Figure 3.30.

When bridge b_2 receives these L1_PDUs, it assembles them into the L3_PDU. Subsequently,

1. bridge b_2 must "convert" the L3_PDU back into the original FDDI frame, place it on FDDI 2, and then remove it from that ring when it returns (note that the source address on the frame is that of user 1 and not b_2);
2. user 2 copies this FDDI frame.

Bridge b_2 is on FDDI 2 and on the SMDS network.

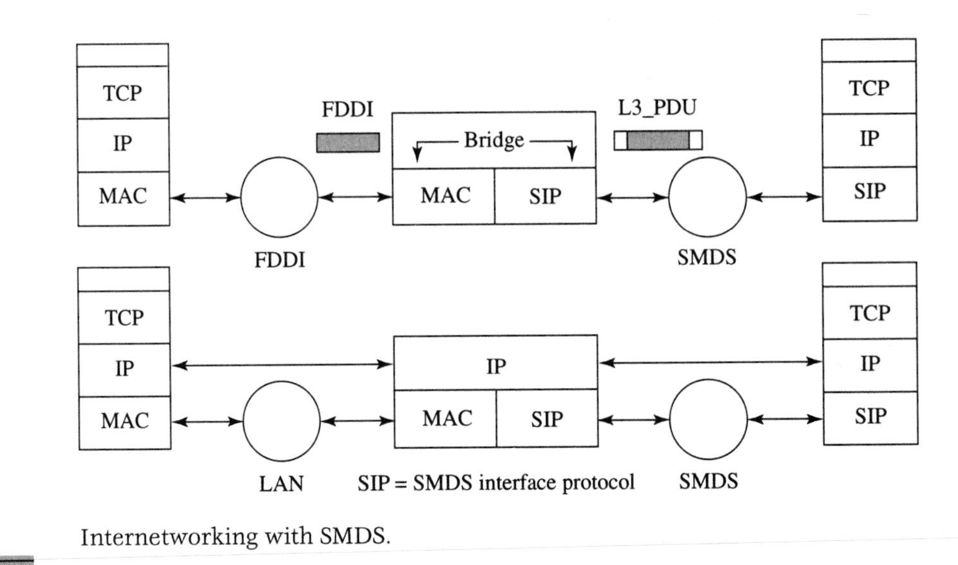

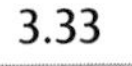

3.33 FIGURE Internetworking with SMDS.

Neither user 1 nor user 2 knows that they are not on the same ring—the FDDI frame is forwarded transparently. Bridge b_1 must carry out two functions: address conversion and frame conversion. When b_1 reads the FDDI frame addressed to user 2, it must recognize that user 2 is on a "remote" ring that can be reached via bridge b_2. This address conversion is carried out with the help of two address tables. The first has entries of the form: (remote user MAC address, remote bridge SMDS address). The second is a list of all MAC addresses on its own ring. The bridge must then convert the FDDI frame into an L3_PDU. The most common way to do this is to encapsulate the entire FDDI frame as a payload in an L3_PDU. This is shown in the top panel of Figure 3.33. We will shortly see how the address table can be constructed.

After bridge b_2 assembles the L3_PDU, it must decapsulate it and recover the FDDI frame. From its address table, b_2 recognizes that the destination address, user 2, is on its "own" ring. It then places the frame on the ring, and removes it when it returns.

The two address tables, one with remote MAC and SMDS bridge addresses, the other with MAC addresses on its own ring, can be built up as follows. Each time b_1 receives an L3_PDU, it notes the SMDS address of the sending bridge, and, from the encapsulated FDDI frame, it obtains the MAC address of the source. In this way it builds the first table. The second table is built simply from the source address on each FDDI frame on its own ring.

In case user 1 uses the IP address of user 2, the stations b_1 and b_2 will have to be routers. This requires additional functions suggested in the lower panel of Figure 3.33.

3.8 SUMMARY OF PACKET-SWITCHED NETWORKS

The packet-switched networks studied above show a steady advance in speed, connectivity, delay, and flexibility or ability to accommodate more kinds of traffic types. We summarize this development in Table 3.1.

All the networks in this table, except Frame Relay and SMDS, implement layers 1 and 2 of the OSI model. In section 3.9 we describe the Internet Protocols or IP, which roughly correspond to layers 3 and above of the OSI model. IP can be implemented on top of layer 2 of these networks. Frame Relay and SMDS also implement layer 3 (the network layer). By means of a router with some additional functionality, these networks can also implement IP.

Name	Speed, connectivity	Delay	Application
Ethernet	10 Mbps, local area	Random, increases with load	Transfer of messages between nearby computers
Token ring	4, 16 Mbps, local area	Random but bounded	Transfer of messages between nearby computers, some real-time traffic
FDDI	100 Mbps, LAN and campus	Random but bounded	LAN interconnections, real-time and CBR applications
DQDB	Unspecified	Random	Unspecified but similar to FDDI
Frame Relay	1.5 Mbps, wide area	Random, increases with load	Transfer of messages between distant computers
SMDS	1.5 to more than 45 Mbps	Random, traffic shaping	LAN interconnections, migration to ATM

TABLE 3.1 Summary of advances in packet-switched networks.

3.9 INTERNET

The Internet today is used to interconnect a large number of computers and local area networks throughout the world. In 1995, according to the Internet Society, there were 4.5 million such computers being used by 30 million people. The Internet has its origin in the ARPANET network sponsored by the U.S. Department of Defense starting in the 1960s. The ARPANET was a datagram store-and-forward network that the Department of Defense liked for its ability to reroute packets around failures. This feature makes datagram networks survivable. However, the technical success of the Internet is due to another feature, namely, the large variety of applications that IP can support on the one hand and, on the other hand, the many different networks that can implement IP. (See Figure 3.34.) We will discuss this key feature at the end of this section. A major factor that contributed to the popularity of the Internet is the exploitation of network externalities through early standardization and free distribution of its protocols and their software implementations that can run on PCs and Macintoshes as well as on workstations and mainframe computers.

The Internet is a network of networks. It comprises a backbone network of point-to-point links that connect regional gateways. Local access is pro-

Telnet	FTP	SMTP	rcp	rsh	rlogin	TFTP	7 6 5
TCP						UDP	4
IP: ICMP ARP RARP							3
X.25, Ethernet, token ring, FDDI, ...							1, 2

3.34 FIGURE

Internet protocols are arranged in a layered hierarchy and compared with the OSI seven-layer model. On top of the basic IP or network layer and TCP/UDP or transport layer are generic applications such as file transfer (File Transfer Protocol or FTP), remote file copy (rcp), remote terminal (Telnet), remote login (rlogin), and e-mail (Simple Mail Transfer Protocol or SMTP). User applications such as Mosaic and gopher are built on top of these generic applications.

vided through routers, to which individual computers are connected via local area networks. Over time, and incrementally, backbone link speeds have increased from 56 Kbps to 1.5 Mbps to 45 Mbps. Local area network speeds have increased even faster to 100-Mbps FDDI rings. The Internet is used for applications that require (relatively) low transmission rates and that can tolerate large delays.

In this section we describe the main components of the Internet software. The Internet protocol layers are shown in Figure 3.34. The figure gives the correspondence between the IP protocols and the OSI seven-layer model. The correspondence is inexact. Layers 1 and 2 are implemented by many networks, including those studied earlier. Layer 3, the network layer, is implemented by IP. Layer 4, the transport layer, is implemented by two protocols, UDP and TCP. There is no direct counterpart to the higher OSI layers.

3.9.1 IPv4, Multicast IP, Mobile IP, IPv6, RSVP

The network layer of Internet, called the *Internet Protocol* (IP), is the most important. The most recent version in use, implemented in 1995, is IPv4. Version 5 was used for some experiments. The next version, IPv6, is to be implemented over the next few years. We start by describing IPv4, the multicast and mobile version of IP, then we explain the modifications of IPv6. We conclude with a discussion of the Resource Reservation Protocol, RSVP.

IPv4

In terms of the Open Data Network model of section 2.7, the IP bearer service is the delivery of messages in the form of datagrams of size up to 2^{16} bytes (64 Kbytes). This delivery service carries no guarantee of service quality in terms of delay or bandwidth. Such service is called *best-effort* service, meaning thereby that the network will attempt to do the best it can.

Each IP packet has an IP header of at least 20 bytes. See Figure 3.35. The header indicates the source and destination network addresses of the message. The IP header also specifies the time to live of the packet. When a router gets an IP message it decrements its time to live and it discards the message if the time to live reaches 0. This procedure prevents packets from floating around for a long time in the network in case of routing errors. The header indicates the type of service requested by the message: either low delay, high throughput, or very reliable. This service type is ignored in

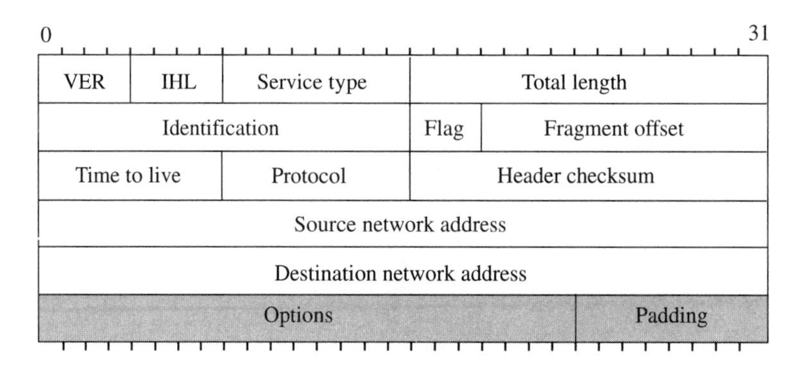

FIGURE 3.35 The IP header specifies the source and destination addresses and control fields described in the text. (VER = version; IHL = header length/32 bits.)

current implementations, although it could be used in the future to provide some service-quality differentiation.

The IP header also specifies the protocol that requested the transmission of the message so that the receiver knows how to handle the message. Finally, the header may request some optional services such as recording the route followed by the packet (the list of nodes is then written in the IP header as the packet progresses), following a route specified by the source, or time-stamping of the message by each node that it goes through.

We explain three important aspects of IP: addressing, segmentation/reassembly, and routing.

Addressing The authors' computer's *network address* is 128.32.110.56. In this address, 128.32 specifies that the computer is on the Berkeley campus of the University of California, 110 designates a specific Ethernet on that campus, and 56 refers to one specific computer interface card (of diva.eecs) on that Ethernet. Note the hierarchical form of the network address.

A directory service is provided by the Internet as an application to find the network address of a computer with a given name. The authors' computer's name is *diva.eecs.berkeley.edu*. This name consists of the institution name (*berkeley.edu*) and the name of the computer (*diva.eecs*). To send a message to one of the authors, you send it to

varaiya@diva.eecs.berkeley.edu or to *wlr@diva.eecs.berkeley.edu*.

You do not need to know where diva.eecs.berkeley.edu is located. The directory service translates the name into the address 128.32.110.56, and the network layer finds a path that leads to that computer. The directory tables are hierarchical, organized in a treelike structure.

If your computer is in France and sends a packet to diva.eecs.berkeley.edu, your local directory recognizes that the address is in the United States. (The addresses in all other countries end with a two-letter country code such as uk for United Kingdom, fr for France, in for India.) Your local directory service sends the request to a directory server in the United States that knows how to find the network address of the directory service for berkeley.edu. The Berkeley directory then finds the network address of the Ethernet to which diva is connected. Thus, after a while, the routers find out that diva.eecs.berkeley.edu is on the Ethernet 128.32.110.

Segmentation/Reassembly The OSI model specifies that layer 4 (the transport layer) perform the fragmentation and reassembly of messages. However, in the Internet, IP fragments messages into IP packets for the data link layer and reassembles IP packets into messages. For instance, if the messages go through an Ethernet, then they must be divided into packets of at most 1.5 KB. The header contains an identification number of the message to which the packet belongs and an offset indication that specifies the position of the data portion of the IP packet inside the original message. The segmentation/reassembly is shown in Figure 3.36.

Routing To route datagrams, the Internet nodes maintain routing tables. A routing table specifies where the node should send a datagram next. If the datagram is destined to a network to which the node is not directly attached, then the routing table specifies the next node for that network. If the datagram is for a computer on the same network as the Internet node, then that node consults the routing table to find the physical (MAC) address of the computer.

To maintain the table with the addresses of computers attached on the same network, the Internet node uses the *Address Resolution Protocol* (ARP). Using this protocol, a node searches its routing table for a given computer name whose MAC address it is trying to locate. If the computer name is not in the table, the node broadcasts a message on the network asking for the computer with the given name to reply. When it sees the reply, the node finds the MAC address as the source address of the packet. The table entries have a time to live, and they are purged when that time is exceeded. That is how the node updates the table entries to reflect changes in the network.

To maintain the routing table to the other networks, the Internet nodes use two versions of the shortest-path algorithm: the Bellman-Ford algorithm

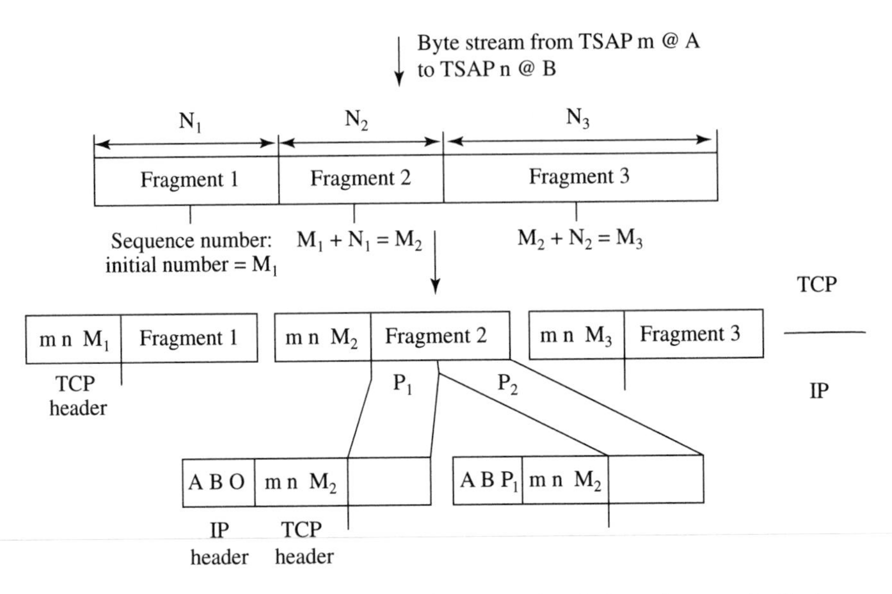

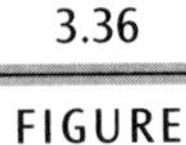
3.36 FIGURE

TCP delivers a byte stream by first decomposing it into fragments that are transported by IP. TCP numbers the fragments, starting with some initial sequence number and then counting the bytes. IP segments the TCP fragments into packets. The IP header indicates the offset of the segment in the fragment.

and Dijkstra's algorithm. We explain these two versions below. To implement that algorithm, the nodes of the Internet exchange control information such as their estimates of the length of a shortest path. These messages and the shortest-path algorithm constitute the *gateway-to-gateway protocol* (GGP). In addition, the Internet nodes use a protocol called the *Internet Control Message Protocol* (ICMP) for exchanging control information. ICMP provides a number of services that Internet nodes use for monitoring the network. One such service is *ping*, which asks a node to echo a packet. Another service is the indication by a gateway to the source of an IP packet that the destination is unreachable. Similarly, a gateway indicates to the source when one of its packets exceeds its time to live. A gateway can also tell a source to redirect its packets in case of difficulties or to slow down transmissions. The protocol also can request time stamps when it needs to synchronize the clocks of the nodes.

Bellman-Ford algorithm Figure 3.37 illustrates a distributed algorithm that finds the shortest paths from all the nodes in a network to any particular node that we call the *destination*. (See RFC 1075.) This is the *Bellman-Ford*

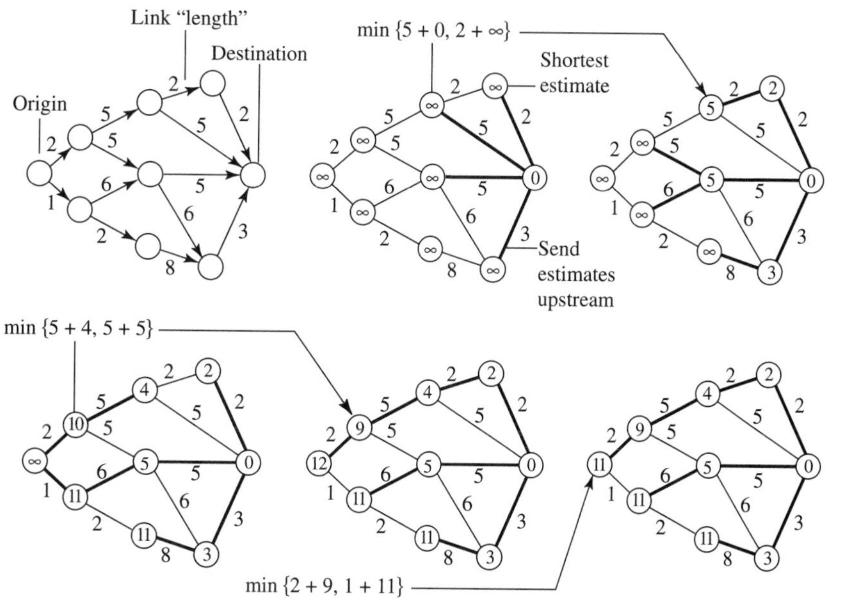

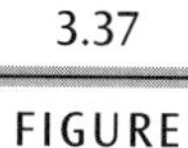

FIGURE 3.37 Using the Bellman-Ford algorithm, the nodes find the shortest path to a destination. The nodes estimate the length of the shortest path to the destination. These estimates are shown inside the circles representing the nodes. When its estimate decreases, a node sends the new value to its neighbors. The figure shows, from left to right and top to bottom, the successive steps of the algorithm for the graph shown in the top-left diagram. Some intermediate steps are not shown.

algorithm. The algorithm assumes that each node knows the *length* of the links attached to itself. These lengths can be a measure of the time taken for a packet to be transmitted along the links, or they can be arbitrarily selected positive numbers (e.g., 1 for every link). This routing strategy is called link-state routing when the length reflects the status of the link. At each step of the algorithm, every node keeps track of its current estimate of the length of its shortest path to the destination. Whenever a node receives a message from another node, it updates its estimate. If the updated estimate is strictly smaller than the previous estimate, then the node sends a message containing the new estimate to its neighbors. A node i estimates the length of the shortest path to the destination by computing

$$L(i) = \min_{j}\{d(i,j) + L(j)\}$$

where the minimum is over all the neighbors j of node i. In this expression, $d(i, j)$ is the length of the link from i to j and $L(j)$ is the latest estimate received from j of the length of its shortest path to the destination.

Initially, all the estimates $L(i)$ are set to infinity. At the first step, the destination reduces its estimate to 0. The destination then sends messages to its neighbors, who then update their estimates, and so on. Eventually, every node i finds out the shortest distance $L^*(i)$ between itself and the destination. Moreover, there is a shortest path from i to the destination that goes first to j if and only if $L^*(i) = d(i, j) + L^*(j)$. Shortest paths need not be unique and the nodes break ties arbitrarily. Figure 3.37 shows the convergence steps of the algorithm. The shortest paths are shown by thicker lines in the last step, at the bottom right of the figure.

The shortest-path algorithm that we just explained is used by bridges for spanning tree routing. It is also used by the network layer of some networks to select good paths along which to route packets. Typically, these networks define the length of a link as a linear combination of the average transmission time and of the recent backlog in the queue of the transmitter of the link. By running the shortest-path algorithm periodically, the nodes can adapt their routing decisions to changing conditions. In this manner, the network reacts to link and node failures, and it also controls congestion.

The Bellman-Ford algorithm is distributed: each node has only a partial knowledge of the topology of the network. It can be shown that the algorithm may fail to converge. (Convergence may fail if the delay between updates is large and if during that time a significant amount of traffic is rerouted.) To avoid this difficulty, the Internet started using a centralized shortest-path algorithm due to Dijkstra.

Dijkstra's algorithm Consider once again the network shown in Figure 3.37. Assume that each node has a complete map of the network. To update the maps, each node sends to all the other nodes a message that indicates the lengths of the links attached to it. Each node then can implement a shortest-path algorithm to find the shortest path to every possible destination. The node could implement the Bellman-Ford algorithm that we discussed earlier. It can also implement the following algorithm due to Dijkstra.

The steps of the algorithm are shown in Figure 3.38. The algorithm, called *shortest-path first* by the Internet community, constructs a shortest spanning tree from any given node. The steps of the algorithm are shown from left to right and from top to bottom in Figure 3.38. To construct a shortest spanning tree rooted at the shaded node, the algorithm first attaches the closest node

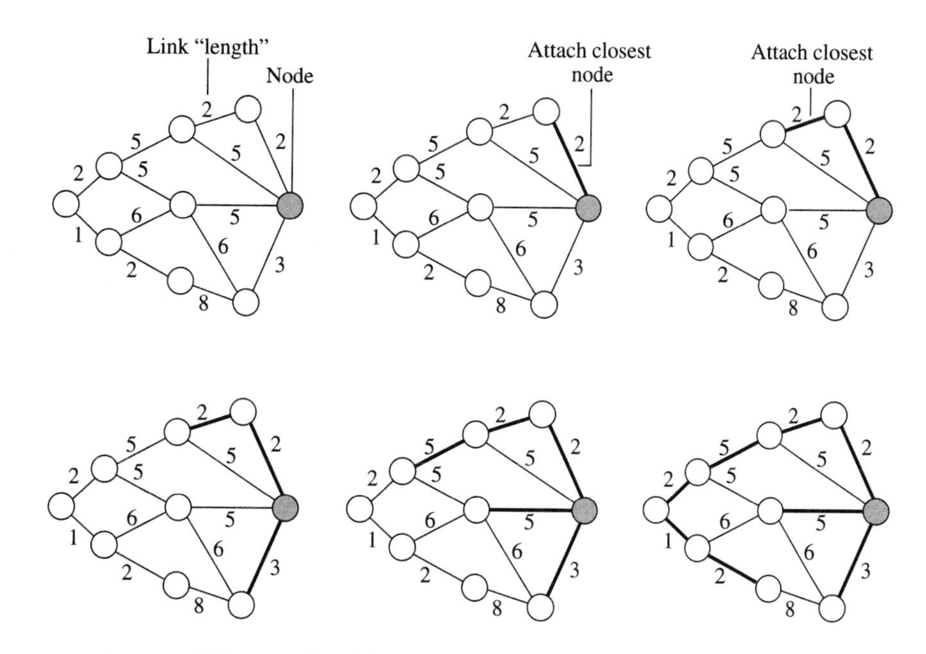

3.38

FIGURE

A node uses Dijkstra's algorithm to construct a shortest spanning tree. The network is shown in the top-left graph. Consider the node labeled *Node.* We assume that this node has a complete map of the network. At each step, shown left to right and top to bottom, the algorithm attaches the closest node to the partially constructed tree. Ties are resolved arbitrarily. The final result is shown in the bottom-right graph. Some intermediate steps are not shown.

to the shaded node. These two attached nodes constitute a (partial) tree. The algorithm then attaches the node that is closest to that tree. The procedure continues until all the nodes are attached. The algorithm breaks ties arbitrarily. The final result is shown in the bottom-right graph. This result is a tree. It is the shortest spanning tree for the graph, the length of a tree being defined as the sum of the lengths of its links.

Border Gateway Protocol In IP, the routing uses a hierarchy. The network is decomposed into *autonomous systems* (ASs). An AS is a subnetwork under a single administration. Within each AS, the routing algorithm (Open Shortest Path First, or OSPF) proceeds as follows: each area router calculates the shortest paths to all the other routers in the areas to which it belongs using three different metrics for the links—delay, throughput, and reliability. Routers do this by exchanging link metrics by flooding using IP.

The routing between ASs uses the Border Gateway Protocol (BGP) (see RFC 1771, 1772). ASs are attached together by one or more routers, called Border Gateways, that share a link layer independent of their AS. Each AS contains a router, called Border Gateway speakers, that implements BGP. BGP speakers use TCP (see below) to exchange routing tables and their updates. These routing tables specify the paths that each BGP speaker currently uses to reach other BGP speakers. BGP speakers exchange keep-alive messages to make sure that the paths are still valid (except when used over switched virtual circuits).

Each BGP speaker compares these paths to construct preferable paths between itself and the other BGP speakers and sends these paths as updates to the other BGP speakers. This selection is left to each autonomous system. The selection must prevent the creation of loops and take into account restrictions that ASs may have about carrying traffic from other ASs. Thus BGP operates on complete paths instead of summarizing them by their distance to the destination as the algorithms by Dijkstra or Bellman-Ford do. BGP enables autonomous systems to refuse to carry traffic originating from other given autonomous systems and also to favor specific paths for their traffic. Each BGP speaker should remember the set of feasible paths, although it advertises only the preferable one.

Multicast IP

Multicast IP is a routing protocol that sends an IP packet to a group of hosts identified by a group address. (See RFC 1112, 1584.)

Multicast routing is implemented by a subset of IP routers called *multicast routers*. Group addresses may be either permanent or transient and are of class D (i.e., have 1110 as their higher order 4 bits). Each host can become or stop being a member of a group by sending join and leave request messages specifying the group address of the group it wants to join or leave. A multicast router learns that information by periodically (every minute) polling the members of its domain. This polling process proceeds hierarchically using a protocol called the Internet Group Management Protocol (IGMP, see RFC 1112).

Specifically, the multicast router sends a query, "which groups do you belong to?" to all the hosts of its local domain (using a group address to which all the multicast-capable hosts belong). On receiving the query, a host starts a count-down timer with a random value between 0 and some T ($T = 10$ seconds is recommended). When the timer expires, the host sends a report with the group address as a destination address for each group it belongs to. The multicast router listens to all those reports destined to group addresses. If a host hears a report from another host—signaling that host's membership to a common group—then the host does not send its own reply corresponding to that group. This procedure limits the number of reports per query. When it wants to join a new group, a host sends a report for that group, without waiting for a query. That report should be sent a few times at random intervals to make sure it gets to the multicast router. Note that the multicast router does not maintain a list of group members. It is recommended that hosts use their individual IP address, and not a group address, as the source address of the IP packets they send. For instance, the IP datagram reassembly algorithm assumes that each host uses a different source address.

The multicast routers then construct a tree of multicast routers from a host to the destinations that belong to the multicast group. This tree is the tree of shortest paths from the source to all the possible destinations pruned back to cover only the destinations of the multicast group. To calculate that tree, the routers run the Bellman-Ford algorithm, explained earlier, where the destination is in fact the source of the multicast. In the pruned tree, the packets are duplicated at the last fork of the tree. Note that the pruned tree does not minimize the sum of the distances covered by the replicated packets. The results of the tree calculations are cached in the routers for subsequent packets.

As hosts join or leave a group, the routers automatically update the tree. If group members belong to the same LAN, the multicast router of the LAN converts the IP group address into a multicast LAN address. If group members belong to a common nonbroadcast multiaccess network (such as Frame Relay), then that network's multicast router must transmit copies to the different group members.

The routers that are not upgraded with the multicast IP software forward the IGMP packets and the IP packets in an oblivious manner.

Mobile IP

The objective of Mobile IP is to deliver packets to mobile nodes automatically. A mobile node is a host or a router that changes its point of attachment from one network or subnetwork to another. Using Mobile IP, a mobile node may change location without changing its IP address. (See [P96].)

A mobile node is associated with a fixed IP address, called its *home IP address*. A router on the mobile node's home network delivers IP datagrams to the mobile node when it is on a foreign network, i.e., away from its home network. That router is called the *home agent*.

Two procedures can be used to deliver the datagrams to the mobile node on a foreign network. In the first procedure, the mobile node gets a temporary *care-of address* on the foreign network from some assignment mechanism and registers that address with its home agent. When the home agent gets a datagram for the mobile node, it encapsulates the datagram in an IP packet with the care-of address as destination address.

In the second procedure, the mobile node uses a *foreign agent.* The foreign agent is a router on the network visited by the mobile node. When it moves to another network, the mobile node registers with a foreign agent and obtains a care-of address from that agent. The foreign agent sends its address to the mobile node's home agent. The home agent encapsulates the datagrams destined to the mobile node in IP packets that it sends to the foreign agent. The foreign agent decapsulates the packets and delivers them to the mobile node with the care-of address.

The encapsulation/decapsulation process is called *tunneling.* In either procedure, when returning home, the mobile node deregisters with its home agent. Also, the mobile node uses its home IP address as the source address of its IP packets.

Agents (home or foreign) advertise their availability by periodically sending *agent advertisement messages*. Moreover, mobile nodes may solicit such a message by sending an *agent solicitation message*. As it receives these messages, the mobile node can determine whether it is on its home network or on a foreign network. On its home network, the mobile node operates without the mobility services of Mobile IP, i.e., by using the standard IP.

The advantage of the first procedure is that it does not require foreign agents. Its disadvantage is that networks must maintain a pool of addresses available for visiting mobile nodes.

The mobility management messages are authenticated to prevent redirection attacks, i.e., a third party stealing messages by giving its address as the new mobile node address.

IPv6

The current version of IP is version 4. Its main limitations are a limited number of addresses, a complex header, a difficult mechanism for introducing extensions and options, a limited number of different services, and poor security and privacy. IPv6 is being designed to overcome these limitations. The transition from IPv4 to IPv6 is expected to take many years. In the meantime, the two versions can coexist by *tunneling*: IPv4 forwards IPv6 packets without

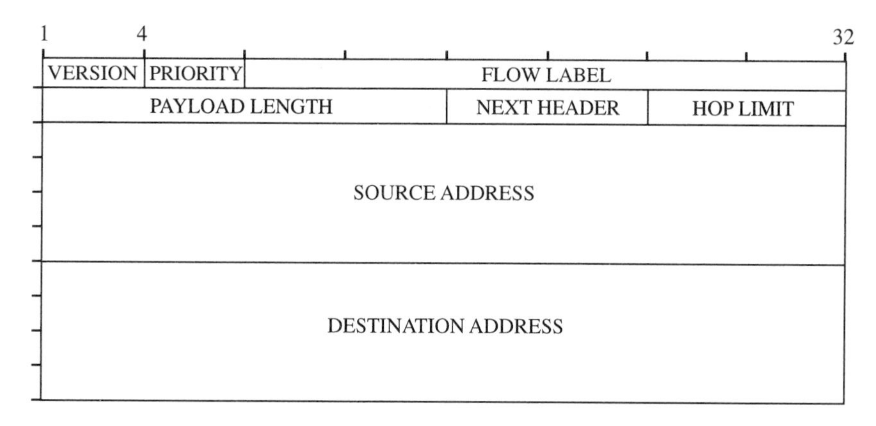

3.39 FIGURE The IPv6 header consists of the 40-byte header shown here, followed by up to six extension headers.

modification. (For details, see RFC 1883, "IPv6 specification.") An IPv6 version of ICMP is designed. (See RFC 1885.) The IPv6 header consists of a 40-byte header followed by up to six optional extension headers.

The 40-byte header is shown in Figure 3.39. The VERSION field contains the binary representation of the number 6, indicating that the packet is an IPv6 packet. The router uses the PRIORITY field to determine the importance and urgency of the packet. FLOW LABEL may be used to decribe the characteristics of the traffic to which the packet belongs. PAYLOAD LENGTH indicates the number of bytes in the packet that follows the header; zero here indicates that the actual packet length is specified in a hop-by-hop extension header (see below). NEXT HEADER indicates which of the six extension headers follows this header, if any, and whether the packet is for UDP or for TCP, otherwise. The extension header has a similar NEXT HEADER field. HOP LIMIT is decremented by one by each router to prevent packets from looping forever in the network; the packet is discarded when the value reaches zero, so that the maximum number of hops is 128. The SOURCE ADDRESS and DESTINATION ADDRESS have 16 bytes (see RFC 1884), enough for 5×10^{28} addresses per human being, which should suffice for a while; there is no explicit support for mobile hosts. Note that there is no error-detection field in the header: IPv6 relies on higher-layer protocols for error control. In IPv6, the routers do not fragment packets. When a packet is too long for a router to handle, the router sends an ICMP message to the source asking it to fragment the packet and resend it.

The extension headers are the following:

- *Hop-by-hop Options:* Information for routers, such as the length of a datagram that exceeds 64 KB (called a *jumbogram*).
- *Routing:* Specifies a set of routers that the datagram must visit (a list of up to 24 IPv6 addresses).
- *Fragmentation:* This header specifies which fragment of a larger packet this IPv6 packet contains. By using this extension header, the destination can reassemble the packet as in IPv4.
- *Authentication:* This header contains a checksum that enables the destination to authenticate the sender.
- *Encapsulating Security Payload:* This header contains the encryption key number and the encrypted payload. The destination can recover the original payload from these two items using a decryption algorithm that is left to the choice of the user.
- *Destination Options:* The possible use of this header, intended only for the destination, has not been defined so far.

This header structure is very flexible in that it leaves the door open for extensions. That structure is also rather efficient in that the minimal header is simpler than in IPv4 by the removal of the checksum and of fragmentation. With time, the Internet designers will develop mechanisms that exploit this format to implement a large range of services.

RSVP

RSVP (Resource Reservation Protocol) is being developed to improve the quality of service in the Internet. In RSVP, the applications signal to the network their requirements, and the protocol reserves resources in the network switches.

A source sends messages (called *path messages*) that describe the specifications of the traffic that the source will generate to a multicast group.

Receivers join a multicast tree by sending messages (called *reservation messages*) with a description of the quality of service they expect. The routers forward upstream a least upper bound of the QoS requirements made downstream. Thus, if a router gets a request from a downstream node for a 200-ms delay version of a stream and another from a different node for a 100-ms delay version, the router forwards the 100-ms delay request upstream. (The precise meaning is admittedly fuzzy.) Once the router gets the 100-ms delay version from upstream, it sends it to the downstream nodes. A similar idea is used for throughput requests. If a router does not find enough spare resources to carry the stream with the requested QoS, it sends a reject message to the receiver. Presumably, a switch algorithm uses these requests for QoS of different streams to determine the resources that should be reserved and how to schedule the transmissions of packets.

To adapt to changes, RSVP specifies that sources periodically send messages that describe their streams and receivers send periodically reservation requests. The downstream messages sent by the source are routed by multicast IP. The reservation messages are sent upstream along the reversed tree. All these messages carry a timeout value that the nodes use to set timers and to delete the corresponding information when the timers expire.

An Internet Engineering Task Force (IETF) working group is refining this basic proposal.

3.9.2 TCP and UDP

Internet uses two different transport layers: TCP and UDP. UDP is the *User Datagram Protocol*. UDP is a connectionless service that provides for error detection.

TCP, the *Transmission Control Protocol*, is a connection-oriented protocol. TCP uses the Selective Repeat Protocol (SRP), discussed in section 2.5.3. Figure 3.40 shows the TCP header. Note that the TCP header contains a packet error-detection field.

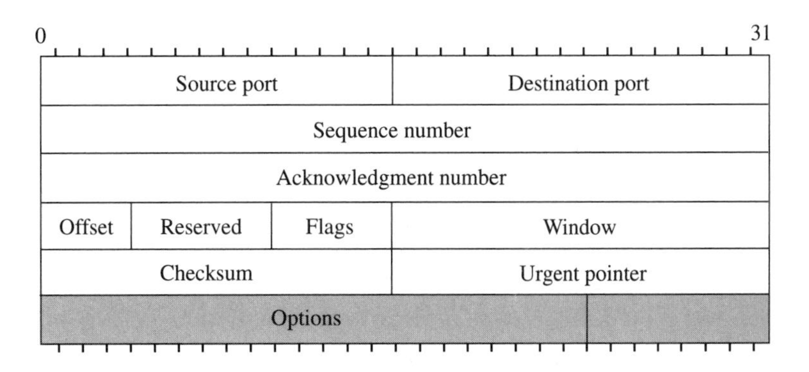

3.40 FIGURE

TCP header. This header indicates the source and destination port numbers. The header also contains the sequence and acknowledgment numbers needed by the SRP that TCP implements. The window size specified by the header is negotiated by the hosts.

The window size can be adjusted by either the source or the destination to control the flow. The sequence numbers have 32 bits and the initial sequence number for a connection is agreed on by a three-way handshake: (1) the source announces the initial number; (2) the destination acknowledges that number; (3) the source sends the first packet with that initial number. This procedure prevents the misunderstanding that would be caused by a delayed packet. Similarly, connections are released by one two-way handshake for each direction.

As we saw in section 2.5.4, flow-control schemes may involve adapting the window size to prevent congestion. A number of such window adaptation schemes for TCP have been proposed, and a few have been implemented. Most of these schemes use timeouts as indication of congestion and reduce the window size quickly when timeouts occur. These schemes typically increase the window size slowly when no timeout occurs.

FTP

FTP, the *File Transfer Protocol*, enables users to transfer files between computers. As Figure 3.41 shows, FTP opens two connections between the computers: one connection for the commands and replies and the other for the data transfers. FTP is interactive. Its commands are *send, get, transfer,* and *cd* (change directory). FTP transfers files in three modes: *stream, block,* and *compressed.* In the stream mode, FTP handles information as a string of bytes without separating boundaries. In the block mode, FTP decomposes the information into blocks of data. In the compress mode, FTP uses the Lempel-Ziv algorithm to compress data.

3.9.3 SMTP, rlogin, and TFTP

We summarize three popular Internet applications. The first two run on top of FTP and the third is based on UDP.

The *Simple Mail Transfer Protocol* (SMTP) is used for e-mail. SMTP accepts messages with a list of destinations. When it does not succeed in delivering a message, SMTP retries a number of times, and it notifies the sender if it cannot deliver the message.

The *remote login service,* rlogin, enables a user to access a remote machine. Rlogin establishes the connection to the remote machine, exchanges the required authorization information, and eventually logs in the user. When rlogin is used, the echoing is performed by the remote machine. That is, the characters typed by the user are sent back by the remote machine before be-

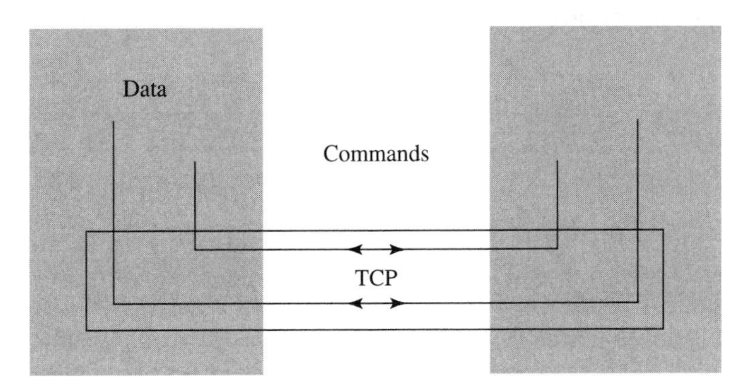

FIGURE 3.41 The File Transfer Protocol (FTP) sets up two connections: one for the commands, the other for the data exchange.

ing displayed. Special control commands such as *stop* and *resume* are sent as urgent data.

The *Trivial File Transfer Protocol* (TFTP) transfers data as blocks of 512 bytes. TFTP sends one block of 512 bytes and waits for an acknowledgment. TFTP retries after a timeout until it succeeds and then proceeds to the next block. TFTP numbers the blocks sequentially from 1. This robust protocol operates even when the transport layer is of low quality.

3.9.4 Real-Time Protocols

A major shortcoming of the Internet protocols TCP/IP is their lack of support for delay or throughput guarantees. Such guarantees are needed for real-time applications.

Many researchers are developing modifications of TCP/IP that attempt to support such guarantees without modifying the protocols so drastically as to make them impossible to implement on the current Internet nodes.

A number of the proposed modifications of TCP/IP attempt to estimate the current delay faced by connections between two hosts. The protocols use such estimates to determine if the requirements of an incoming call request can be met. The proposals differ in how the estimates are calculated. Assuming that the Internet nodes are not instrumented to measure their delay, most proposed protocols involve sending test packets to the destination and measuring the time taken for the acknowledgments of the packets to come back.

A major unknown about such approaches is the possible obsolescence of the estimates.

Other strategies propose to reserve bandwidth and buffer space for specific connections.

Such variations on the current TCP/IP protocols can probably improve the quality of the service offered to some connections. However, TCP/IP protocols were designed without provisions for delay/throughput guarantees, and many researchers are convinced that TCP/IP and the hardware at the nodes (routers) will have to be significantly changed to provide such guarantees efficiently.

3.9.5 Faster Transport Protocols

TCP was designed for relatively slow links (56 Kbps). With such links, the amount of data of a given TCP connection that can be in transit in the network can be kept small. For instance, the propagation time from coast to coast in the United States is less than 20 ms, and during that time a 56-Kbps link transmits fewer than 1,200 bits. By contrast, if the network uses links at 155 Mbps, then up to 3 Mb can be in transit in the fibers. We explained in section 2.5.3 that a window error-control protocol is efficient only if the window size is sufficiently large to fill up the transmission path. However, if that window size is very large, then the error-control protocol is no longer an effective flow-control mechanism since the large window size can overwhelm the destination computer or saturate some nodes of the network.

In view of this shortcoming of window flow-control protocols for high-speed links, network engineers have proposed rate-control protocols. Examples of such protocols include XTP and VMTP.

The operating principle of rate-control protocols is to limit the rate at which the transport protocol sends data by using timers. Although such rate control can be performed without feedback from the network, such feedback can also be used to reduce the rate when nodes detect congestion.

We discuss rate control in detail for ATM networks in Chapters 6 and 7.

3.9.6 Internet Success and Limitation

We studied the Internet suite of protocols in terms of the variety of services that it supports. We also saw that those protocols are organized in a multilayered architecture resembling the seven-layer OSI model. Figure 3.34 summa-

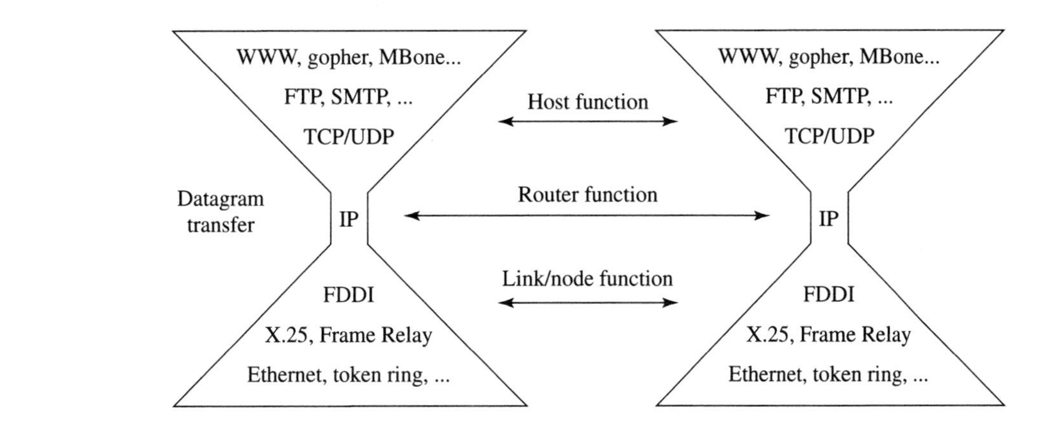

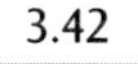

3.42

FIGURE

The Internet implementation of the Open Data Network model shows a simple bearer service, supported by many networks and supporting a variety of applications. The bearer service is provided by simple routers, facilitating network interconnection. Applications are provided by hosts, facilitating experimentation and incremental evolution.

rizes both of these aspects. It lists the different protocols in a manner that shows their layered dependence and how those layers are related to the OSI model.

In order to explain the technical basis for the astounding success of the Internet and to foresee its limitations, it will be more revealing to view the Internet protocols as an implementation of the Open Data Network model of section 2.7. Such a view is presented in Figure 3.42.

The narrow "waist" of the figure represents the single and simple IP bearer service: the end-to-end transport of IP datagrams. This service can be provided by routers capable of the basic tasks of storing packets and routing them to appropriate outgoing links after consulting a routing table. Most significantly, there are no performance requirements, so that slow and old routers can coexist with new, high-performance routers. This "backward compatibility" is enormously valuable and explains why networks with widely differing performance can be interconnected (see Figure 1.13). It also explains why network growth is opportunistic rather than planned.

The brilliance of the IP design lies in its simplicity: because IP datagrams are self-contained, routers do not need or keep any state information about those datagrams. As a result, the network becomes very robust. If a router fails, datagrams in the router may be lost, but new datagrams would automatically get routed properly, with no special procedures. (By contrast, if the bearer service were connection-oriented like ATM, routers would have to maintain connection-state information, and if a router were to fail, that state information would be lost, making it very difficult to restore the connection.) Simple rules in routers can help route traffic around congested parts of the network, giving the network the capability to adapt to changes in traffic.

The third feature of the figure—its wide top—represents the rich variety of applications that the IP bearer service can support. The last decade of successful, sophisticated applications like the World Wide Web has shown that datagrams can serve as building blocks for complex services, provided that the end hosts are sophisticated.

The most remarkable aspect of this feature is that the application software resides entirely in the end hosts and not in the routers. This means that the same basic service, implemented by simple routers, can support these sophisticated applications. Thus the network hardware and software have a much longer technical and economic life than does the end host. Indeed, in the mid-1990s, significant numbers of Internet routers are 15 to 20 years old. This also implies that parts of the Internet can experiment with new applications on advanced hosts using the real network, while other parts of the Internet continue undisturbed to run old applications on primitive hosts. This ability to experiment with new applications has greatly helped the proliferation of new applications.

Thus the technical basis for the Internet's success is its reliance on simple routers to transfer individual datagrams and on advanced end hosts to run sophisticated applications. The simple infrastructure is very cheap compared to the end hosts, and so it makes economic sense to subsidize the infrastructure for collective use. The developers of successful applications can distribute them for fame or profit, without requiring any change from the infrastructure. The contrast with the telephone network could not be more striking. There the network "intelligence" is located in its expensive switches, while the end hosts (the telephone sets) are primitive, with little functionality. The introduction of new services requires changes in the infrastructure—changes that are slow and expensive. Hence experiments are costly and infrequent, and new services are introduced after much deliberation and planning.

The limitations of the Internet can be foreseen from the figure. The IP bearer service cannot provide any guarantees in terms of delay or bandwidth or loss. Routers treat all packets in the same way. (This "equal service for all" is, perhaps ironically, called *best-effort service.*) This is an innate feature: the absence of state information means that packets cannot be differentiated by their application or connection, and so routers would be unable to provide additional resources to more demanding applications.

The technical challenge is to expand the IP bearer service to provide guarantees in a way that preserves the Internet's accommodation of backward compatibility and incremental change. We have discussed some of the proposals to modify IP to meet this challenge.

3.10 SUMMARY

Over its 100-year history, the services provided by the telephone network seem hardly to have changed. The quality of voice is much improved, connections can be set up through "direct dialing" to virtually anywhere in the world, service is very reliable, and costs have steadily declined. Nonetheless, users must find these improvements slow, cumulative, and unremarkable. To appreciate the progress of the telephone system, one has to study the changes in its infrastructure: the increased throughput and capabilities of its switches and its links.

The 20-year history of packet-switched networks presents a sharp contrast. From the viewpoint of user applications, its progress is dramatic. Starting out in its humble beginnings to interconnect a computer with a printer or another computer, these networks enable users around the world to communicate in the form of data, text, images, movies, and animation.

Several technical advances cooperated in bringing about this dramatic progress. The acceptance of the principles of layered architectures for developing new services encouraged modularity and reuse of existing services and permitted network engineers and application developers to work independently. The choice of transporting packets as the basic service turned out to be ideally suited for interconnecting computers and other devices, since it isolated the rapid technical advances in link speed from the advances in software. Finally, the invention of local area networks such as Ethernet quickly reduced the cost of access at the same time as their speed increased, and the capabilities of workstations and personal computers kept up with the improvements in access and speed.

These cooperating technical advances all contributed to the success of the Internet. However, the choice of transporting datagrams as the basic service, which may be the Internet's most important feature, may also turn out in the long run to limit its growth to applications that do not require performance guarantees in terms of delay or bandwidth. (The potential for such applications, however, is enormous as the growth of WWW suggests. A 1995 CommerceNet/Nielsen survey finds that 17% of the total population of the United States and Canada 16 years of age or older have Internet access, 11% have used the Internet in the past three months, and 8% have used the WWW.) It is easy to design new Internet protocols and routers to meet the needs of these applications. The real technical challenge is to do that in such a way that the new protocols and routers can coexist with the Internet's vast legacy. If that challenge can be successfully met, the Internet could claim to provide universal networking. At the same time, advances in circuit-switched networks are enormously extending their service capability. We study those networks in the next chapter.

3.11 NOTES

The OSI model is described in detail in [T88, W91]. Standards for local area networks such as Ethernet, token ring, FDDI, etc. are published by the IEEE. Frame Relay, SMDS, and SONET (discussed in Chapter 4) are described in [B95].

TCP/IP protocols are discussed in [T88, W91, C95]. Most TCP/IP documents are published in a series called Request for Comment or RFC. Those documents are available on-line at a Web site

http://www.cis.ohio-state/htbin/rfc/rfc-index.html

or at the ftp site

ftp://nic.merit.net/documentsrfc/INDEX.rfc

The documents are cited as RFC ####, where #### is the RFC number.

For details about the software implementations of TCP/IP in UNIX, see [C95]. The RSVP protocol is described in [ZDESZ93]. A comparison of IP-based and ATM-based service models that provide some real-time guarantees is carried out in [CWSA95]. XTP is discussed in [X92], and VMTP is described in [CW89].

3.12 PROBLEMS

1. There are many cases in which a link is shared by different devices. In such cases there is some explicit or implicit protocol that regulates the access to the link. Discuss the following examples.
   (a) Computer backplane bus is used by many devices (e.g., CPU, memory, I/O devices) to communicate. How is access to this common bus regulated?

(b) In a cellular phone system, a fixed number of channels is available for use by all the mobile phones within the same cell. (The ratio of number of phones to the number of channels is the "pair gain.") How does a mobile phone user get access to an idle channel or learn that no channel is idle?

(c) There is a fixed number of seats in a theater or bus or airplane. Many users may wish to occupy a seat. How is access regulated? Airlines sell more tickets than seats because people cancel their flight. How is access regulated when there is "overbooking"?

(d) When a driver wishes to change into an adjacent lane in which there are other cars, the lane-change maneuver needs to be coordinated to prevent an accident. What "protocol" do drivers use to achieve coordination?

2. Figure 3.43 shows how network managers can partition an Ethernet network to accommodate more nodes. The figure assumes that each node transmits R bps, on average, during a representative period of time. There are N nodes to be connected. If the total transmission rate $N \times R$ is larger than the rate that can be handled by one Ethernet, then the network manager can try to partition the network. For instance, if the efficiency of one Ethernet connecting the N nodes is 80% and if $N \times R > 8$ Mbps, then one Ethernet cannot handle all the nodes. Let us assume that the N nodes can be divided into two groups that do not exchange messages frequently. For simplicity, say that each group sends a fraction p of its messages to the

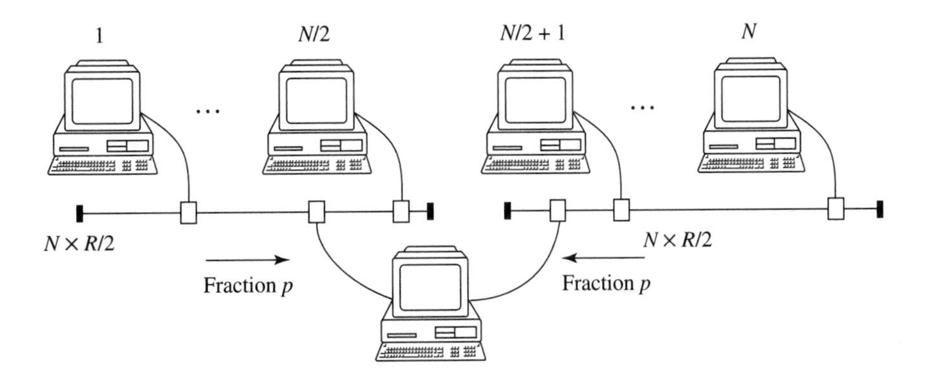

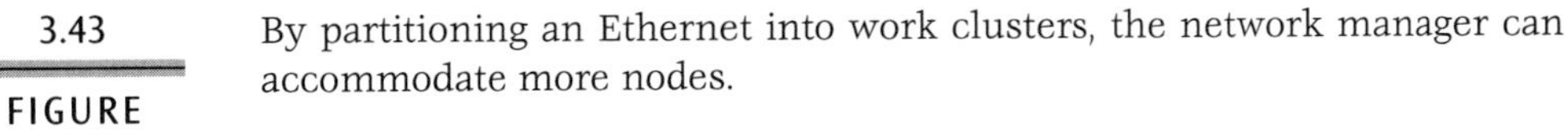

3.43 FIGURE

By partitioning an Ethernet into work clusters, the network manager can accommodate more nodes.

other group. Let us connect the computers in each group with a dedicated Ethernet, as shown in the figure. The two Ethernets are connected by a bridge or by a device called a *switching hub* that performs the same function between Ethernet segments. Calculate the traffic on each Ethernet. The arrangement can handle the N nodes if this rate is less than the efficiency of each Ethernet times 10 Mbps. Of course, the bridge must be able to handle the traffic.

3. The OSI transport layer segments a message into packets and appends a sequence number to each packet so that the receiver can reassemble the original message. In practice, the sequence number is selected modulo N, for some integer N, and consecutive packets are numbered $0, 1, \ldots, N-1, 0, \ldots$. (For example, in SMDS the sequence number has 4 bits, so $N = 16$.) With this choice of sequence numbers, the receiver would be unable to detect the loss of N consecutive packets of the same message. How large should N be to keep the probability of missed detection small? How large is N for IP, ATM, SMDS?

4. Modify Figure 3.20 to reflect the release after transmission protocol used by the 16-Mbps token ring network. Show that its efficiency is approximately $(1 + a/N)^{-1}$, which approaches 100% as N becomes very large.

5. The physical layer of a *token bus network* is a broadcast coax cable like Ethernet, but the MAC layer is like the token ring. Each station on the bus has a number 1, 2, . . . , say. After station i transmits its packets, it releases a token, indicating that the token is intended for station $i + 1$. When the last station, say N, finishes transmission, it releases the token, indicating that the next station is station 1. Analyze the efficiency of the token bus network.

6. Suppose there are N stations on an Ethernet. Suppose the probability is p that any of them has a packet ready for transmission and suppose that different stations are independent. What is the probability p_m that m stations have a packet ready for transmission, $m = 0, 1, \ldots, N$? The average number of packets ready for transmission is $\lambda := pN$. Suppose that $N \to \infty$, $p \to 0$ in such a way that $N \times p \to \lambda$. For this asymptotic case, show that the probability that there are m packets to transmit is Poisson with mean λ.

7. If your computer is connected to an Ethernet, find out what the network utilization is, how many collisions have occurred, and other network statistics that are available. What is the maximum sustained rate at which your computer can generate network traffic? What is limiting this rate? Is it the disk, I/O bus, etc.?

8. Recall the FDDI MAC protocol.
   (a) Show that the maximum time between successive token arrivals in the FDDI network is TTRT + TRANS.
   (b) Show that this time is 2 TTRT (neglecting packet transmission time) when synchronous traffic is included.
   (c) Show that when provision for synchronous traffic is made, the token arrivals are not periodic. What is the maximum deviation in the token interarrival time for synchronous traffic? How much buffering would be needed to transfer constant bit rate synchronous traffic?
   (d) Give a definition of fairness, and argue that FDDI provides fair access to all stations.
   (e) The DQDB MAC protocol attempts to regulate access as if all the stations placed their packets in a single queue that is served on a first-come, first-served (FCFS) basis. Would you call FCFS access fair? Why?

9. We want to analyze the efficiency of DQDB similar to the analysis of the token ring network in section 3.3.2. Suppose there are N equally spaced stations, indexed $1, 2, \ldots, N$, arranged from left to right, and suppose *PROP* is the propagation time between adjacent stations. Suppose T is the transmission time at each station of a 53-byte frame. (For example, if the link speed is 100 Mbps, then $T = 53 \times 8 \times 10^{-8}$ s.)
   (a) Suppose the head station transmits F idle frames per second, back to back. What is F (in terms of T)?
   (b) We define utilization as the fraction of frames that carry data. Suppose that at all times at least one station has some information to send to its right. Show that the utilization of bus A in Figure 3.26 is 100%.
   (c) Consider two stations i and j with i to the left of j. Suppose both are transmitting to a station k on their right. Suppose i and j receive packets to transmit at times t_i and t_j. Suppose $t_i > t_j$, i.e., i's packet arrived later than j's packet. Show that the DQDB protocol works in such a way that j will transmit its packet before i. Now suppose $t_i < t_j$. Construct an example such that the protocol will transmit i's packet after j. What is the maximum amount of time $t_j - t_i$ for which this could happen? (*Note:* In a true first-come, first-served system, i should transmit before j whenever $t_i < t_j$.)

10. Consider the expressions shown in Figure 3.27. Show that if transmission from each node to the next corrupts a packet with probability $1 - p$, and if each frame is retransmitted until it is successfully received, then each

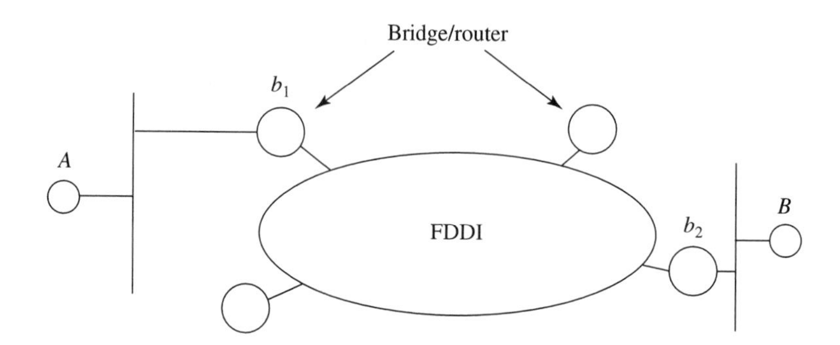

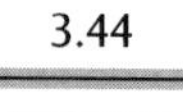
3.44 FIGURE

Bridges b_1 and b_2 must be designed so that computer A can transparently send packets to B connected to a different Ethernet. Note that the MAC addresses on FDDI and Ethernet are different.

frame is transmitted $1/p$ times on average before being successfully transmitted. Calculate the packet error rate $1 - p$ for packets of size 100 and 1,000 bytes and bit error rates of 10^{-4} and 10^{-8}.

11. Suppose in Figure 3.27 that $\sigma_1 = 50$ ms and $\sigma_2 = 3$ ms. Suppose $k = 10$, so there are 10 nodes. How small should $(1 - p)$, the packet error rate per link, be before Frame Relay has less delay than X.25?

12. Suppose an FDDI network connects two Ethernets as shown in Figure 3.44. Bridge b_1 is attached to one Ethernet and the FDDI, while b_2 is attached to the other Ethernet and the FDDI. These bridges are supposed to provide transparent routing. Explain how computer A can send packets to B using b's MAC address? Explain how the bridges can obtain their address conversion tables.

13. Show that the shortest-path first algorithm of Figure 3.38 produces a spanning tree with the shortest length, where the length of a tree is defined as the sum of the lengths of all its links.

14. Define the three-way handshake of the TCP initialization protocol in more detail, using timeouts. Explain how it avoids misunderstandings due to a delayed packet.

15. Propose a TCP window flow-control scheme that reduces the window size when congestion is high and increases it when it is low. Also propose a method to measure congestion. Propose a simple mathematical model that can be used to analyze the performance of your scheme.

4

CHAPTER

Circuit-Switched Networks

The packet-switched networks that we studied in Chapter 3 are well suited for message traffic. Virtual circuit-switched networks, designed for variable bit rate traffic, are discussed in Chapter 5. Constant bit rate traffic is better transported by circuit-switched networks, which we study in this chapter. The most important circuit-switched networks are the telephone, cable TV or CATV, and some cellular telephone and satellite networks. We discuss only the telephone and CATV networks.

This chapter presents the underlying concepts and the technologies deployed in the 1980s and 1990s that changed the telephone system from a voice-only network to one that supports all types of high-performance traffic. You will become familiar with the most important technologies: SONET, optical local loop, ISDN, and INA. You will also learn how CATV systems are changing from a one-way video distribution system into a network that offers limited communication in the upstream direction. The telephone and CATV networks, which until 1990 served separate markets, have since begun to compete and collaborate. (It is too early to assess the impact of the 1996 Telecommunications Bill.) This chapter explains the technological basis underlying this change.

Today's telephone network consists of a high-speed digital backbone network. Subscribers gain access to this network using dedicated low-speed analog or 64-Kbps digital links, called the *local* or *subscriber loop*. The network provides circuit-switched connections for carrying constant bit rate (64-Kbps) voice traffic. (Higher bit rate connections are available to a limited extent.) We introduce in section 4.1 some simple concepts that are used to describe circuit-switched networks. More elaborate mathematical models are developed in Chapters 6 and 7.

Over time, as advances in optical transmission systems have become incorporated into the backbone network, its speed has grown. However, the speed of subscriber access has remained unchanged. Thus the cost of "long-distance" telephone calls has declined steadily, while the cost of "local" calls has declined much less. (Whether this shift in cost of local compared to long-distance calls is directly experienced by subscribers is a different matter having to do with regulatory tariffs as well as cost.)

Four innovations have dramatically expanded the capabilities of telephone networks: SONET (Synchronous Optical Network), optical local loop, ISDN (Integrated Services Digital Network), and INA (Intelligent Network Architecture).

SONET provides higher speeds than the current transmission system. SONET also simplifies multiplexing equipment. SONET is discussed in section 4.2. In terms of the Open Data Network model, SONET implements a "bit way," as indicated in Figure 2.32. SONET provides end-to-end fixed-rate channels at speeds that are multiples of 55.84 Mbps. Those channels can be used in a flexible manner to support a variety of bearer services. SONET is "backward compatible" because it can carry current telephone bit streams. It is also "forward compatible," since it can be used to transport ATM cells. Indeed, the transport of ATM cells over SONET is the basis of the Broadband Integrated Services Digital Network or BISDN, which some regard as the "universal" network.

The second innovation in the telephone network is the replacement of the existing copper subscriber loop by optical fiber. The replacement will eliminate the current 64-Kbps speed bottleneck, and users then will be able to send and receive all forms of information (video, image, text, voice) requiring high speed. Only when high-speed access is widely deployed will BISDN or the "information superhighway" become a reality that many can share. In section 4.3 we study three technologies for optical local loop: the AT&T subscriber loop carrier, British Telecom's TPON (Telephony over Passive Optical Network), and an as yet impractical scheme that depends on wave-division multiplexing (WDM). Recent advances in modulation techniques such as ADSL may permit access speeds of 1 Mbps or more over the existing copper subscriber loop. Combined with better compression techniques, these modulation techniques may extend the economic life of the existing copper plant.

The third innovation, ISDN, gives users a combination of data and voice channels multiplexed and made available at a single interface. To an extent that is circumscribed by the limited access speed, ISDN integrates constant bit rate (voice) traffic and message traffic. We describe ISDN in section 4.4.

SONET, optical local loop, and ISDN enhance the telephone network's information transfer services. INA is an innovation of a different sort. To appreciate INA, we must think of the telephone network as a collection of (hardware and software) resources, each of which can execute a number of activities. For example, circuits can be combined to form a path and used to transfer a constant bit rate voice stream; a recording device can be activated to record a message or to play back a previously recorded message; a switch can send a "ring" signal to a telephone; the access-line interface at the switch can record the digits dialed by a subscriber; and so on. These activities can be combined into sequences to produce valuable new services such as the 800 number service, call forwarding, voice mail, caller identification, etc. INA facilitates the specification and implementation of these activity sequences. INA is described in section 4.5.

The second most important circuit-switched network is CATV. Until the mid-1990s CATV was primarily a one-way analog distribution network with a large bandwidth going "downstream" to users, little or no bandwidth going "upstream" from users, and with no switching capability. However, recent technological innovations in networking and signal processing may reduce or eliminate these handicaps. If that happens, CATV will become a formidable competitor to the telephone companies because, unlike the telephone network, CATV already has a high-speed "local loop"; because it has a customer base that is comparable to that of the phone companies; and because it can use its large subscriber fees to finance the necessary investment. (The telephone companies are taking this potential competitor seriously. Some of them are adopting for their own local loop the same technology that CATV will adopt in order to offer video programs and compete with CATV.) CATV is discussed in section 4.6.

A summary of this chapter is given in section 4.7.

4.1 PERFORMANCE OF CIRCUIT-SWITCHED NETWORKS

Circuit-switched networks consist of switches interconnected by links. The capacity or bandwidth of each link is divided into fixed-rate *circuits* (e.g., 64-Kbps circuits for the telephone network). Alternatively, we can view each link as comprising several multiplexed circuits. Telephone networks employ time-division multiplexing, while CATV, some cellular telephone, and satellite networks today employ frequency-division multiplexing.

Users connect to a network switch by dedicated or shared access links. A user wishing to exchange information makes a request to the network for a *connection* or *call* to another user at a specified address. (Different networks have different addressing conventions. For example, the North American telephone network has a 10-digit address.) On receiving the call request, the network switches exchange information according to specified algorithms in order to make two related decisions: the *call admission* decision determines whether to admit or reject the request; and if the decision is to admit, the *routing* algorithm selects a route or sequence of idle circuits to connect the two users. Once the connection is established, the users exchange information. (The time taken by the switches to establish a connection is the connection *setup time*. Packet-switched networks have no setup time.) When the exchange is over, the user requests connection termination, and the switches return the circuits to the idle state.

A connection request that is rejected by the network is said to be *blocked*. Requests may be blocked because the switches are unable to find a route with idle circuits or because idle circuits are kept in reserve for future connection requests. The most important performance measure of the network is the *blocking probability*, defined as the chance that a call request is rejected during the busiest traffic hour.

During the information exchange period, user data is transferred in frames at a fixed rate. The frames carry a connection identifier and the switches maintain a state table relating the connection identifier to the next link in its route. Therefore, frames in the same connection travel over the same route. Frames in different connections that go through the same link are multiplexed. At the switch, the different connections on each incoming link are demultiplexed, and different combinations are remultiplexed to form the stream for each outgoing link.

Because the network allocates a fixed bandwidth to each connection, the frames face no queuing delay. Hence the total end-to-end delay is the propagation time plus a small processing delay in each switch along the route. This total delay is both small and constant. That is why circuit-switched networks are well suited for real-time, constant bit rate traffic. In terms of the Open Data Network model of section 2.7, the bearer service offered by the network is the real-time transfer of information with specified peak rate.

This bearer service may also be used for variable bit rate traffic whose peak rate is smaller than the connection bandwidth. This may lead to a low

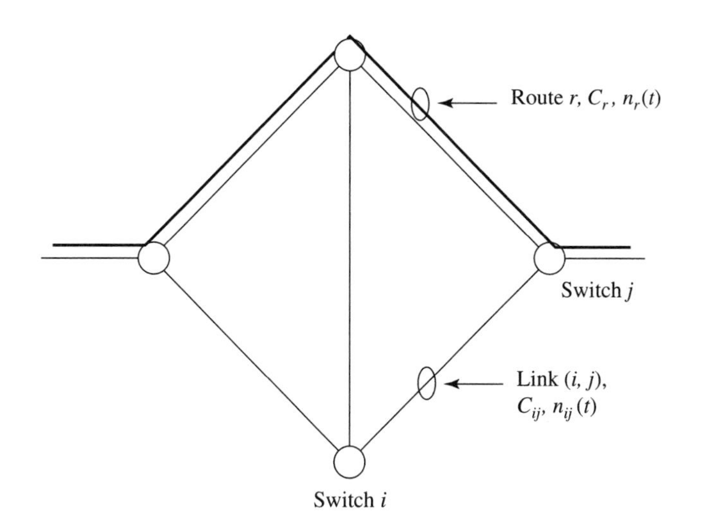

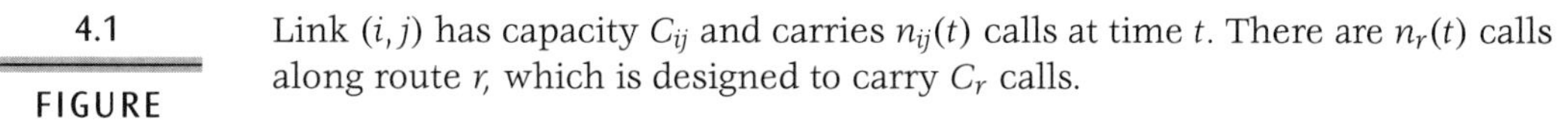
4.1 FIGURE

Link (i, j) has capacity C_{ij} and carries $n_{ij}(t)$ calls at time t. There are $n_r(t)$ calls along route r, which is designed to carry C_r calls.

utilization, defined as the ratio of the average rate of information transfer divided by the bandwidth of the connection. If the utilization is significantly below one, then link capacity is very poorly utilized. For example, in Telnet applications, the user's typing speed limits the information transfer to 40 bps. If this application is run over a 64-Kbps telephone line, the utilization is 40/64,000, which is less than 0.1%. The bearer service may also be used to transfer messages. In this case, the large setup time may make this use of the service noncompetitive relative to message transfer through packet-switched networks with no setup time.

Several questions of network planning, route assignment, and blocking probability can be formulated using the mathematical model of the network described next. See Figure 4.1. (We study stochastic models in Chapter 7.)

We represent a circuit-switched network as a graph whose nodes $i = 1, \ldots, n$ denote the switches, and there is an edge between nodes i and j if there is a transmission link connecting switches i and j. We suppose that the network provides connections with a fixed bandwidth. (For example, the telephone network provides 64-Kbps connections.) The capacity of the link joining i and j is denoted by C_{ij}, the number of simultaneous (two-way) connections or calls that this link can accommodate. (For example, a T-1 link can accommodate 24 voice calls; see Table 1.2.)

A route r is a sequence of connected links starting at an origin switch and ending at a destination switch. We write $(i,j) \in r$ if link (i,j) is part of the route r. Let R denote the set of all routes. At any given time t there are a number of simultaneous active connections through different routes. We denote these connections by the vector $n(t) = \{n_r(t), r \in R\}$, where $n_r(t)$ is the number of connections through route r at time t. The number of calls through link (i,j) is the number of calls through all routes that pass through (i,j):

$$n_{ij}(t) = \sum_{\{r|(i,j)\in r\}} n_r(t).$$

The number of connections through a link is limited by its capacity and so $n(t)$ must satisfy the constraint:

$$n_{ij}(t) = \sum_{\{r|(i,j)\in r\}} n_r(t) \leq C_{ij}, \text{ for all } (i,j). \tag{4.1}$$

Relation (4.1) determines the maximum possible set of connections that can be accommodated by the network. The vector $n(t)$ changes randomly over time. When a call along route r terminates, $n_r(t)$ decreases by one; if a new call is placed along route r, $n_r(t)$ increases by one.

Suppose that the network is designed to carry C_r simultaneous calls along route r, $r \in R$. Then the link capacities C_{ij} must be so large that

$$\sum_{\{r|(i,j)\in r\}} C_r \leq C_{ij}, \text{ for all } r \in R. \tag{4.2}$$

The system of inequalities (4.2) establishes a relation between the capacity invested in the transmission system, the choice of routes, and the maximum number of simultaneous calls that the network can accommodate.

The random amount of time that a connection endures is the call *holding* time. The random time between the arrival of two consecutive connection requests is the *interarrival* time. It is customary to denote the average holding time by μ^{-1} and the average interarrival time by λ^{-1}, so λ is the average rate of call requests (calls per second).

Fix a route r, and let λ_r, μ_r be the call parameter values for that route. Suppose p_r is the blocking probability for these calls. Then the rate of *carried* calls is $(1-p_r)\lambda_r$ calls per second. Each of these calls lasts μ_r^{-1} seconds on average. So the average number of active connections along route r is

$$(1-p_r)\frac{\lambda_r}{\mu_r} =: (1-p_r)\rho_r,$$

where ρ_r is the number of call requests during one call holding time. ρ_r is an important measure of traffic intensity. Its unit of measurement is named after Erlang for his pioneering work in traffic engineering.

Evidently, we must have $(1 - p_r)\rho_r \leq C_r$, all $r \in R$, which places a lower bound on the blocking probabilities $\{p_r\}$. This lower bound, however, is quite optimistic, since it is based on average values and ignores the statistical fluctuations in the request arrivals and call holding times. In Chapter 7 we compute more accurate estimates of blocking probabilities.

The model described above, together with related problems presented in section 4.9, can be used as examples of how to formulate and resolve questions of planning and operations of circuit-switched networks. The parameters of costs and capacity that enter in those questions depend on technology. We now describe the main technological innovations in circuit-switched networks, beginning with SONET.

4.2 SONET

We view SONET (Synchronous Optical Network) as a bit way implementation, providing end-to-end transport of bit streams. As its name suggests, the development of SONET was spurred by advances in the accuracy of clocks and in optical transmission.

SONET is an ITU standard. It encodes bit streams into optical signals that are propagated over optical fiber. SONET's high speed and its frame structure permit it to support a very flexible set of bearer services. The standard specifies the frame structure as well as the characteristics of the optical signal. We will discuss only the frame structure.

The most important feature of the standard is that all clocks in the network are locked to a common master clock, so that the simple time-division multiplexing (TDM) scheme of Figure 2.9 can be used. Multiplexing in SONET is done by byte interleaving, as depicted in Figure 4.2. As a result, as we see in the figure, if each of N input streams has the same rate R, the multiplexed stream has rate NR. Because sources are synchronized, the buffers at each input line will be very small as they have only to accommodate the effect of jitter. (Jitter is the inevitable, small variation in the successive clock ticks recovered from a periodic signal corrupted by some noise.)

In North America and Japan, the basic SONET signal is STS-1 (Synchronous Transport Signal-1). It has a bit rate of 51.84 Mbps. Higher rate signals are multiples of this rate; see Figure 4.3. In Europe, the basic rate is STS-3 or

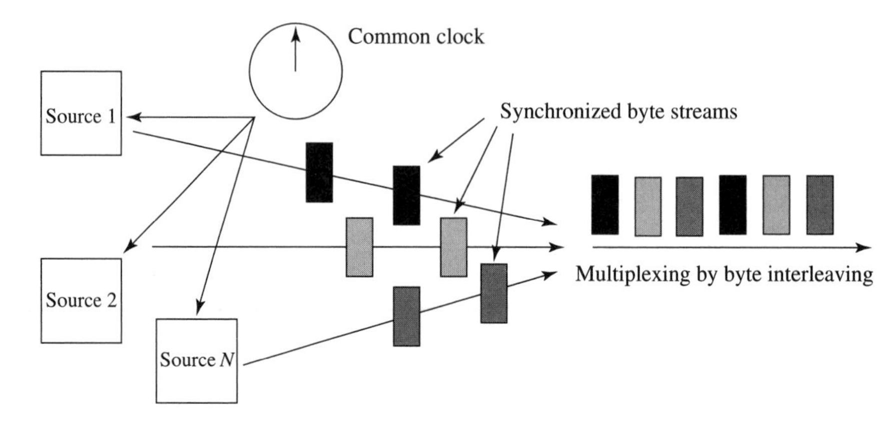

4.2 FIGURE SONET sources are synchronized to a common master clock. Different streams are multiplexed by byte interleaving.

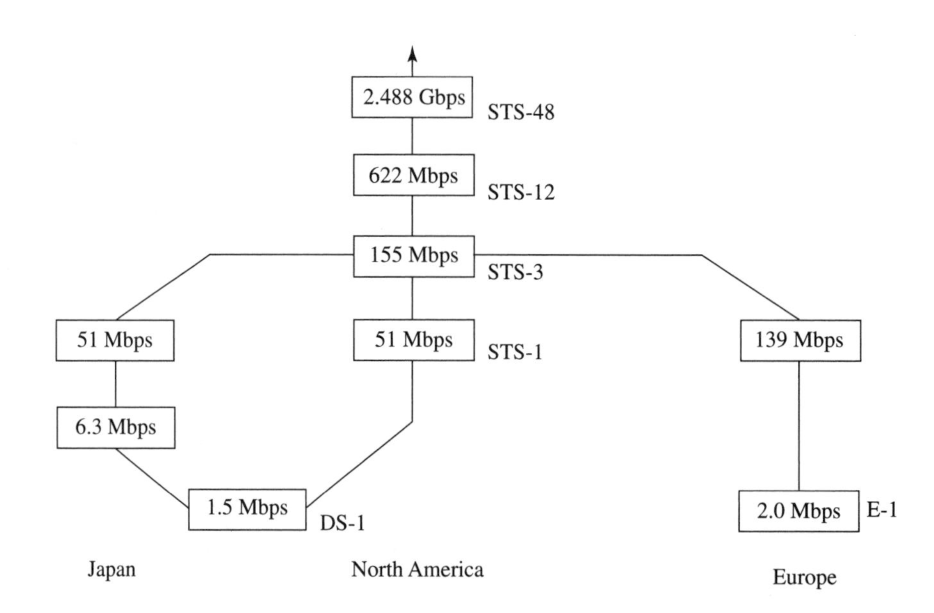

4.3 FIGURE The STS-n signal has a rate equal to $n \times 51.84$ Mbps. In Europe the hierarchy starts at 155.52 Mbps. All the standards become compatible at speeds of 155 Mbps.

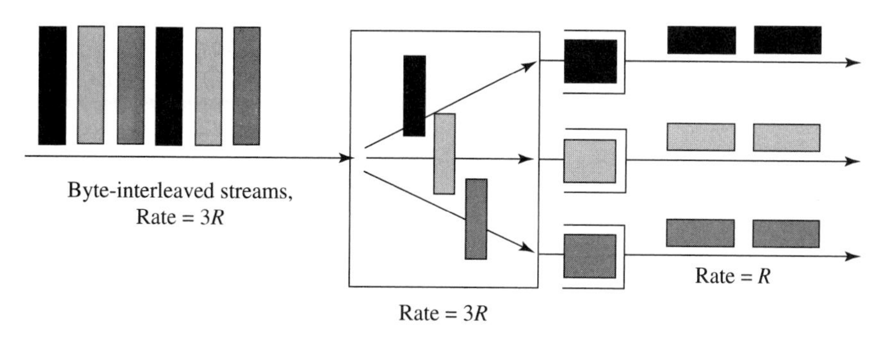

4.4 FIGURE — SONET streams are demultiplexed byte by byte by counting.

155.52 Mbps, and the STS hierarchy is called the SDH or Synchronous Digital Hierarchy, starting at 155.52 Mbps.

The simplicity of the STS hierarchy makes a contrast with the current digital signal (DS) hierarchy shown in Table 1.2, which, because of the way it accommodates asynchronous streams, makes multiplexing much more complex. As Figure 4.3 suggests, moreover, SONET is "backward compatible" in the sense that it can transport current telephone signals: the 1.544-Mbps DS-1 signal in North America and Japan and the 2.0-Mbps E-1 signal in Europe. SONET's frame structure is also "forward compatible" in that it can support the transport of ATM cells. The frame also provides channels for organization, administration, and management in a uniform manner.

SONET's frame structure and multiplexing methods also simplify equipment. The byte-interleaved time-division multiplexing makes demultiplexing easy, because as suggested in Figure 4.4, individual streams can be determined simply by position in the frame.

A SONET ADM (add/drop multiplexer) is also less difficult to design and build. The function of an ADM is to drop one of the incoming multiplexed streams and replace it with another stream; see Figure 4.5. (In the current digital signal hierarchy of Table 1.2, an ADM requires a full demultiplexer-multiplexer pair.) Thus, for several functions, SONET equipment is cheaper than current equipment.

Telephone networks need to have enough redundancy and automatic procedures that use this redundancy to restore quickly the integrity of the network in the event of failures. Two examples will illustrate the magnitude of the losses that can occur when restoration procedures fail. A power outage in an unmanned central office in Hinsdale, Illinois, on Mother's Day, 1988, affected 500,000 customers who made three and a half million calls per day.

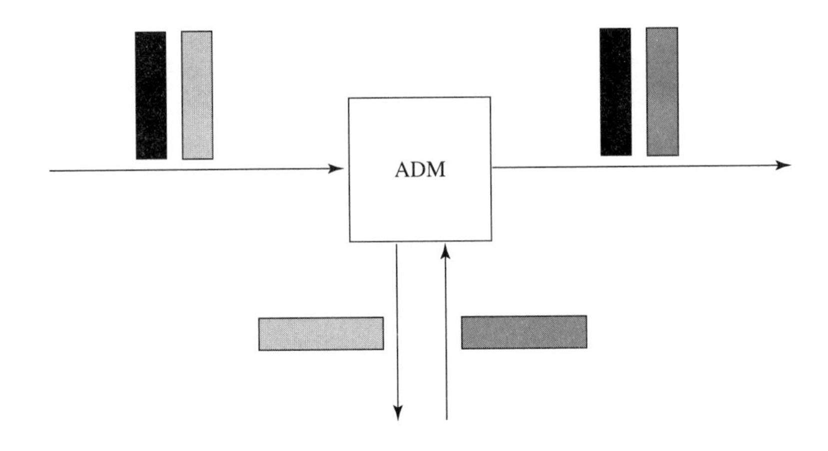

4.5 FIGURE The add/drop multiplexer (ADM) replaces one of the incoming streams with another.

Full service was not restored for one month. In November 1988, a construction crew accidentally severed a major fiber-optic cable in New Jersey, disrupting much of the long-distance traffic along the East Coast, blocking more than three million call attempts. There are many approaches to redundancy, depending on the network. We shall briefly discuss the increased reliability possible when SONET systems are deployed as dual rings. It is believed that dual rings may restore service within 50 ms of a network failure.

The basic idea is illustrated in the left panel of Figure 4.6. One fiber carries the working ring in which the signal moves counterclockwise. In the standby or protection fiber, the signal moves in a clockwise direction. Normally, there is no signal in the protection ring. If a cable between any two central offices is cut or if there is a node failure, the failure would be detected by the automatic protection system, which would perform a loop-back function using the standby fiber. (This is very similar to the FDDI recovery procedure.)

In the panel on the right in the figure, an ADM is replaced by a digital cross-connect system (DCS), which permits an incoming link to be directly connected to an outgoing link. The DCS has created a logical direct link from ADM1 to ADM2, and another logical link from ADM2 to ADM1, so that the configuration of the right panel is equivalent to a ring among the three ADMs. In this way, by setting DCS interconnections appropriately, one can effectively change the topology of the network. Such changes are carried out in order to meet shifts in demand over a 24-hour period or some other period.

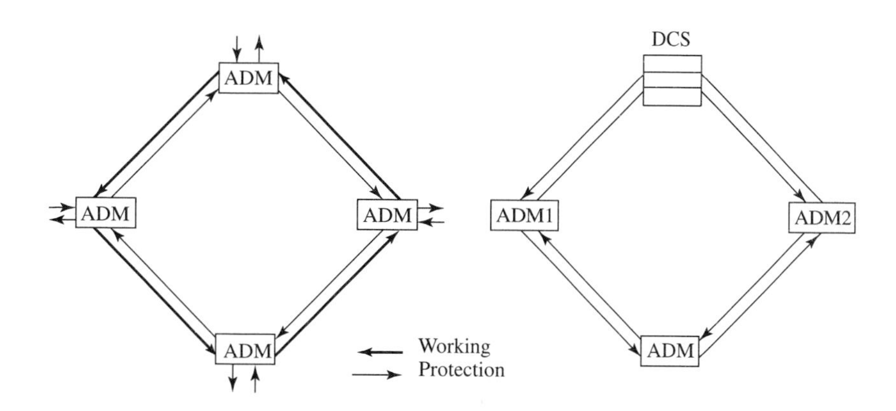

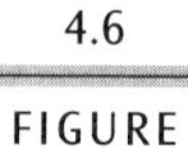
4.6 FIGURE

The panel on the left shows the configuration of a SONET ring. The panel on the right shows how digital cross-connect systems (DCSs) can be used to create logical links and reconfigure the network topology.

4.2.1 SONET Frame Structure

We now describe the main features of the SONET frame structure. Figure 4.7 depicts typical transmission equipment connecting a network element (on the left) where a SONET frame is assembled to another element (on the right) where that frame is disassembled.

(Two terms in the figure may be obscure. A *regenerator* (REG) is a device that boosts the power of the optical signal. The device has three components:

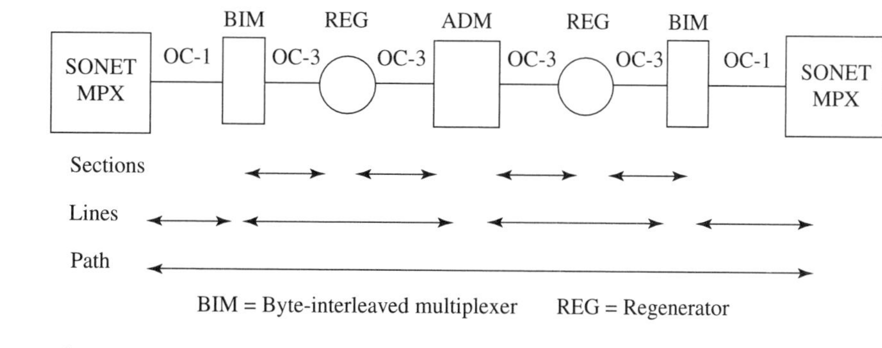

4.7 FIGURE

The main SONET network elements.

the first component converts the optical signal into an electrical signal, the second amplifies the electrical signal, and the third converts the amplified electrical signal back into a more powerful optical signal. The second possibly obscure term is *OC-n* or *Optical Carrier-n*. It is the name of the facility that transmits the STS-*n* signal.)

Figure 4.7 also gives certain definitions used in the SONET frame overhead structure. *Section* is the portion of the transmission facility between a terminal network element and a regenerator or between two regenerators. A *terminal point* is the point after signal regeneration at which performance may be monitored. *Line* is the transmission medium (optical fiber) and associated equipment required to transport information between two consecutive network elements, one of which originates the line signal and the other of which terminates it. (Line corresponds to what we call a link.) A line has a certain bit rate. *Path* at a given rate is a logical connection between a point at which a standard frame for the signal is assembled and a point at which the standard frame is disassembled. (Path corresponds to an end-to-end route.)

The layered overhead structure associated with these definitions is given in Figure 4.8. The four layers of the structure are the path, line, section, and photonic layers. Their main functions are summarized in Table 4.1.

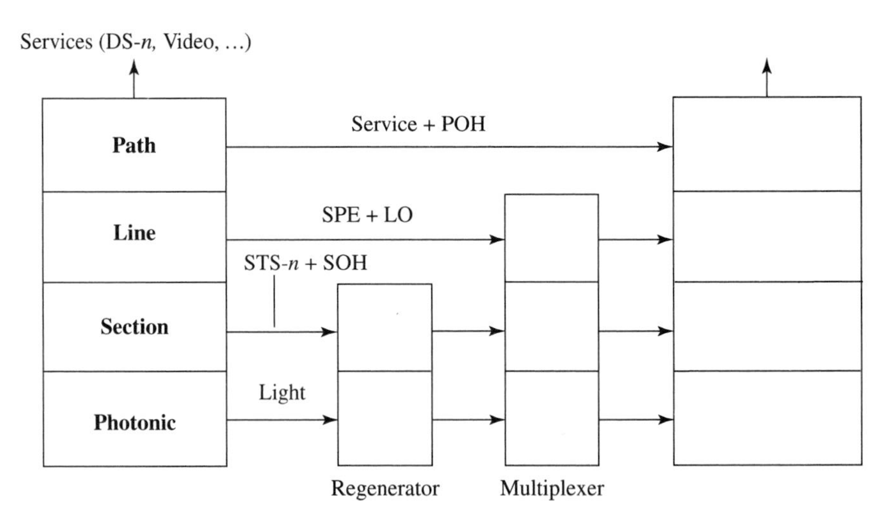

FIGURE 4.8 The four layers of the SONET overhead.

Layer	Function
Path	Services; end-to-end error detection
Line	Multiplexing, with frame and frequency alignment; protection switching; data links
Section	Framing, scrambling, data links
Photonic	Electrical to optical and optical to electrical signal conversion

4.1 TABLE The layers of SONET and some of their functions.

Consider an example where two path layer processes are exchanging DS-3 frames. (The 45-Mbps DS-3 signal belongs to the current telephone signal hierarchy.) The DS-3 frames, plus POH (path overhead), are mapped into an STS-1 SPE (synchronous payload envelope), which is then given to the line layer. The line layer may multiplex several different payloads (in addition to the STS-1 SPE containing the DS-3 signal) from the path layer (frame and frequency aligning each one) and add LOH (line overhead). In addition to multiplexing, the line overhead provides other functions, e.g., protection switching. Finally, the SOH (section overhead) provides framing and scrambling prior to transmission by the photonic layer. The photonic layer converts the electrical bit stream from the section layer into an optical signal.

Going bottom up, each layer builds on the service provided by the layer below. The photonic layer provides optical transmission at some rate; the section layer provides framing and scrambling for the bits being transmitted; the line layer provides line maintenance and protection and multiplexing of STS-1 signals; the path layer provides a mapping from services (e.g., DS-3 frames, ATM cells, digital voice streams) into STS-1 SPEs. In terms of the Open Data Network model, the SONET path layer implements the bearer service consisting of end-to-end transport of bit streams in the format of the SPE.

Certain network equipment may not participate in some of the upper-layer functions. For example, the multiplexer in Figure 4.8 operates at the line layer, and the regenerator operates at the section layer. A photonic amplifier (which directly amplifies the optical signal without converting it into an electrical signal as a regenerator does) operates at the photonic layer.

Figure 4.9 presents overall features of the SONET frame. The 810-byte STS-1 frame is organized into nine rows of 90 bytes (transmitted left to right,

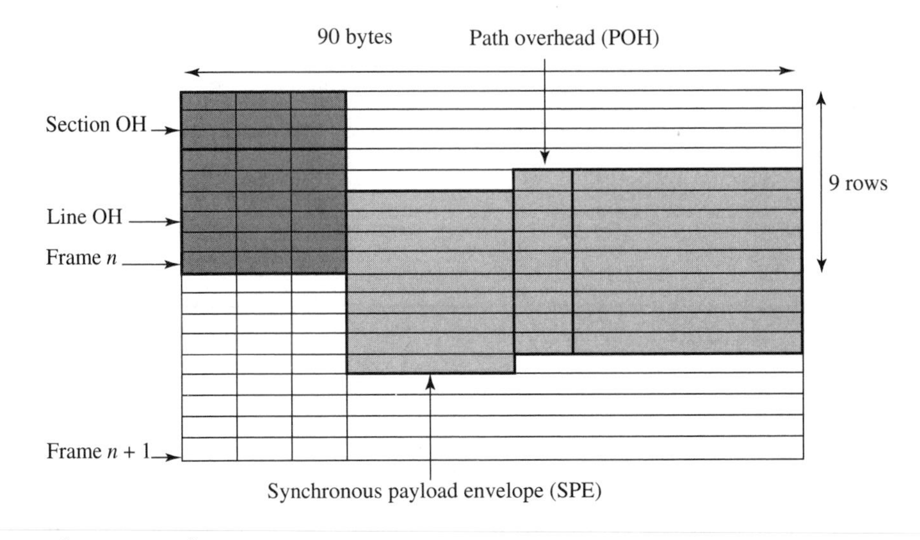

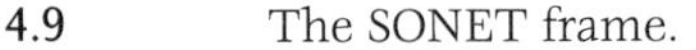

FIGURE 4.9 The SONET frame.

top to bottom). A frame is 125 μs in duration, corresponding to one 8-kHz voice sample. (This gives the STS-1 rate of 810 bytes/frame $\times$ 8,000 frames/s or 51.840 Mbps.) The first three columns (27 bytes) are for SOH and LOH. Thus an SPE is 9 rows $\times$ 87 columns, of which the first column (9 bytes) is devoted to POH.

The most unusual feature is that an SPE does not need to be aligned to a single STS-1 frame: it may "float" and occupy parts of two consecutive frames as shown in the figure. Two bytes in LOH are allocated to a pointer that indicates the offset in bytes between the pointer and the first byte of the SPE.

Figure 4.10 explains the function of some individual overhead bytes. The standard defines the function of each byte and assigns it to a layer and position in the frame. Each of most of the unidentified bytes provides a 64-Kbps data link for use by the transmission equipment for sending messages, commands, and alarms for transmission functions. Examples of these functions are testing, protection switching, and fault isolation.

The first three rows are assigned to SOH. Two bytes (called A1, A2 in the standard) are for framing each STS-1. Another byte (C1) is the STS-1 channel ID. It is a unique number assigned to each channel prior to byte interleaving into an STS-n signal and stays with that channel until it is de-interleaved (demutliplexed). It is used to determine the position of the other STS-1 signals. Byte B1 is a bit-interleaved parity code 8 (BIP-8) using even parity. The

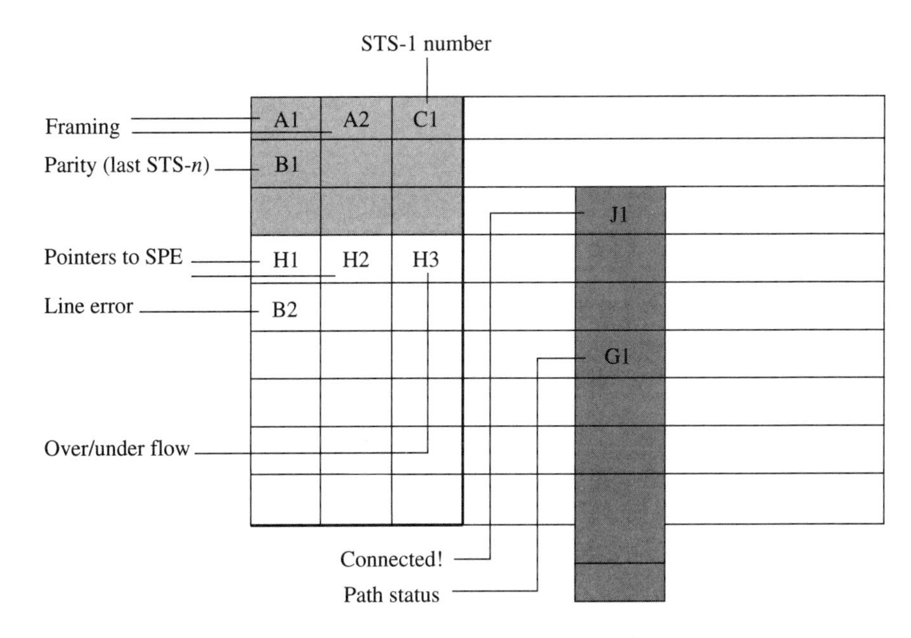

4.10 FIGURE Function of some overhead bytes.

section BIP-8 is calculated over all bits of the previous STS-n after scrambling. It is defined only for the first STS-1 tributary of an STS-n signal.

The remaining six rows are for LOH. Byte B2 is another parity code used for line-error monitoring. Bytes H1, H2 are allocated to a pointer indicating the offset in bytes between the pointer and the first byte of the STS SPE. Byte H3 is used for frequency justification, that is, to compensate for clock deviations. Even though all sources are synchronized to the same master clock, deviations do occur. When the rate of the source (payload) is higher than the local STS-1 rate, H3 is used to add an extra byte to the SPE; when the source is slower, a byte is deleted from the SPE. We explain how this works in detail using Figure 4.11.

For convenience, compatibly number the frames and the SPE, as in the figure. By definition, in any frame (H1, H2) points to the start of the SPE in that frame. Suppose that the payload speed is higher than the frame speed. Then, on occasion, one extra byte is added to the payload. This is done by decrementing the pointer (in frame $n + 1$) by one, so that SPE $n + 2$ starts sooner. The extra byte is placed in H3.

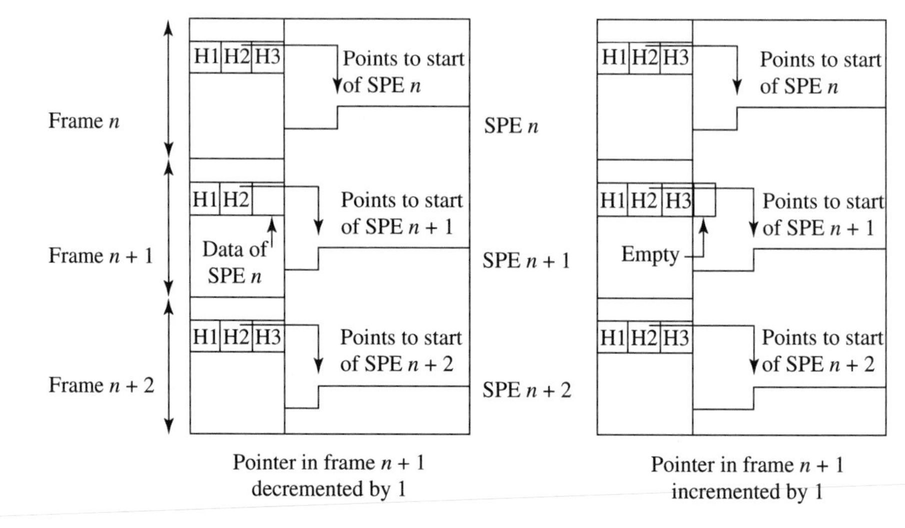

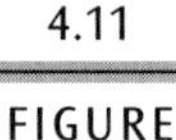

4.11 FIGURE Frequency justification: when the payload rate is higher than the frame rate, H3 is used to provide an extra byte as on the left; when the payload rate is lower, a byte is left empty or "stuffed" as on the right.

Suppose instead that the payload bit rate is slower than the frame rate. Then, on occasion 1 byte is not made available to the payload. This is done by incrementing the pointer (in frame $n + 1$) by one, so that SPE $n + 2$ starts later. The byte next to H3 is now left empty or "stuffed" and not allowed to be part of the payload envelope.

The POH is assigned and remains with the payload until the payload is demultiplexed. The first byte of POH, J1, is used to repeat a (user-programmable) 64-byte, fixed-length string so that a path-receiving terminal can verify its continued connection to the intended transmitter. The third byte, G1, is used to convey back to an originating STS path terminal the path status and performance. This feature permits the status and performance of the complete duplex path to be monitored at either end or at any point along that path.

Figure 4.12 illustrates the ITU standard on how an STS-3 frame would be used to carry (approximately) 44 ATM cells of 53 bytes each. An interesting feature is that no framing bits are provided to delimit the boundary of the ATM cells within the payload. We will see in Chapter 5 how the CRC bits of the ATM cell header are used to find the cell boundary.

In summary, SONET is a transmission standard that assumes a synchronous mode (all signals synchronized to the same clock). Its flexible frame

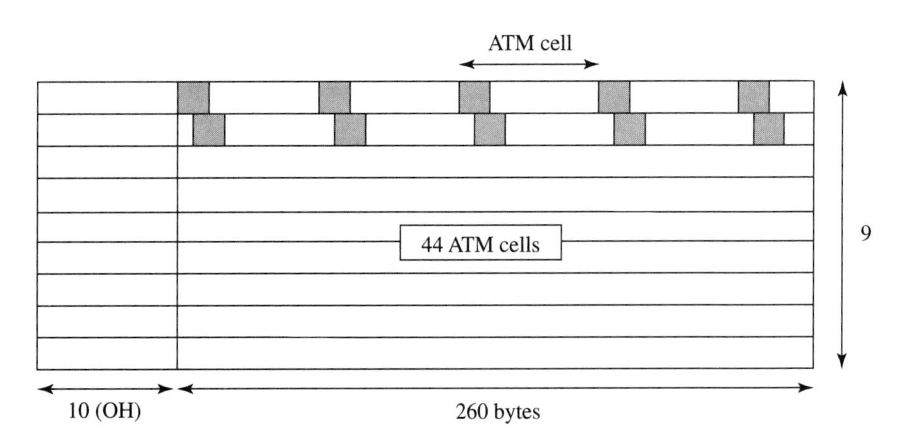

4.12 FIGURE

An STS-3 frame accommodates 44 ATM cells. No framing bits are provided to delimit the cell boundary.

structure accommodates both asynchronous traffic such as ATM and synchronous transfer mode (STM) traffic such as voice. Presumably, at a switch these two kinds of traffic would be separated and recombined, as suggested in Figure 4.13. Whether a particular service will follow an STM or ATM mode will depend on the traffic characteristics.

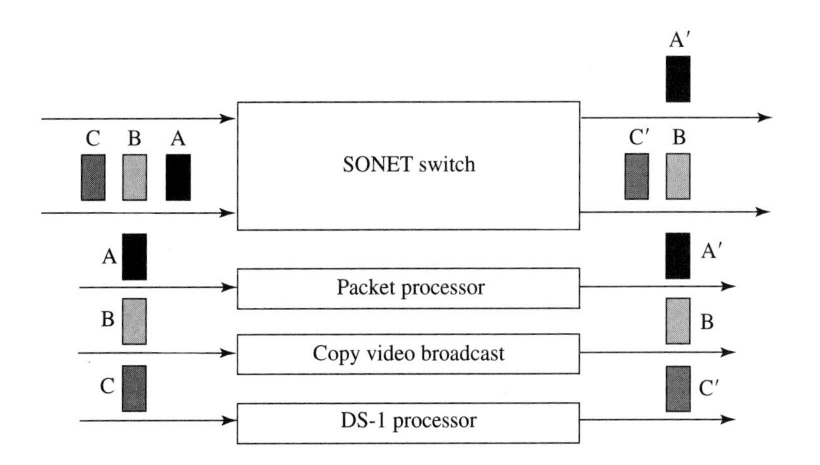

4.13 FIGURE

At a SONET switch, streams of different traffic types are demultiplexed, appropriately processed, and then remultiplexed.

4.3 FIBER TO THE HOME

In this section we consider three systems designed to address the "last mile" problem: "Assuming that a high-speed backbone network is available, how can residential or business customers be connected to it in a cost-effective manner?" The optical subscriber loop systems connect customer premises to a central office with optical fiber. The proposed systems allocate a channel with a fixed bandwidth to each customer location. Typically, the bandwidth of the channel is 64 Kbps, or 2B + D (2×64 Kbps $+ 16 = 144$ Kbps), or 1.544 Mbps. The AT&T SLC Series 5 system uses a fiber to replace a copper line. The British Telecom TPON system multiplexes and demultiplexes the signals optically. Other proposals (PPL and hybrid systems) use wavelength-division multiplexing.

There is an economic justification for using fibers in the subscriber loop when fibers are less expensive than copper or when new services require a larger bandwidth than that provided by copper. The cost of installing a fiber loop is comparable to that of a copper loop. Moreover, optical loops are less expensive because they have a longer lifetime. As a result, local operating companies often deploy optical loops for new installations. Some are replacing old copper loops with fibers when maintenance is needed or anticipated. (The use of fibers does entail additional cost of optical to electric conversion. Also, as we will see later, because it is difficult to amplify optical signals, only a small "pair gain"—the number of subscriber lines per feeder line—can be achieved with an all-optical distribution system.)

Most observers believe that the installation of fiber will be justified by the anticipated new services that require high-speed access. Examples of these services are ISDN, TV distribution, home shopping, and library access. Working at home (telecommuting) and distance learning may contribute greatly to the demand for high-bandwidth services.

4.3.1 The AT&T Subscriber Loop System

Figure 4.14 gives a block diagram of the AT&T Subscriber Loop Carrier (SLC) Series 5 system. In this system, the signals to and from different subscribers are multiplexed electronically (TDM) at a remote terminal (RT). Fibers connect the remote terminal to the customer premises. Thus the optical subscriber signal is converted to an electric signal at the RT, the different electric signals are multiplexed, and the multiplexed signal is converted to an optical

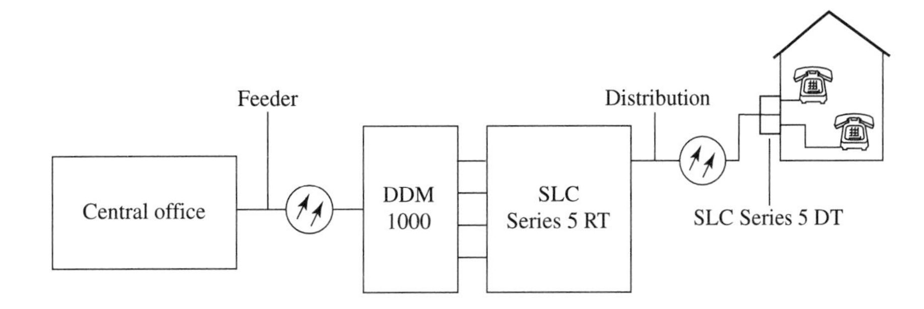

4.14 FIGURE The AT&T subscriber loop system.

signal and sent to the central office. (By contrast, the subscriber signals in the British Telecom system are multiplexed in the optical domain, saving equipment needed for electrical/optical signal conversion.) The characteristics of the optical subscriber loops are indicated in Figure 4.15. (These characteristics may not seem meaningful. In Chapter 9 we study what they mean.)

The system allocates a DS-1 (1.544-Mbps) channel to each customer location. As a result, additional lines and ISDN services are easily provided to the customers. There is a significant economy in the cost of local access. In current telephone systems, there is a wire pair from each subscriber phone that terminates at a switch line interface card. However, if several subscriber

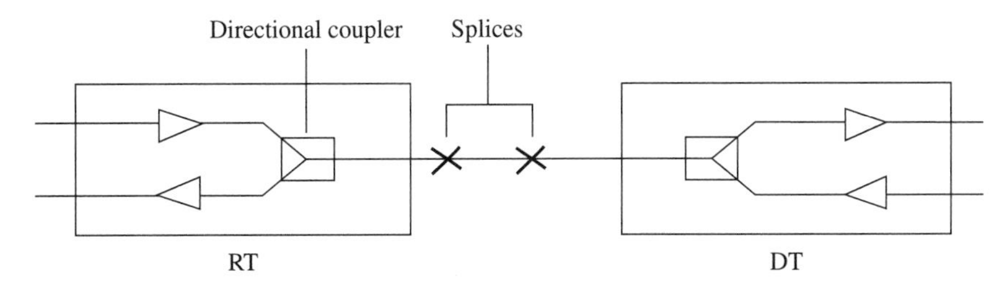

Single-mode fiber, 3-dB directional couplers

InGaAs laser diode, 1.3 mm, $P_T = -20$ dBm

InGaAs PIN diodes with -46 dBm sensitivity at 1.5 Mbps

4.15 FIGURE Details of the AT&T subscriber loop system.

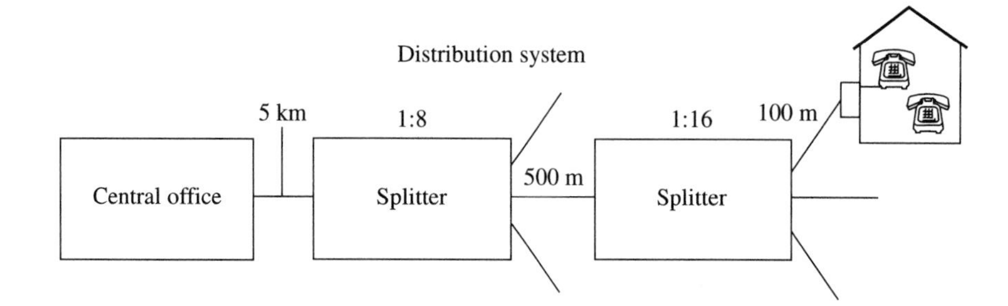

4.16 FIGURE The British Telecom TPON system.

phone signals are multiplexed together as happens with the AT&T Subscriber Loop Carrier Series 5 system, a separate interface card is not needed for each signal, so that the cost per line is reduced.

4.3.2 The British Telecom TPON System

Figure 4.16 depicts the British Telecom TPON (Telephony on a Passive Optical Network) system. This system uses passive (optical) multiplexing to reduce the cost per customer. The TPON topology is a "double star" that attaches 128 (16 × 8) customers on each fiber from the central office. The two splitters in the figure are optical (contrast with Figure 4.14, where multiplexing and demultiplexing are done electronically, requiring electrial/optical conversion).

Figure 4.17 shows the frame structure of TPON for the signals from the central office to the customers. The frames are repeated 100 times per second, and each frame contains 80 slots. Thus, slots are repeated 8,000 times per second, so that each byte (8 bits) repeated in all the slots constitutes a 64-Kbps channel, which may be used for voice or data. Each 2,496-bit slot thus provides 2,352 8-Kbps channels for 128 users plus overhead channels. These channels can be packaged as voice channels, ISDN services, data channels, or their combinations. The system uses the additional bits in each slot and each frame for synchronization and maintenance. Some of these bits are also used for ranging measurements needed for multiplexing of the signals coming from the customers. As shown in Figure 4.18, the signals from the customers are multiplexed in time. The 128 subscriber units time their periodic transmissions so that their signals are interleaved without overlapping when they reach the center of the star (the central office). (Signal overlap is analogous to

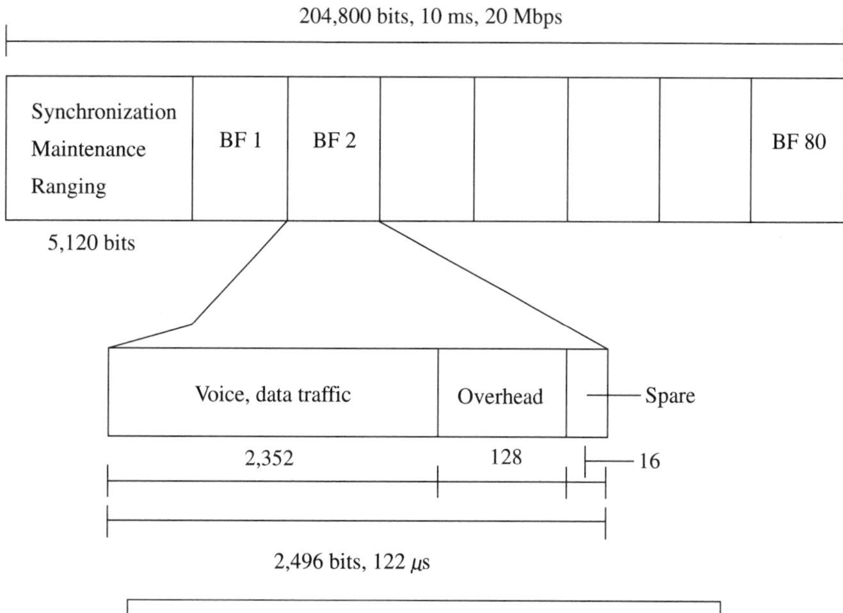

4.17 FIGURE The TPON frame structure.

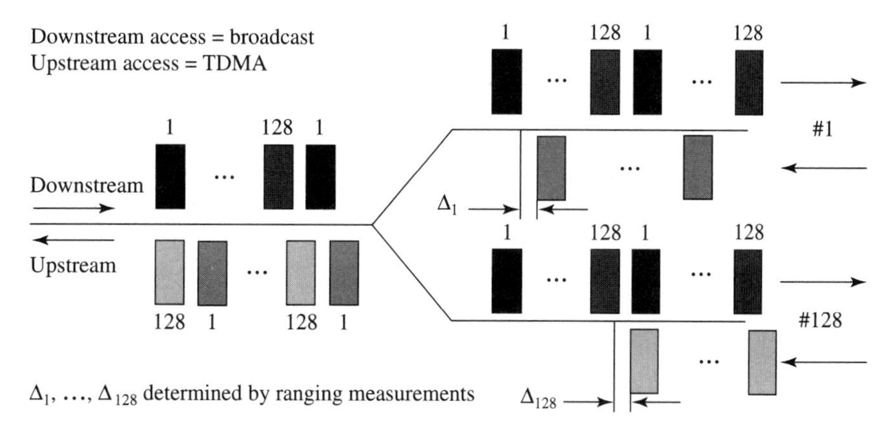

4.18 FIGURE Streams are multiplexed, without overlapping.

a collision on the Ethernet.) To reduce the chance of overlap, a "guard band" must be provided around each individual transmission, which wastes potential bandwidth.

As in the case of the AT&T SLC system, the multiplexing of signals of the 128 subscribers eliminates the need for individual line interface cards.

4.3.3 Passive Photonic Loop

The passive photonic loop (PPL) uses wave-division multiplexing or WDM instead of TDM. PPL allocates one pair of wavelengths of light to each subscriber. Thus n subscribers need $2n$ different wavelengths. A subscriber transmits using one wavelength and receives the other wavelength. A block diagram of the PPL system is shown in Figure 4.19. In the figure, subscriber 1 is allocated wavelengths λ_1, λ_{n+1}, subscriber 2 is allocated wavelengths λ_2, λ_{n+2}, and so on. The subscriber can modulate light of one wavelength to send information to others; and others can modulate light of the second wavelength to send information to the subscriber. There is a modulation bandwidth around each wavelength, which defines the maximum transmission speed. In this sense, WDM is similar to frequency-division multiplexing (FDM), described

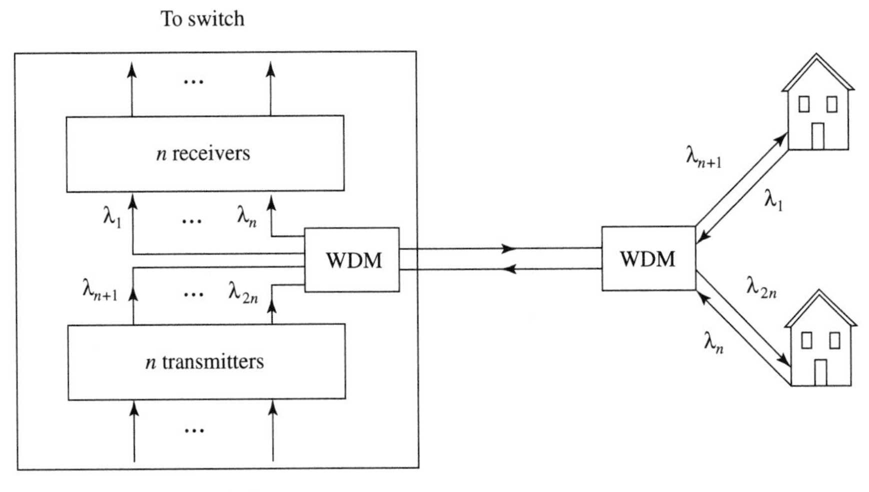

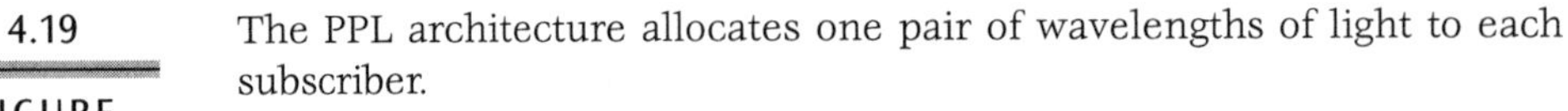

4.19 FIGURE

The PPL architecture allocates one pair of wavelengths of light to each subscriber.

in section 2.5.1. The bandwidth available to each subscriber is huge, on the order of hundreds of GHz, although current technology limits utilization to a few Gbps. (The technology is discussed in Chapter 9.)

The number of subscribers that PPL can accommodate is limited by the number of wavelengths that can reliably be differentiated using tunable lasers and optical filters. Current technology places an upper bound of 50.

PPL and other WDM-based networks today exist only as laboratory experiments. Many advances are needed before such networks will be commercially viable. However, as we will see in Chapter 9, WDM seems to be the only feasible modulation scheme that can effectively utilize the enormous bandwidth of optical fiber.

4.3.4 Hybrid Scheme

Hybrid schemes that use different forms of multiplexing are also possible. One such system is shown in Figure 4.20. In this system, a number of bit streams are multiplexed by ATM. The resulting ATM cell streams are multiplexed with some TV signals either by subcarrier multiplexing (described in Chapter 9) or by WDM. Such a system can be used to superpose CATV and ATM networks on the same fiber system. Since the systems can make use of

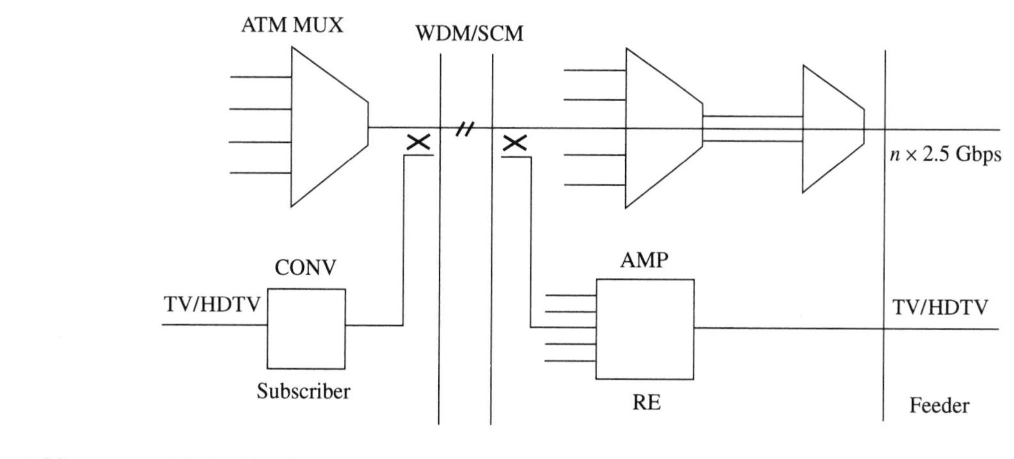

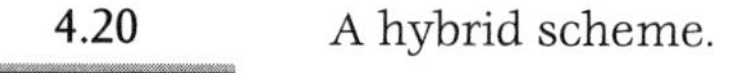
4.20 FIGURE A hybrid scheme.

parts of existing distribution systems for CATV and telephone, these systems may prove to be very cost-effective to deploy in the first stage of service integration.

4.4 ISDN

Telephone companies are currently implementing the Integrated Services Digital Network (ISDN) according to standards defined in successive recommendations culminating in [I120]. The objective of ISDN is to offer new digital transmission services to subscribers. The telephone network uses digital transmission for voice and a packet-switched X.25 network for the transport of signaling information. ISDN makes these internal transport facilities available to users as new services.

ISDN offers a variety of bearer services built on top of the first three OSI layers, higher-level services (called teleservices), and supplementary services. The teleservices are application layer services in terms of the Open Data Network model. The supplementary services are concerned with call control functions rather than communication per se, and do not fit directly in the Open Data Network model.

The main bearer services are the transport of audio and voice (at 64 Kbps), circuit-switched digital channels at rates that are multiples of 64 Kbps, packet-switched virtual circuits, and connectionless service (datagrams). The teleservices include telex, facsimile, videotex, and teletex transmissions with specific coding and end-to-end protocols. A message-handling service and a directory service are also being defined for ISDN. The supplementary services include telephone services such as caller identification, call forwarding, call waiting, and conference calling. Figure 4.21 gives a schematic of the ISDN architecture. Circuit-switched, packet-switched, dedicated point-to-point, and call-control (common channel signaling or CCS) services are brought together at an ISDN switch and accessed by the user through a common terminal equipment (TE).

The user interfaces to ISDN are defined as combinations of three types of channels: B, D, and H. (See Figure 4.22.) The B channel is a 64-Kbps channel that transports a circuit-switched connection, an X.25 service (packet-switched, virtual circuit), or a permanent digital point-to-point connection. The D channel is a 16-Kbps or a 64-Kbps channel used for signaling information (call control) and for low bit rate packet-switched services. An H channel is a 384-Kbps, 1,536-Kbps, or 1,920-Kbps channel used like a B channel but

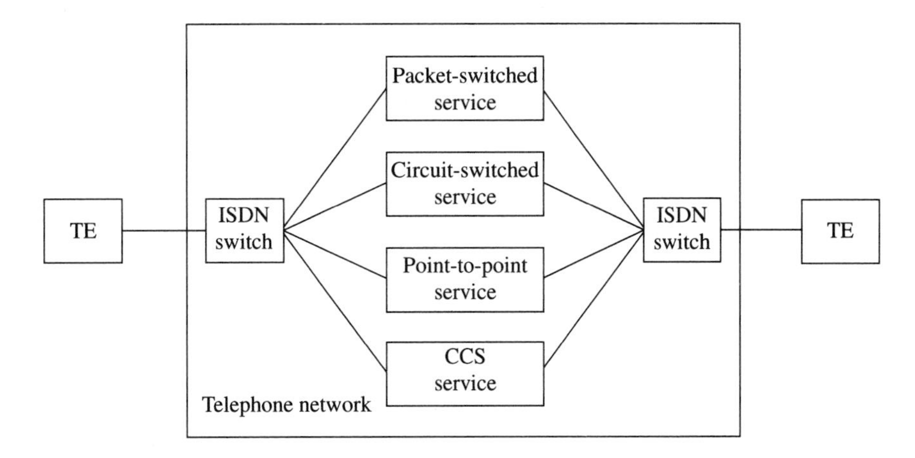

4.21 FIGURE ISDN architecture. Circuit, packet, unswitched point-to-point, and call-control services are accessed through a common interface.

for higher-rate services. The ISDN standards specify the basic access and the primary access for users. The basic access is 2B + D; it consists of two full-duplex B channels and a full-duplex 16-Kbps D channel. The primary access is 30B + D (64 Kbps) in Europe, and it is 23B + D (64 Kbps) in the United States, Japan, and Canada.

Channel types:

- B: 64 Kbps CS, X.25 (PS, VC), or permanent signaling and low rate PS
- D: 16 Kbps or 64 Kbps
- H: 384 Kbps, 1,536 Kbps, or 1,920 Kbps as B

Basic access:

- 2B + D (16)

Primary access:

- 30B + D (64) in Europe
- 23B + D (64) United States, Japan, Canada

Basic interface:

- Pseudo-ternary (1 = 0 V, 0 = ± 0.75 V alt.)
- Frame level: 192 Kbps, sync., DC balancing (144 Kbps)
- Data link:
  - B/PS: LAPB (GBN, ACK + NACK)
  - B/CS and Permanent: user's choice
  - D: LAPD: acknowledged = GBN, VC unacknowledged: datagram with discard

Network: routing, mpx, congestion control, call control

4.22 FIGURE ISDN services and standards.

The ISDN standards specify a network-user interface that can be accessed directly by ISDN terminal equipment such as digital telephones and, via terminal adapters, by other devices such as computers.

We now summarize some of the ISDN standards for implementing the lower three OSI layers. For basic access, the physical layer of ISDN specifies an 8-pin connector to attach to the network, a pseudo-ternary encoding (1 is represented by 0 volt and 0 by alternatively +0.75 volt and −0.75 volt), a frame format that includes synchronization and DC-level balancing bits, and a line rate of 192 Kbps corresponding to the 144 Kbps of user data rate ($2 \times 64 + 16$ Kbps) plus the overhead bits. In addition, the physical layer specifies a contention-resolution protocol for access to the D channel by up to eight terminals attached to a common (multidrop) line. (See Figure 4.22.)

The data link layer of ISDN is LAPD for the D channel and LAPB for packet-switched connections on the B channel. For circuit-switched or permanent connections on the B channel, the users can choose the data link protocol and can use the I.465/V.120 protocol defined by the CCITT for such connections.

LAPD provides unacknowledged and acknowledged information-transfer services. The frame structure is essentially that of X.25: bit-oriented frames that start and end with an 8-bit flag that is avoided inside the frame by bit-stuffing; a 16-bit CRC is used for error detection; 16-bit addresses are used to distinguish users connected to the same interface and different connections with a given user (i.e., different service access points). The unacknowledged service is implemented as a datagram; erroneous frames are discarded. The acknowledged service is implemented as a virtual circuit with Go Back *N* link error control. The receiver can turn the sender off and on by sending it "receiver not ready" and "receiver ready" frames.

The I.465/V.120 data link protocol is a modified version of LAPD that provides asynchronous data transfer, HDLC synchronous data transfer, and bit-transparent asynchronous transfer. To use this transfer protocol, the users first set up a circuit on the B channel by using the D channel. When the transfer is complete, the users release the circuit also by using the D channel.

The network layer of ISDN specifies the routing, multiplexing, and congestion control, in addition to the call-control messages.

In summary, ISDN is an attempt to diversify the bearer services offered by the telephone network. The different services are provided by different networks (rather than in one single network) accessed through a common ISDN switch. Diversity is limited because these services are built on top of the traditional 64-Kbps channels, which constrains the bit stream rates that can be supported.

4.5 INTELLIGENT NETWORKS

In our historical summary in section 1.1.1 we noted that modern switches in the telephone networks are programmable computers, which makes them very flexible. By sending instruction to a switch, one can modify its configuration. This contrasts with the earlier electromechanical switches, whose functions were built into their hardware. In modern switches, the control is separated from the hardware that executes the elementary switching operations.

This separation of control and basic operations is also present in the other network elements. As a result, these elements, too, are programmable. The separation enables telephone companies to develop their own services and implement them on switches and other network elements obtained from different equipment vendors.

Intelligent Network or IN is the name given to this network of programmable elements, organized to facilitate introduction of new services. In this section we explain an architecture model for intelligent networks proposed by telecommunication engineers. Since 1985, the IN concept has evolved through several proposed architectures. A number of telephone companies have implemented their own version of IN. Some wireless network operators are also implementing IN for mobile subscribers. Finally, the capability to implement new services that IN offers to a telephone company can in part be delegated to customers who can use the capability to design their own services. In section 4.5.1 we examine services that the telephone companies provide. We explain how the provision of a service can be decomposed into a sequence of basic steps. That is, service provision can be viewed as a script, or program, whose elementary steps are operations performed by the network elements. In section 4.5.2 we examine an architecture model for intelligent networks. In section 4.5.3 we list the basic actions the network performs when it implements a service.

4.5.1 Service Examples

Figure 4.23 illustrates the plain old telephone service or POTS. POTS is the basic telephone call service. When the network implements a telephone call, its elements execute the sequence of steps listed in the figure. In the old telephone network, with electromechanical switches, the call functions of the network elements are wired into those elements. For example, when a user

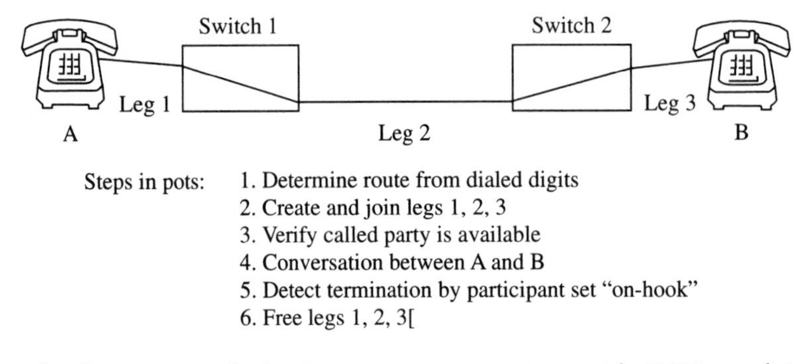

4.23 FIGURE

The figure notes the basic steps necessary to provide POTS or plain old telephone service.

dials a telephone number, the successive digits generate a string of impulses that activate a sequence of rotary switches in the central offices of the telephone company. When an electrical current is sent to an on-hook phone, this current activates the bell. This arrangement of the hardware is inflexible. For instance, a customer cannot instruct the switches to block calls dialed from specific numbers or to forward a call to another number at certain hours of the day.

Figures 4.24 and 4.25 illustrate the operations performed by the modern network when it implements the *call forwarding* service. A customer using call forwarding can instruct the telephone network to forward an incoming call to another telephone set when the normal phone is not picked up after it rings (say) three times.

Figure 4.24 shows two telephone sets A and B connected via two switches 1 and 2. The customer at set A calls set B. The customer at B has instructed the telephone network to forward a call to some set B′ after three rings. This instruction is stored in a database of switch 2 to which set B is attached. Moreover, the phone number of set B is stored in a *trigger table*. When a call for a telephone set reaches switch 2, the switch checks its trigger table. If that set is not in the trigger table, then the call is handled in the standard way. If the set is in the trigger table, then the switch places a request to its database. The database responds to the request by providing the instructions that the switch must follow to handle the call. Thus, when the call for set B reaches switch 2, the switch is triggered, and it requests the instructions from its database. Note that the telephone network can implement such services because the controls are separated from the actual operations of the switches.

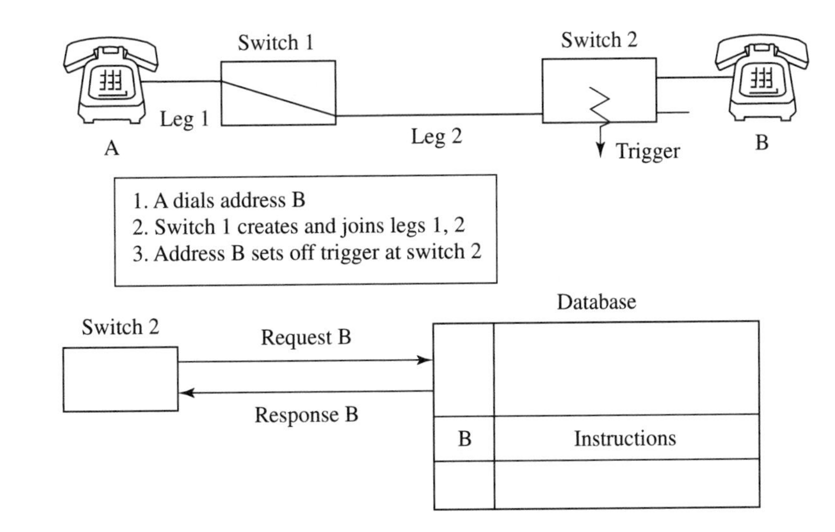

4.24 FIGURE

Call forwarding: when a call for B reaches switch 2, the switch searches a database for instructions.

4.5.2 Intelligent Network Architecture

The *Intelligent Network Architecture* (INA) is a model for organizing the programmable network elements and the communication between those elements. The objectives of that model are in part the same as those of the OSI

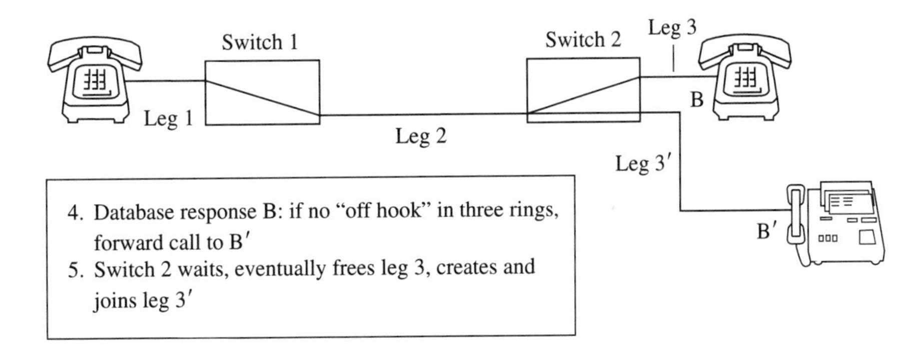

4.25 FIGURE

Call forwarding, continued: if the set B is not picked up after three rings, the call is forwarded to B′.

reference model for computer networks: decomposition and standardization. INA is more complex in that it involves not only communication protocols and data elements as OSI does, but other network resources, including hardware elements. On the other hand, the OSI model is quite abstract, whereas the INA architecture is specifically designed for telephone networks.

As we discussed above, the main feature of an intelligent network is that the *control functions* and the *network resources* are separated. The network has transmission and other resources, including subscriber lines, trunk lines, switch ports, databases, and voice recorders. The control functions are call-control functions and resource-control functions. Examples of these functions are: connecting three network users in a conference call, playing a recording, and collecting digits dialed by a user. To implement a control function, the network executes a sequence of atomic operations called *functional components.*

INA classifies the network elements into three types: *service switching points* (SSPs), *service control points* (SCPs), and *intelligent peripherals* (IPs). (See Figure 4.26.) These elements exchange information over the *common channel*

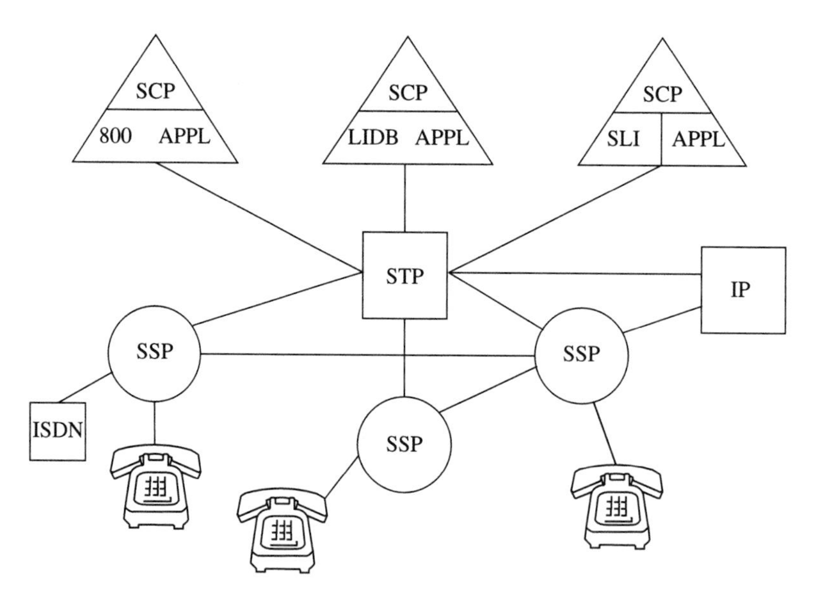

4.26 FIGURE

INA is a model for organizing network elements and the communication between them.

signaling (CCS) packet-switched network, using the *signaling system 7* (SS7) protocol. INA calls the CCS network switches *signal transfer points* (STPs).

The figure shows three telephone sets and one IP attached to the network of SSPs. The figure represents the links used by the voice signals by regular lines. The bold lines represent links of the CCS network. The SCPs contain databases and instructions for special applications.

The network elements are capable of performing different functions. The SSP can detect the need for IN service based on an originating line trigger such as off-hook, triggers applying to all calls such as for 800 or 911 calls, and terminating line triggers such as for call forwarding users. When such a trigger is activated, the SSP sends a request to the SCP for instructions. The SSP can identify, monitor, and allocate transmission resources (legs) connected to it. An IP can identify, monitor, and allocate nontransmission resources connected to it. The SCP detects IN service requests forwarded by the SSP. It then interprets the request according to a *service logic program* (SLP). A *service logic interpreter* (SLI) executes the SLP. A given SLI can execute multiple SLPs concurrently. The execution of the SLP involves invoking functional components and monitoring resources. Finally, at completion of the execution, the SLI notifies the SSP.

4.5.3 Functional Components

We briefly describe the functional components, the atomic actions on network resources, in five types of network operations.

Control of Processing

The functional components for control of processing are of two types: to provide instructions when the SSP asks the SCP to take control of the call processing and to effect the release of control when the SCP returns the control to the SSP after servicing the request.

Connection Request

A connection request involves the following functional components: creating a leg between an SSP and another network element, joining a leg to an ongoing call, splitting a leg from an ongoing call, and freeing a leg to release the resource.

User Interaction Request

Two types of functional components are invoked in user interactions: sending information such as a prerecorded announcement to a call participant and receiving specific information such as dialed digits from a call participant.

Network Resource Status Request

Network resource status requests are used by the SLP in processing some call control. Monitoring is a functional component that instructs the SSP for notification of a particular event on a specified leg, such as on-hook, flash-hook, and off-hook.

Network Information Revision Request

Network information revision requests enable the SLP to change the data stored in the SSP tables.

In summary, INA is a culmination of a long development in which network element functions or operations are separated from the control of those functions. This separation permits the creation of new services as programmable sequences of functional components. Sophisticated customers can program these sequences by themselves. A very important example is 800 number services. Companies, such as credit card and direct order companies that provide direct customer services over the 800 number phone, can route customer calls to different parts of the country or to different operators depending on the time of day, the subscriber's location, and other information provided by the subscriber via the telephone keypad.

4.6 VIDEO DIAL TONE

The standard CATV distribution system is illustrated in Figure 4.27. The head station distributes analog video programs. Several programs are frequency-division multiplexed at the head end. Amplifiers are inserted into the distribution cable in order to boost the CATV signal. Every subscriber receives the same programs.

A video dial-tone network provides its subscribers access to video connections and a signaling system to select video programs. In a likely implementation, the users select programs by browsing menus on the screen of their television set with a remote control. The programs consist of video clips of news items, music videos, educational material, or other programs. The sys-

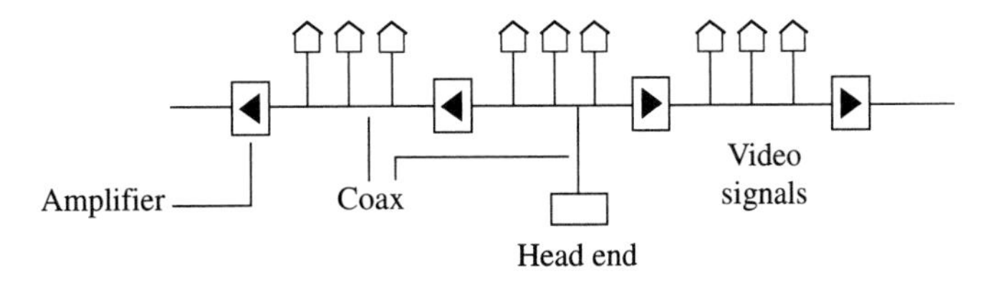

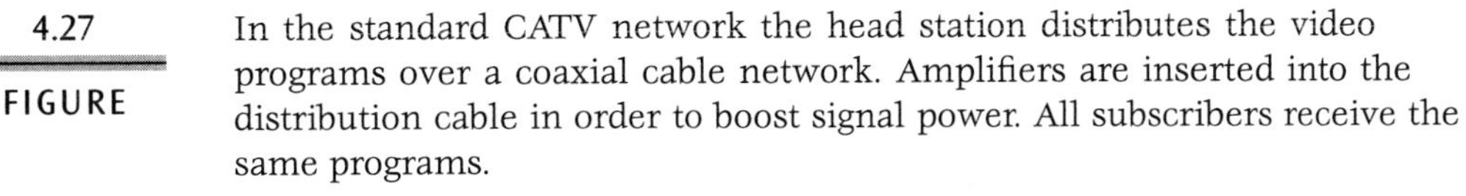

4.27 FIGURE In the standard CATV network the head station distributes the video programs over a coaxial cable network. Amplifiers are inserted into the distribution cable in order to boost signal power. All subscribers receive the same programs.

tem can also be used to deliver CD-quality audio programs and videotex pages. The most straightforward implementation of the video dial-tone network will involve a distribution system with fiber to the curb and a local coaxial network.

In this section we discuss some features of proposed implementations of a video dial-tone network. We start with an overview of the physical layout of such a network in section 4.6.1. We then discuss the control and video networks in section 4.6.2. In section 4.6.3 we elaborate on the MPEG standards used for video compression.

4.6.1 Layout

The physical layout of a video dial-tone network is sketched in Figure 4.28. The network between the head stations and the users consists of fiber to the curb attached to a local coaxial network. The head stations of the service providers are connected by a backbone network, not shown in the figure. The lower part of the figure shows two generations of technology. The second generation proposes a much larger bandwidth in the reverse direction.

The head stations are equipped with video servers controlled by a computer. The video servers consist of a combination of VCR, laser disk, and CD-ROM changers, and hard disk arrays. User control devices are set-top boxes.

The video and control signals are both digital and travel in opposite directions, as illustrated in Figure 4.29. The video and control signals occupy different frequency bands, as shown in the lower panel of Figure 4.28. This separation of frequency bands prevents the amplifiers from amplifying their own output, which would lead to saturation of the amplifiers and to oscillations.

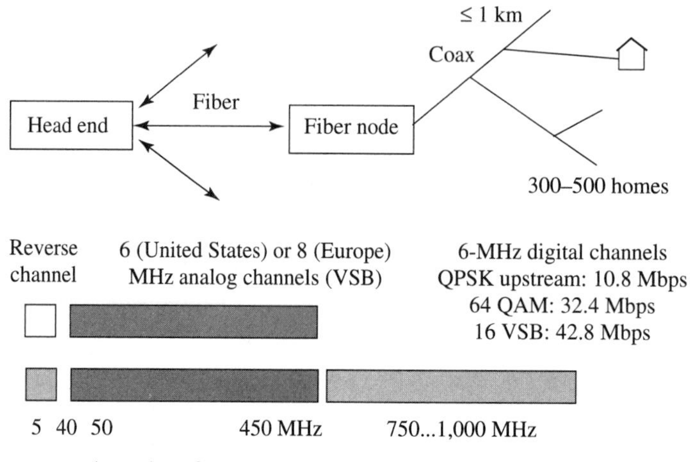

4.28 FIGURE

The network connects the head end stations of the service providers to the user equipment. The video signal is carried over an optical fiber to the curb, where it is converted into an electrical signal and distributed over coaxial cable. User control signals use the same physical network.

Figure 4.30 gives an alternative distribution technology in which all or part of the distribution system is wireless. In the wireless cable system, subscribers directly access the signal broadcast from the head end station. In the hybrid/fiber wireless system a digital video signal is sent to the curb over optical fiber (as in Figure 4.28), and the local coaxial distribution system is replaced by a local wireless system. The wireless portion of these networks would extend over short distances. These systems may be less expensive than cable when there is a high geographical concentration of users.

4.6.2 Control and Video Networks

The network combines two types of services: the exchange of control messages between the users and the head stations and the transmissions of video programs (and possibly other programs).

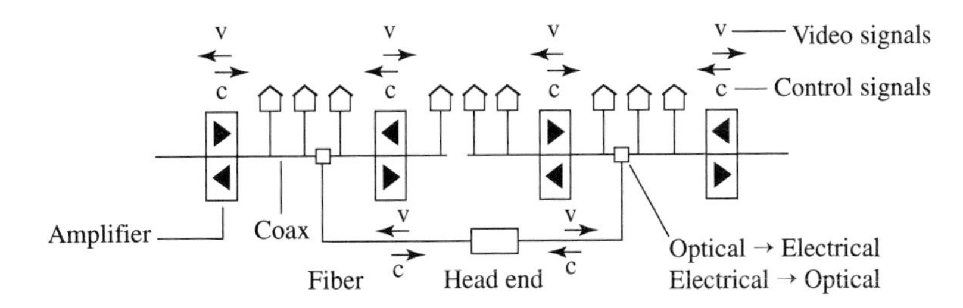

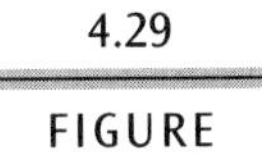
4.29
FIGURE

The network connects the head stations of the service providers to the user equipment. Video signals are sent downstream from head stations to users, and user control signals travel upstream to head stations.

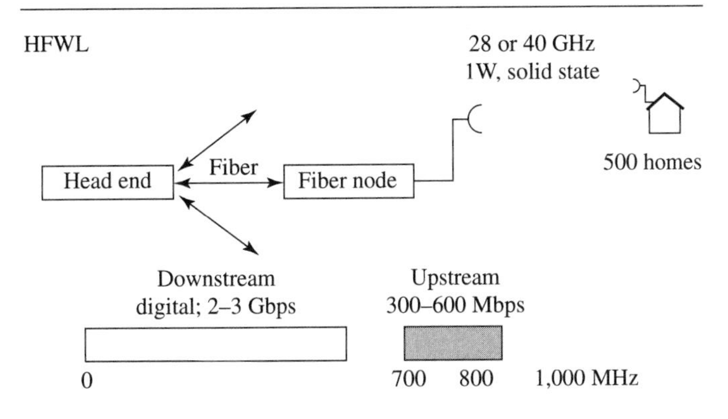

4.30
FIGURE

The wireless cable (WLC) system replaces the current distribution system. The hybrid/fiber wireless (HFWL) system replaces the local coaxial distribution.

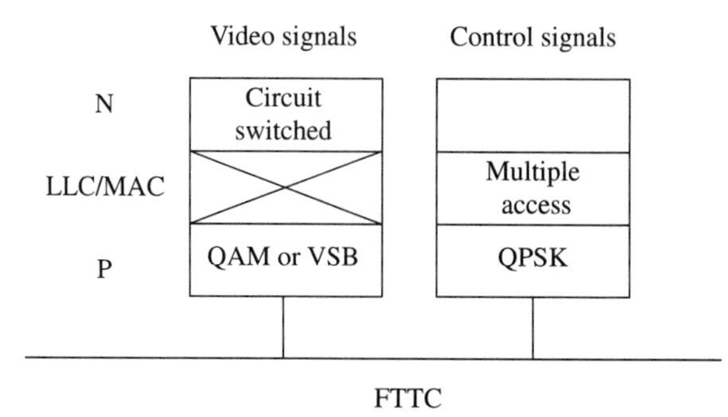

4.31 FIGURE Plausible three-layer decomposition of network functions. The video signals are carried by a circuit-switched network. The control signals are sent using a multiple access scheme, e.g., Ethernet.

Figure 4.31 sketches a plausible three-layer decomposition of the functions of the fiber-to-the-curb (FTTC) network. The transmissions of control messages and of programs involve different operations. The video signals are circuit switched. At the physical layer either QAM (quadrature amplitude modulation) or VSB (vestigial side band) modulation may be used. The control signals are exchanged using some multiple-access network, e.g., Ethernet. The bits are modulated using QPSK (quadrature phase-shift keying). QPSK is a modulation scheme that converts groups of n bits into sine waves with a different phase for each of the 2^n n-bit words.

The advantage of QPSK is that by choosing the carrier frequency of the sine waves judiciously, the network engineer can make sure that the modulated bit stream does not interfere with the video signals. Another advantage is that this modulation scheme requires only a small range of frequencies to transmit the bits. Only a small range of frequencies is generated because the modulated signal changes only once every n bits.

The head station first compresses the video signals for the different users according to the MPEG standards. The head station then modulates the different bit streams on different frequencies, using QPSK or similar modulation methods. The resulting signal modulates a laser diode, which produces an optical signal. Optical fibers carry the optical signal to a number of optical receivers, which convert the optical signal back into an electrical signal and transmit the latter over the local coaxial cable distribution plant. The set-top box of a user demodulates the signal intended for that user, decompresses it, and generates the NTSC or HDTV signal for the TV set. Such MPEG decompression equipment is already used by direct-broadcast satellite TV systems. We discuss MPEG compression in the next section.

The combination of modulation of different signals on different frequencies and transmission by an optical fiber that this network uses is called subcarrier modulation (SCM). We discuss SCM in more detail in Chapter 9.

The video dial-tone network permits users to send and receive messages; the network can also be used to carry videotex, data, digital audio, and other programs. The main limitation of the system is that the transmission rate available to each user is much smaller than the rate the user can receive. The key advantage of the video dial-tone network is that it uses the CATV technology with the relatively minor modification of adding an Ethernet or some other multiple-access network for the control messages. (Recent reports suggest, however, that CATV companies are encountering difficulties with these modifications.) The rate of the information streams that a user receives is orders of magnitude larger than the rate Internet users receive.

A backbone network connects different head stations. Live events and other programs can be broadcast over that network and made available by the local head stations. A program or information that is not available locally can be made available by the backbone network. The users can search remote databases via control messages carried by the backbone network.

4.6.3 MPEG

The Moving Pictures Expert Group has defined a number of standards for video compression. The MPEG compression algorithms use the redundancy within each frame and the similarity of successive frames.

To compress video, an MPEG algorithm first calculates the discrete Fourier transform of a frame (discrete cosine transform or DCT). The algorithm then encodes the resulting cosine coefficients using Huffman encoding. Huffman encoding assigns shorter codes to values of the coefficients that occur more frequently so as to minimize the average number of bits required to encode a frame. These encoded frames are called I frames.

The second method that MPEG uses to reduce the number of bits is motion compensation, illustrated in Figure 4.32. The figure shows two successive frames of a video. The algorithm decomposes frame n into blocks. The algorithm then examines frame $n + 1$ and finds the translation of the blocks of that frame that makes them most similar to the corresponding blocks of the previous frame. Instead of transmitting all the blocks, the algorithm transmits the displacement vectors and the differences between a block in a frame and the corresponding block in the next frame after translation. The frames that contain this motion-compensation information are called P frames. Finally, an

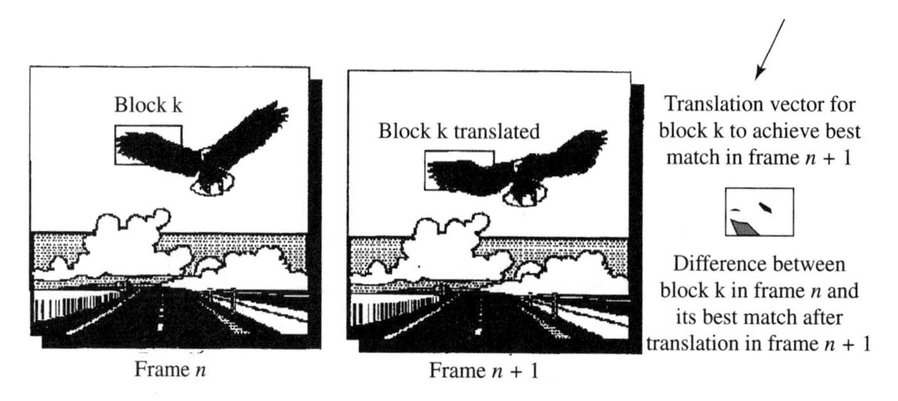

4.32 FIGURE

The motion-compensation algorithm compares successive frames. A frame is divided into blocks and the corresponding blocks in the next frame are moved until they match those in the previous frame best. The algorithm transmits the block displacement vectors and the residual differences between the blocks.

MPEG algorithm can use interpolation frames, called B frames. The B frames perform an interpolation between P frames.

Figure 4.33 shows the sequence of frames produced by MPEG2. Refresh frames are sent twice every second to avoid error propagation. The number of B and P frames can be selected to produce different qualities and resulting bit rates.

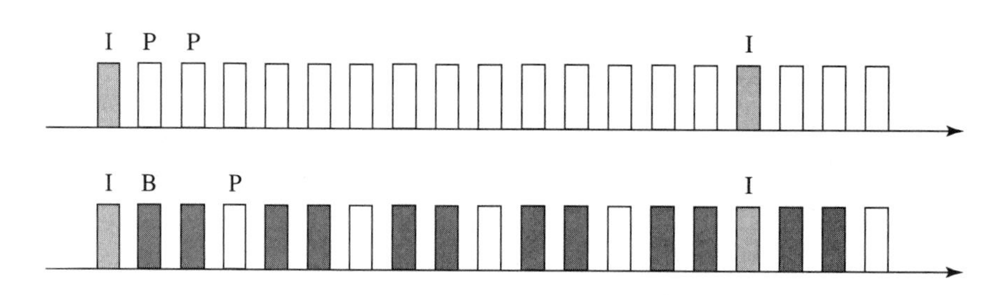

4.33 FIGURE

Sequence produced by an MPEG2 algorithm. This sequence consists of refresh frames (I), motion-compensation frames (P), and interpolation frames (B).

4.7 SUMMARY

The bearer service offered by circuit-switched networks is fixed bit rate end-to-end connections with a delay equal to the propagation delay. This bearer service is well suited for constant bit rate traffic. The service can of course be used to support both variable bit rate and message traffic. However, used in that way the utilization of the network link capacities may be very low.

The telephone and CATV networks are the most important networks in terms of number of customers and ability to finance new investment. Recent innovations will enable these networks to provide new services that can propel them well beyond their traditional markets.

Innovations in the telephone network have led to (1) higher bandwidth links in the backbone network through SONET; (2) higher speed local loop using optical fiber; (3) integrated services, providing a common interface for message transmission over packet-switched networks and constant bit rate circuit-switched connections; and (4) Intelligent Network Architecture, permitting a very flexible means to create new user services.

SONET implements the bit way layer of the Open Data Network model, taking advantage of the economies of scale offered by optical communication. The optical local loop, perhaps more expensive than today's copper wire, will bring to the user the high-speed access necessary for video applications. Lastly, ISDN and INA exploit the economies of scope by integrating packet- and circuit-switched services and other communication-related services such as call forwarding.

Innovations in CATV have made possible (1) a more efficient use of bandwidth by digital compression; (2) increased bandwidth by use of optical fiber for transmission to the curb; and (3) transmission of information by the subscriber. With these innovations CATV can begin to compete for local telephone services.

Despite their innovations, packet-switched networks are best suited for message traffic, and circuit-switched networks are best suited for constant bit rate traffic. Both types of network are still not well suited for variable bit rate traffic. That traffic is best carried by ATM networks, which we study in the next chapter.

4.8 NOTES

The SONET/SDH framing conventions are described in [BC89, S92, B95]. A discussion of SONET rings and other architectures that improve network survivability appears in [WL92, W95]. The AT&T SLC 5 is briefly described in [C89]. The British TPON system is described in [R91]. The passive photonic loop is described in [WL89]. Promising new local loop technologies that can provide high-speed access are described in [CM91]. ISDN standards are described in [S92]. See [F95, P95] for a recent appraisal of the market penetration of ISDN. An issue of *Communications Magazine* [CM92] contains good discussions of IN in the United States and several other countries. For a description of video dial tone see [CM94] and [IN95]. The MPEG standards are discussed in [S96].

4.9 PROBLEMS

1. For a network like the one in Figure 4.1, let $\{C_{ij}\}$ denote link capacities. Suppose $C_R = \{C_r\}$ is a set of route calls, and say that C_R is feasible if (4.2) holds. If there are N routes in R, then C_R is an N-dimensional vector.
   (a) Show that the set of feasible vectors is convex.
   (b) Suppose that a connection over route r brings a revenue of p_r per unit of time. Show that the set of route calls C_R^*, which maximizes revenue, can be obtained as a solution of a linear programming problem.
   (c) As each call request arrives, the call admission algorithm decides whether to admit or reject the call. We would like to design an algorithm that maximizes revenue. This problem illustrates the difficulties in designing such an algorithm. Consider the network shown in Figure 4.34 with three links, with capacities as shown, and with two routes. Show that the set of feasible calls is given by the convex region on the right. Design an admission procedure so that the system will operate near the desired point (4,6). Consider two cases. In the first case, existing calls can be terminated (so-called preemptive case); in the second case, existing calls cannot be terminated (nonpreemptive case).
   (d) Planning problems are concerned with expanding the network capacity to meet growing demand. Suppose that this growth is described as an increase of Δ_r calls along route r, $r \in R$. Suppose this increase is to be met by expanding the capacity of each link (i, j) by amount $\delta_{ij} \geq 0$ at a unit cost of p_{ij}. Formulate the minimum cost expansion as a linear programming problem.
2. In the preceding question, it is assumed that the set of routes has already been selected and the number of simultaneous calls C_r on each route

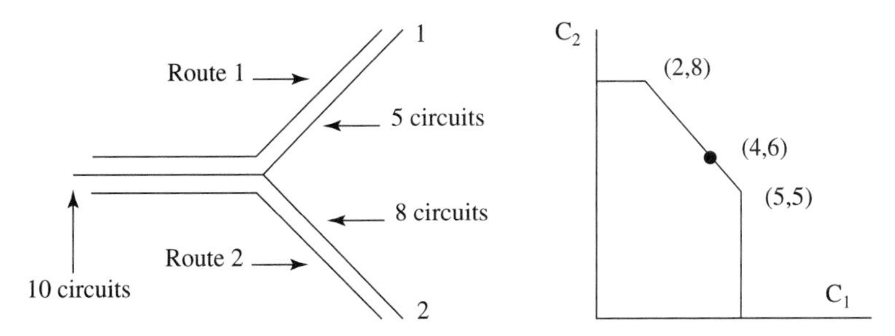

4.34 FIGURE

The panel on the left shows a network with three links and two routes. The set of feasible calls is the convex region shown on the right.

$r \in R$ is given. But the route selection and assignment may be changed. Suppose the link capacities C_{ij} are given. Suppose we wish to route N_{xy} simultaneous calls from every originating switch x to destination switch y. Formulate a linear programming problem whose solution gives a routing assignment.

3. A cellular telephone network consists of a number of base stations (switches) connected by wired links. A user within a cell served by a particular base station gains access to that station using an idle radio or wireless link. If all the access links are busy, the user's request is blocked. See Figure 4.35. The number of access links to each base station is fixed by government regulations. The number of mobile users varies randomly.
   (a) Formulate a model to determine the blocking probability.
   (b) Suppose a call is placed from a mobile in one cell to another mobile in another cell. The route for such a call would have three parts: a radio access link in the first cell, a route over the wired links from the first base station to the second, and a radio access link in the second cell. What is the blocking probability of the call in terms of the blocking probabilities for each part of the route?

4. The textbook time-division multiplex scheme assumes that all the multiplexed signals have identical bit rates or frequency (see Figure 2.9). Since each signal comes from a separate source, this assumes that the clocks at all those sources are perfectly synchronized. This is impossible in practice, and so, over time, the clocks will drift apart. The more accurate the clocks, the smaller will be the drift. If we assume that the clock accuracy is on the order of 10^{-n} (i.e., they drift apart by one out of every 10^n seconds), and if we assume a bit rate of 10^m bps, then these signals will

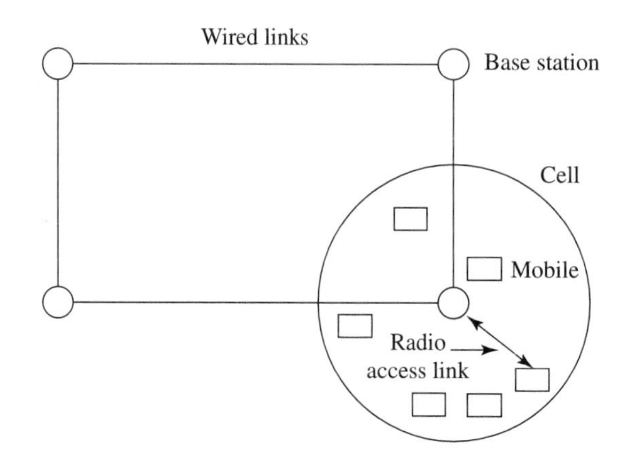

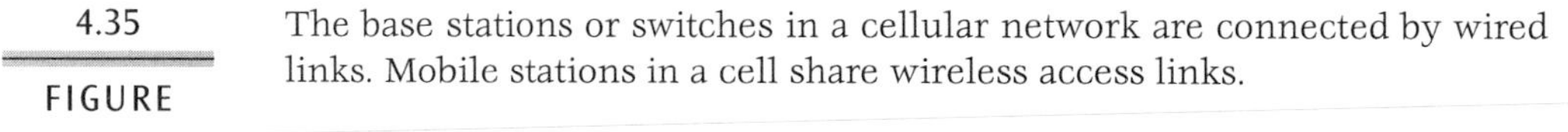

4.35 FIGURE

The base stations or switches in a cellular network are connected by wired links. Mobile stations in a cell share wireless access links.

drift by 1 bit every 10^{n-m} seconds. For example, if we assume a clock accuracy of 10^{-12} (very high) and bit rate of 10^9 bps, there is a drift of 1 bit every 10^3 seconds (about 1 bit per hour). If we assume a clock accuracy of 10^{-4} and bit rate of 10^6, there is a drift of 100 bits every second. Use those concepts to analyze how frequently it will be necessary to use the frequency justification procedures of Figure 4.11.

5. The frame structure in X.25 uses an 8-bit flag. Suppose it is 01111110. This 8-bit pattern may be present in the data, in which case it may be interpreted as the flag. To avoid this confusion, whenever a pattern of six 1s appears in the data it is followed by a seventh stuffed bit of 1. Show that the data will never carry the flag pattern. Also explain how the original data can be recovered.

6. Design a caller ID service using the functional components of section 4.5.3.

5 CHAPTER Asynchronous Transfer Mode

In Chapters 3 and 4 we studied packet- and circuit-switched networks. Those networks are suitable for message and constant bit rate traffic, respectively. In this chapter we examine the Asynchronous Transfer Mode or ATM networks. ATM networks combine the good features of both types of networks, making them suitable for variable bit rate (VBR) traffic. ATM networks potentially can provide bearer services with a specified quality of service to meet the needs of all traffic types. Whether this potential can be realized depends on how well the problems of management and control of ATM networks can be solved. Those problems are discussed in Chapters 6 and 7.

This chapter presents the concepts of ATM. It offers a simple model to calculate the delay of an ATM network. By the end of this chapter you will understand the ATM layered architecture, the addressing and routing standards, and the formats for different services. You will also know the important proposals for ATM LANs and for IP service over ATM. With these proposals, ATM becomes backward compatible with existing LAN equipment and IP software.

We saw in section 2.1 that different applications impose different performance requirements on the network bearer services in terms of delay, bandwidth, and loss. If all these applications are to share the same network resources (links, buffers, switches)—and this is very desirable to gain the economies of scale and service integration—the network must be able to allocate its resources differently to different applications. Because switches or routers in datagram networks do not have connection state information, they cannot differentiate among packets by application. Therefore, datagram networks cannot discriminate among applications by quality of service.

Circuit-switched networks do maintain connection information. But since they provide a fixed set of resources to every connection or call, they, too, cannot differentiate among connections.

In the mid-1980s some engineers argued that virtual circuits were ideal for the efficient utilization of network resources when applications have widely different performance requirements. In a virtual circuit network, the nodes can set aside resources for specific connections and they can also discriminate among connections in order to meet their different requirements. Those arguments culminated in the development of a new set of standards for a class of virtual circuit networks called ATM networks. ATM networks seek to provide the end-to-end transfer of fixed-size packets or cells over a virtual circuit and with specified quality of service (in terms of delay, speed, and error rate).

We describe the main features of ATM networks in section 5.1. In section 5.2 we present the structure of the ATM header. In section 5.3 we discuss the ATM adaptation layer, which converts information streams in a variety of forms into a sequence of ATM cells. In order to satisfactorily serve a wide range of applications, ATM networks must be appropriately controlled. Standards relating to the management and control of ATM networks are discussed in section 5.4. ATM networks are designed to support both real-time applications such as video connections and telephone services and non-real-time applications such as e-mail and file transfers. In section 5.5 we study how ATM may support Broadband Integrated Services Digital Networks (BISDNs). In section 5.6 we discuss internetworking with ATM. Our objective is to explain how the familiar TCP/IP applications can be supported by an ATM network. Section 5.6 also covers LAN emulation by ATM networks.

We provide a summary in section 5.7.

5.1 MAIN FEATURES OF ATM

Four features distinguish the ATM bearer service from other bearer services:

1. the service is connection-oriented, with data transferred over a virtual circuit (VC);
2. the data is transferred in 53-byte packets called *cells;*
3. cells from different VCs that occupy the same channel or link are statistically multiplexed;
4. ATM switches may treat the cell streams in different VC connections unequally over the same channel in order to provide different qualities of service (QoS).

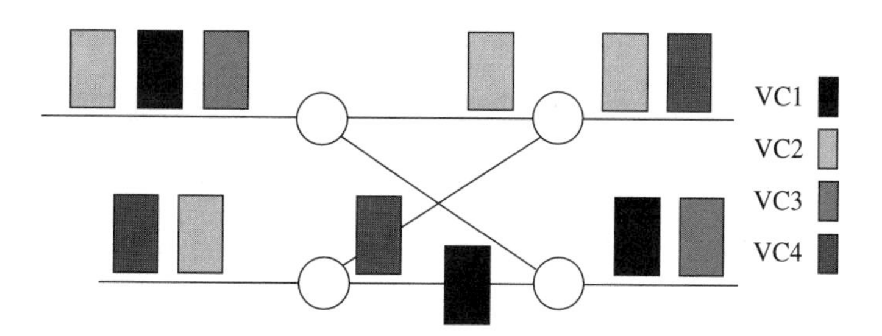

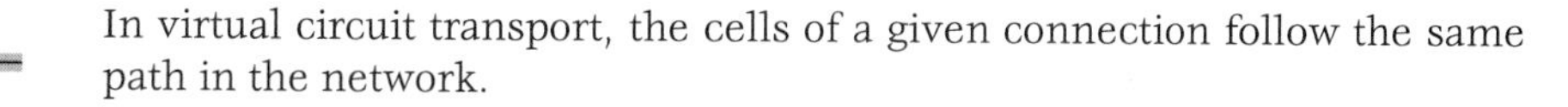

5.1 FIGURE In virtual circuit transport, the cells of a given connection follow the same path in the network.

We emphasize that ATM is a *bearer* service and *not* a bit way in terms of the Open Data Network model of section 2.7. The actual transfer of the bits that constitute the cell may be carried out in very different bit ways, ranging from SONET to DS-3 channels to proprietary bit ways. We now discuss the advantages and disadvantages of the four features listed above.

5.1.1 Connection-Oriented Service

In a connection-oriented service over a virtual circuit, the data stream from origin to destination follows the same path. See Figure 5.1. Data from different connections is distinguished by means of a virtual channel identifier (VCI). A connection over a virtual circuit is called a *virtual channel* in the ATM terminology. Thus each cell incurs an overhead corresponding to the length (number of bits) of the VCI, which is generally much smaller than the length of a full source/destination address needed in a datagram service. Second, cells in the same connection reach the destination in the order they are sent from the source, thus eliminating the need for sequence numbers and for buffering packets at the destination if they arrive out of order. (However, sequence numbers are necessary if the application layer must detect cell loss. We see this in our discussion of the adaptation layer, in section 5.3.) More importantly, ATM switches can identify different connections by their VCI. Consequently, the switches potentially can discriminate among different connections.

This potential can be used in many ways: admission control (refusing certain connections if sufficient network resources are unavailable), congestion control (limiting the amount of traffic accepted from a connection), resource allocation (negotiating the bandwidth and buffers allocated to a connection), and policing (monitoring the burstiness and average rate of traffic in a connection). We will study these different modes of control in Chapters 6 and 7.

The main disadvantage of connection-oriented service is that the network must incur the overhead of connection setup even when only a few cells are to be transferred, which could be done more efficiently by a datagram service. Another disadvantage is that a link or node failure terminates the virtual channel, whereas such a failure affects only a few packets in a datagram network.

The ATM Forum specifies five categories of services that an ATM network can provide: constant bit rate (CBR), variable bit rate–real time (VBR-RT), variable bit rate–non-real time (VBR-NRT), available bit rate (ABR), and unspecified bit rate (UBR). These services differ in the parameters of the quality of service and of the traffic that they specify. The parameters of the traffic are defined by an algorithm—called the generalized cell rate algorithm (GCRA)—that controls the arrival times of cells. These parameters are the following:

- peak cell rate (PCR),
- sustained cell rate (SCR),
- cell delay variation tolerance (CDVT),
- burst tolerance (BT),
- minimum cell rate (MCR).

We examine these parameters in Chapter 6. For now, here is a brief discussion of their meaning. PCR is the reciprocal of the minimum time between two cells. SCR is the long-term average cell rate. CDVT measures the permissible departure from periodicity of the traffic. BT measures the maximum number of cells in a burst of back-to-back cells. MCR is the reciprocal of the maximum time between two cells.

For CBR traffic, PCR measures the maximum rate, and CDVT specifies the acceptable jitter, i.e., departure from strict periodicity at rate PCR. For VBR traffic, the key parameters are SCR and BT, which measure the long-term rate of the traffic and its burstiness. For ABR traffic, the cell rate is between MCR and PCR. For UBR, no parameters are controlled, but no QoS is guaranteed.

Thus, CBR is an essentially periodic stream of cells with some acceptable jitter that may be unavoidable because of framing or packetization. VBR is a bursty traffic that may be generated by a variable bit rate codec. ABR and UBR are irregular data traffic.

Attribute	ATM Layer Service Characteristics					Parameter
	CBR	VBR(RT)	VBR(NRT)	ABR	UBR	
CLR	Specified	Specified	Specified	Specified	Unspecified	QoS
CDT, CDV	CDV and Max CTD	CDV and Max CTD	Mean CTD only	Unspecified	Unspecified	QoS
PCR, CDVT	Specified	Specified	Specified	Specified	Specified	Traffic
SCR, BT	n/a	Specified	Specified	n/a	n/a	Traffic
MCR	n/a	n/a	n/a	Specified	n/a	Traffic
Congestion control	No	No	No	No	Yes	

5.1 TABLE

ATM service classes and applicable parameters. Source: [AL95].

The quality of service (QoS) parameters (attributes) are the following:

- cell loss ratio (CLR),
- cell delay variation (CDV),
- maximum cell transfer delay (Max CTD),
- mean cell transfer delay (Mean CTD).

The service categories specify the traffic and QoS parameters according to Table 5.1.

The UBR service is best effort and requires only the selection of one path from the source to the destination. To provide CBR, VBR, and ABR services, the ATM switches must reserve resources when the call is set up. In the case of ABR, as we discuss in Chapter 6, the flow of cells is regulated based on feedback information about the network congestion. CBR and VBR traffic are regulated at the source (by the GCRA) without any network feedback.

To find a path from the source to the destination that can provide the service, the ATM network uses a routing algorithm. This algorithm is still largely unspecified, except for its general mechanism that works as follows. The user indicates to the network that it desires a given service, say VBR-RT with given parameters. That indication is carried on a specific VCI between the user end system and the ATM switch it is attached to—we call this switch the border switch. That switch then looks into its routing database, which maintains an image of the network state. Using a possibly precomputed routing algorithm, the switch identifies a suitable next hop to carry the service and forwards the request to that next hop while reserving the resources it will need to carry the service. The path creation then continues from hop to hop. If all goes well, eventually the switch attached to the destination forwards the service request to that destination. The destination then accepts or rejects the request. Messages are then sent along the reverse path to indicate the request acceptance or rejection. The switches then update their routing databases accordingly. It is possible that a switch along the path rejects the request because the routing databases are never completely up to date. When that happens, that switch sends rejection messages along the reverse path all the way to the border switch. The border switch then may try to repeat the procedure along an alternate path. This procedure for finding an alternate path is called *crankback* by the ATM Forum. It is similar to the dynamic alternate routing of circuit-switched networks.

This routing procedure assumes that the nodes exchange state information to update their routing databases. Such an exchange is conceivable for a private ATM network but more doubtful when part of the network is public. Accordingly, the ATM Forum is specifying the routing algorithm for private networks. That algorithm is called P-NNI, for private network node interface. The network state is summarized by a set of link parameters: maximum cell transfer delay, maximum cell delay variation, maximum cell loss ratio, administrative weight (desirability as defined by network administrator), available cell rate, cell rate margin, and variance factor. These parameters are still tentative and their usage is yet to be specified.

In addition, the routing procedure requires that the nodes can determine the resources they need to carry a service. This determination is simple for CBR traffic but substantially more complex in the case of VBR traffic, as we explain in Chapter 6.

To be scalable, a routing algorithm must be hierarchical. The P-NNI routing algorithm defines a hierarchy of nodes. At each level, nodes are clustered in peer groups. The nodes of the same peer group elect a peer group leader, which maintains an aggregate description of the group. This aggregate description indicates the characteristics of the service across the group. The precise procedure for deriving such aggregate description is not fully specified yet. The group members also maintain routing tables to reach each other and their peer group leader. The peer group leaders exchange information to identify their neighbors and to calculate routing tables to reach each other. At the next level up, these peer group leaders are then clustered into a parent peer group, and so on. The addresses of the nodes are designed to identify their group memberships, as the telephone numbering does (country code, area code, zone, number in zone).

To route from A to B, the routing algorithm can then look at the addresses to find out the group leaders that should be involved. Thus, if the address of A is XYZ and that of B is XVW, then A and B belong to the *same group or groups of nodes.* The routing will then go from A to its peer group leader, say C, to the parent peer group leader of C, say D, to the peer group leader of B, say E, to B itself. The routing table at A specifies how to get to C. The table at C specifies how to get to E, and the table at E indicates how to reach B. At each level, the routing is decided by an algorithm such as a shortest-path algorithm. Note however, that the algorithm must combine the service charactersitics such as the maximum transfer delay, the maximum cell loss ratio, and so on, that we listed earlier in our description of the link parameters.

If the connection must go through a connectionless public service, such as Frame Relay or SMDS, then it becomes almost impossible to guarantee the quality of service of the connection. This is the case even though one can transmit the desired characteristics of the connection across the public service to other private ATM networks.

The ATM addressing has been defined by the ATM Forum. Each ATM system is assigned an ATM address that is independent of the higher protocol addresses (such as IP addresses). Routers implement an address-resolution protocol that maps the higher-level address into an ATM address. This decomposition (called *overlay model*) enables the higher-level protocols to be developed independently of the ATM protocols. The ATM addresses have 20 bytes, contain a 48-bit MAC address, and may contain an E.164 (standard telephone number) as a subaddress or another hierachical address. Through signaling, an end system can specify its MAC address to the ATM switch it is attached to and get back its full ATM address. Group addresses are also defined. For example, to send an IP packet to a given IP address over an ATM network, routers use the address-resolution protocol to determine the ATM address of the destination. The routing algorithm (e.g., P-NNI) then sets up the connection to the destination. We discuss IP over ATM in more detail later.

5.1.2 Fixed Cell Size

In order to recognize packet or cell boundaries in the bit stream transmitted by the physical link, it is customary to delimit packets by a distinguished bit pattern. However, if cells are of fixed length and if they contain a fixed-length error-checking sequence such as a CRC (which ATM cells do), then the node can use these features to determine cell boundaries implicitly. The basic idea is illustrated in Figure 5.2.

Suppose that each cell of length N contains a CRC sequence of length n computed over the preceding $m = N - n$ bits. In ATM cells, this CRC field is the header error-control field (HEC). The HEC field is the last header byte,

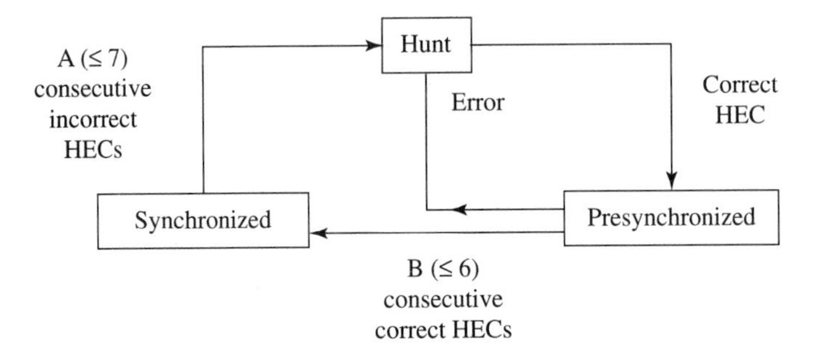

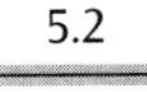

5.2 FIGURE

The figure summarizes the algorithm that a node uses to locate the cell boundaries. The node verifies the header error-control field (HEC) and searches for the cell boundary when it detects errors. Note the hysteresis of the algorithm designed to prevent it from hunting for a new boundary after every transmission error.

and it is calculated over the previous 4 bytes. The algorithm starts with any n-bit "window" as a tentative CRC and matches it with the CRC sequence computed over the preceding m bits. If a match occurs, the cell boundary has been correctly identified; otherwise, the algorithm shifts the window by 1 bit and repeats the procedure. A match occurs in at most N steps if the cell contains no errors. Once the cell boundary is identified, subsequent boundaries are found by counting bits. Conversely, if CRC errors are detected in several consecutive cells, this can be taken to indicate loss of synchronization (loss of cell boundary location).

The relatively short size of the ATM cells implies a large overhead that takes up approximately 5/53 or 9.4% of the bandwidth of every link, which is a disadvantage. (Were the cell size doubled, this overhead would be under 5%.) The short size of ATM cells was selected to reduce packetization delay for real-time voice data. Suppose voice is sampled 8,000 times per second, or once every 125 μs, and suppose each sample is encoded into 1 byte. It then takes $125 \times P$ μs to fill up a cell containing data of P bytes. Thus the packetization delay for ATM cells with 48 bytes of data is $125 \times 48 = 6{,}000$ μs. (The packetization delay for higher bit rate real-time data such as video is proportionately less.) To place this delay in perspective consider all the delays encountered by a cell as it traverses the ATM network. See Figure 5.3.

The cell encounters five types of delay:

1. packetization delay (PD) at the source,
2. transmission and propagation delay (TD),

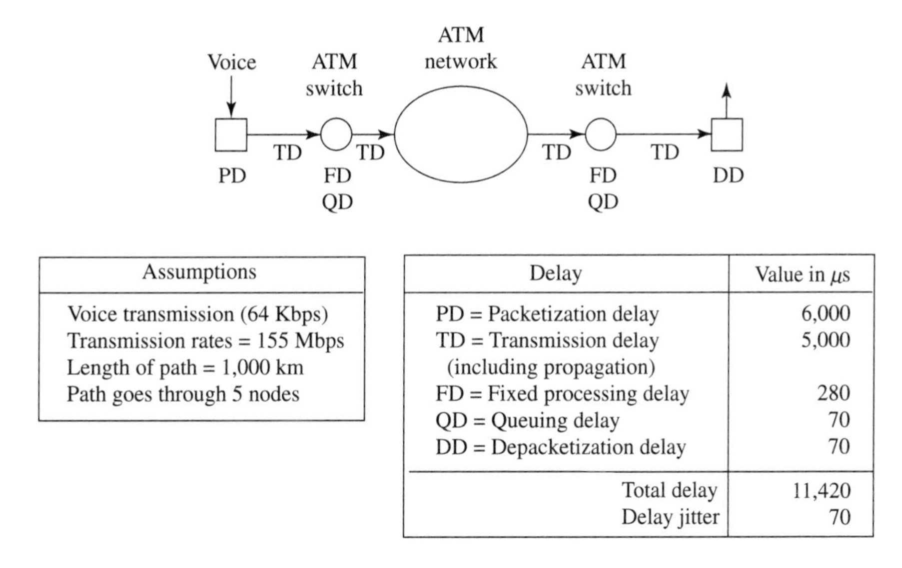

Assumptions
Voice transmission (64 Kbps)
Transmission rates = 155 Mbps
Length of path = 1,000 km
Path goes through 5 nodes

Delay	Value in μs
PD = Packetization delay	6,000
TD = Transmission delay (including propagation)	5,000
FD = Fixed processing delay	280
QD = Queuing delay	70
DD = Depacketization delay	70
Total delay	11,420
Delay jitter	70

5.3 FIGURE Five types of delays are encountered by ATM cells. The table gives typical values of these delays for a voice conversation.

3. queuing delay (QD) at each switch,
4. a fixed processing delay (FD) at each switch, and
5. a jitter compensation or depacketization delay (DD) at the destination.

We already discussed packetization delay. The propagation delay for electric and optical signals is between 4 and 5 μs/km, so for a 1,000-km path from source to destination, TD is about 5,000 μs. The transmission delay at a switch is the time needed to transmit one cell, or 53 × 8 bits. If the transmission speed is 155 Mbps, this time is about 3 μs, which is negligible. Hence TD is equal to propagation delay.

A cell suffers queuing delay when there are other cells that arrived earlier or simultaneously and that have not yet been processed by the switch. This delay is random and depends on the traffic load and on the switch architecture. Consider, for example, an output-buffered switch such as that discussed in Chapter 10. Arriving into the buffer of output link 1 (say) are the cells in all the virtual channels that go through link 1. If there are many such virtual channels, and if we assume that the cell streams in these channels are statistically independent, we may approximate the cell arrivals into the output buffer as a Poisson process with rate ρ per unit of time, where one unit of

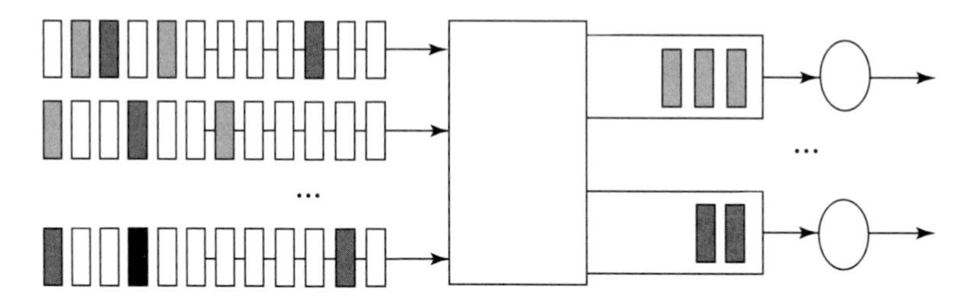

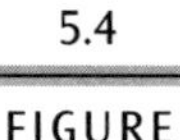

5.4 FIGURE

In an output buffer switch, the cells that arrive into a buffer are a subset of the cell streams at the input lines of the switch.

time is the amount needed to transmit one cell over the output link. See Figure 5.4. (For example, if the link's transmission speed is 155 Mbps, the speed of a SONET STS-3 signal, and a cell contains $53 \times 8 = 424$ bits, this unit of time is $424/155 = 2.74\ \mu s$.) Thus the output buffer is modeled as an M/D/1 queue, with service rate equal to one cell per unit time. The average number of cells in the buffer is given by

$$N = \frac{2\rho - \rho^2}{2(1 - \rho)},$$

where ρ is the traffic intensity or load or average link utilization. (See below for a derivation of this queuing delay formula.) For a load of 80%, this gives $N = 2.4$ cells, which for a 155-Mbps transmission rate yields QD of about 6.6 μs. If a cell goes through 10 such output buffers as it travels from source to destination, QD is about 70 μs. (This is average queuing delay; it may be more useful to define QD at the 90 or 99 percentile level, i.e., such that Probability{queuing delay $<$ QD} $= 0.9$ or 0.99.)

In addition to the queuing delay, a cell undergoes a fixed (almost deterministic) processing delay (FD) as it goes through a switch. This delay is due to the cell being copied into the switch memory one or more times and the time taken for computing the CRC and for translating the VCI into a route through the switch, etc. The time taken for the memory copies is proportional to the cell size, whereas the time taken for looking up the VCI routing table depends on its size and on how it is organized. Some of these operations may be done in parallel. A reasonable value of FD is about 10 cell times, which for a bit rate of 150 Mbps amounts to 28 μs. If the cell goes through 10 switches, total FD is 280 μs.

Finally, although voice and other real-time sources generate cells at a fixed rate, those cells arrive at the destination with random intercell delay, called *jitter*. Jitter is mainly due to the random queuing delay. To eliminate jitter, cells arriving at the destination are copied into a buffer that, in turn, is read out at a constant rate. This introduces a depacketization delay (DD) equal to the queuing delay (QD).

The table in Figure 5.3 summarizes the various delays calculated above for a 64-Kbps voice signal transported over a virtual circuit that goes though 10 nodes with 155-Mbps links. We see that for voice the packetization delay is a significant fraction of the total delay. If the cell size were doubled, so would that packetization delay.

Queuing Analysis

This section may be skipped by those unfamiliar with probability calculations. We calculate the average delay through an ATM switch. Our model of the switch is an M/D/1 queue. That is, during the nth cell transmission time, a random number $A(n)$ of cells arrive at the buffer. The random variables $A(n)$ are independent and Poisson distributed with mean ρ. Thus, $E\{A(n)\} = \rho$, and one can show from the properties of the Poisson distribution that $E\{A(n)^2\} = \rho + \rho^2$. Let X_n be the number of cells in the buffer at the beginning of the nth slot time (one slot time is equal to one cell transmission time). Then,

$$X_{n+1} = (X_n - 1)^+ + A_n = X_n + A_n - 1(X_n > 0), \qquad n \geq 0, \tag{5.1}$$

where we use the notation that for any number z, $z^+ = \max\{z, 0\}$ and $1(\cdot)$ is the indicator function, so $1(z > 0) = 1$ if $z > 0$, and 0 otherwise. The term $(X_n - 1)^+$ accounts for the fact that if $X_n > 0$, then one cell will be transmitted, leaving $(X_n - 1)$ cells in the buffer. Assume that we have reached statistical steady state, so that $E\{X(n+1)\} = E\{X(n)\}$ and $E\{X(n+1)^2\} = E\{X(n)^2\}$. Taking expectations on both sides of (5.1) we get

$$E(X) = E(X) + E(A) - P(X > 0),$$

so

$$P(X > 0) = E(A) =: \rho.$$

Next we square both sides of (5.1) and take expectations to get

$$E(A^2) + \rho + 2\rho E(X) - 2E(X) - 2\rho^2 = 0.$$

Since $E(A^2) = \rho^2 + \rho$, we find

$$E(X) = \frac{2\rho - \rho^2}{2(1-\rho)}. \qquad (5.2)$$

5.1.3 Statistical Multiplexing

A virtual circuit specifies a path from source to destination going through several links and switches. Many virtual circuits occupy the same link. A switch has ports terminating several incoming and outgoing links. (A large switch such as in a telephone central office may terminate thousands of links; such a switch is most likely built in modular fashion following the principles presented in Chapter 10. A local area ATM switch, which interconnects terminal equipment within a local area, may terminate tens of links; that kind of switch is likely to be built in a single stage.)

We now describe the five tasks that a switch carries out. The tasks are: demultiplexing, routing through the switch, multiplexing, buffering, and discarding.

The switch demultiplexes the cell stream arriving over each incoming link into "tributary" streams belonging to different virtual channels. The switch does this on a cell by cell basis, using the VCI in each cell.

The switch then routes the cell stream in each virtual channel to the appropriate output port. For reasons explained in the next section, the switch may change the VCI assigned to a particular channel. Routing is carried out with the help of a table with entries of the form

$(\text{VCI}_{in}$, input port, VCI_{out}, output port).

(Routers in datagram networks do not have such connection state information.) An entry in the routing table is created at the time the virtual channel is set up, and it is deleted when that virtual channel is torn down. This routing of the cell stream is internal to the switch; the mechanism that implements routing depends on the switch architecture, as we will see in Chapter 10.

Next, the switch multiplexes the cell streams directed to the same output port. This is statistical multiplexing. (The cell stream in a virtual channel may carry constant bit rate (CBR), variable bit rate (VBR), or message traffic.) The capacity of the output transmission link exceeds the average bit rate of the incoming virtual channels, but it may not exceed their peak rate. Hence, the switch has to buffer excess cells. This need arises when for a short time interval the number of cells to be transferred over an output link exceeds the capacity of that link. Finally, if the buffer is full, the switch must discard cells of lower priority. (ATM specifies two priorities; see section 5.2.)

5.1.4 Allocating Resources

ATM networks are expected to offer to transfer cell streams from source to destination, under a range of quality of service, to meet the varying needs of applications, as we saw in section 5.1.1. One may think of the service parameters as establishing a contract: the network guarantees specified bounds on delay and cell loss if the cell stream emitted by the user conforms to the bounds on average rate and burstiness.

In order to meet its obligations under such contracts, the network

1. exercises admission control,
2. selects the route (virtual channel path) of admitted connections,
3. allocates bandwidth and buffers separately to each connection,
4. selectively drops low priority cells, and
5. asks sources to limit the cell stream rate.

These network decisions are considered in depth in Chapters 6 and 7. Admission control requires the network to determine whether, given existing connections, there are sufficient idle resources to meet the QoS requirements of a new connection request. Once a new connection is admitted, the network must assign to it a virtual circuit with a route that has sufficient resources. The network must then allocate those resources to the connection so that it can meet the QoS. The first three decisions are taken during the connection setup phase. The last two decisions are taken during the data transfer phase. It happens on occasion that buffers at a switch become full, and some cells must be dropped. If there are cells of different priorities in the buffer, it is preferable to drop those with low priority. Finally, the switch may need to signal to a source to reduce or increase its rate (flow control) depending on how well it conforms to the QoS contract.

In order to implement the routing and resource allocation decisions, the network needs to distinguish between cell streams of different connections. That distinction can be made on the basis of each cell's VCI. Of course, each switch has to create a table of the relation among VCI, its QoS parameters, its route, and its allocated resources and consult this table as needed. The information in the table is said to be implicit in the VCI. (Since referencing the table takes time, we shall see that some of the implicit information is, in fact, explicitly present in each cell of a connection.) By contrast, different cells with the same VCI may have different priorities, and so priority information must be indicated explicitly in each cell. Similarly, the switch may wish to signal to the source the necessity for flow control, and so explicit provision must be made within the cell for representing this signal. We now see how the ATM cell provides this information.

5.2 ATM HEADER STRUCTURE

In this section we will examine the ATM cell structure, and we will see how an ATM switch can obtain the information it needs from the cell header. As indicated in Figure 5.5 the AAL (ATM adaptation layer) produces a data stream of 48-byte cells or PDUs (protocol data units). (We will see later the function of the AAL layer.) The ATM layer adds a 5-byte header and forwards the 53-byte cell to the physical layer. The 53-byte cell is converted into a serial bit stream by reading the cell from right to left (most significant bit first) and top to bottom (byte 1 first).

The figure shows the header structure for both the user-network interface (UNI) and the network-network interface (NNI). The abbreviations used in the figure are

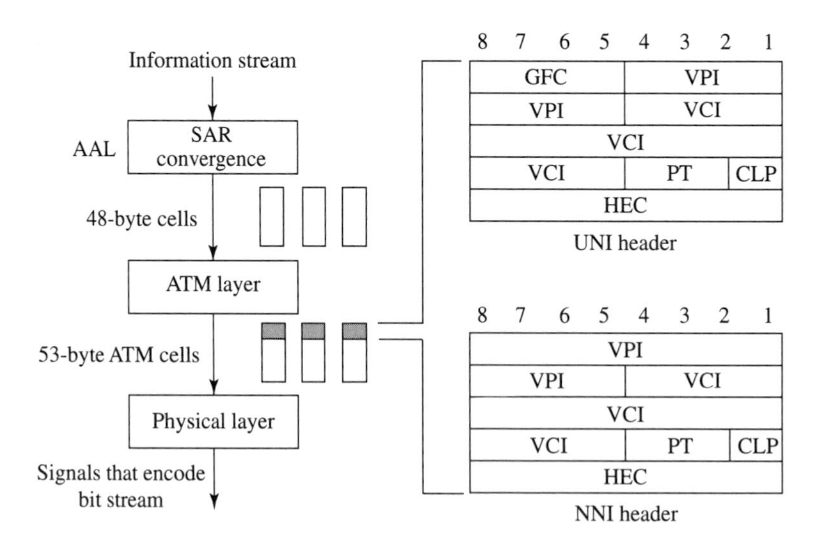

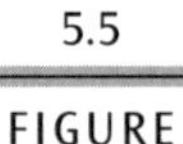

5.5 FIGURE

The figure shows the headers of ATM cells across the user-network interface (UNI) and across a network-network interface (NNI).

- GFC, generic flow control,
- VPI, virtual path identifier,
- VCI, virtual channel identifier,
- PT, payload type,
- CLP, cell loss priority,
- HEC, header error control.

The only difference between the two headers is that the UNI header has a 4-bit field that can be used to signal to the user the need for flow control. The NNI uses those bits to expand the VPI field.

5.2.1 VCI and VPI

Consider first the 16-bit VCI. The most important feature is that the VCI is *local* to each link. More precisely, different simultaneous connections that share a common link must have different VCIs, but connections that have no common link may have the *same* VCI.

Because VCIs are local to a link, connections coming into a switch from different switches (or sources) may have the same VCI. However, if these connections share the same outgoing link, they must be assigned different VCIs on that link. This works as in Figure 5.6, which shows four user nodes (A, B, C, D) and two switches. Initially there is a connection over VCI #1 from

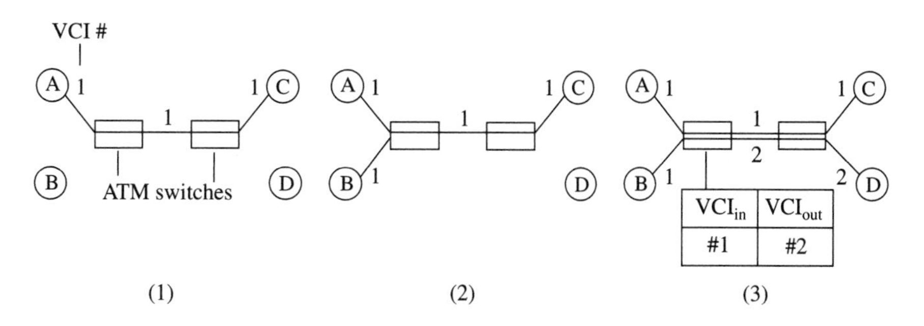

5.6 FIGURE

Virtual channel identifiers must be unique per connection on any given link. When the network sets up the second virtual channel, it assigns it a different VCI on the link that the second virtual channel shares with the first. (1) VCI #1 from A to C has been set up. (2) Request from B for connection to D is initially assigned VCI #1. (3) Because of common link, new connection is given VCI #2. The switch creates and updates the VCI translation table.

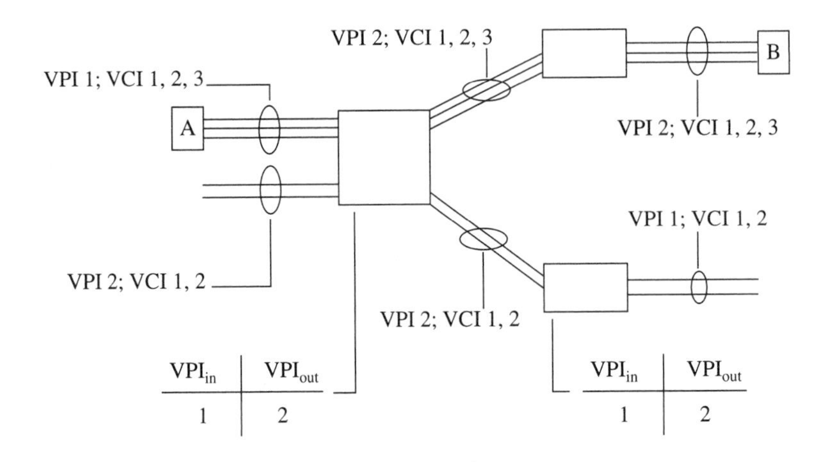

5.7 FIGURE A virtual path is a group of virtual channels that the network routes together.

A to C. At some later time there is a request to establish a connection from B to D. (This VCI assignment is carried out by separate control cells during the connection establishment phase.) Since B is not currently using VCI #1, it assigns that VCI to the connection. When the packet establishing this connection reaches the first switch, that switch knows that VCI #1 is assigned to another connection that shares a link with the proposed connection. The switch therefore changes the VCI from #1 to #2, noting the change in a table. (During the data transfer phase, this switch must change the VCI on each cell on the B–D connection from #1 to #2.)

The 16-bit VCI field permits 64,000 simultaneous connections through each link, and since the same identifier may be reused by connections with disjoint paths, the network can support orders of magnitude of more simultaneous connections.

Some purposes may be better served by treating several virtual channels together as a group. Suppose, for example, as shown in Figure 5.7, that a user wishes to establish three virtual channels from A to B.

It would then make sense to give these channels the same path and permit the user to assign VCIs to those channels arbitrarily. That is the function of the VPI, or virtual path identifier. Virtual channels with the same VPI form a group. They are assigned the same path and switched together, i.e., routing and switching decisions are based only on the VPI. More interestingly, the bandwidth and buffer resources allocated to a VPI may be done statically (at the time of service subscription) and shared only by virtual channels with that

VPI. This use of VPIs permits the creation of "virtual private networks": a multilocation firm can rent several virtual paths to form its own private network whose resources are then shared by its virtual channels.

From Figure 5.5 we note that the UNI header has an 8-bit VPI field, while the NNI has an additional 4 bits. The network may use these additional bits to create certain fixed routes, similar to what is done in today's long-distance telephone networks. This may allow resources allocated to virtual paths to be statically assigned and shared dynamically among component virtual channels. The use of VPIs may speed up processing, since switches only need to consult a table with entries indexed by shorter VPIs.

5.2.2 Other Fields

The 4-bit GFC (generic flow control) may be used by the network to signal to a user the need for momentary changes in the instantaneous cell stream rate. The functionality of the GFC has not yet been established.

The 3-bit PT (payload type) permits networks to distinguish between different types of information such as user data (PT = 000) and maintenance (PT = 100 or 101). This field permits network equipment to introduce and remove special cells that are routed as ordinary cells but that carry special information for control purposes.

The 1-bit CLP (cell loss priority) distinguishes among cells that the network may not discard (CLP = 0) and cells that it may discard (CLP = 1) if necessary. Of course, the network may introduce a low-priority service (as a possible QoS) at the level of a connection—all cells in such a connection are subject to discard. Since that information is implicit in the VCI, the CLP field is not needed, although it may be used to make the information explicit. The CLP field is needed, however, if different loss priority cells are present in the same connection. One example would be if voice is encoded into equal numbers of high and low order bits and sent over cells that alternately carry the high and low order bits. The low order bit cells would be assigned a CLP = 1, and be subject to discard, while the cells with high order bits would be assigned CLP = 0 and would not be discarded. By encoding voice in this way, the statistical multiplexing gain can be increased considerably. Also, if a border switch notices that cells do not conform to the service parameters, then that switch can set CLP = 1 in those cells so that these nonconformant cells are eligible for discard. The border switch detects such nonconformant cells by using the GCRA with the traffic parameters, as we explain in Chapter 6.

Finally, the 8-bit HEC (header error control) field is equal to the sum of the byte 0101'0101 and the CRC (cyclic redundancy check) calculated over the rest of the header with the generator polynomial 1'0000'0111. (The functioning of the CRC is discussed in Chapter 2.) As explained in section 5.1.2, the HEC is also used to delineate the cell boundary. The HEC can correct single bit errors and detect multiple bit errors. The error-control algorithm has two states: error detection (D) and error correction (C). The algorithm is initially in state C. If the algorithm is in state C and detects a single bit error, it corrects the error and moves to state D. If it detects a multiple bit error when in state C, the algorithm discards the cell and moves to state D. The algorithm discards cells that contain errors when in state D and remains in state D. The algorithm moves from state D to state C when it gets a cell without error.

5.2.3 Reserved VCI/VPI

Some VCI/VPI combinations are reserved. One set of combinations transports so-called unassigned cells. These may be introduced by the transport layer at the source, but they do not contain user information. These cells are removed by the destination transport layer; they are not forwarded to the application layer.

Sets of VCI/VPI combinations are reserved for cells that can be introduced and removed by the physical layer. These cells are invisible to the ATM layer. One set is assigned to "idle" cells that may be introduced periodically to help synchronization (cell boundary recovery). If the bit stream generated by the physical layer is synchronous (as in SONET), then "idle" cells may be used to fill up a frame whenever there is no user traffic to be sent. Another set of VCI/VPI combinations is assigned for OAM (organization, administration, and management) functions at the physical layer. In essence, these form connections dedicated to assist in various monitoring and physical layer control functions.

5.3 ATM ADAPTATION LAYER

As shown on Figure 5.8, the network converts the information stream into a stream of 48-byte data cells. This conversion is performed by the ATM *adaptation layer* or AAL. The AAL is divided into two sublayers: the CS or convergence sublayer and the SAR or segmentation and reassembly sublayer.

The CS converts the information stream into packet streams of data in a variety of convenient formats, broadly divided into five types of traffic. The individual packets are called SDUs (service data units), following OSI terminology. Since a user information stream may be encoded into packet

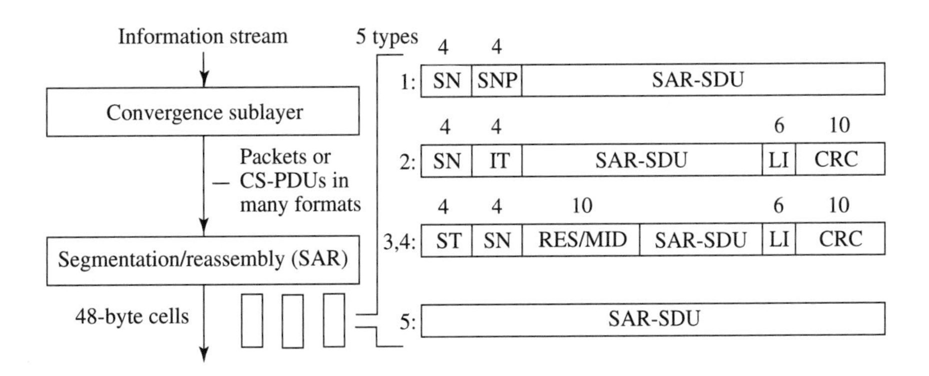

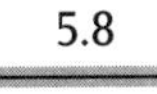
5.8 FIGURE

The ATM adaptation layer (AAL) converts the information stream into 48-byte cells. The AAL is decomposed into the convergence sublayer and the segmentation/reassembly sublayer.

streams in many different ways (e.g., video may be encoded into a constant or variable bit rate stream), and since some of this encoding information must be included in the SDU, CS tasks are likely to be application dependent.

By contrast, SAR tasks are quite standard, depending only on traffic type. The SAR must segment the SDUs received from the CS, add the necessary overhead, and convert the result into 48-byte cells or PDUs (protocol data units) that are then handed over to the ATM layer. As we will explain, the overhead is needed by the SAR at the destination to reassemble the SDUs from the 48-byte cells.

As mentioned, the CS converts the information stream into one of five types of packet streams. The streams are designed to match the requirements of five traffic types: constant bit rate–real time, variable bit rate–real time, connection-oriented packet streams, datagrams, and IP packets. We describe each type, its typical intended applications, and the corresponding SAR functions.

5.3.1 Type 1

This traffic is generated at a constant bit rate, and it is required to be delivered at the same rate (with a fixed delay). Intended applications are voice and constant bit rate video or audio. Since the cells may suffer variable delays, the CS at the destination must compensate for those delays. This is done in one of two ways. Incoming cells are buffered at the destination and the buffer is read out at a fixed rate.

Alternatively, the CS at the source inserts an explicit time stamp into the packet stream. The destination CS extracts this time stamp and attempts to read cells with a constant latency. (A cell that arrives with a delay that exceeds this latency may be discarded.) The source and destination CS sublayers must agree on the scheme to be used for timing information. That agreement must be reached during the connection setup phase.

Figure 5.8 displays the structure of the 48-byte PDU for Type 1 traffic. The SAR sublayer takes the (periodic) packet stream generated by the CS sublayer, segments it into 47-byte SDUs, and prepends to each SDU a 4-bit sequence number (SN) protected by a 4-bit sequence number protection (SNP) field. The SNP corrects single bit errors and detects multiple bit errors in the sequence number. The eight possible sequence numbers permit the SAR sublayer at the destination to determine if fewer than eight consecutive SDUs are lost. (The source and destination must agree on what to do when cell loss is discovered.)

5.3.2 Type 2

This traffic is generated at variable bit rate. As with Type 1 traffic, there are timing considerations. These are indicated by a time stamp. As before, the CS sublayers at source and destination must agree on how to deal with timing and cell loss.

The SAR function is more complicated than that for Type 1 traffic, since it must segment variable-sized packets generated by CS. When the SAR sublayer segments a variable-sized packet into SDUs, it appends to each SDU a 4-bit IT (information type) field to indicate whether that SDU is at the beginning of message (BOM), is a continuation of message (COM), or is at the end of message (EOM) and also whether the SDU is a component of the audio or video signal. In the case of EOM, an LI (length indicator) field is added if the cell is only partially filled by the message.

5.3.3 Type 3

This traffic is suitable for a connection that occupies one or more variable-sized packets. (It is well suited for transfer of SMDS L3_PDUs.) Two kinds of error quality are envisaged. In the first kind the CS sublayer guarantees error-free delivery. In the second kind no such guarantee is made. Error-free delivery is ensured by the destination CS requesting retransmission when an error is detected. To ensure reconstruction of the packets and to detect errors, the overhead fields appended by the SAR sublayer are as in Figure 5.8. ST or segment type is like the IT of Type 2: ST = 10 for BOM, ST = 00 for COM, ST = 01 for EOM, and ST = 11 if the packet fits inside a single cell. The 4-bit SN (sequence number) is used as before to detect loss or cells inserted by network routing errors. The 10-bit RES field is reserved for future use. The 6-bit LI field is needed when the cell is only partially filled. Finally the 10-bit CRC is a check over the SAR-SDU used to detect errors in the CS SDUs.

5.3.4 Type 4

This is almost the same as Type 3 traffic except that it emulates connectionless (datagram) service. The CS layer generates a single variable-sized packet. Each such packet is segmented into one or more cells by the SAR sublayer. The 10-bit MID field now plays a special role. It would be common for a user to multiplex several different datagrams over the same virtual channel connection. The MID (multiplexing identifier) field can be used to distinguish among the cells originating from different datagrams. (This, too, is similar to the use of the MID or message identifier field in SMDS.)

5.3.5 Type 5

This traffic carries IP packets. The frame structure of Type 5 eliminates the overhead present in Types 3 and 4. The IP packet is packaged into a CS packet that contains a length indicator and a 32-bit CRC calculated over the complete CS packet with the generator 1′0000′0100′1100′0001′0001′1101′1011′0111. The CS packet contains a padding field so that the length of this CS packet is an exact multiple of 48 bytes. The CS packet is sent as a sequence of SAR packets of AAL Type 5. The PT of the ATM cell header indicates whether the cell is the last one of the CS packet or not.

5.4 MANAGEMENT AND CONTROL

A very important feature of ATM networks is that they can take a number of management and control decisions to discriminate among connections and to provide the variety of QoS that different applications need. The decisions are divided into three groups. When a request is made for a connection with a particular QoS, the network must determine whether to accept or reject the request, depending on the resources then available. (Recall that QoS involves three sets of parameters: delay, cell loss, and source traffic rate.) If the resources are insufficient to meet the request, the network may negotiate with the user for a connection with a different QoS.

Once the connection is admitted, the network must assign a route or path to the virtual channel that carries the connection. It must inform the switches and other network elements along the path that this virtual channel must be allocated certain resources so that the agreed-on QoS is met.

Lastly, the network must monitor the data transfer to make sure that the source also conforms to the QoS specification and to drop its cells as appropriate. The network may also ask a source to slow down its transmissions.

In addition, the network carries a number of information flows to monitor its operations and to detect and identify the location of congested or failed devices.

The BISDN standard is silent about how these decisions are to be carried out. We shall discuss potential solutions in Chapter 6. The ATM Forum has proposed specific frame formats that the network should use to carry its monitoring information and to interact with users. We review those proposals next.

The network uses operation and maintenance information flows for the following functions:

- fault management,
- traffic and congestion control,
- network status monitoring and configuration, and
- user/network signaling.

These functions, like the other network functions, are organized into layers. Figure 5.9 shows the layer arrangement of all network functions, including those of operation and management. The layers in the *user plane* comprise the functions required for the transmission of user information. For instance, for an Internet Protocol over ATM, these layers could be Telnet/TCP/IP/AAL5. The layers in the *control plane* are the functions needed to set up a virtual circuit connection. These functions, which include the signaling protocols, are needed only for switched virtual connections and are absent in a network that implements only permanent virtual connections. (In a permanent virtual circuit connection, the path or route assigned to a source and destination and the VCI for that route are fixed.) The *layer management plane* contains management functions specific to individual layers. Finally, *plane management* consists of the functions that supervise the operations of the whole network.

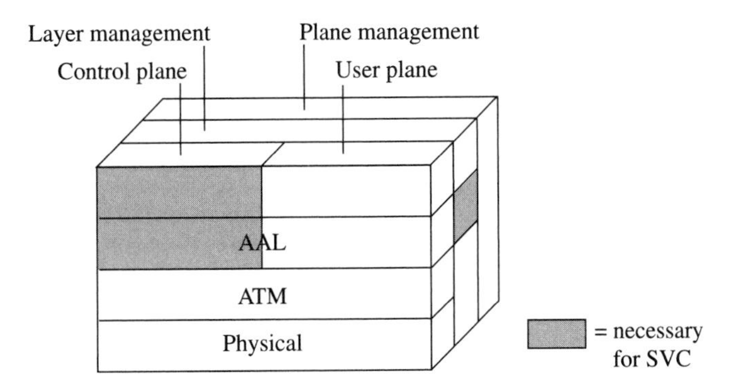

5.9 FIGURE

Layer arrangement of network functions, including the operation and management functions.

5.4.1 Fault Management

Consider a virtual circuit connection over an ATM network and assume that the connection is implemented by a SONET network. We know from section 4.2 that SONET establishes transmission *paths* for the ATM layer. The transmission is over optical fibers. The transmitters in SONET are all synchronized to the same master clock. This synchronization enables the time-division multiplexing of different bit streams. This multiplexing is done byte by byte.

The physical layer (SONET) is decomposed into three sublayers: section, line, and path. The section layer transmits bits between any two devices where light is converted back into electronic signals or conversely. For instance, there is a section between two successive regenerators or between a regenerator and a multiplexer. The line layer transports bits between multiplexers where SONET signals are added to or subtracted from the transmission. Finally, the path layer transports user information. Thus, a path goes across a number of lines (or links) that are switched by the SONET demultiplexers and multiplexers, and a line consists of a number of sections. Each layer inserts and strips its own overhead information, which it uses to monitor the transmission functions for which it is responsible. (See Figure 5.10.)

Each of the three sublayers uses overhead bytes in the SONET frames to supervise its operations. The overhead bytes are said to carry a *flow* of operation and maintenance information. The flow carried by the section overhead bytes is called F1. The flow carried by the line and path overhead bytes are F2 and F3, respectively. The virtual circuit connection is carried by a virtual path connection. Accordingly, the network uses a flow of cells to supervise

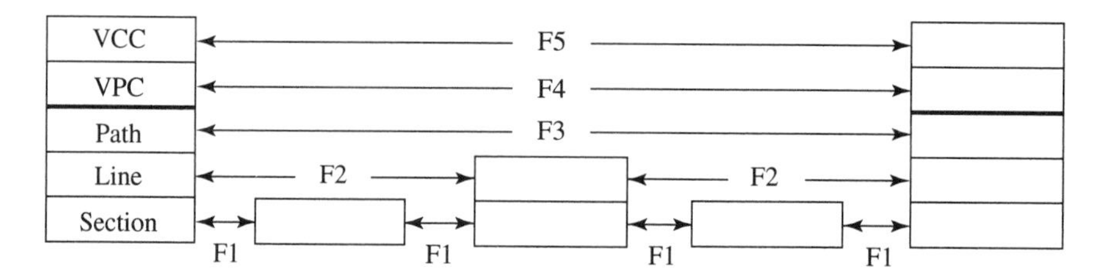

5.10 FIGURE

Operation and maintenance flows for a virtual circuit connection over SONET.

the virtual path connection and a flow of cells to supervise the virtual circuit connection. These two flows are called F4 and F5, respectively.

The format of the F4 and F5 cells depends on whether the cells monitor the segment across the user-network interface or the end-to-end connection (see Figure 5.11). The cell formats are shown in Figure 5.12. Note that the F5 cells have the same VPI/VCI as the user cells of the connection they monitor. The F5 cells are distinguished from the user cells by the PT field. Similarly, the F4 cells have the same VPI as the user cells and are distinguished by their VCI.

The main function of the OAM cells is to detect and manage faults. Fault-management OAM cells have the leading 4 bits of the cell payload set to 0001. The next 4 bits, the function type (FT) field, indicate the type of function

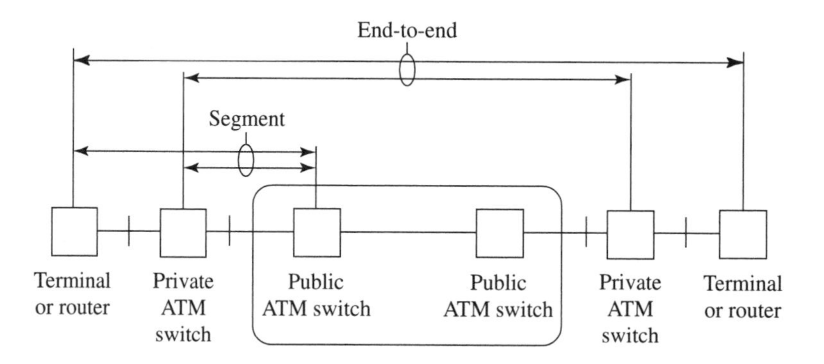

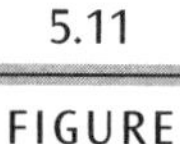

5.11 FIGURE

A segment indicates a connection across the user-network interface. An end-to-end connection is between the source and destination user equipment.

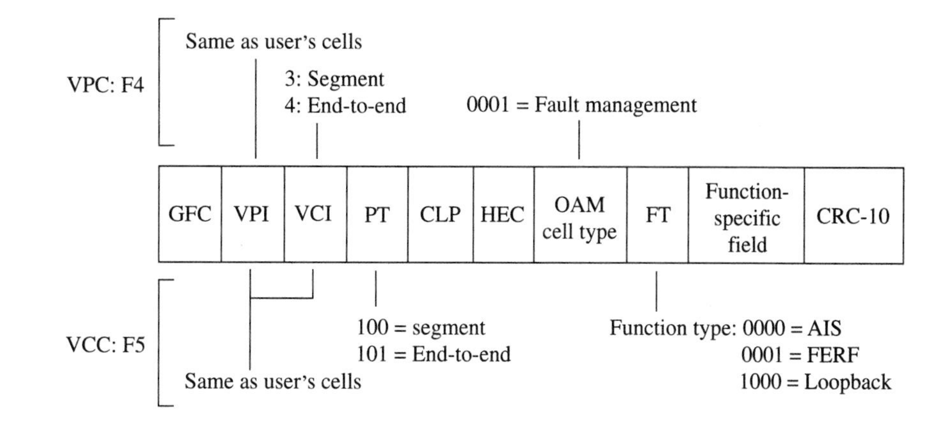

5.12 FIGURE Format of OAM cells.

performed by the cell: alarm indication signal (AIS) signaled by FT = 0000, far end receive failure (FERF) signaled by FT = 0001, and loopback cell, signaled by FT = 1000. The AIS cells are sent along the VPC (virtual path connection) or VCC (virtual circuit connection) by a network device that detects an error condition along the connection. Those cells are then sent along to the destination of the connection. When the equipment at the end of that connection receives the AIS, it sends back FERF cells to the other end of the connection. As shown in Figure 5.13, the AIS and FERF cells specify the type of failure as well as the failure location.

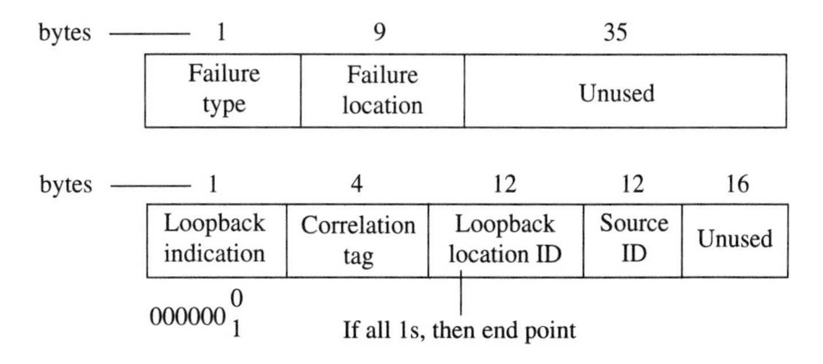

5.13 FIGURE Function-specific fields in AIS and FERF cells (above) and in loopback cells (below).

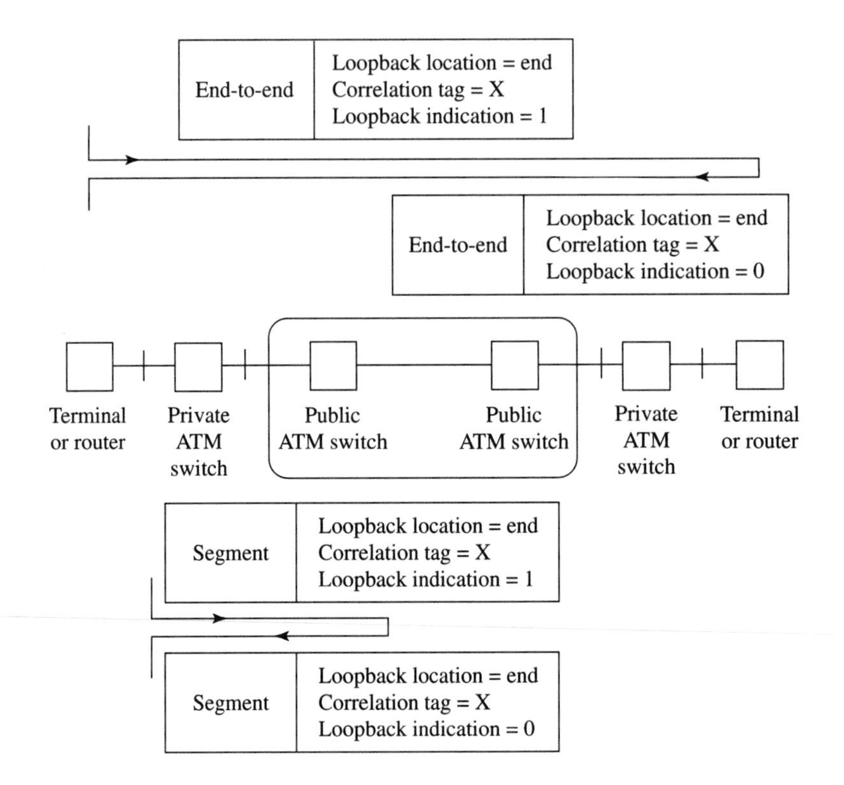

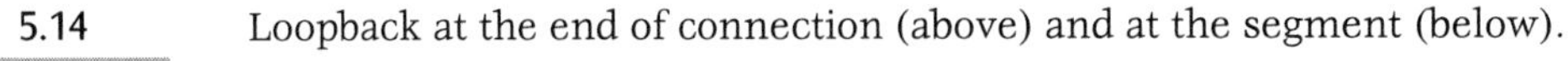

FIGURE 5.14 Loopback at the end of connection (above) and at the segment (below).

A loopback cell contains a field that specifies whether the cell should be looped back, a correlation tag, a loopback location identification, and a source identification. These loopback cells are used as shown in Figure 5.14.

The device that requests a loopback (we call it the *source*) inserts a loopback cell and selects a value for the correlation tag. The device can specify where the loopback should take place. The device sets the loopback indication field of the cell to 1 to indicate that the cell must be looped back. When the device where the loopback must occur receives the cell, it sets its loopback indication field to 0 and sends the cell back to the source. The source compares the correlation tag of the cell it receives with the value it selected. This correlation tag prevents a device from getting confused by other loopback cells.

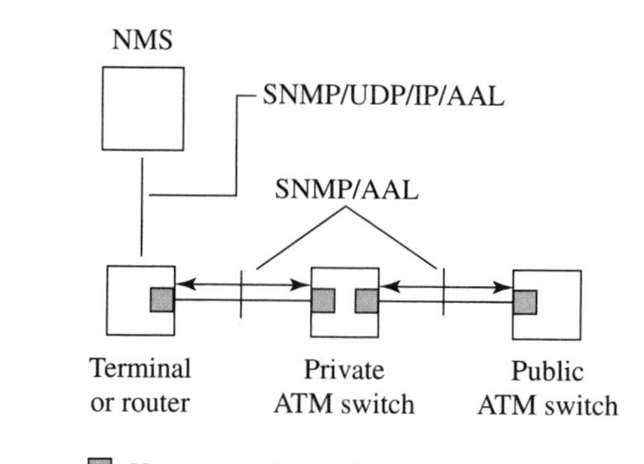

5.15 FIGURE

The Intermediate Local Management Interface (ILMI) protocol is designed to supervise the connections across user-network interfaces.

5.4.2 Traffic and Congestion Control

The objectives of traffic and congestion control are to guarantee the contracted quality of service to virtual connections. The operations that the network performs are the subject of Chapters 6 and 7.

5.4.3 Network Status Monitoring and Configuration

A protocol is being defined to facilitate the monitoring and management of an ATM network. This protocol is a version of SNMP, the Simple Network Management Protocol, adapted for ATM networks. The 1996 version of this protocol is called the *Intermediate Local Management Interface* or ILMI protocol. The objective of the ILMI protocol is to supervise the connections across the user-network interfaces. The situation is illustrated in Figure 5.15.

The figure shows a private ATM network connected by a private ATM switch to a public ATM network. Each connection across a user-network interface (UNI) is supervised by two UNI management entities: one for each of the ATM devices. Two such management entities are said to be *adjacent,* and the ILMI specifies the structure of the management information base (MIB) that contains the attributes of the connection supervised by the adjacent entities.

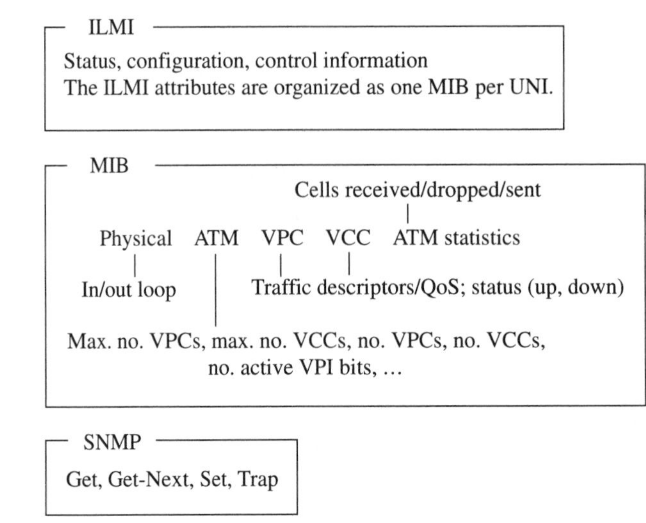

5.16 FIGURE Structure of the ILMI management information base (MIB) that contains the attributes of the connection supervised by adjacent UNI management entities.

The structure of the ILMI MIB is summarized in Figure 5.16. As the figure indicates, one MIB is defined per UNI. The contents of the UNI MIB are the attributes of the physical layer (which implements the bit way), the ATM traffic, the VPCs, and VCCs that go across that UNI. The figure indicates representative attributes.

5.4.4 User/Network Signaling

The basic signaling functions between the network and a user follow:

- the user requests a switched virtual connection,
- the network indicates whether the request is accepted or not, and
- the network indicates error conditions with a connection.

In the current version (called *phase 1*) of the user/network signaling protocol, the signaling takes place over a single permanent virtual circuit connection between the user and the network. When the user requests a new connection, the request specifies the address of the connection and the quality of service requested for the connection. The network accepts or refuses the request. No negotiation is accommodated by phase 1 of the specifications.

For details on the format of the signaling messages, see [A93].

5.5 BISDN

The ATM bearer service is the transport of cells with a variety of quality of service. We have seen that together with the formats defined for the ATM adaptation layer, this bearer service can be used to support a wide range of applications. In this section we explain how the ATM bearer service in turn can be implemented using the telephone companies' SONET networks. The result is an implementation of a Broadband Integrated Services Digital Network, or BISDN.

BISDN is the proposed standard for the fully integrated universal future network—the "information superhighway." The standard specifies both transport and physical layers. The network and data link layers are the same as the ATM standard described above. The physical layer standard is that of SONET. (We are using the terminology of the standard, which is closer to that of the OSI model. In terms of the Open Data Network model, the bearer services are as described in the previous sections, whereas the bit way standards are the same as those of SONET.)

At this time the physical layer standard for BISDN has been established for STS-3C signals. This is a bit rate of 155.52 Mbps. The STS-3C frame, shown in Figure 5.17, is arranged in a 9 × 270 byte matrix.

The first nine columns are devoted to section and line overhead (SOH, LOH). This leaves 9 × 261 bytes, of which one column is devoted to path overhead (POH). The resulting 9 × 260 byte SPE (synchronous payload envelope)

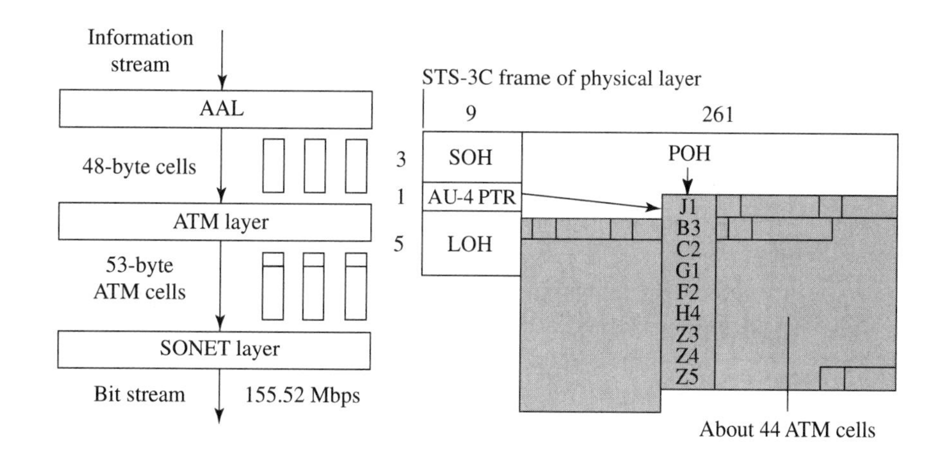

5.17 FIGURE BISDN over STS-3C SONET frames.

contains consecutively arranged 53-byte ATM cells. (Details of SONET can be found in section 4.2.)

The most important features of this frame structure are the following. The ATM bit rate is $155.52 \times 260/270 = 149.76$ Mbps. The $9 \times 260 = 2{,}340$ byte SPE holds about 44.15 53-byte ATM cells. Thus ATM cell boundaries bear no relationship to the SPE or STS-3C frame boundaries. (The H4 byte in POH is a pointer to the next occurrence of the cell boundary and may be used by the destination to recover cell boundaries. However, the latest version of the standard requires recovery of the cell boundary from the header CRC, as explained before.) If the transport layer does not provide a sufficient number of ATM cells, the physical layer inserts idle ATM cells into the frame that are removed by the physical layer at the destination. As with SONET traffic generally, the SPE is not aligned with the STS-3C frame. The SPE location is obtained from the AU-4 pointer in the LOH.

Individual subscribers to BISDN receive an STS-12 (622.08-Mbps) signal, and they can send a STS-3C signal. The subscriber can share this bandwidth among many simultaneous connections of varying QoS: constant and variable bit rate video and audio traffic with real-time constraints, data traffic with guaranteed retransmission in case of error for file transfer traffic, datagrams traffic for short transactions such as database queries or remote procedure calls, etc. The ATM/BISDN standard is sufficiently flexible to accommodate this variety of traffic service. However, the standard is silent about the ways in which network resources should be managed and controlled in order to meet this variety.

5.6 INTERNETWORKING WITH ATM

An ATM network can be used to carry internetwork traffic. For instance, an ATM network can be used to interconnect various LANs or IP subnetworks. Such internetworking can take place at the data link layer (bridging) or at the network layer (routing). For more details on the material of this section, we refer the reader to [AL95].

Two tasks are required for internetworking over ATM. The first is encapsulation of the protocol data units and the second is the routing or bridging of these PDUs. The routing itself consists of the route calculation and the switching of packets. The route calculation necessitates the address resolution plus a routing algorithm. The address resolution maps the protocol address (such as

IP, MAC, FR, or SMDS) into an ATM address. The routing algorithm calculates the routes through the network. That routing algorithm for the private ATM network is P-NNI, as we learned in section 5.1.1. The non-ATM network can either use its own routing algorithm or an extension of P-NNI.

In the following subsections we explain encapsulation, LAN emulation, IP over ATM, a more general multiprotocol over ATM, and FR or SMDS over ATM.

5.6.1 Multiprotocol Encapsulation over AAL5

The encapsulation of internetwork traffic is defined in RFC 1483. The RFC specifies two methods depending on whether one ATM VCC carries a single or multiple protocols. In either case, the protocol data units are carried by the payload of the convergence sublayer of AAL5. That is, a PDU of up to 64 KB is first padded to an exact multiple of 48 bytes. To these bytes, the convergence sublayer adds an 8-byte trailer that specifies the length of the PDU and a CRC.

If the VCC carries a single protocol, this protocol is identified implicitly by the VPI/VCI. Otherwise, the convergence sublayer PDU starts with a header that specifies the protocol (e.g., routed IP or bridged 802.3-6).

If the protocol is bridged, then the MAC address of the destination must be specified in the convergence sublayer PDU. Note that the ATM interface must perform the usual functions of a bridge with dynamic learning by looking into the MAC addresses of the encapsulated PDUs.

We explain how such a bridging function is implemented in LAN interconnections in the next section.

5.6.2 LAN Emulation with ATM

LAN emulation is a glue that enables ordinary LAN software to operate over ATM and also to interconnect LANs and ATM. (See [A95].) This emulation enables the connection of Ethernets through an ATM network or of token rings through an ATM network (not a mixture).

Figure 5.18 shows the LAN emulation in the protocol suite. A packet destined to B invokes the LAN emulation layer that contacts the server to find the ATM address of the bridge. The packet is segmented and sent to the bridge, whose LAN emulation software reconstructs the packet before sending it on the LAN. The reverse transfer, from B to A, is similar.

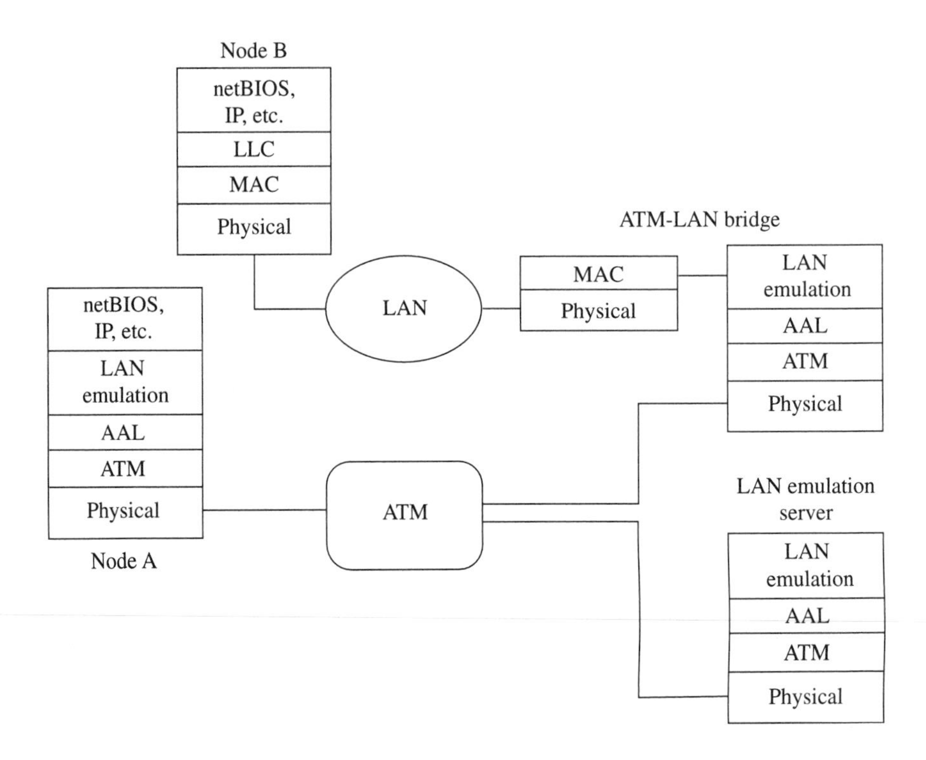

5.18 FIGURE

A LAN emulation layer is inserted between the network layer and the AAL layer in ATM nodes.

Broadcast packets and packets with unknown destination addresses are flooded by the server.

This solution enables ATM networks to interoperate with existing "legacy" LANs.

5.6.3 IP over ATM

As we learned in Chapter 4, IP is a powerful and widely used internetworking protocol. With IP we can interconnect IEEE 802 networks easily. IP is a datagram network layer. In this chapter, we discussed the ATM technology and how it can be used to build local area networks and wide area networks with a good control on the quality of service it provides to applications.

For the ATM technology to have any chance of being widely implemented, it must interoperate with the Internet protocol suite. In this section we explain how ATM networks can transport IP packets. This possibility enables a progressive upgrade of the Internet to the ATM technology. The advantage such an upgrade would provide is that applications requiring the tight control of QoS can be supported by ATM and not easily by the TCP/IP protocols. Thus, progressively, the Internet would evolve into a BISDN network while remaining compatible with the installed base of best-effort services.

We explain three strategies: the classical IP model, the shortcut models, and the integrated models. We then explain multicast IP over ATM. These strategies are being evolved by the IETF working group IP over ATM. (See RFC 1754.)

Classical IP

Consider the situation shown in Figure 5.19. In the so-called classical IP model, the nodes attached to an ATM network are grouped into logical IP subdomains. Routing between logical IP subdomains is via routers, as shown in the figure. Note that AAL5 is used so that the router reassembles the packet before forwarding.

Within one given logical IP subdomain, a node uses an address-resolution protocol (ARP) server. The stations all know the ATM address of their ARP server. Thus, to find a particular destination, instead of broadcasting a request *are you node IP.address* to find the physical address, here a node sends a request to the ARP server of the subdomain asking *what is the VCI of a particular IP.address?*

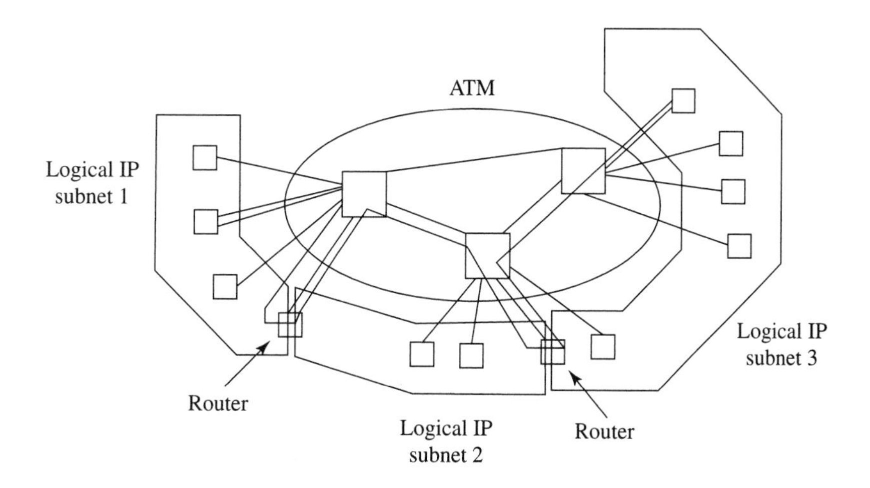

5.19 FIGURE

Nodes attached to an ATM network are grouped into logical IP subdomains. The routing is via routers between subdomains. Within one domain, the routing uses ATM ARP with an ARP server.

In the case of SVCs (switched virtual circuits), the nodes need to register with the ARP server. They do so by calling the ARP server (using the ATM addresses). The server then asks *what is your IP.address?* and enters that information in its table [IP → ATM].

The IP and ARP packets are encapsulated over AAL5. Two alternatives exist: either one VC is allocated per protocol (one for IP, one for ARP), or multiple protocols are multiplexed over a single VC per subnetwork attachment point (SNAP). The maximum transmission unit in IP over ATM is fixed to 9,180 bytes. Other sizes (up to 64 KB) can be agreed on by configuration (in the case of PVC or permanent virtual circuit) or signaling (for SVC). The type of ATM connection is either CBR or VBR with specified peak rates forward and backward. Other encapsulations are being proposed to eliminate or reduce the redundant IP header. (See RFCs 1577, 1483, 1626, 1755.)

Shortcut Models

Instead of retransmitting via routers as in the classical IP model, the idea of the shortcut models is to go directly from source to destination ATM nodes. In the accepted terminology, the ATM network is called a nonbroadcast multiaccess (NBMA) link layer. In Figure 5.20 we indicate how a node S finds the NBMA address.

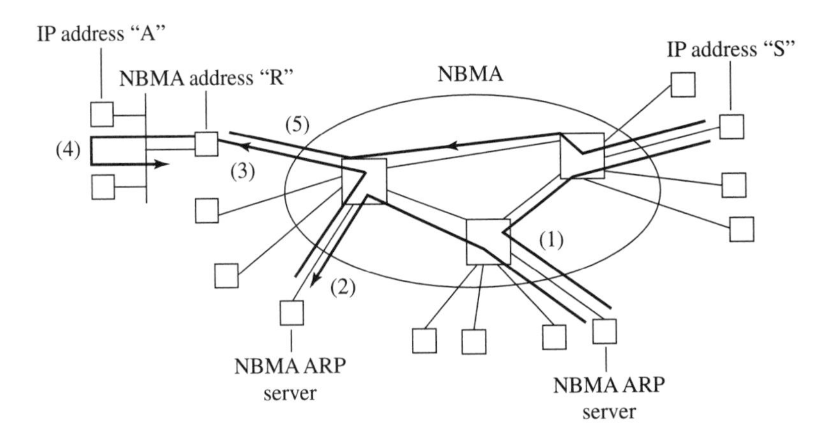

5.20 FIGURE

To transmit to node A, node S finds out the nonbroadcast multiaccess (NBMA) link address of the router R according to the following steps: (1) What is the NBMA address of A? (2) Forward request; (3) Knows router of network of A; (4) ARP to locate A; Then, in reverse, (3), (2), (1) give NBMA address R; a connection is then made as shown in (5).

Note that this method can also be used for Frame Relay, ISDN, and X.25 networks. (See RFC 1735.)

Integrated Model

The integrated model proposal aims to simplify the routing by integrating addressing and routing of IP and ATM. In this model, the ATM address could be a superset of the IP address. (This approach does not work with networks that use the E.164 addresses.) The IP router then maps the destination IP address into the ATM address of the destination if it is directly reachable or of the best router otherwise. The selection of the best router may be load dependent.

Multicast IP over ATM

The main difference between ATM multicast VCs and multicast IP is that the source must add new destinations in the case of ATM.

The mechanism uses a Multicast Address Resolution Server (MARS) that maps an IP multicast address to the set of all the individual ATM addresses. A cluster is a set of hosts that use the same MARS. The communication between clusters works as the regular IP multicast routing. We describe the communication within one cluster.

There are two approaches to multicast connections within one cluster. In the direct approach, each sender sets up one VC per member of the multicast group. In the indirect approach, the sender sends to a multicast server, which then sets up VCs to the group members.

In the direct approach, the hosts send join and leave requests to the MARS. The MARS maintains a point to multipoint VC to all the sources (called the ClusterControlVC) to inform them of group changes. The join and leave messages are retransmitted over the ClusterControlVC. Before sending, the source asks MARS for list of ATM addresses. Then the sender sets up the VCs.

In the indirect approach, the sender knows only the MCS as a member of a group. The hosts register with MCS, and every MCS registers with MARS.

5.6.4 Multiprotocol over ATM (MPOA)

To increase the likely penetration of ATM technology, the industry is developing mechanisms to run a large class of protocols over ATM networks. The objective of these mechanisms is to enable the internetworking of existing network layer protocols and LANs.

As we indicated earlier, two major tasks must be performed: address resolution and routing. The MPOA working group of the ATM Forum is developing three approaches.

In the first approach, there is an algorithmic map from the network layer address into the ATM address. As a result, the signaling requests can be routed using the ATM routing algorithm (P-NNI), and there is no need for an external address-resolution scheme. The difficulties of this scheme lie in the integration of ATM and non-ATM routing protocols.

The second approach assumes that the non-ATM networks adopt the P-NNI routing algorithm. This approach would permit a unified routing algorithm across the complete network. The difficulty is the need to migrate the current routing algorithms to P-NNI.

The third approach is an extension of the LAN emulation that replaces the layer 2 emulation by a layer 3 emulation. The advantage is that the routing of connections can then exploit the QoS of the ATM network. One proposal for making such layer 3 emulation economical is to share a route computation engine across a network of packet switches, preferably using a standardized protocol.

5.6.5 FR and SMDS over ATM

The transport of Frame Relay packets over ATM requires the conversion of the FR data link connection identifier into a VPI/VCI for the ATM connection and the segmentation of the FR packet payload into AAL5 packets. The FR congestion and discard eligibility fields are mapped into the ATM EFCI bit and the ATM CLP bit, respectively. Moreover, the FR committed information rate is mapped into VRB traffic parameters. (See ATM Forum 94-0996.)

SMDS packets are mapped into AAL3-4 cells and transported over a well-known VPI/VCI. A connectionless server within the ATM network receives the cells and forwards the packet.

5.7 SUMMARY

In the two previous chapters we studied packet- and circuit-switched networks. Packet-switched networks handle well message traffic that imposes very loose delay constraints. Circuit-switched networks are well suited for constant bit rate traffic with hard delay constraints. Of course, both networks can accommodate all types of traffic, but they may do this inefficiently.

Engineers invented ATM networks with the objective that they would provide a highly flexible bearer service capable of supporting in an efficient manner applications that range from those that need guaranteed delay and loss bounds to those that need only best-effort service provided by datagram networks such as Internet. At the end of the 1980s, ATM networks were exotic topics of discussion at research conferences. By the mid-1990s, tens of vendors were providing ATM equipment, although with limited functionality.

This chapter was devoted to a discussion of the functionality of ATM. In principle ATM can offer the full range of services needed to make the claim that ATM networks can function as the "universal network." The great interest among telephone companies in providing ATM over SONET suggests that ATM will be a serious contender for this title, in the name of BISDN. However, BISDN is still a considerable distance in the future. A more immediate goal is the migration path for IP over ATM. We saw two proposals that offer such a path. The migration can permit an "upgrade" of IP in the sense that IP/ATM may be able to provide service guarantees that IP cannot. If this turns out to be the case, IP and ATM may gain through economies of scope and service integration.

Tremendous progress in ATM has been made over the period 1990–95. But the potential of ATM remains to be realized until the difficult issues of resource allocation and control are satisfactorily resolved so that the same ATM network can efficiently provide a variety of service qualities. In Chapters 6 and 7 we discuss these issues.

5.8 NOTES

The basic ATM formats and some protocols are discussed in [P94]. The all-important management functions that guarantee the service quality level, however, are not yet sufficiently defined to constitute a proposal. Nevertheless, there is important and continuing discussion within the ATM Forum that seeks to develop those proposals and standards. Those discussions are published by the ATM Forum [A93, A95]. Proposals for IP over ATM are discussed in various IETF RFC, notably, RFC 1483, 1577, 1626, 1735, 1755.

ATM Forum contributions are available only to principal members of the ATM Forum. Published ATM Forum specifications can be purchased through the ATM Forum (e-mail: af-info@atmforum.com). IETF RFC publications are available via

http://www.cis.ohio-state/htbin/rfc/rfc-index.html

or

ftp://nic.merit.net/documentsrfc/INDEX.rfc

5.9 PROBLEMS

1. Suppose the transport layer transfers fixed-size packets of N bits either as datagrams or over a virtual circuit. In the former case n_d bits are used to encode the destination address. In the latter case $n_c < n_d$ bits are used to encode the VCI. In addition, in the latter case, there is a fixed delay incurred to set up the connection. We measure this delay in terms of the time needed to transmit D bits. Suppose the source wants to transmit a message of M bits. Will the datagram or the connection-oriented service incur greater delay?

2. How many sequence numbers are needed to detect cell loss? How many are used in SMDS?

3. The packetization delay depends on the speed of the information transfer. Calculate the packetization delay for (a) 53-byte ATM cells and (b) a 1,000-byte packet transfer service for (1) voice samples that are sampled 8,000 times per second and encoded into a 64-Kbps stream; and (2) MPEG1, which takes 30 video frames per second and encodes them into a 1-Mbps stream.

4. The size of the ATM cell affects cell error rate. If the transmission system has a bit error rate (BER) of p, and there are N bits per cell, show that the cell error rate (CER) is

$$CER = 1 - (1 - p)^N \approx Np.$$

Suppose that a cell is retransmitted whenever it contains an error. Then, the average number of cells transmitted per error-free cell reception is $(1 - CER)^{-1}$. Suppose the overhead per cell is fixed at n bits, independent of cell size. Show that the average number of bits transmitted per error-free reception of one bit of information is

$$\frac{N + n}{N(1 - p)^{N+n}}.$$

For a given BER, p, and overhead n, the optimal cell size N minimizes this number. Taking $n = 40$ (5-byte overhead), find the best N for two extreme cases: $p = 10^{-9}$ and $p = 10^{-3}$.

5. How many simultaneous connections are needed to support all foreseeable needs, including telephone, video, text, etc. Will the 16-bit VCI be enough?

6. Compare the number of bits devoted for IP addressing with the requirements for ATM. Suppose you want to transmit voice in small IP packets so that the packetization delay is not more than X ms. How large a packet can you tolerate, and what is the overhead rate (ratio of number of overhead bits to packet size)?

7. Suppose voice is sampled and transported over ATM. Suppose the cell stream is subject to random delay. How would you characterize the resulting distortion?

6

CHAPTER

Control of Networks

In the preceding chapters we described the trends in packet- and circuit-switched networks that, together, culminate in the design of ATM networks and in improvements of the Internet protocols. We saw in Chapter 5 that a network that combines an ATM transport layer with a high-speed physical layer such as SONET can potentially provide the large range of quality of service (QoS) necessary to support most applications. However, in order to provide this range of QoS, the network's resources (bandwidth and buffers) must be properly managed or controlled.

In this chapter you will learn the concepts and fundamental techniques used to control circuit-switched, datagram, and ATM networks in order to achieve efficient use of network resources. You will understand how different control techniques affect different network performance measures, and you will acquire the skills to evaluate those performance measures, including blocking probability, delay, and loss. You will learn the deterministic proposals of the ATM Forum for admitting new connections and for allocating resources to those connections. You will see that those proposals are based on worst-case scenarios and that more sophisticated proposals based on effective bandwidth can realize the gains from statistical multiplexing. From Chapter 2 you already know the requirements imposed by various applications; now you will be able to calculate how well a specific application can be supported by the bandwidth and buffers allocated by the network for that application.

In section 6.1 we describe the objectives and the methods of control. We explain that with good control strategies the network can carry more traffic with the same quality of service. We discuss the meaning of quality of service, and we compare deterministic and statistical guarantees. We then classify the methods available to control the operations of circuit-switched, packet-switched, and virtual circuit-switched networks. There are four principal methods: admission control, routing, flow and congestion control, and resource allocation. A fifth method, control via pricing, is discussed in Chapter 8.

In section 6.2 we explain admission control and routing mechanisms for circuit-switched networks such as the telephone network. We point out the trade-off between accepting all calls that can be carried and the negative impact on future calls. A compromise is achieved by trunk reservations.

We discuss flow- and congestion-control procedures for datagram networks in section 6.3. Flow-control procedures attempt to limit the number of packets in transit in the network. An excessive number of packets in transit results in long queues in network nodes. Long queues cause large delays without increasing the throughput of the network. An effective flow-control procedure prevents such queues from building up while maintaining a steady flow of packets. The congestion-control mechanism is used to prevent some nodes of the network from becoming much more congested than other nodes. Whereas flow control regulates the flows of packets, congestion control adjusts the routing to circumvent congested nodes. Datagram networks, such as the Internet, do not guarantee delay or throughput. Their flow control and congestion control attempt to reduce the average delay per packet for any given throughput.

Section 6.4 introduces the formulations and some of the results on the control of ATM networks. Since ATM networks use virtual circuits and guarantee delays, throughput, and loss rates of individual connections, these networks must exercise admission control, and they must reserve bandwidth and buffer capacity for connections. One key question therefore is to determine how much of these resources the network must reserve. Resource allocations procedures determine which cells the network nodes should buffer or transmit. For instance, a node may transmit cells of a video connection before e-mail cells. The node may also discard audio cells to make room for data cells.

Chapter 7 carries out the mathematical analysis that justifies the results that we describe here.

Some of the material in this chapter relies on concepts of probability theory. Although we have attempted to provide intuitive discussion, readers with no probability background may not fully appreciate the argument in some places. Those readers can skim that material.

6.1 OBJECTIVES AND METHODS OF CONTROL

We provide an overview of the objectives of control and the principal means available or likely to be available to exercise control. These means involve taking decisions at very different time scales and based on different information. Some examples are given to illustrate the ideas.

6.1.1 Overview

Let us consider an ATM network that transfers information between users in cell streams over virtual circuits. A virtual circuit specifies a route of links and switches that connects the users. The cells in different virtual circuits share the transmission bandwidth and buffers that their routes have in common. The way in which those resources are shared is determined by the network's control strategy.

Because of fluctuations in the cell streams, there are periods of time when the cells arrive in a buffer faster than the transmitter empties that buffer. When this situation occurs, the buffer stores the excess cells. This temporary storage results in delays and, in case the buffer is full, in cell loss. Delays and losses affect the quality of service provided to the user.

With better control strategies, the network can carry more virtual circuits, while maintaining the quality of service promised to the users. This possibility has long been known in the case of the telephone network, where improved routing algorithms enable the network to carry more phone calls with the same blocking probability. Thus, by improving the control algorithms, network managers can provide better service without additional hardware. In a public network that sells services to users, better control leads to greater revenue. In a private network, better control results in lower cost for service.

There are four basic means of control: admission control, routing, flow and congestion control, and allocation control. As indicated in Table 6.1, the means available depend on whether the transport method used by the network is circuit, datagram, or virtual circuit switching.

6.1.2 Control Methods

Admission control determines which circuit or virtual circuit connection requests are accepted by the network. This is similar to the admission of telephone calls by the telephone network. When a subscriber places a request for

Network type	Admission	Routing	Flow	Allocation
Circuit switched	Free paths and costs	• Static (list) • Dynamic (least full)	None	None
Datagram	None	• Static (random) • Dynamic (shortest)	• Link window • End-to-end window	
Virtual circuit	Current VCs	Dynamic (largest spare capacity)	• Window • Rate	Priority, fairness, and QoS based

TABLE 6.1 Means of control of different types of network.

a new connection or call, the network can notify the user that it is too busy to accept that request. The subscriber can then decide to place the request at some later time or to use a competitor's network, if one is available. Normally, a datagram network always accepts the packets (datagrams) submitted by a user and does not exercise admission control. A number of issues are related to the admission of calls. For instance, a user may ask the network to schedule a connection at some later time. A user may ask the network to call back when it can set up a less-expensive connection. The problems of multiparty connections, such as conference calls, are also related to admission control. Can new parties join in? What happens if one party drops out?

When the network accepts a circuit or virtual circuit call request, it must decide the path or route to be followed by the bit stream of the call or the packets. This decision is called *routing* and remains fixed for the duration of the connection. In multiparty connections, the network must decide where to copy a stream it delivers to multiple destinations. What if different destinations require different versions of the packet stream? What if the characteristics of the path from the source to the destination change over time, as in mobile wireless networks?

A datagram network can modify its routing decisions to avoid congested parts of the network. This possibility is called *congestion control*.

For datagram and virtual circuit switching, the network can also decide whether bit streams should be forwarded along their routes quickly in order to reduce delay or whether they should be slowed down (throttled) in order to prevent congestion downstream. This decision is called *flow control*. A related method, called *traffic shaping*, is used by a traffic source to regulate its stream of packets before sending them to the network. Because a circuit-switched

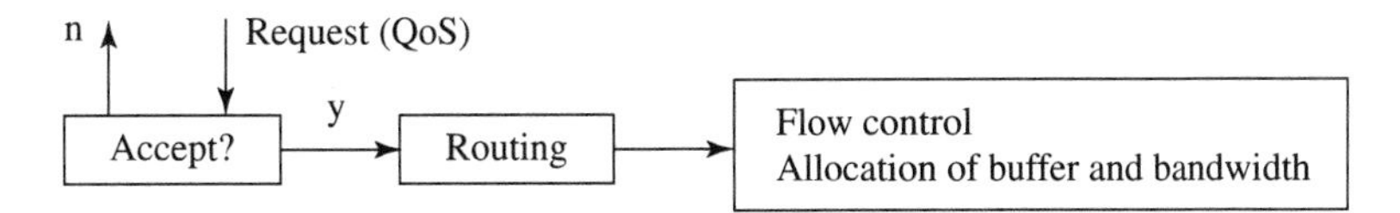

Notes: No admission control in datagram networks.
No flow control or dynamic allocation in circuit-switched networks.

FIGURE 6.1 Sequence of control actions for a given call.

network provides a constant bandwidth (and no buffers) to each connection, it does not exercise flow control.

Lastly, in virtual circuit switching, the network can control the bandwidth and buffers allocated to each virtual circuit. This is called (resource) *allocation control*. The allocation can be static, i.e., fixed at the beginning of the call request, or it may be dynamic and changed over the duration of the call. It is this flexibility that permits the network to provide connections with different qualities of service. Circuit-switched networks do not exercise allocation control.

We will return to discuss the remaining entries of Table 6.1. Figure 6.1 illustrates the order in which the control decisions are taken. The figure is almost self-explanatory. In the case of a virtual circuit network, for example, when a request for a connection with a particular QoS is received, the network must first decide to accept or reject it. If the answer is "yes," a route must then be assigned. Moreover, during the lifetime of the call, the network will exercise flow control and decide how much resources to allocate to the virtual circuit.

6.1.3 Time Scales

The four types of control decisions are taken at different time scales, and they are based on different types of information. We illustrate this for virtual circuit networks using Figure 6.2. The decision to accept a new call is based on the number of virtual circuits currently active in the network and on the typical statistics of the bit streams carried by these virtual circuits. The point is that the admission-control decision should not be based on the instantaneous traffic conditions; rather, it must be based on the typical conditions that are likely to prevail for the duration of the calls. Indeed, the decision to accept a call is irreversible and cannot be revoked if currently active calls suddenly become more busy than they were when the call was accepted.

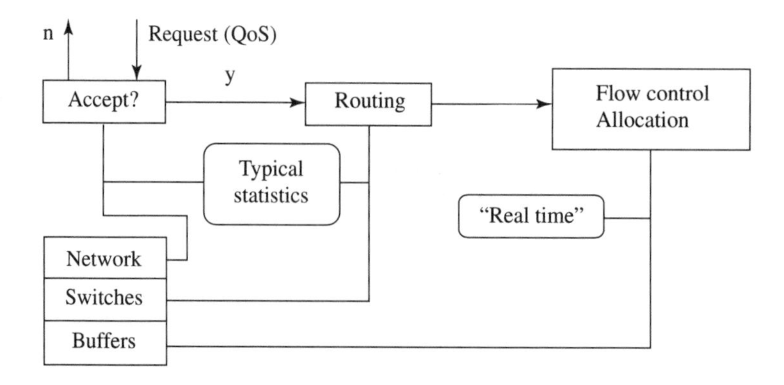

6.2
FIGURE
The different control actions are taken at different levels and based on statistics that are relevant on different time scales.

The same considerations apply to routing decisions, and so these must also be based on typical traffic statistics. Thus, call admission and routing are long-term decisions that have to be maintained for the duration of the calls. By contrast, flow- and allocation-control decisions can be based on instantaneous conditions and can be modified as these conditions change; hence the label "real time" in Figure 6.2. Voice calls, for example, typically last several minutes. If voice is transmitted over a high-speed ATM network, the voice cells must be switched within a few microseconds. Thus the time scales involved in the different control actions range over several orders of magnitude.

6.1.4 Examples

We now return to Table 6.1, discussing each row in turn. For a circuit-switched network, the decision to admit a new call is based on the paths that are then free or, equivalently, on the circuits that are busy (not free) when the call request is made. Thus, the network keeps track of which circuits are busy and uses that information to decide whether to accept a new call. The routing is also based on the circuits that are busy when the new call is placed. There are two types of routing algorithms: static and dynamic. A static algorithm uses precomputed decisions whereas a dynamic algorithm bases its decision on the current state of the network. There is no flow control and no allocation control in these networks, as we noted before.

A datagram network normally accepts all packets, so there is no admission control. The routing algorithm can be static or dynamic. The flow control in a datagram network typically uses windows, as explained in Chapter 2. The windows may correspond to individual links or to end-to-end paths through the network. For instance, the HDLC (High-Level Data Link Control) used by X.25 networks limits the number of unacknowledged packets between two successive nodes. This is a form of link-level window flow control. Specifically, HDLC uses the Go Back *N* protocol. The Selective Repeat (SRP) and Automatic Request (ARQ) protocols are other window flow-control protocols.

The TCP (Transmission Control Protocol) of Internet uses an end-to-end window flow control (SRP). Some datagram networks use a crude form of allocation control. For instance, TCP has a provision for expedited data transfer. A packet sent as expedited data is made to jump to the head of the queues that it goes through in the sending and receiving computers (not in the routers, since they are unaware of TCP and know only IP). RSVP, a proposed modification of IP, uses a more sophisticated resource allocation in the routers.

A store-and-forward network can also be used as a virtual circuit network (not separately indicated in Table 6.1). IBM's SNA (System Network Architecture) is an example of a virtual circuit store-and-forward network. Such a network accepts all requests for virtual circuits, so there is no admission control. The routing decisions are made for individual virtual circuits rather than for each packet, as in datagram networks. The networks use window flow-control methods, as in datagram networks. These networks also use a priority bandwidth allocation for expedited data packets.

ATM is the most important example of a virtual circuit-switched network. The QoS requirements in an ATM network can be more stringent than in datagram networks, and so an ATM network does not accept all virtual circuit requests. Were it to accept all the virtual circuit requests, it would be unable to deliver the cells with small loss and delay. Thus, ATM networks must control the admission of new virtual circuits. Some methods are being proposed for such admission control and also for routing, flow control, and allocation. We will discuss these proposals in section 6.4.

Another set of questions arises when different types of networks are interconnected. For instance, consider a datagram network attached to a virtual circuit network. If individual packets are sent from the datagram network to users through the virtual circuit network, then a virtual circuit needs to be set up for transporting the packets in the second network, as we explained in our discussion of IP over ATM in Chapter 5. If some form of quality of service is provided by the different networks, then they should collaborate to provide an end-to-end quality of service.

6.1.5 Quality of Service

We said earlier that with better control strategies the network can provide more connections or calls with the same quality of service. We will make the concept of QoS more precise.

In the best case, the QoS guarantees very small cell (or packet) loss rate and delay. By very small cell loss rate, we mean a loss rate comparable to the loss rate due to unavoidable transmission errors. For example, suppose the bit error rate along the fibers is about 10^{-12} and that a cell has 424 bits (53 bytes). Then the fraction of cells lost because of transmission errors is on the order of $424 \times 10^{-12} \approx 10^{-10}$. This is the least cell loss rate that could be promised to users. The smallest delay that could be promised would be comparable to the propagation delay. For example, the propagation time of a cell from San Francisco to Boston is on the order of 10 ms. Thus, in the best case, the network could promise a cell loss rate of about 10^{-10} and a delay on the order of 10 ms for cells going from San Francisco to Boston.

The network could propose inferior QoS. The worst quality is the so-called best-effort service, where the network promises to deliver the cells only if it finds the resources to do so. The range of QoS can be arranged from best to worst as in Figure 6.3. These different levels of QoS are similar to the different grades of service offered by the postal system. The users can choose overnight delivery, express mail, first-class mail, and so on, down to the lowest grade of bulk mail. The user selects the QoS based on the application. For instance, a user mailing a large number of advertisements might be satisfied if 95% of the customers receive them within a few weeks.

Just as in the postal system, users will select from the menu of QoS offered by the network that service which best meets the needs of their application. In Figure 6.4, the needs of several applications are displayed in a QoS parameter space of three dimensions: rate (bandwidth), loss, and delay. There are two points to remember. First, quality of service can have many aspects: in addition to the three listed in the figure, one may include security, reliability, and availability of the connection. However, rate, loss, and delay are the critical aspects for most applications. Second, the network must be controlled so that it will provide different QoS to different virtual circuits (users or applications) at the same time. Providing the best QoS for all the connections is wasteful, like sending all junk mail by express delivery, and it is therefore important that the network be able to handle different connections differently. The benefits of supporting many services on one common network (due to economies of scale and scope and network externality) appear to outweigh the costs of the increased complexity.

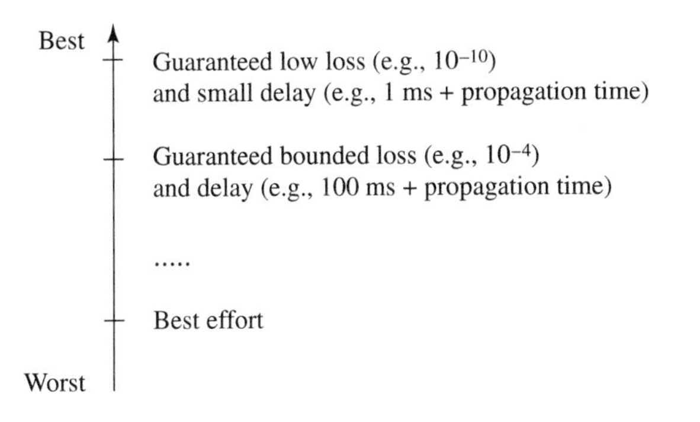

6.3 FIGURE

Future networks will be able to offer a wide range of qualities of service from low loss and low delay to best effort.

The guarantee of quality of service across a set of interconnected networks is an important but complicated issue. In a first approximation, the rate through a series of networks is the minimum of the rates, the delay is the sum of the delays through the individual networks, and the loss rate is also approximately the sum of the individual loss rates. The actual situation is more complicated because of the needed interfaces between networks and because the delay, loss, and rate are not measured by scalar quantities.

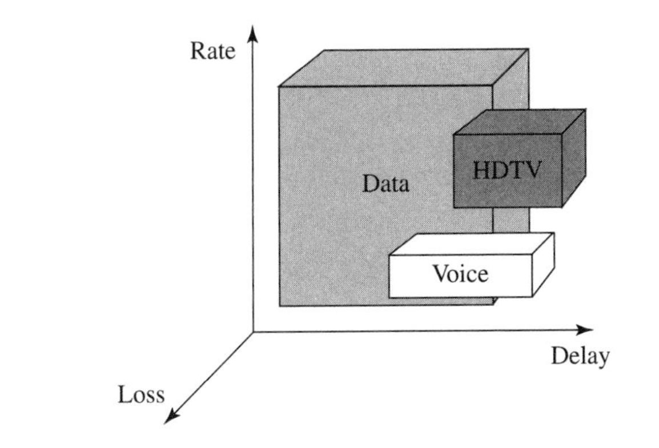

6.4 FIGURE

The quality of service specifies a number of parameters for the service. The figure illustrates three of the parameters and sketches their acceptable ranges of values for different types of services.

It is useful to think of the relation between the network and a user as being ideally defined by a (service) contract. The contract obligates the network to transfer the user's information with a defined QoS (delay, loss, rate, etc.) provided that the user's traffic conforms to specified limits (bit rate, burstiness, etc.). Thus the contract says that the network will provide a particular QoS provided the user's traffic conforms to specified conditions. With this view, the objective of network control is to fulfill the largest set of contracts.

6.2 CIRCUIT-SWITCHED NETWORKS

We first explain, using a very simple network, that the most important QoS for circuit-switched networks is the blocking probability. We then show how the blocking probability for circuit-switched networks depends on routing and admission control.

6.2.1 Blocking

Figure 6.5 illustrates a simple circuit-switched network. The network consists of two groups of telephone sets in two buildings. The telephone sets in each building are connected to a switch (a PBX, or private branch exchange) and the two PBXs are connected by two lines.

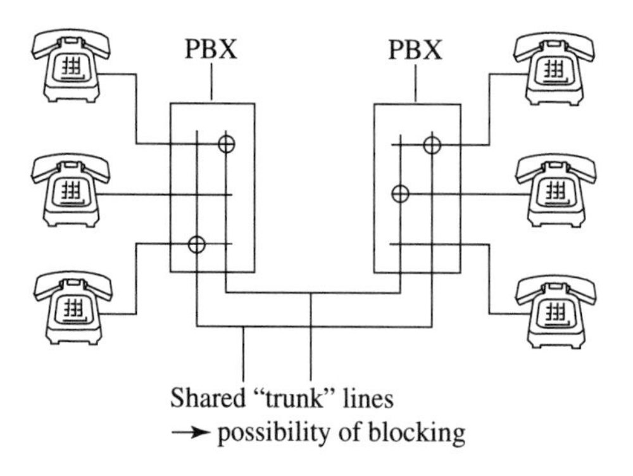

6.5 FIGURE

A small telephone network that consists of two switches and a few telephone sets. It may happen that all the lines between the switches are busy when a user attempts to place a new call. When this occurs, the new call is blocked.

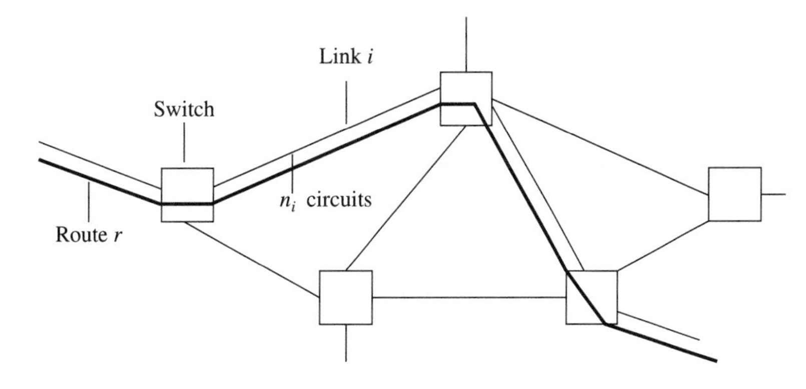

6.6 FIGURE Circuit-switched network. Switches are connected by links. Link i consists of n_i circuits. A route is a set of links that form a path.

You see that it is possible for a call to be requested between the two buildings when the two lines between these buildings are busy. Using a smaller number of lines between the buildings than might be needed in the worst case leads to the possibility of calls being blocked. Of course, in any realistic situation it would be impractical to use the maximum possible number of lines. For instance, if the two buildings had 100 telephone sets each, then 100 lines would be needed between the two buildings to allow all the telephone sets to be used at the same time for calls between the buildings. In practice, the likelihood of such a situation is exceedingly small, so that one may reasonably design the telephone system with a smaller number of lines.

The problem faced by the network designer is to select the number of lines between the buildings so that the blocking probability is small enough. We sketch the main features of the methods that the network designer uses. We present the underlying mathematical derivations in the next chapter.

Consider the network of Figure 6.6, which summarizes the basic model that network engineers use to analyze routing in circuit-switched networks. Switches are connected by groups of circuits called links (or trunks). For instance, link i is composed of n_i circuits. A circuit provides the bandwidth to transmit one phone call. A route r is a path in the network from one user to another user. A call in progress along a route uses exactly one circuit on each link along the route. There are R routes.

As customers place telephone calls, the network uses some routing algorithm to select the routes for the calls. Consequently, calls are placed along the various routes at random times. These calls have variable durations. For instance, assume that the network routes N_r calls along a particular route r every minute, on average. Assume also that the average duration of a call is T minutes. With these assumptions, we would expect that $\nu_r := N_r \times T$ calls are in progress at a typical time. Indeed, in T minutes, about ν_r calls are placed along route r, and these calls then terminate and are replaced by other calls. Because of the variability in the times when calls are placed and in the call durations, the actual number X_r of calls along route r may be somewhat larger or smaller than ν_r. Consider now a particular link i with its n_i circuits. The number Y_i of calls carried by link i is the total number of calls along routes that go through link i. Denote by R_i the set of routes r that go through link i. Then,

$$Y_i = \sum_{r \in R_i} X_r.$$

We expect Y_i to be approximately equal to $\sum_{r \in R_i} \nu_r$, but we know that Y_i has some probability of being larger than that value. When Y_i exceeds n_i for some link i along route r, the call is blocked.

This discussion shows that the probability that a call placed along route r is blocked because all the circuits of one of the links along the route are busy is some function of the rates $\{\nu_1, \ldots, \nu_R\}$. We denote that function by $B_r(\nu_1, \ldots, \nu_R)$. We explain algorithms for calculating the function B_r in Chapter 7.

6.2.2 Routing Optimization

We now formalize the best call-routing decision as a solution to an optimization problem. We are given the network topology, i.e., the number of circuits n_i of each link i between a pair of switches. We are also given the rate λ_{AB} of call requests (per unit of time) between every pair (A, B) of customer locations.

The routing problem is to select a route for each call so as to maximize the network revenues. We will explain several formulations of the routing problem depending on the information available and the computational effort one is willing to spend.

Static Routing

One possible formulation of the routing problem is to decide that a call between users at A and B is routed along route r_1 with probability $p(AB, r_1)$, along route r_2 with probability $p(AB, r_2)$, and so on. Thus, a route r is selected with probability $p(AB, r)$ for a call from A to B, and the call is routed along that route if there is a free path along that route. Otherwise, the call is blocked. With this specific routing assignment, we can calculate the rate ν_r of call requests along every route r from the rates λ_{AB}. We can then calculate the rate of network revenues

$$W := \sum_r w_r \nu_r [1 - B_r(\nu_1, \ldots, \nu_R)],$$

where w_r is the cost rate for calls along route r and $\nu_r[1 - B_r]$ is the rate of calls that route r actually carries because they are not blocked.

Thus, given the static routing decisions—the probabilities $p(AB, r)$, for every pair AB and route r—we can calculate the network revenues W. A good static routing algorithm must come up with routing decisions that yield large network revenues. One method that can be used to develop a good routing algorithm is to design an improvement step that specifies how the probabilities $p(AB, r)$ can be modified to increase W. If the improvement step is well designed, then its successive application should lead to routing probabilities that result in a large W.

Improvement steps based on properties of the gradient of W with respect to the routing probabilities have been proposed in the literature.

Dynamic Alternate Routing

In static routing, a call is blocked if there is no idle path along the assigned route, even if there is an idle path along a different route. The dynamic alternate routing algorithm takes the network congestion into account in deciding how a call should be routed.

In its basic form, the alternate routing algorithm uses a table that proposes two routes $r_1(A, B)$ and $r_2(A, B)$ for every pair (A, B) of customer locations. For instance, these could be the shortest two routes between the locations. An incoming call of type (A, B) is first assigned route r_1. If there is no idle path along this route, the call is assigned route r_2. If there is no idle path along r_2, the call is blocked.

The advantage of this algorithm over the static routing algorithm is that route selection depends on network congestion. (Of course, in order to implement this algorithm, the state of network congestion must be known.) One disadvantage of the basic dynamic alternate routing algorithm is metastability, as we explain next.

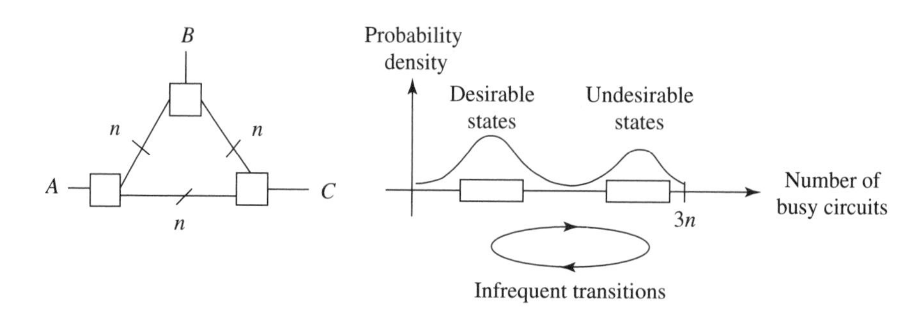

6.7 FIGURE

Illustration of metastability. The network on the left is symmetric and calls are placed with rate λ from each node. Each call is equally likely to be for each of the two other nodes. If the network routes a call whenever it finds either a direct path or an indirect (alternate) path, then the probability density of the number of busy circuits is as shown in the figure on the right. This figure illustrates the persistence (metastability) of undesirable network states.

Metastability and Trunk Reservations

Metastability is the property of a set of states of a dynamic system that can persist for a long time. When a circuit-switched network uses dynamic alternate routing, a set of states corresponding to a congested network can be metastable. This possibility is illustrated in Figure 6.7.

Consider the network shown in the left part of the figure. It is a symmetric network with three user locations, A, B, C, and three switches connected by links with n circuits each. Assume that the call requests are also symmetric. Specifically, assume that calls between any pair of locations are made with rate λ. The network uses the alternate routing algorithm. For every pair the preferred route r_1 is the direct route through one link, and the alternate route r_2 is the indirect route through two links.

Assuming that the call durations are independent and exponentially distributed, the vector of number of circuits busy in the three links is a Markov chain. (See Chapter 7 for the theory of Markov chains.) We can calculate the steady-state probabilities of that Markov chain by solving the balance equations. The result of this calculation is shown in the right part of the figure. The plot shows the steady-state probability distribution of the total number of busy circuits. This total number can range from 0 to $3n$. The plot shows two groups of values with a large probability: one group of small values and one group of large values. The analysis (not shown here) of the time evolution of the network reveals that the number of busy circuits can remain large for a long time before it gets smaller. That is, the state of the network can remain for a long time in a set of undesirable states where many circuits are busy. This set of states is almost stable and is said to be metastable.

We can explain the metastability of congested states as follows. Assume that the network is very congested and that a new call is requested. Since the network is congested, it is unlikely that the new call can be routed along the shortest route. As a result, it is likely that this new call will require two circuits and thereby increase the congestion even further.

Trunk reservation provides an effective protection against the metastability of undesirable congested states. When the routing algorithm uses trunk reservations, a number of circuits are reserved for routing calls along shortest routes. Let us describe how trunk reservation would be implemented in the network of the figure. First one chooses a number m smaller than n. When a new call between A and B is requested, the routing algorithm first tries to route the call along the shortest route, using the direct link. If this is not possible, then the algorithm attempts to route the call along the indirect route. That route will be used as long as there are at least m circuits between A and C and between C and B that are either used by shortest-route calls or that are free. Thus, m circuits of each link are reserved for routing calls along their shortest route. This reservation avoids the network reaching a state where most calls are routed along indirect routes, thereby using the network links inefficiently.

See the references cited in section 6.6 for a more detailed discussion of routing in circuit-switched networks.

Separable Routing

Separable routing is also a dynamic routing algorithm, since the route for a new call is chosen on the basis of the network congestion.

The separable routing algorithm calculates its routing decisions by performing one policy improvement step starting with the best static routing algorithm. To understand how this step is performed, consider the following situation. Say that the network is initially in state N. Here N is a vector that specifies how many calls are being carried along every possible route in the network. Denote by $V(N, t)$ the expected network revenues between time 0 and time t when the best static routing algorithm is used. Now assume that a new call is requested and that two routes, say 1 and 2, can be used for this call. If route 1 is used, then the state will jump from N to some other value N' to indicate that there is a new call being carried along route 1. If route 2 is used, the state jumps from N to N''. Which is the better decision? The policy improvement step decides that route 1 is preferable if $V(N', t) > V(N'', t)$ for all large t and that route 2 is preferable otherwise. Note that the policy improvement step assumes that the static routing algorithm is used at all subsequent times in order to compare the initial decisions.

The crucial question is how we can compare $V(N', t)$ and $V(N'', t)$. That is, we have to evaluate the impact of an additional call on the network revenues. This additional call has a finite holding time. During its holding time, the additional call increases the likelihood that subsequent calls are blocked, which reduces the total network revenues. The effect of one call on the blocking of other calls can be evaluated.

Admission Control

So far we have considered the routing decision, assuming a call is accepted. That is, we have shown how the blocking probabilities of the calls depend on the routing decisions. How should the network decide which calls should be accepted? The future networks will carry many different types of calls: audio, voice, data, video, etc. These calls differ in terms of their resource needs, revenue generated, request rate, and call duration. Consequently, a good admission policy must base its admission decision on the set of calls currently carried by the network and on the type of call being requested.

We examine a simple model of this admission control problem in section 6.4. We also present a more complex model in Chapter 7.

6.3 DATAGRAM NETWORKS

When a network transports data as datagrams, it first decomposes the data into packets of variable size. The packets are then sent one by one from node to node along a path to the packet destination. Each packet contains a special field that specifies its destination address. Thus, in a datagram network, the nodes store and forward the individual packets, one at a time. The routes taken by successive packets may be different, even if they go from the same source to the same destination. Also, because the packet sizes may be different, as suggested in Figure 6.8, the transmission times of the packets are different, since they are equal to the packet lengths divided by the transmission rate. In our study of datagram networks we begin by describing a model that we will use to formulate the control questions.

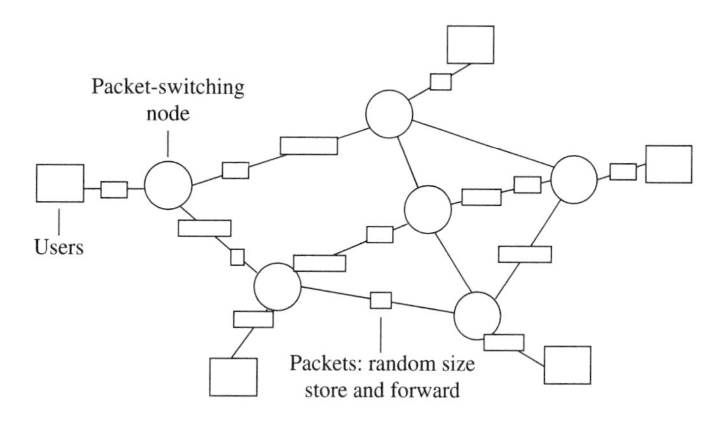

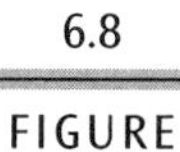

6.8 FIGURE Datagram networks. Packets of different length are transported in store-and-forward manner.

6.3.1 Queuing Model

The queuing model of Figure 6.9 can be used by the designer to predict the transmission delays and to design good routing and flow-control algorithms. The top part of the figure shows one node with its components: a receiver converts the optical signal on the incoming fiber into packets. The packets are stored into memory and are then retransmitted on one outgoing fiber.

The bottom part of the figure shows an abstract representation of the same node. The packets are viewed as "customers" or "jobs," in the language of queuing theory, arriving at random times into a queue. The customers wait for their turn to be served and the service times are random. The service time

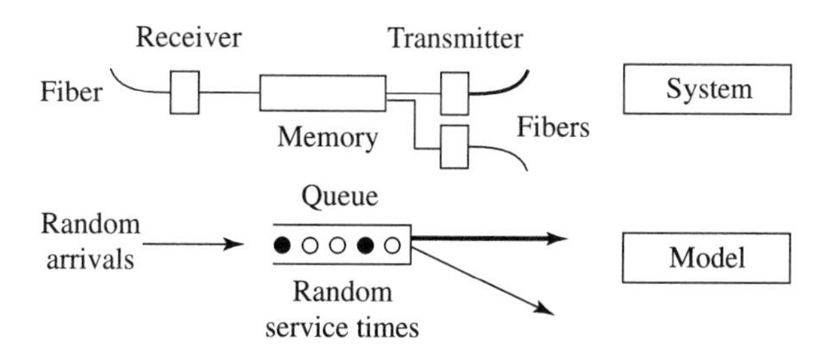

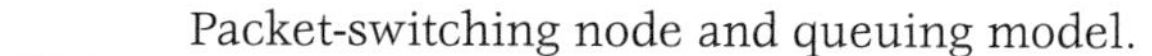

6.9 FIGURE Packet-switching node and queuing model.

is the length of the packet in bits divided by the transmission bit rate. The service time is random when the packet length is random. If the packets all have the same size, as in ATM, the service time is deterministic. Thus, the fluctuations of the arrival times and of the transmission times are modeled by random variables. The specific assumptions about the distributions of these random variables depend on the precise model being used.

6.3.2 Key Queuing Result

The most useful result of queuing theory for the analysis of datagram networks concerns the network shown in Figure 6.10. That result is the formula for the average delay per packet in such a network.

The network consists of a collection of nodes connected by links. The packets that arrive from outside of the network are assumed to form Poisson processes and the packet transmission times in the various nodes are assumed to be independent and exponentially distributed. We use the following notation. At each node j, packets arrive at rate λ_j packets/s, and the service rate is μ_j packets/s. Thus μ_j is the transmission rate in bits per second divided by the average packet size in bits. It is assumed that $\mu_j > \lambda_j$ for all j. Then the average delay faced by a packet entering the network is

$$T = \frac{1}{\gamma} \sum_j \frac{\lambda_j}{\mu_j - \lambda_j}, \tag{6.1}$$

where γ is the total rate at which packets arrive into the network.

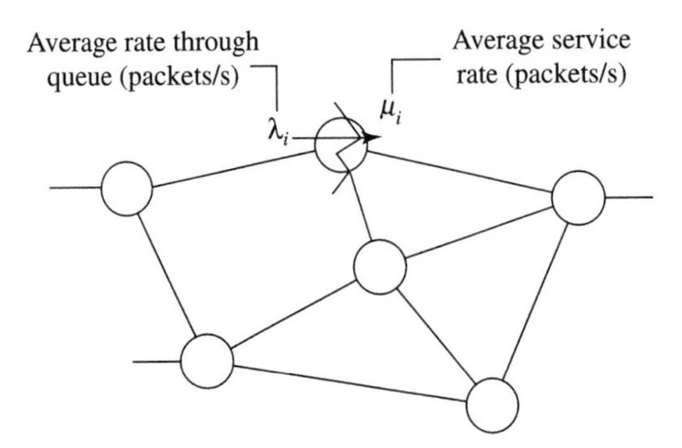

6.10 FIGURE

Queuing network. The symbol λ_i indicates the average rate of packets going through node i (in packets per second) while μ_i is the average transmission rate of that node, also in packets per second, when the node is nonempty.

The assumptions are not exactly satisfied in actual networks. For instance, the packet lengths are not exponentially distributed. Typically, the packets that travel on the Internet tend to have a bimodal distribution: most packets are either short or long, and the fractions of short and long packets are not consistent with an exponential distribution. Also, since the length of a packet does not change as the packet travels through the network, the transmission times of a packet at the different nodes are not independent. In fact, if one knows the transmission time of a packet at one node, then one can determine the length of that packet and therefore its transmission times in all the other nodes.

Although these assumptions are not always valid, the formula for the average delay per packet provides a reasonably good estimate of the actual value of that average delay in a real network. This simple formula is the starting point for the construction of routing and flow-control algorithms. (The formula is usually conservative, i.e., the actual delay is smaller than that predicted by the formula.) We derive the delay formula (6.1) in section 7.3.3.

6.3.3 Routing Optimization

We explain how the delay formula is used to design good routing strategies. We first consider static routing.

Static Routing

Consider the simple network illustrated in Figure 6.11. Packets arrive with rate λ. The network can send these packets along two different routes to their common destination. Each route consists of one node, i.e., of one buffer

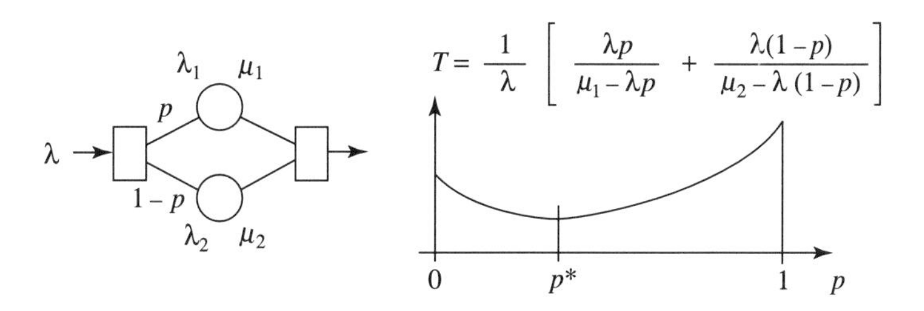

FIGURE 6.11 Static routing optimization. Packets are sent to node 1 with probability p independently of one another and to node 2 otherwise. The right part of the figure shows the average delay T per packet through the network as a function of p.

equipped with a transmitter. Our objective is to choose the fraction p of packets that should be sent along the first route so as to minimize the average delay per packet in the network.

We assume that the packets arrive as a Poisson process and that their lengths are independent and exponentially distributed. The two links use transmitters with different rates. Consequently, the two nodes 1 and 2 are modeled as queues with exponential service times with different rates.

To use formula (6.1), we need to identify the parameters in that formula. The parameter γ is the total rate of packet arrivals into the network. That rate is λ. Thus, $\gamma = \lambda$. The delay formula contains a sum over all the network nodes. For each node j, λ_j designates the average rate of packets going through that node, and μ_j is the average service rate of that node. Here, $\lambda_1 = \lambda p$ and $\lambda_2 = \lambda(1-p)$. The service rate μ_1 of the first node, in packets per second, is equal to the rate of the transmitter in that node, in bits per second, divided by the average length of a packet, in bits. The rate μ_2 is obtained in a similar manner. Substituting these values in the delay formula gives

$$T = \frac{1}{\lambda}\left[\frac{\lambda p}{\mu_1 - \lambda p} + \frac{\lambda(1-p)}{\mu_2 - \lambda(1-p)}\right].$$

The right part of Figure 6.11 is a plot of the average delay T as a function of p for typical values of λ, μ_1, and μ_2. That plot shows that the average delay per packet T is minimized for some value p^* of p. Thus, there is an optimal way of splitting the traffic between the two routes.

We now consider the general case. The optimal static routing problem for a general datagram network is to minimize the average delay per packet T with respect to all routing probabilities.

The network sends a packet leaving node i to node j with probability p_{ij}. Given these routing probabilities p_{ij}, we can calculate the average rates of flow λ_i through the nodes i by solving the following flow-conservation equations:

$$\lambda_i = \gamma_i + \sum_j \lambda_j p_{ji}, \quad \text{for all } i.$$

In these equations, γ_i denotes the rate of arrivals of packets from outside the network into node i. The equations say that, for each i, the rate of flow λ_i through node i is equal to the external arrival rate into that node γ_i plus the sum over all nodes j of the fraction p_{ji} of the rate λ_j of flow leaving that node j and being sent to node i. If the network is open, i.e., if all the packets that enter the network can eventually leave it, then the flow-conservation equations have a unique solution $\{\lambda_i,\ i = 1, \ldots, J\}$.

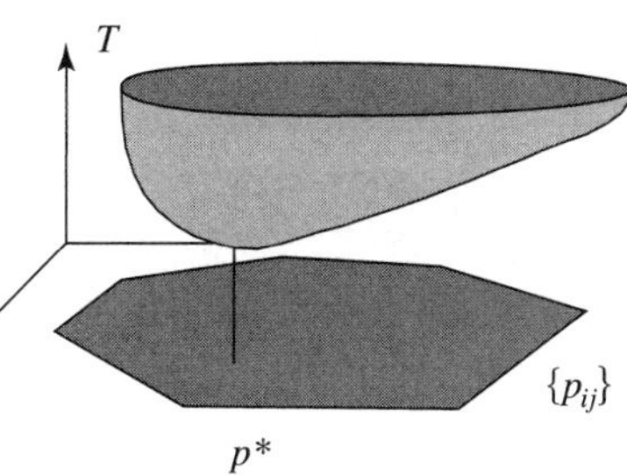
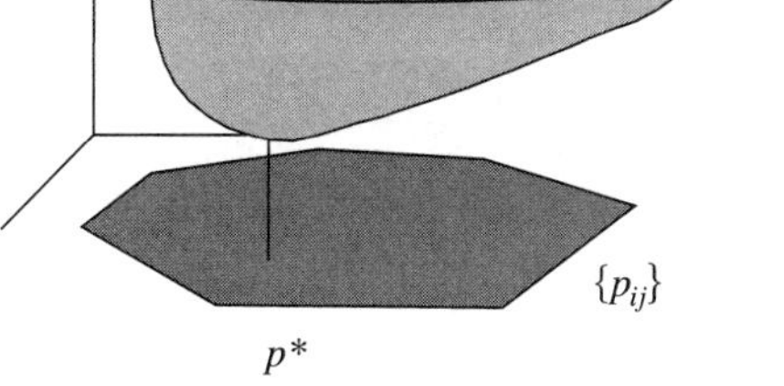

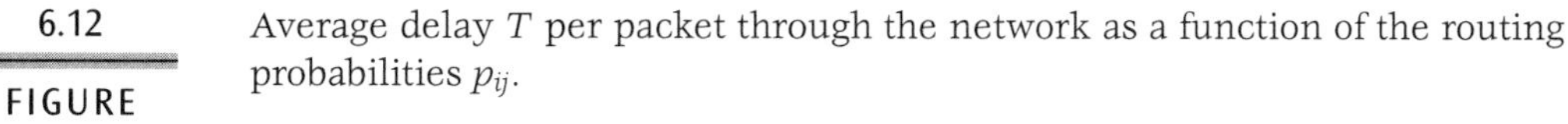

6.12 FIGURE

Average delay T per packet through the network as a function of the routing probabilities p_{ij}.

Thus, given the external rates $\{\gamma_i,\ i = 1, \ldots, J\}$, the flow-conservation equations enable us to calculate the rates $\{\lambda_i,\ i = 1, \ldots, J\}$ as a function of the routing probabilities $\{p_{ij}\}$. Once these rates are determined, we can use our formula (6.1) to compute the average delay T. Figure 6.12 sketches the delay T as a function of the routing probabilities p_{ij}.

The delay T is a complicated function of the routing probabilities p_{ij}. The minimization of T does not result in a closed-form expression for the optimum routing probabilities. Instead, one must use a numerical minimization algorithm. For instance, a gradient projection algorithm can be used to obtain the optimal routing probabilities.

Dynamic Routing

Instead of selecting (fixed) probabilities to route the packets when they leave the nodes, we can devise a dynamic routing algorithm that bases the routing decisions on the actual backlogs of the nodes. We illustrate such an algorithm for the simple network of Figure 6.13. The left part of the figure shows a network in which packets arriving as a Poisson process with rate λ can be sent along one of two routes. Each of the two routes is modeled as a single node with exponential service times.

We assume that the routing probability p can be based on the queue lengths x^1 and x^2. That is, when a packet arrives, the routing controller looks at the two queue lengths x^1 and x^2 and sends the packet along route 1 with probability $p(x^1, x^2)$ and along route 2 otherwise.

The designer of the routing algorithm must find the function $p(x^1, x^2)$ that, when used by the routing algorithm, minimizes the average delay per packet in the network. The solution, i.e., the best function, is illustrated in the right part of Figure 6.13. This function specifies that when x^1 is large and x^2 is

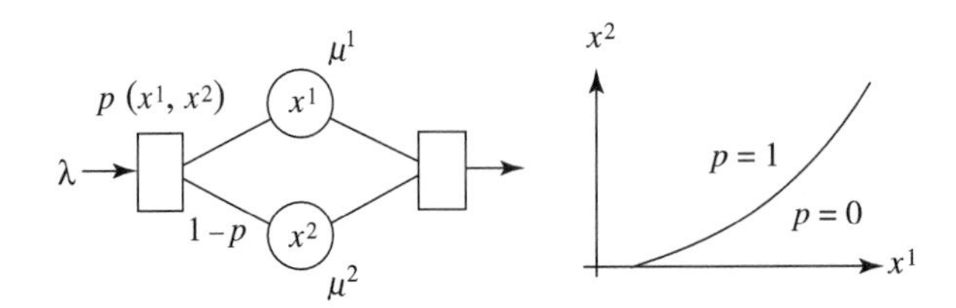

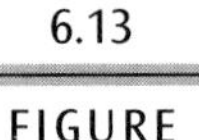
6.13 FIGURE

Dynamic routing. When the queue lengths are x^1 and x^2, an arriving packet is sent to queue 1 with probability $p(x^1, x^2)$. The graph in the right-hand part of the figure shows the function $p(. , .)$ that minimizes the average delay per packet through the network.

small, the packets should be sent along route 2, and vice versa. Moreover, if a packet should be sent along route 2 when the queue lengths are x^1 and x^2, then the same routing decision should be taken when x^1 is larger or when x^2 is smaller.

The proof of these intuitively obvious structural properties of the function $p(x^1, x^2)$ turns out to be rather involved, even for such a simple network. In fact, very few structural results of this type are known for more complicated networks. Moreover, such structural results do not appear to yield improved procedures for calculating the optimal dynamic routing algorithm.

We now turn to the case of a general network. The derivation of the optimal dynamic routing algorithm for a network with many nodes is a formidable problem that is still beyond the reach of current approaches. Consequently, approximations are necessary. Moreover, the state of the network is not known instantaneously, so that, even if it could be derived, the optimum dynamic routing algorithm would not be implementable. As a result of these limitations, network engineers have developed simple heuristics that can be implemented. Two such heuristics are the Bellman-Ford algorithm and the distributed-gradient algorithm.

Bellman-Ford algorithm Figure 6.14 summarizes the setup of the Bellman-Ford algorithm. The model is a network of nodes connected by links. The average delay on each link is estimated by the corresponding transmitter. One possible estimation method is for the transmitter on each link to keep track of the backlog in its buffer and to calculate the average delay by dividing the total number of bits stored in the buffer by the transmission rate. The propagation time of signals along the link can be added for improving the estimate.

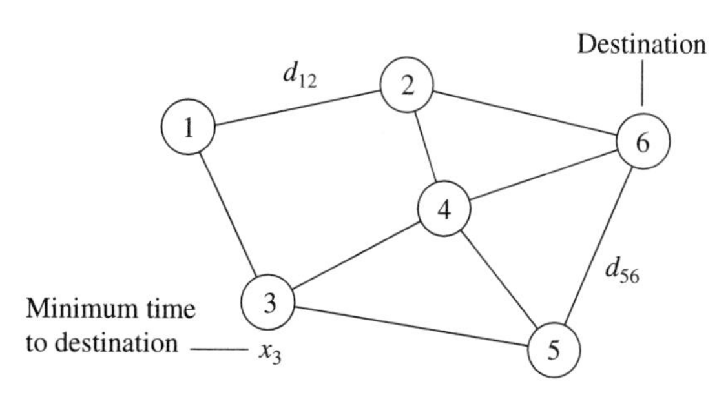

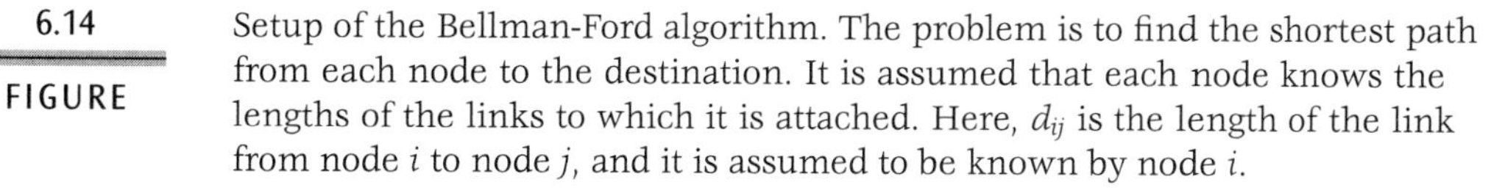

6.14 FIGURE Setup of the Bellman-Ford algorithm. The problem is to find the shortest path from each node to the destination. It is assumed that each node knows the lengths of the links to which it is attached. Here, d_{ij} is the length of the link from node i to node j, and it is assumed to be known by node i.

To explain the calculations performed by the Bellman-Ford algorithm, let us assume that the delay d_{ij} on the link from node i to node j has been estimated for all pairs of nodes. Let us denote by x_i the minimum delay between node i and some fixed destination. The minimum delay x_i must satisfy the equation

$$x_i = \min_j \{d_{ij} + x_j\}. \tag{6.2}$$

These equations are of the form $x = F(x)$, where x designates the vector with components x_i. Thus, the vector x satisfies fixed-point equations. These fixed-point equations can be solved by the recursion

$$x^{n+1} = F(x^n).$$

It can be shown that this recursion will converge to the vector of minimum delays for any nonnegative initial vector x^0. The resulting algorithm is called the Bellman-Ford algorithm.

Once the minimum delays have been calculated, the fastest path to the destination is easily identified: a packet that leaves node i should be sent to node j^* if j^* is the value of j that achieves the minimum in equation (6.2).

Distributed-gradient algorithm The delays through the links of a network depend on the traffic along those links. Consequently, a routing algorithm can be improved by taking into account the effect of routing decisions on the traffic and therefore on the delays. The distributed-gradient algorithm estimates that effect.

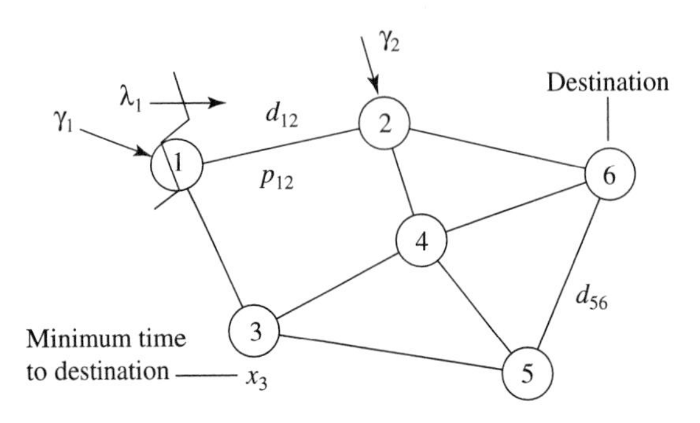

6.15 FIGURE Setup of the distributed-gradient algorithm. The problem is to find the routing probabilities that minimize the average delay from each node to the destination. Packets arrive from outside of the network into node i with rate γ_i. The average delay from node i to node j, d_{ij} is a function of the rate of packets through that link.

The situation is illustrated in Figure 6.15. Denote by γ_i the rate of traffic entering the network via node i and by λ_i the average rate of flow through node i. Also, let d_{ij} represent the delay along link ij and x_i the average delay from node i to a specified destination. By p_{ij} we designate the fraction of traffic leaving node i that is sent to node j. The problem is to determine the values of these routing probabilities that minimize the average delays from the nodes to the destination. In this algorithm, in contrast to the Bellman-Ford algorithm, the link delays are assumed to depend on the rates of traffic.

The key idea of the algorithm is expressed in the following formula:

$$\frac{\partial x_i}{\partial p_{ij}} = \lambda_i \left[\frac{\partial d_{ij}}{\partial \lambda_{ij}} + \frac{\partial x_j}{\partial \gamma_j} \right], \qquad (6.3)$$

where λ_{ij} designates the rate of traffic through link ij. This formula calculates the derivative of the delay between node i and the destination with respect to p_{ij}. This formula may be understood by multiplying both of its sides by ϵ. When p_{ij} is increased by a small value ϵ, the rate through the link ij increases by $\lambda_i\epsilon$. This rate increase has two effects. First, it increases the delay along link ij by $\lambda_i\epsilon$ multiplied by the derivative of d_{ij} with respect to the traffic rate along the link ij. Second, it increases the traffic rate through node j. The increase in the traffic through node j also increases the delay from node j to the destination and will therefore increase the delay from node i to the destination by the same amount. The formula expresses these two effects.

To appreciate how the formula can be used, let us assume that each transmitter can estimate the derivative of the form

$$\frac{\partial d_{ij}}{\partial \lambda_{ij}}.$$

Equation (6.3) then provides a recursive procedure for evaluating the terms $\partial x_i / \partial \gamma_i$. Suppose node i is attached to the destination s by one link is. The corresponding equation (6.3) is then simply

$$\frac{\partial x_i}{\partial p_{is}} = \lambda_i \frac{\partial d_{is}}{\partial \lambda_{is}}$$

(since $x_s = 0, x_i = d_{is}$, because a packet originating at s and destined for s encounters no delay), and that formula enables us to calculate

$$\frac{\partial x_i}{\partial \gamma_i} = \frac{1}{\lambda_i} \frac{\partial x_i}{\partial p_{is}}.$$

In this way, the terms on the left-hand side of (6.3) can be evaluated for the nodes that are directly attached to the destination. These values can then be used together with equation (6.3) to evaluate the terms corresponding to nodes that are two links away from the destination, and so on.

To implement this algorithm, the nodes must broadcast the values of their gradient estimates to their neighbors. The readers should consult the references for details on this algorithm. It should be noted that this algorithm is subject to undesirable oscillations. Remedies for these oscillations have been devised.

6.3.4 Flow Control

Flow control is the name of control procedures that throttle the flow of packets along its path to avoid parts of the network from becoming excessively congested.

Figure 6.16 illustrates the usefulness of flow control for an elementary example. Consider a node that is used by two traffic streams represented by the differently shaded arrows. Assume that the traffic flowing on the two top arrows is guaranteed a delay T less than some specified value T_{max}.

To meet this guaranteed bound on the delay, the source of the traffic represented by the lower arrows must be prevented from entering the node when more than $c \times T_{max}$ Mb are already buffered in that node. Here, c denotes the transmission rate in Mbps.

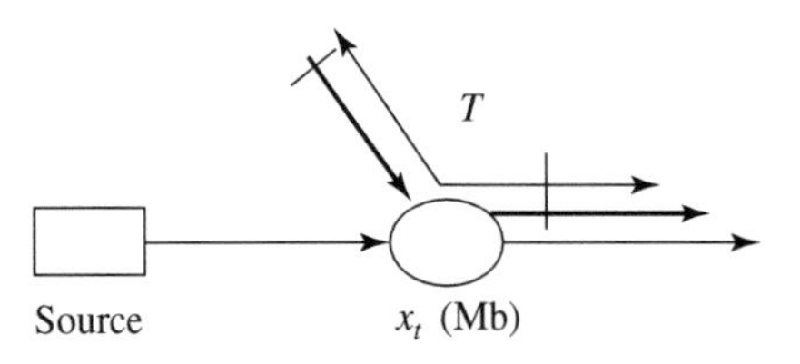

6.16 FIGURE

Illustration of flow control. Two types of packets go through the same node. We assume the packets are served in their order of arrival. If the packets of one type are guaranteed a bounded delay, then the packets of the other type must be stopped when the backlog exceeds a given size.

The resulting flow-control procedure for the traffic flowing along the lower arrows is called a window flow control. A more involved example is explained next.

Window Flow Control

The left part of Figure 6.17 shows a source and a destination connected by a path that consists of three fibers and two intermediate nodes (packet switches). There is also another path for transmitting acknowledgments from the destination to the source. The switches are also used by other streams that are not shown in the figure.

The network uses an end-to-end window flow control that limits the total number of packets along the path from the source to the destination to a max-

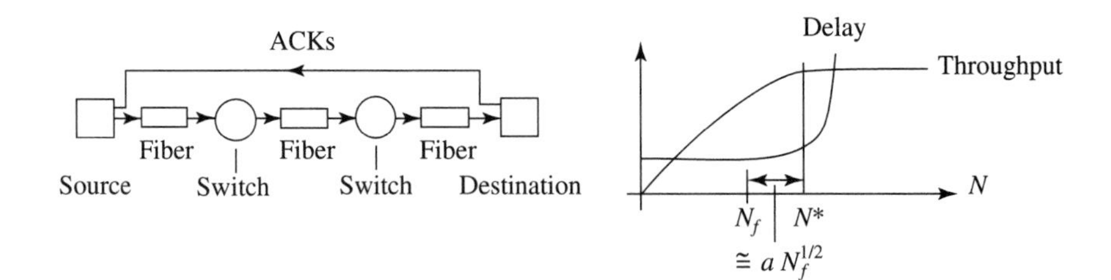

6.17 FIGURE

Window flow control. At most N packets are allowed to be unacknowledged. Assuming that the packet transmission times are independent and exponentially distributed, the graph on the right shows the delay and the throughput as a function of N. These graphs suggest that a reasonable choice for N is N^*.

imum value N. The source implements that window flow control by stopping its transmissions when there are N unacknowledged packets in the network.

The network designer must choose the value of N. The network throughput, i.e., the rate at which the source can transmit packets, and the delay of packets both increase with N. Under specific statistical assumptions, analysis reveals that the delay starts increasing fast with N when the throughput begins to level off as a function of N. The right part of Figure 6.17 illustrates this phenomenon. Thus, once N has a value sufficiently large, a further increase of N results in a small increase in the throughput and in a large increase in the delays. One can then argue that the most suitable value of N is the value N^* that corresponds to the "knee" of the delay curve.

Under specific assumptions, the value N^* can be shown to be slightly larger than the number N_f of packets that can be stored in the fibers. This value of N_f is calculated by taking the product of the propagation time of light through the fibers times the transmission rate and by dividing this product by the average number of bits per packet. The figure gives a more precise estimate of N^* derived under the same assumptions.

Rate Flow Control

Window flow control is not suitable when the capacity of the fiber is utilized by a small number of sources. This is because by the time the acknowledgment of the first packet by the destination reaches the source, a very large number of packets will have been transmitted if the source does not stop in the meantime. Thus, by the time the destination can signal the source that some congestion is occurring, it is probably too late for the source to slow down its transmissions. (See problem 10.)

In addition, window flow control necessitates the establishment of two-way connections to transmit acknowledgments, and this complicates the operations of the network.

To avoid these problems, network researchers are proposing the use of rate-based flow control. Instead of limiting the number of packets in the network sent by each source, a rate-based flow control limits the average rate at which sources transmit packets. This control is easier to implement, since it only requires each source to monitor its transmission rate.

One particular implementation of rate-based flow control being recommended for ATM networks is the *leaky-bucket* controller. This control mechanism regulates the traffic by smoothing out the bursts of packets that the source would otherwise transmit. We explain this next.

6.4 ATM NETWORKS

In this section we attend to the control of virtual circuit networks in general and of ATM networks in particular.

We first outline the problems of control of virtual circuit networks. We explain that the user must describe its traffic in a way that can be enforced and monitored. Moreover, the traffic description must enable the network to control the traffic efficiently.

We then study results obtained by using deterministic models. Such deterministic models form the justification of most current recommendations for the control of ATM networks. In our view, this approach is unnecessarily conservative. Its main, and important, merit is that it is simple to implement.

We conclude the chapter by discussing statistical procedures for specifying traffic characteristics and for controlling the network. We believe that by adopting such procedures the network can derive significantly higher revenues. However, these statistical procedures require further study. Our presentation outlines a few possibilities. We hope that this section will invite network engineers to study such methods further.

6.4.1 Control Problems

We explained in section 6.1 that the user and network enter in a contract in which the network guarantees a quality of service while the user assures that the traffic will obey specific bounds. Implicit in any contract are provisions for verifying that the terms of the contract are observed by the various parties. Thus, the network must be able to verify that the traffic obeys its descriptors and the user that the quality of service is acceptable. The objective of the control is to maximize the network revenues while meeting the contracted quality of service of the connections. The quality of service specifies the loss rate and delay of the transfer of bits (or cells, packets).

Other descriptors of the quality of service include security and reliability. Specific temporal characteristics of losses and of delays may be important in particular applications. In multimedia applications, for instance, two users may be connected by a collection of virtual circuit connections. The quality of service might specify bounds on the synchronization offset between the virtual circuit connections. As you see, even the formulation of the problem lends itself to many variations. In this chapter, we limit our attention to loss rate and delay.

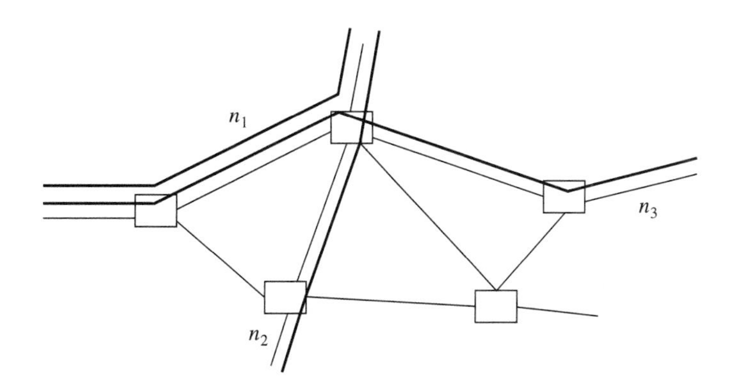

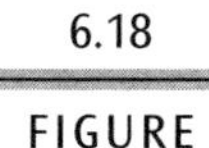
6.18 FIGURE

Network control. The network engineer needs to determine whether a given set of connections can be carried by the network.

We also noted in section 6.1 that a virtual circuit network can exercise four types of control action: admission, routing, flow control, and allocation of buffers and bandwidth.

To properly design the control procedures of the network, the network engineer must be able to evaluate the performance of specific control procedures. That is, if we consider for example the network shown in Figure 6.18, the network engineer should be able to determine whether the loss rates and delays of the different virtual circuit connections are acceptable. The answer to that question depends on the characteristics of the traffic generated by the different connections, on the flow-control and allocation strategies, on the buffer capacities in the switches, on the rates of the transmitters, on the propagation delays and bit error rates of the links, and on the error-correction procedures that the network uses.

The engineer faces three major difficulties in trying to determine the acceptability of given connections. First, the users are often interested in very small loss rates and relatively hard bounds on delays or delay jitter. These loss rates and delay bounds are sensitive to the details of the traffic and are difficult to calculate even for a single node with precise models of the traffic. Second, the network operator never knows precisely the characteristics of the traffic of connections. For instance, even the *mean* rates of compressed video streams vary substantially across movies. Third, the characteristics of the traffic of the virtual circuit connections are modified when they go through nodes and interact with other connections.

One approach to these complex questions is to play it safe and to adopt conservative procedures. This is the approach followed to date by the ATM Forum, as we explain in the next section. These procedures are based on deterministic bounds on the traffic transmitted by the user. The user can enforce these deterministic bounds by using a mechanism called the *leaky bucket*, which we describe also in the next section. The network can verify that the bounds are met by using the same mechanism. Moreover, the network can guarantee deterministic delay bounds and also that no loss occurs when buffers overflow. These procedures based on deterministic bounds may lead the network to carry only a subset of the calls it could carry with statistical procedures.

Deterministic procedures are based on a worst-case analysis and they do not take into account the likelihood of such worst-case occurrences. Such procedures cannot use the fact that many connections are very unlikely to behave in the worst possible way at the same time. Consequently, deterministic procedures do not take advantage of the benefits of statistical multiplexing. To take advantage of these benefits, the network must adopt statistical procedures. When using a statistical procedure, the user specifies statistical descriptors of the traffic.

Three questions arise with such a statistical approach. The first question is how the user can enforce statistical characteristics. The second question is how the network can verify that the traffic satisfies these statistical characteristics. The third question is how the network can use the statistics to decide which calls to accept and which to reject.

We explain later practical statistical methods for specifying and verifying traffic characteristics and sketch procedures for controlling the network.

6.4.2 Deterministic Approaches

In this section we review the ATM Forum recommendations and network control procedures based on these recommendations.

ATM Forum Recommendations

The ATM Forum recommendations for the user-network interface specify a mechanism for describing the traffic flowing through a virtual circuit connection. This mechanism is the *generalized cell rate algorithm* (GCRA). That mechanism defines the five service categories specified by the ATM Forum: constant bit rate (CBR), variable bit rate (VBR) either real time or non-real time, available bit rate (ABR), and unspecified bit rate (UBR) as we explain below.

The GCRA has two parameters, T and τ, and it times the arrivals of cells as follows. The algorithm defines a *theoretical arrival time*, *tat*, of a cell. If the next cell arrives before $tat - \tau$, then the algorithm GCRA(T, τ) declares that cell to be *nonconformant*, and *tat* is unchanged. If it arrives at time $t \geq tat - \tau$, then the cell is *conformant* and the algorithm resets the value of *tat* to $\max\{t, tat\} + T$.

For instance, consider GCRA(10, 3) with the initial value of $tat = 0$ and assume that the arrival times of cells are 1, 6, 8, 19, 27. The first cell is conformant since $1 \geq 0 - 3$. The algorithm updates *tat* to $\max\{1, 0\} + 10 = 11$. The second cell is nonconformant since $6 < 11 - 3$. The third cell is conformant since $8 \geq 11 - 3$, and the algorithm sets $tat = \max\{8, 11\} + 10 = 21$. The fourth cell is conformant since $19 \geq 21 - 3$, and the algorithm sets $tat = \max\{19, 21\} + 10 = 31$. The fifth cell is nonconformant since $27 < 31 - 3$. Summarizing, the decisions of the algorithm are C, NC, C, C, NC where C means conformant and NC nonconformant.

This algorithm is equivalent to a leaky bucket, which we describe next and which we will use to study admission control. Fluid accumulates at a given rate in a bucket that can store up to $T + \tau$ units of fluid. Fluid that arrives when the bucket is full is lost. A cell that arrives when the bucket contains less than T units of fluid is nonconformant. A cell that arrives when the bucket contains at least T units of fluid is conformant, and it removes T units of fluid from the bucket. Let us denote by $F(t-)$ the amount of fluid in the bucket just before time t and by $F(t+)$ the amount of fluid just after time t. The leaky-bucket algorithm is such that a cell is nonconformant if $F(t-) < T$ and then $F(t+) = F(t-)$. Otherwise, the cell is conformant and $F(t+) = F(t-) - T$. If the next cell arrives s time units later, then $F(t + s-) = \min\{F(t+) + s, T + \tau\}$.

In our previous example, assume that the bucket contains $T + \tau = 13$ units of fluid at time 0. Then, $F(1-) = 13, F(1+) = 3, F(6-) = 8 = F(6+), F(8-) = 10, F(8+) = 0, F(19-) = 11, F(19+) = 1, F(27-) = 9 = F(27+)$. Consequently, the successive decisions are C, NC, C, C, NC.

The ATM Forum recommends that *constant bit rate* (CBR) traffic on a line with rate R should specify its peak cell rate *PCR* and its cell delay variation tolerance *CDVT*. The meaning of these parameters is that the cells should be conformant for the

$$\text{GCRA}\left(\frac{R}{PCR}, CDVT\right)$$

algorithm.

Thus, if $R/PCR = 5$ and $CDVT = 0$, then the peak cell rate is 20% of the line rate, and the cells arrive at multiples of five cell transmission times. If $R/PCR = 5$ and $CDVT = 1$, then the fastest arrival times are $-1, 4, 9, 14, 19$, and so on. This sequence corresponds to a periodic stream with rate equal to 20% of the line rate: one arrival every fifth cell transmission time. If $R/PCR = 4.5$ and $CDVT = 1$, then cells can arrive at times $-1, 4, 8, 13, 17, 22, 26, 31$, and so on. This sequence has rate $R/4.5$. An easy way to see why the maximum rate of a CBR stream is indeed PCR is to recall that a cell takes away $T = R/PCR$ from the leaky bucket, which is filled at a specified rate. Thus, over a long time interval with duration t, t units of fluid enter the bucket and at most t/T cells can arrive. The maximum rate is therefore $1/T = PCR/R$ cells per cell transmission time $1/R$ second, or PCR cells per second.

The intuitive meaning of GCRA(R/PCR, $CDVT$) is that the cells can arrive at their peak cell rate PCR but do not have to be exactly periodic. The $CDVT$ measures the departure from exact periodicity. Such departure may be necessary because of the framing structure that carries the cells. For instance, imagine a CBR stream that corresponds to 3.5 cells every frame time of a physical layer. In one implementation, the successive frames might carry 3 or 4 cells, and such framing introduces a cell delay variation. Similarly, multiplexing and the injection of maintenance cells introduce a cell delay variation. Note that a late cell delays the set of all future acceptable arrival times. Thus, it is not correct to think of a CBR stream as having arrival times being multiples of R/PCR with a shift of $CDVT$.

For *variable bit rate* (VBR) traffic, the ATM Forum specifies the PCR, $CDVT$, burst tolerance (BT), and the sustained cell rate (SCR). The meaning of these parameters is that the cells should be conformant for both

$$\text{GCRA}\left(\frac{R}{PCR}, CDVT\right)$$

and

$$\text{GCRA}\left(\frac{R}{SCR}, BT + CDVT\right)$$

algorithms.

The motivation for this definition of VBR is that such a stream might be very bursty and send a number of back-to-back cells separated by idle periods. The BT parameter makes such bursts acceptable.

For instance, consider a VBR stream with $R/PCR = 1, R/SCR = 20, CDVT = 0, BT = 57$. If we think back about the leaky-bucket interpretation of the

GCRA($R/SCR, BT + CDVT$) = GCRA(20,57) algorithm, we see that a conformant cell takes away 20 units of fluid. With an initial content of $T + \tau = 77$ units of fluid, we find that four cells can arrive back to back, at times $0, 1, 2, 3$. Indeed, $F(0-) = 77, F(0+) = 57, F(1-) = 58, F(1+) = 38, F(2-) = 39, F(2+) = 19, F(3-) = 20, F(3+) = 0$. A new group of four cells can then arrive at time 80 because $F(80-) = 77$. Thus, bursts of four cells can arrive at times $0, 80, 160$, and so on. Such a stream is also conformant for the GCRA($R/PCR, CDVT$) = GCRA(1, 0) since this controller makes all the streams conformant. The bursty stream that we constructed has a long-term average rate equal to 4/80 = 1/20, since it has bursts of size 4 every 80 time units. Thus, the sustained cell rate of the stream SCR is indeed such that $R/SCR = 20$.

As another example, consider the parameters $R/PCR = 5, R/SCR = 10, CDVT = 16, BT = 20$. This stream can have bursts of five consecutive cells with bursts at time 0, 1, 2, 3, 4, 50, 51, 52, 53, 54, 100, 101, 102, 103, 104, 150, 151, and so on. The sustained cell rate is 1/10 of the line rate (five cells every 50 cell transmission times).

The ATM Forum is currently developing specifications for *available bit rate* (ABR) transmissions. The operating principle is that an ABR connection may have a guaranteed minimum cell rate and imposed peak cell rate. The actual rate available to the connection varies between these two bounds on the basis of feedback information about the congestion in the network and destination. The rate adjustment scheme is a time-varying GCRA control that operates end to end and is rate-based. The ATM Forum does not specify the rate-control algorithm that end systems and switches must use but defines general mechanisms. The ATM cells contain a flow-control field that the switches fill to indicate the congestion level they experience. Also, the sources and destinations can send resource management (RM) cells.

In a typical scheme, the sources periodically send RM cells to indicate their desired rate. The switches then compare the requests of various VCs and distribute their spare bandwidth accordingly. The switches indicate their bandwidth allocation in returning RM cells that go back to the source. In addition, the switches set congestion indication bits in the forward cells and in the returning RM cells. The destination monitors the congestion indication bits of the arriving cells and marks the corresponding bit of the RM cells accordingly. The source uses the congestion indication bits of the RM cells and the bandwidth allocation values of these cells to modify the parameters of its GCRA. Many variations are possible and are being explored. For more information, the reader should keep track of the revised versions of the ATM Forum recommendations. Thus, ABR is an attempt to use feedback to better utilize the network.

The last service being specified by the ATM Forum (in addition to CBR, VBR, and ABR) is *unspecified bit rate* (UBR). UBR is best-effort service with no guarantee on the quality of service.

Admission Control

How many GCRA(T, τ) connections can go through a buffer equipped with a transmitter that sends C cells per unit of time if the delay through the buffer must be less than D units of time? Here, one unit of time is a cell transmission time on each one of the incoming connections.

To answer this question we first study the following problem. We want to find the *fastest* sequence of cells that is conformant to the GCRA(T, τ). We say that a sequence of arrival times $\{x_1, x_2, \ldots\}$ is faster than another sequence of arrival times $\{y_1, y_2, \ldots\}$ if $x_k \leq y_k$ for all $k \geq 1$. We construct this fastest sequence by filling up the leaky bucket with $T + \tau$ units of fluid at time 0 and by sending a cell and removing T units of fluid as soon as the bucket contains T units of fluid.

Assume now that a GCRA(T, τ) stream goes through a buffer equipped with a transmitter that transmits C cells per unit of time. We want to analyze the maximum backlog of the buffer, assuming that it is initially empty. The maximum buffer backlog will occur if the cell stream is the fastest. If $C > 1/T$, then the buffer is faster than the arrival stream (since $1/T$ is the maximum long-term rate of a GCRA(T, τ) stream) and the maximum backlog is finite.

Figure 6.19 shows the evolution of the buffer occupancy for the fastest streams possible with given GCRA parameters. Case (1) is unstable. This instability is caused by C being smaller than $1/T$. Case (2) is stable since $C > 1/T$. By comparing cases (3) and (4) we observe the effect of the burstiness of the stream.

If N streams that conform to GCRA(T, τ) share the same buffer, then the maximum occupancy of the buffer occurs again when the streams are the fastest. We present a few examples in Figure 6.20. By comparing these cases we observe that multiplying the number of sources N and the service rate C by the same factor V results in multiplying the worst-case occupancy by V. This should be expected since both the number of arrivals and the number of departures in each time step are multiplied by V.

A good approximation of the maximum backlog in the buffer with rate C and N GCRA(T, τ) streams can be derived as follows. Assume that $\tau \gg T$. Then $K + 1$ cells can be sent back-to-back by a single stream if

$$\tau + T - KT + K \geq T.$$

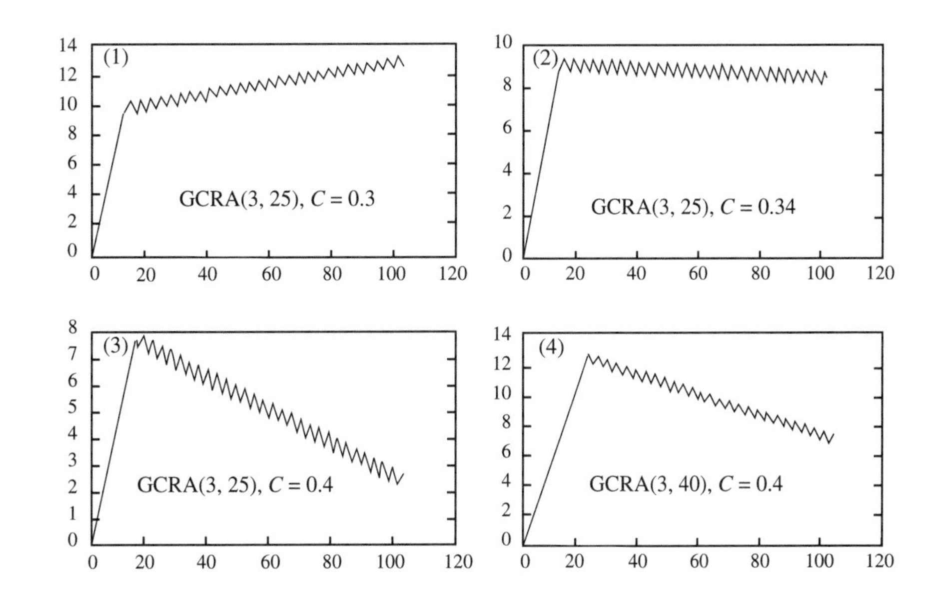

6.19 FIGURE

Buffer occupancy for the fastest streams with the parameters indicated. In (1), the system is unstable because C is too small. The other cases show the effect of the burstiness of the stream on the backlog.

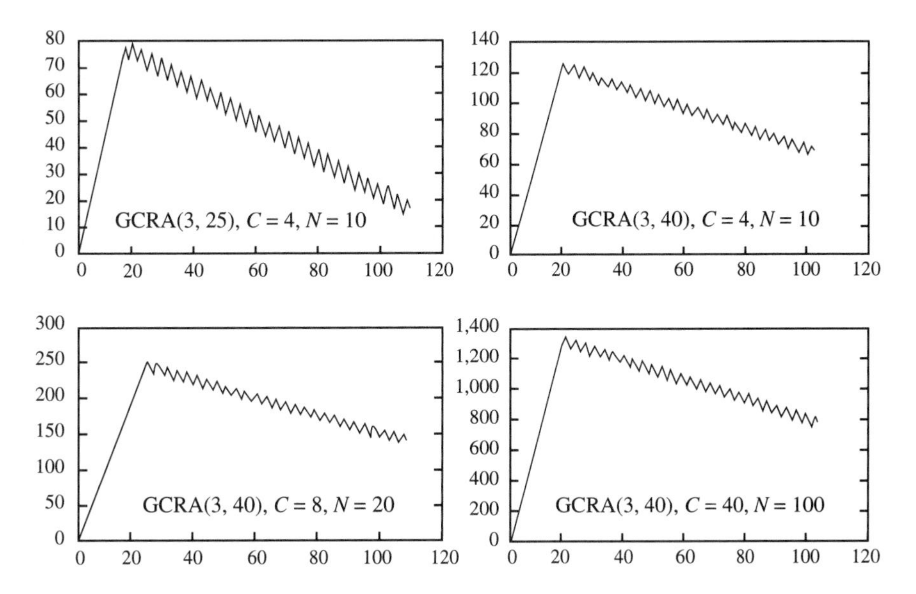

6.20 FIGURE

Buffer occupancy when N fastest streams possible with GCRA(T, τ) share a buffer with service rate C.

Indeed, during the time interval $[0, K)$, K units of fluid enter the leaky bucket and KT units of fluid are removed by the first K cells. The left-hand side of the above inequality is the amount of fluid that is left at time K, and this amount should be larger than T if the $(K+1)$th cell is conformant. The maximum number of back-to-back cells is therefore approximately equal to

$$M := \frac{\tau}{T-1} + 1 = \frac{\tau + T - 1}{T-1}. \tag{6.4}$$

After these back-to-back cells, the other cells follow each other approximately every T time units, the time required to collect enough fluid in the leaky bucket to send a new cell. Our description neglects round-off effects that are not significant for our analysis. If each of N streams produce these M back-to-back cells during the interval $[0, M-1]$, then the buffer accumulates $N \times M$ cells and can serve only $(M-1) \times C$ cells, so that the backlog is approximately $B(T, \tau, C, N)$ where

$$B(T, \tau, C, N) := N \times M - (M-1) \times C \approx N\frac{\tau + T - 1}{T-1} - \frac{C\tau}{T-1}.$$

If $T \times C > N$, as must be assumed for stability, then the cells stop accumulating after the burst of back-to-back cells, so that B is the maximum backlog.

For the numerical examples of Figures 6.19 and 6.20 this formula gives

$$B(3, 23, 0.34, 1) = 8.59, B(3, 25, 0.4, 1) = 8.5, B(3, 40, 0.4, 1) = 13$$
$$B(3, 25, 4, 10) = 85, B(3, 40, 4, 10) = 130$$
$$B(3, 40, 8, 20) = 260, B(3, 40, 40, 100) = 1300.$$

Comparing these numbers with the figures shows that this approximation is satisfactory.

Let us go back to the question that we asked at the beginning of this section. Assume that the maximum acceptable delay through the buffer is D cell transmission times. We want to estimate the maximum number N of GCRA(T, τ) streams that the network can accept.

To answer the question, we observe that the maximum backlog must be at most $D \times C$ cells. Indeed, a cell that faces a backlog of $D \times C$ experiences a delay equal to D. Consequently, the maximum number N is obtained by solving

$$B(T, \tau, C, N) = D \times C,$$

i.e.,

$$N \times M - (M-1) \times C \approx N\frac{\tau + T - 1}{T - 1} - \frac{C\tau}{T - 1} = D \times C.$$

By solving this equation we find

$$N \approx C\frac{D(T-1) + \tau}{T + \tau - 1}.$$

Recall that these derivations assume that $C > N/T$. An equivalent way to look at the above equation is to write the constraint on N as

$$N \times \alpha_G(T, \tau, D) \leq C, \tag{6.5}$$

where

$$\alpha_G(T, \tau, D) := \max\left\{\frac{T + \tau - 1}{D(T-1) + \tau}, \frac{1}{T}\right\}. \tag{6.6}$$

We can interpret the inequality (6.5) as stating that each GCRA(T, τ) connection with maximum delay D requires a bandwidth $\alpha_G(T, \tau, D)$ given by (6.6). We can call $\alpha_G(T, \tau, D)$ the *effective bandwidth* of a GCRA(T, τ) connection with maximum delay D.

If we recall the formula (6.4) for the maximum number M of back-to-back cells, we find that

$$\alpha_G(T, \tau, D) := \max\left\{\frac{M}{D + M - 1}, \frac{1}{T}\right\}. \tag{6.7}$$

Thus, if D is small compared with M, the effective bandwidth is close to one. That is, the switch must treat such a source as a constant bit rate source with a rate equal to the line rate R. At the other extreme, if D is much larger than M, then the effective bandwidth is close to M/D and about D/M can be accommodated (so long as $T > D/M$).

For instance, one finds that

$$\alpha_G(3, 40, 30) = 0.42, \alpha_G(3, 40, 20) = 0.52, \alpha_G(3, 40, 10) = 0.70.$$

Using the ATM Forum parameters for a VBR connection, we see that

$$\alpha_G(R/SCR, BT + CDVT, D) = \max\left\{\frac{\beta + 1}{\beta + D}, \lambda\right\}, \tag{6.8}$$

where we introduce the parameters

$$\lambda := \frac{SCR}{R} \text{ and } \beta := \frac{BT + CDVT}{R/SCR - 1}. \tag{6.9}$$

This effective bandwidth is a measure of the cost of carrying such a VBR connection. Note that the cost increases as the acceptable delay decreases. The cost also increases with the burstiness measured by β.

The delay constraint becomes binding if D is small enough or if β is large enough. Otherwise, the bandwidth is determined by the average rate λ.

Summarizing this section, we have explored the implications of the GCRA control mechanism recommended by the ATM Forum. We have analyzed the maximum number of connections that can go through a buffer subject to a maximum delay constraint. The analysis is based on the worst-case behavior of the connections and ignores the likelihood of such behavior. The result can be viewed as requiring an effective bandwidth per connection. The effective bandwidth increases with the burstiness of the connection and decreases with the acceptable delay.

Pricing Calls

In this section we explore the pricing implications of deterministic approaches. We study a simple model that highlights some features of the problem.

Consider calls of two types: 1 and 2. Each call requires resources of two classes A (e.g., buffer space) and B (e.g., bandwidth) in order to be carried by the network. Assume that a call of type i requires a_i units of resources of class A and b_i units of class B (for $i = 1, 2$). The total amount of resources of classes A and B available are α and β, respectively.

Denote by n_i the number of calls of type i carried by the network for $i = 1, 2$. The set of admissible pairs (n_1, n_2) is

$$S := \{(n_1, n_2) \in \{0, 1, 2, \ldots\}^2 | a_1 n_1 + a_2 n_2 \leq \alpha \text{ and } b_1 n_1 + b_2 n_2 \leq \beta\}.$$

This set is illustrated in Figure 6.21. Assume for the time being that the parameters of the problem are such that the two oblique boundary lines cross each other at some point X, as shown in the figure.

We consider that the network will charge p_i for a call of type i $(i = 1, 2)$ per unit of time, and we examine the unit prices (p_1, p_2) that the network can charge in a competitive environment. To simplify the problem, we assume that the demand is larger than the supply, so that we can attract as many customers as we can carry, if our price for the service is competitive. Let us call the network **N**.

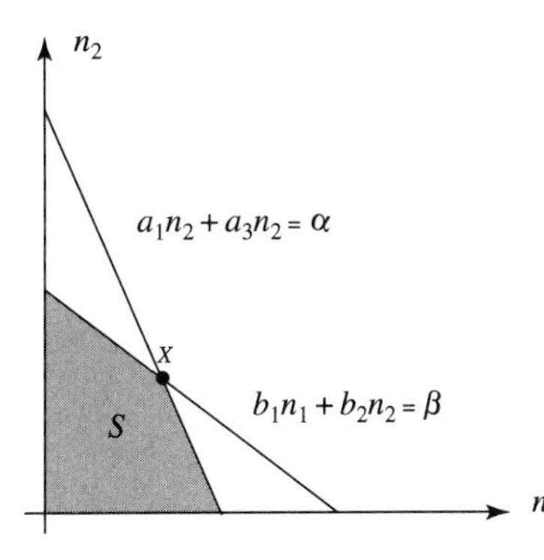

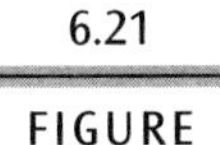
6.21 FIGURE

Set S of acceptable numbers of calls of two types that require resources of two classes. We show that when the constraint lines intersect as here at a point X, a competitive network should operate at the intersection point.

We define the competitive environment as follows. Assume that **N** offers the prices (p_1, p_2) and that some other network **N**$'$ offers the prices (p'_1, p'_2). (**N**$'$ has the same resources, α and β.) If the revenues per unit of time R of **N** and R' of **N**$'$ are such that $R' \geq R$, then it must be that $p_i \leq p'_i$ if **N** carries calls of type i, for $i = 1, 2$. The interpretation of this condition is that if **N**$'$ is as profitable as **N**, to stay in business **N** cannot charge more than **N**$'$; otherwise, **N** would lose its customers.

Network **N** wants to find the prices it can charge to collect revenues at rate R. (R may equal the cost of the resources plus return on investment.)

Our first observation is that if the network chooses to carry only calls of type 1, then it can obtain the rate of revenues R by charging $p_1 = Ra_1/\alpha$ since the network can carry α/a_1 calls of type 1 (see figure). Consequently, it must be that

$$p_1 \leq Ra_1/\alpha.$$

Indeed, if **N** charges (p_1, p_2) with $p_1 > Ra_1/\alpha$, then it cannot compete with a network **N**$'$ that charges $(Ra_1/\alpha, \infty)$ and carries only customers of type 1.

Similarly, we find that

$$p_2 \leq Rb_2/\beta.$$

Some algebra shows that there is a pair of prices (p_1^*, p_2^*) that satisfy the above two conditions with strict inequalities that yields the revenues R when the network operates at the operating point X of the figure. Moreover, when

N offers these prices, the only networks that can be competitive with N must offer the same prices and operate at the same point X.

Thus, when the boundary lines cross, it is optimal (necessary) for the network to carry both types of calls. To make the discussion more concrete, consider that calls of type 1 are CBR with rate λ and that calls of type 2 are VBR with GCRA(T, τ) where $\tau \gg T$.

Imagine that n_1 calls of type 1 and n_2 calls of type 2 go through a buffer with capacity B and with transmitter rate C. No losses are acceptable. The first condition required is

$$n_1\lambda + n_2\frac{1}{T} \leq C,$$

so that the transmitter can keep up with the average rate of the calls.

The second condition concerns the buffering. In the worst case, calls of type 2 send M back-to-back cells followed by periodic cells every T time units, where

$$M = \frac{\tau + T - 1}{T - 1},$$

as we explained in the previous section. The maximum backlog in the buffer occurs at the end of the M back-to-back cells and is equal to $M(n_1\lambda + n_2 - C)$. Thus, the second condition is

$$n_1 M\lambda + n_2 M \leq B + CM.$$

The two constraint lines cross each other if and only if

$$B < CM(T - 1) = C(\tau + T - 1).$$

6.4.3 Statistical Procedures

In this section we describe statistical procedures for specifying traffic and quality of service and for controlling the network.

Traffic Models

Network engineers use a few different stochastic models for specifying traffic in high-performance networks. Stochastic fluid models view traffic as a fluid with a randomly varying rate. Queuing models view packets or cells as discrete entities that arrive one at a time or in finite batches.

Markov-Modulated Fluids

A Markov-modulated fluid (MMF) is a nonnegative function of a finite continuous-time Markov chain. Accordingly, an MMF is given as $r(z_t)$, where $\{z_t, t \geq 0\}$ is a continuous-time Markov chain on some finite state space $\mathbf{Z}$ with rate matrix $Q = \{q(i,j), i,j \in \mathbf{Z}\}$ and $r : \mathbf{Z} \to [0, \infty)$ is a given function. (r may be measured in bits or cells per second.)

The interpretation of the definition is that $r(z_t)$ is the rate at time t of some bit stream. The rate fluctuates randomly. This model is motivated by observing the bit stream that a variable bit rate video compression algorithm generates.

Assume that the MMF $r(z_t)$ goes through a buffer equipped with a transmitter with rate c. We want to analyze the evolution in time of the buffer occupancy process x_t. The analysis enables us to determine how large the buffer capacity should be and the delays that the MMF faces through the buffer.

We show in Chapter 7 that the buffer occupancy process has an invariant distribution of the form

$$P(x_t > x) = \sum_{z \in \mathbf{Z}} a(z) e^{-\beta_z x}, x \geq 0.$$

In this expression, $\{\beta_z, z \in \mathbf{Z}\}$ are the eigenvalues of the matrix $A = [A(i,j), i,j \in \mathbf{Z}]$, where

$$A(i,j) = q(i,j)/(r(j) - c).$$

To determine the coefficients $a(z)$, one must calculate all the eigenvalues and eigenvectors.

To simplify the analysis, one usually approximates this sum of exponentials by a single exponential,

$$P(x_t > x) \approx Ke^{-\beta x}, \text{ for } x \gg 1, \tag{6.10}$$

where β is the smallest of the eigenvalues $\{\beta_z, z \in \mathbf{Z}\}$ and K is the corresponding coefficient. The calculation of K requires computing all the eigenvalues and eigenvectors.

A useful example is when $r(z(t))$ is the sum of the rates of N independent and identically distributed sources, each modeled by on-off Markov fluid. That is, each source is modeled by a Markov chain on $\{0, 1\}$ with rate matrix such that $q(0, 1) = \lambda > 0$ and $q(1, 0) = \mu > 0$. When the source is in state 1, it produces fluid at rate 1. When it is in state 0, the source does not produce fluid. Assume that the superposition of these N sources goes through a buffer with transmission rate C. Then one finds

$$\beta = \frac{1 + \lambda - N\lambda/C}{1 - C/N} \tag{6.11}$$

and

$$K = \left(\frac{N\lambda}{C(1+\lambda)}\right)^N \Pi_{z=1}^{N-[C]-1} \frac{\beta_z}{\beta_z - \beta}. \tag{6.12}$$

To estimate K one must solve for all the eigenvalues β_z, which can be done numerically easily as long as N is not too large.

This approach has the merit of yielding complete results for simple models. However, the analysis of networks with Markov-modulated fluid sources using this approach appears beyond reach.

In the following section we discuss a different approach.

More General Model

The Markov-modulated traffic model describes the random fluctuations of the rate of a bit stream. A more general traffic model is a random process $\mathbf{A} := \{A(t), t \geq 0\}$ that specifies the number $A(t)$ of bits carried by the traffic during the interval $[0, t]$ for $t \geq 0$. Without further assumptions, this model is too general for us to be able to analyze how the network can transport such traffic. Somehow we must specify the average rate of the traffic and some measure of burstiness.

The average rate is $[A(t+T) - A(t)]/T$ for large T. Under suitable assumptions, this average rate is a well-defined quantity λ. That is, $[A(t+T) - A(t)]/T$ approaches λ as T increases, and that value λ does not depend on the realization of the stochastic process $\mathbf{A}$ nor on t. The process $\mathbf{A}$ has these properties if it is stationary and ergodic.

Averaging Rate Fluctuations

The rate of a traffic stream fluctuates. For instance, the peak rate of a video stream may be 10 times larger than its average rate. To prevent losses, a network node could allocate to each stream a bandwidth equal to the peak rate of that stream. However, such an allocation is overly conservative and it is sometimes possible for the transmitter to allocate a bandwidth much closer to the average rate than to the peak rate. There are two fundamentally different methods that a network can use to reduce the bandwidth it must allocate to each connection. The first method is multiplexing many sources. The other method is buffering.

The multiplexing method exploits the fact that different sources fluctuate independently so that when the rate of a source is larger than average, the rate of another may be smaller than average. Consequently, the rate of the superposition of many sources tends to be close to its average value. This observation is similar to the fact that if one throws 1,000 fair coins, about 500 of them land on heads. The probability that more than 600 coins land on heads is very small, about 10^{-10}. Thus, if each of 1,000 sources is off 50% of the time and transmits at rate α the other 50% of the time, then the total rate of the sources rarely exceeds $600 \times \alpha$. The analysis of the multiplexing method determines how many sources must be multiplexed and the transmission rate per source required so that the probability that the total rate of the sources exceeds the transmitter rate is smaller than some specified value, say 10^{-10}.

The buffering method uses the fact that over a long duration $[t, t+T]$ a stream A with rate λ produces a total number of cells close to λT. Consequently, if $c > \lambda$, then $A(t+T) - A(t) \leq cT$ with a large probability. Now, if it were true that $A(t+T) - A(t) < cT$ for all $t \geq 0$, then a node that transmits the stream with rate c would delay it by at most T. To see this, note that the buffer cannot remain nonempty for T consecutive time units. Indeed, if the buffer is empty at time t and nonempty during $[t, t+T]$, then during that interval it has output cT bits and it must be that more than cT bits entered the buffer, i.e., $A(t+T) - A(t) \geq cT$, a contradiction. If the buffer cannot remain nonempty for T consecutive time units, then every bit that enters must leave before T time units. This argument shows that, by using a buffer, a node can transmit at a rate c only slightly larger than the average rate λ. The node delays the bit stream by at most T if $A(t+T) - A(t) \leq cT$ for all $t \geq 0$. In the statistical approach, we do not insist that this inequality hold all the time, but only that it hold with a large probability. We then conclude that the node delays the bits by at most T with a large probability. The analysis of this method estimates the transmitter rate needed so that the node delays the bit stream by more than T seconds with some specified small probability, say 10^{-10}.

How do these methods compare? Multiplexing is effective for real-time traffic and buffering is effective for non-real-time traffic. The intuitive justification for this statement is that bursty real-time traffic cannot be buffered long enough to average out its rate fluctuations. For instance, a video connection has a rate close to the peak rate for as long as a few seconds, say 10 s. These long periods of high bit rate correspond to active scenes in the video, say a car chase in an action movie. The motion compensation video compression algorithm is not effective during fast-changing scenes and, consequently, produces a large bit rate to reflect the rapid modifications of the successive frames. If the node buffers the video bit stream for less than a few seconds, then it still must transmit the stream at a rate close to the peak rate. One might think that by superposing a large number, say 100, of such video bit streams, buffering would become much more effective. However, analysis and simulations show that not to be the case. We show below that buffering is ineffective for real-time streams such as video bit streams. However, multiplexing can be effective for such streams. In contrast to the real-time traffic case, buffering is effective for streams that can be delayed by a few seconds in the network, such as streams produced by interactive applications. Obviously, buffering is the way to handle best-effort traffic (UBR or ABR).

Multiplexing

We now present the main results on the multiplexing of many sources. Remember that the objective of multiplexing is to use a transmission rate per source close to the average rate of each source instead of requiring a rate close to the peak rate. Network engineers call this possibility of reducing the required rate per source the *multiplexing gain*.

Consider N sources. For $n = 1, \ldots, N$, let Y_n be the rate at some time t of source number n. We assume that the sources are stationary, independent, and identically distributed. That is, the rates $\{Y_1, \ldots, Y_N\}$ are independent random variables that have a common distribution that does not depend on t. We want to find the rate c such that

$$P\{Y_1 + \cdots + Y_N > cN\} \le 10^{-9}.$$

In words, if a node transmits the superposition of the N sources with that rate c, then it drops at most a fraction 10^{-9} of the bits.

As we may expect, the rate c is slightly larger than the average rate, say λ, of each source. The precise value of c depends on N and on the distribution of the random variables Y_i. Using the Bahadur-Rao theorem (see section 7.4.6), we find

$$P(Y_1 + \cdots + Y_N > Nc) \approx \frac{1}{\sqrt{2\pi}\sigma\theta_c\sqrt{N}} e^{-NI(c)}. \tag{6.13}$$

In the above expression, θ_c achieves the maximum in

$$I(c) = \sup_{\theta}[\theta c - \varphi(\theta)]$$

where

$$\varphi(\theta) = \log E[\exp(\theta Y_1)]$$

and

$$\sigma^2 = \varphi''(\theta_c) = \frac{\partial^2 \varphi}{\partial \theta^2}(\theta_c).$$

($\varphi(\theta)$ is called the logarithmic moment generating function.) In the case of on-off sources with $P(on) = p$ and peak rate a, the coefficients of (6.13) have the following form:

$$\theta_c = \frac{1}{a}\log\left(\frac{c(1-p)}{p(a-c)}\right), I(c) = \frac{c}{a}\log\left(\frac{c(1-p)}{p(a-c)}\right) - \log\left(\frac{a(1-p)}{(a-c)}\right), \sigma^2 = c(a-c).$$

As a numerical example, we use the following values for the homogeneous on-off sources: $\lambda = 1/20s$, $\mu = 1/5s$, $a = 18$ Mbps, $c = 8$ Mbps. These parameters correspond to $P(on) = \lambda/(\lambda + \mu) = 0.2 = p$ and therefore to a mean rate equal to $p \times a = 3.6$ Mbps.

We find that $\theta_c = 0.0646, I(c) = 0.1523, \sigma^2 = 80$. We can then calculate the value of N needed so that $P\{Y_1 + \cdots Y_N > cN\} \leq 10^{-9}$. The computer finds that $N \geq 118$ is the required condition.

Note that the number of sources that are in the on state has a binomial distribution. Thus the probability that the aggregate input rate exceeds the output rate can be represented exactly as

$$\sum_{k \geq Nc/a}^{N} \binom{N}{k} p^k (1-p)^{N-k}.$$

We can evaluate this expression directly and find $P\{Y_1 + \cdots Y_N > cN\} \leq 10^{-9}$ for $N \geq 114$. You will note the remarkable accuracy of the Bahadur-Rao approximation. The Bahadur-Rao approximation can be used for complex distributions of the random variables Y_k where a direct calculation is very complex. Note that for such distributions, the evaluation of the parameters that enter (6.13) must be performed numerically.

The Bahadur-Rao theorem enables us to analyze the case of multirate sources. That theorem can also be used to analyze the overflows of mixtures of different types of sources. To do this, say that we have 50% of sources of type Y and 50% of sources of type Z. We can then construct a hybrid source that is of type $Y + Z$. The analysis of such a situation reveals that the value of c required to achieve a small loss probability cannot be written as the sum of the necessary rates for the Y sources and the Z sources. Thus, unfortunately, for small buffers there is no additive result similar to the effective bandwidth.

On-Line Estimation

The above discussion shows that the network needs detailed information about the statistics of the sources in order to determine the capacity that it should allocate to connections. In practice it may not be realistic to expect the users to know that information when they set up the connection. These contradicting aspects seem to make statistical procedures impractical. We believe that this conclusion is not correct. The network could guarantee a quality of service by being conservative in its initial admission control and measure the traffic to determine the actual resources that the traffic requires. For instance, the admission control could be based on the peak rate of the traffic. Once the connection is in progress, the network can monitor its actual requirements. (The network also could calculate the price from the actual resources that the connection utilized.) Such a procedure presents a risk that all the ongoing connections might, after a while, suddenly become much more active and require more bandwidth. However, such an event is very unlikely.

According to the above description, we propose four methods for estimating the actual bandwidth that connections require. The methods differ in their numerical complexity and in their efficiency.

Assume that identical and independent sources with known mean rate m and peak rate a want to be transmitted by a transmitter with rate C. We want to design an on-line admission procedure that accepts the maximum number of sources subject to a loss probability of 10^{-9}. We assume that nothing is known about the sources other than their mean and peak rates.

We first assume that the sources are on-off with $P(on) = m/a$. On-off sources are easily seen to be the most bursty sources with given mean and peak rate in that fewer of them can be accepted for a given loss rate. We use the Bahadur-Rao formula to determine the maximum number N_0 of such sources that can be accepted, as we did in the previous section. We then accept the N_0 sources, and we measure their statistics.

The four methods we propose differ in how they infer the actual number of sources that can be carried by the transmitter. In methods 1 and 2, we calculate the bandwidth needed to carry the N_0 sources with a loss rate of 10^{-9}. In methods 3 and 4, we estimate the parameters of the Bahadur-Rao formula, and we calculate the maximum value N that can be carried.

Method 1 We divide the interval between the mean rate $m \times N_0$ of the N_0 sources and C into a number of equal parts. That is, we calculate $C_0 = m \times N_0, C_1, C_2, \ldots, C_K = C$ so that $C_1 - C_0 = C_2 - C_1 = \cdots = C_K - C_{K-1}$. We then monitor the instantaneous rate of the N_0 ongoing connections and determine the loss rate L_k that these connections would face if the service rate were C_k instead of C, for $k = 0, 1, \dots, K$. This monitoring is performed in parallel, by a device that does not perturb the connections. We then find $C(N_0) = \min\{C_k | L_k \leq 10^{-9}\}$.

We could decide that $C(N_0)$ has been determined if its value has stopped fluctuating for some time. More research is required to determine satisfactory stopping rules.

For instance, if $C = 155$ Mbps and $C(N_0) = 120$ Mbps, we are led to think that 25% more calls could be accepted. We can then accept a few more calls and repeat the above procedure.

This method has the advantage of being simple to implement.

Method 2 This second method is a modified version of Method 1 and results in faster estimation. We define N_0 and C_k as in method 1, and we accept N_0 calls. We group the N_0 calls into two subgroups. Subgroup 1 has 40% of the calls, and subgroup 2 has the remaining 60%. In parallel, we measure for each value of $k = 0, 1, \dots, K$ the loss rates L_k^1 and L_k^2 that the two groups would face if they were transmitted with respective bandwidth $0.4C_k$ and $0.6C_k$. We use these numbers to estimate the loss rate L_k that N_0 calls would face with bandwidth C according to the formula

$$L_k = 1.1619(L_k^1)^{-2}(L_k^2)^3.$$

This formula is derived from the Bahadur-Rao formula (6.13), which shows that the loss rate $L(N)$ for N calls has the form

$$L(N) = \frac{A}{\sqrt{N}} \exp\{-NG\}$$

so long as the bandwidth per call c is constant. By measuring $L_k^1 = L(0.4N_0)$ and $L_k^2 = L(0.6N_0)$ with $c_k = C_k/N_0$, we can determine the two unknown parameters A and G and calculate $L_k = L(N_0)$.

The motivation behind this approach is that the subgroups have fewer calls and benefit less from statistical multiplexing. Consequently, loss rates L_k^1 and L_k^2 are substantially larger than L_k and are faster to estimate.

Method 3 The third method accepts the same number, N_0, of calls as the previous two methods and monitors these calls to estimate the coefficients of the Bahadur-Rao formula. The estimation is based on the approximation

$$\varphi(\theta) := \log E[\exp(\theta Y_1)] \approx \frac{1}{N_0} \log \frac{\int_0^T \exp\{\theta X(N_0, t)\} dt}{T}, \text{ for } T \gg 1,$$

where $X(N_0, t)$ is the total instantaneous rate of the N_0 calls.

This approximation assumes that the processes are ergodic and stationary, so that

$$\frac{\int_0^T \exp\{\theta X(N_0, t)\}dt}{T} \to E\exp\{\theta X(N_0, t)\} = E\exp\{\theta(Y_1 + \cdots + Y_N)\}.$$

Once $\varphi(\theta)$ has been estimated in parallel for a large number of different values of θ, we can determine θ_c, $I(c)$, and the other required parameters. We can then find out the value of N that would result in the desired loss rate.

Method 4 This final method is similar to method 3 but uses the independence of the calls in a different way. The estimation is based on the approximation

$$\varphi(\theta) := \log E[\exp(\theta Y_1)] \approx \log \frac{\sum_{n=1}^{N} \int_0^T \exp\{\theta X_n(t)\}dt}{N_0 T}, \quad \text{for } T \gg 1,$$

where $X_n(t)$ is the instantaneous rate of the nth call.

This estimation uses the fact that the random variables

$$\frac{\int_0^T \exp\{\theta X_n(t)\}dt}{T}, n = 1, \ldots, N$$

are independent and identically distributed and converge (by ergodicity and stationarity) to $E\exp\{\theta(Y_1)\}$.

The method then proceeds as method 3. By better exploiting the independence of the calls, this method obtains estimators with a lower variance.

Our experiments suggest that the second method represents a good trade-off between complexity and speed. The fourth method is the fastest but requires complex computations.

Buffering

In this section, we consider a situation where the loss rate is kept small by using a large buffer. The buffer stores bursts of cells that arrive faster than they can be transmitted. It is unlikely that the bursts are frequent enough to make the buffer overflow.

Except for very simple source models (e.g., Poisson or a Markov-modulated process with a small number of states), it is difficult to analyze exactly the small loss rate at a large buffer. The cause of the difficulty is that the state space of a Markov model of the source and buffer system is large, which makes the numerical analysis complex. Because of that complexity, and with the objective of deriving tractable results, we turn to an asymptotic analysis of the loss rate as the buffer becomes large. Not surprisingly, the loss rate becomes smaller as the buffer increases. When the buffer is large, the loss rate is well approximated by an exponential function of the buffer size, as we already saw in (6.10).

Roughly, the loss rate is approximately $\exp\{-BI(C)\}$ where B is the buffer size (in cells, say) and $I(C)$ is some increasing function of the transmitter rate C and, obviously, of the statistics of the traffic. We argue that we should choose C large enough so that $\exp\{-BI(C)\}$ is small enough. For a video source, we might want $\exp\{-BI(C)\} \approx 10^{-10}$ when $B = 1{,}000$. Thus, we want $I(C) \approx 1\%$. For a database source, we might want $\exp\{-BI(C)\} \approx 10^{-8}$ for $B = 20{,}000$, so that $I(C) \approx 0.1\%$. We designate this target value of $I(C)$ by δ. Thus, $\delta = 1\%$ for video and $\delta = 0.1\%$ for database. (Once again, recall that these are working hypotheses and not standards.)

Now suppose that there are J types of traffic, and n_j sources of type j are multiplexed onto an output link. We want

$$\lim_{B \to \infty} \frac{1}{B} \log P(W \geq B) \leq -\delta,$$

where W is the buffer occupancy. Under appropriate assumptions, this constraint can be satisfied when

$$\sum_{j \in J} n_j \alpha_j(\delta) \leq C, \tag{6.14}$$

where C is the total output link rate, and $\alpha_j(\delta)$ is the effective bandwidth for the type j source corresponding to δ.

Equation (6.14) allows a simple policy for call acceptance that is analogous to that of the traditional circuit-switched networks, since the effective bandwidth for each call can be determined independently of the other types of calls. Furthermore, since $\alpha_j(\delta)$ lies between the mean and peak rates of the source, the difference between the peak rate and $\alpha_j(\delta)$ is the bandwidth saving through multiplexing.

The effective bandwidth $\alpha(\delta)$ of a source that produces a random number $A(t)$ of cells in t seconds can be calculated as

$$\alpha(\delta) = \frac{\Lambda(\delta)}{\delta}, \tag{6.15}$$

where

$$\Lambda(\delta) = \lim_{t \to \infty} \frac{1}{t} \log E\{e^{\delta A(t)}\}. \tag{6.16}$$

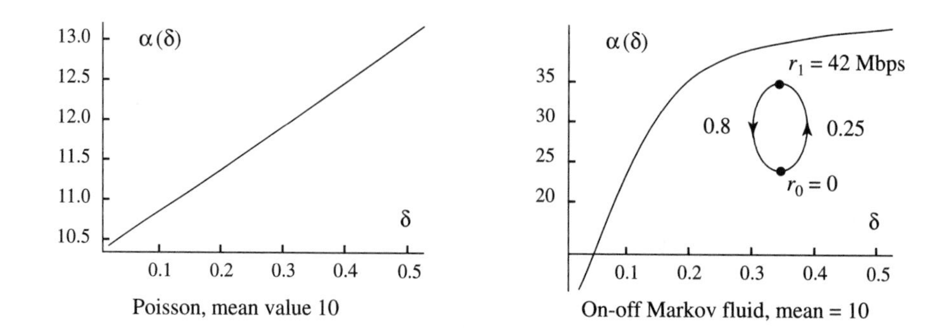

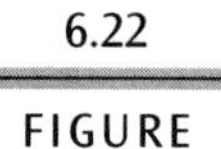

6.22 FIGURE

Effective bandwidth of iid Poisson and of on-off Markov-modulated sources with the same average rate. Note that these effective bandwidths differ substantially.

Figure 6.22 shows the effective bandwidth for two types of sources. The left-hand source produces independent identically distributed (iid) batches of bits that are Poisson distributed with mean rate 10 per unit of time. The right-hand source is an on-off Markov-modulated fluid with mean rate 10 and with the parameters shown in the figure. The on-off source has a larger effective bandwidth than the Poisson source.

When δ is small enough, one may be able to justify the following approximation:

$$\log E\{e^{\delta A(t)}\} \approx \log\left[1 + \delta E\{A(t)\} + \frac{\delta^2}{2}E\{A(t)^2\}\right] \approx \delta E\{A(t)\} + \frac{\delta^2}{2}E\{A(t)^2\},$$

which results in

$$\alpha(\delta) \approx \lambda + \frac{1}{2}\delta D^2, \tag{6.17}$$

where

$$\lambda := \lim_{t\to\infty}\frac{1}{t}E\{A(t)\} \text{ and } D^2 := \lim_{t\to\infty}\frac{1}{t}E\{A(t)^2\}.$$

In these definitions and in (6.17), λ is the average rate of the stream, and D^2 is called its *dispersion*. This simple approximation for $\delta \ll 1$, shows that the effective bandwidth increases with the burstiness of the stream and indicates that an approximate measure of burstiness (for large buffers and small δ) is the dispersion. When $\delta \ll 1$, we are willing to lose quite a few cells, and the second moments of the stream are good predictors of the loss rate, as one might guess from a functional central limit theorem. When δ is larger, the losses are determined by the tail behavior, and the higher moments cannot be neglected in the calculation of the effective bandwidth.

Formulas or algorithms are available for calculating the effective bandwidth of a large class of models. Methods for on-line estimation of the effective bandwidth are the subject of current research and so are adaptive techniques for selecting a suitable value of C.

These results deal with a single buffer and may be usable for a local ATM network. When traffic goes through multiple buffers, the situation is more complex. When streams share a buffer, they interact and modify each other's statistics and effective bandwidth. At first the problem appears intractable: the statistics of a stream depend on those of all the streams it interacted with, and the same is true for the latter streams. Fortunately, a simplification occurs. One can show that if the transmitter rate C of a buffer is large enough, then a stream preserves its effective bandwidth as it goes through the buffer. Specifically, stream j preserves its effective bandwidth if C is larger than the sum of $\alpha_j^*(\delta)$ and the average rate of all the other streams that share the buffer. Here, $\alpha_j^*(\delta)$ is the *decoupling bandwidth* of stream j. Formulas for calculating that decoupling bandwidth are given in Chapter 7 where the applications of that result to call admissions are discussed.

The approach above works well only if the buffer is large. Numerical and simulation experiments show that admission control based on the notions of effective and decoupling bandwidth may be too conservative. What is happening is that the method is based on the estimate of the exponential rate of decay of the loss probability, and it ignores the preexponential factor, which may be very small.

Statistical Multiplexing and Buffering

In many networks, a large number of sources are multiplexed and losses are further reduced by buffering. In such networks, the two effects that we discussed in the previous sections are combined.

The analysis of the combined effect of statistical multiplexing and buffering is rather complicated and does not yet carry over to more than a single queue. Nevertheless, the methods help understand the behavior of queues.

We present two approaches. The first approach yields satisfactory approximations for small buffers. The second approach is more suitable for larger buffers, but it is not as accurate in estimating the effect of statistical multiplexing.

Approach 1: Small Buffer To analyze the loss rate at a buffer of size B that serves N sources with rate Nc, we argue that losses occur when two events happen: first, the aggregate rate of the N sources must reach the value Nc; second, the rate must remain large enough until the buffer overflows. The probability of the first event is given by the Bahadur-Rao formula (6.13). The probability of the second event was obtained by Weiss,

$$P[\text{Buffer overflow} \mid \text{total rate} > Nc\,] = \exp\left\{-\left(N\frac{B}{\beta}\right)^{1/2} K(nc)\right\}$$

where β is the burst size of one source and $K(Nc)$ is a constant that depends on Nc. Combining the two results, the probability of overflow can be estimated as

$$\frac{1}{\sqrt{2\pi}\sigma\theta_c\sqrt{N}}\exp\left\{-NI(c) - \left(N\frac{B}{\beta}\right)^{1/2} K(Nc)\right\} \tag{6.18}$$

where

$$\sigma^2 = \frac{M''(\theta_c)}{M(\theta_c)} - c^2, \qquad M(\theta) := E[\exp(\theta Y_1)]$$

and θ_c achieves the maximum in

$$I(c) = \sup_{\theta}[\theta c - \varphi(\theta)].$$

The burst size of a source is defined as the product of the peak rate of the source times the mean holding time of that peak rate.

Figure 6.23 illustrates this formula. In that figure, 10 sources are served by a buffer with rate 150 Mbps. The sources are modeled by on-off Markov-modulated fluids with the parameters shown in the figure. The graphs show the loss rate as estimated by the small-B asymptotics, by the combination of small-B and Bahadur-Rao estimate, and the exact loss rate that can be calculated for this simple model.

Approach 2: Many Sources We limit the discussion to homogeneous sources. Let N be the number of incoming traffic streams. Assume that all sources are independent of each other. In particular, we consider the case that the arrival process from each source is modeled by a Markov-modulated fluid with two states. The off ($=0$) and on ($=1$) states have exponentially distributed holding times with parameters λ and μ, respectively. The output rate of a source is a in the on state and 0 in the off state. This on-off Markov model has received much attention in the research community because of its

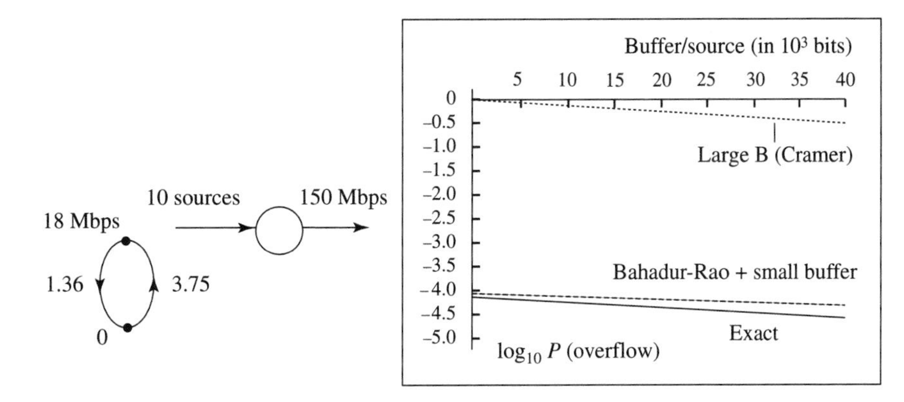

6.23

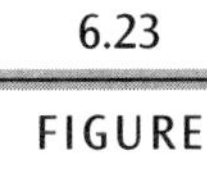

FIGURE

The figure compares three methods for calculating the loss rate at a small buffer. The first method is based on Cramer's theorem and ignores the statistical multiplexing gain. The second method combines the Bahadur-Rao estimate for 0-buffer and Alan Weiss's analysis of the overflow of small buffers. The third method, possible only for simple systems, is an exact calculation of the loss rate.

ability to model bursty processes such as sampled voice and its usefulness in queuing analysis.

Assume that the output buffer has first in, first out (FIFO) service discipline. Let b and c denote respectively the amounts of buffer space and bandwidth per source. Assume that $\frac{\lambda}{\lambda+\mu}a < c < a$, where the first inequality ensures stability and the second inequality allows a nonzero probability for the aggregate input rate to exceed the output rate.

We now explain the approximations to be made. Assume that time is discretized into epochs, with X_n equal to the number of cell arrivals from a single source in epoch n. Define

$$\varphi_m(\theta) = \frac{1}{m} \log E\left[\exp\left(\theta \sum_{n=1}^{m} X_n\right)\right].$$

Assume that the asymptotic logarithmic moment generating function, defined as

$$\varphi(\theta) = \lim_{m\to\infty} \varphi_m(\theta),$$

exists. Denote $\Phi(c, b, N)$ as the proportion of time that the buffer is full.

Theorem 6.4.1 Under appropriate assumptions,

$$\lim_{N\to\infty} \frac{1}{N} \log \Phi(c, b, N) = -I(c, b),$$

where

$$I(c, b) = \inf_{m} \sup_{\theta}[\theta(b + mc) - m\varphi_m(\theta)]. \tag{6.19}$$

Proof The intuitive explanation of this result is that if an overflow occurs at time 0, then there must have been some time $-m$ at which the buffer was last empty and since when at least $N(b+mc)$ cells have arrived. The probability of this many arrivals decays exponentially with N. The most likely way for an overflow to occur corresponds to the duration m with the smallest exponential decay rate. ■

Multiclass Case

We explained in the previous discussions that real-time traffic can be handled without buffering. The network determines the number of connections that it can accept so that the probability that the total rate of the connections exceeds the available bandwidth is very small. The interactive connections are handled with buffering. The network also estimates the bandwidth required per connection so that the probability of buffer overflows is very small. In the following we explain how the network can handle these two traffic types simultaneously. The basic idea is that the nodes give priority to real-time traffic and low priority to interactive traffic. Thus, the real-time traffic is not affected by the interactive traffic. However, the interactive traffic only gets the bandwidth that the real-time traffic does not use. The key question for us is to determine how the network should take that effect into account.

The real-time traffic with instantaneous rate $\mathbf{v} := \{v(t), t \geq 0\}$ that goes to a transmitter with rate C occupies a bandwidth equal to the effective bandwidth of $\mathbf{v}_C := \{\min\{v(t), C\}, t \geq 0\}$. Thus, if the real time traffic $\mathbf{v}$ and an interactive traffic $\mathbf{d} := \{d(t), t \geq 0\}$ go through a transmitter with rate C, the transmitter gives priority to $\mathbf{v}$ over $\mathbf{d}$ and makes sure that

- $P\{v(t) \geq C\}$ as estimated by Bahadur-Rao is acceptably small;
- $EB(\mathbf{v}_C) + EB(\mathbf{d}) \leq C$;
- the decoupling conditions are satisfied.

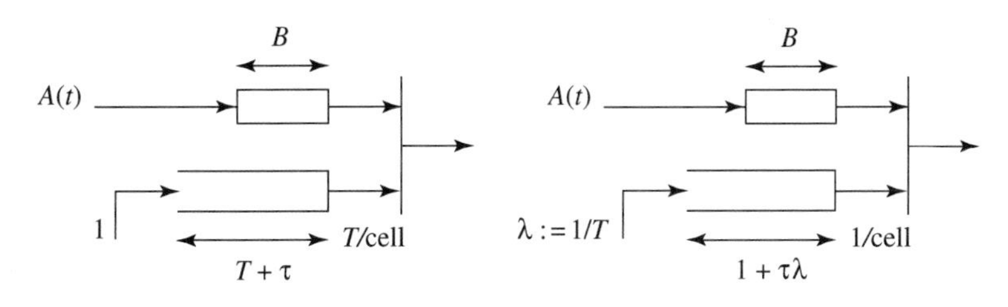

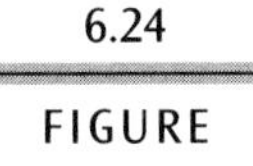

6.24 FIGURE

We use a leaky bucket GCRA(T, τ) to regulate the stream $A(t)$ in the left-hand side figure. The bottom buffer is the fluid buffer that provides the permits to transmit cells. We redefine units in the figure on the right-hand side so that one cell takes away one unit of fluid. Fluid enters at rate $\lambda = 1/T$.

Choosing a GCRA

Consider a user application that generates a random stream that must be transported using a service characterized by GCRA. What parameters of GCRA should the traffic stream request? In practice, the user will try various combinations among the values made available by the network until the cheapest acceptable parameters are identified. With some experience, equipment and service providers will be able to advise users and to recommend specific parameters for a given application.

In this section we look at the question from a theoretical angle and try to identify the set of parameters that is acceptable to carry a given traffic. The objective of the exercise is to understand the key statistics that affect that selection.

Consider using a GCRA(T, τ) controller with $\tau \gg T$ to shape some process $\{A(t), t \geq 0\}$. That leaky bucket is equipped with a cell buffer as shown in the left-hand side of Figure 6.24.

We redefine the units of fluid so that one cell requires one new unit of fluid. With this new unit, the fluid enters at rate $\lambda = 1/T$. The modified system is shown in the right-hand part of Figure 6.24.

We want to determine the values of B, T, and τ so that the probability that the buffer overflows is some small value, say 10^{-8}. These parameters will also give us a bound on the delay $B/\lambda = BT$ caused by traffic regulation. Alternatively, we could specify the maximum acceptable delay D, then calculate $B = \lambda D = D/T$ and find parameters T and τ that result in an acceptable loss probability.

To analyze the loss probability, we note that the system shown in Figure 6.24 is equivalent to the system shown in Figure 6.25. The interpretation

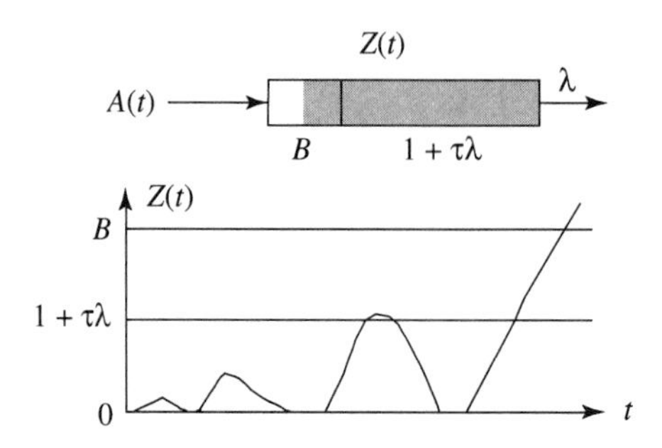

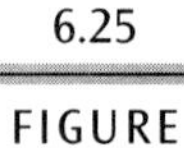
FIGURE 6.25 A backlog less than $\lambda\tau + 1$ means that there is a supply of tokens available. A backlog larger than $\lambda\tau + 1$ means that cells are backlogged.

of a backlog z in this system is that if $z < \lambda\tau + 1$, then $\lambda\tau + 1 - z$ is the amount of "token fluid" available in the bottom buffer of the right-hand part of Figure 6.24; if $z > \lambda\tau + 1$, then $z - \lambda\tau - 1$ is the amount of cells backlogged in the top buffers of Figure 6.24.

To analyze the probability of overflowing the cell buffer, we can analyze the probability that $Z(t)$ exceeds $B + \lambda\tau + 1$. If $\lambda > \alpha(\delta)$, the effective bandwidth of A, then this probability is approximately $\exp\{-\delta(B + \lambda\tau + 1)\}$. Thus, if we want this probability to be approximately $10^{-8} \approx e^{-18.3}$, then we need

$$\delta(B + \lambda\tau + 1) = 18.3$$

and

$$\lambda \geq \alpha\left(\frac{18.3}{B + \lambda\tau + 1}\right),$$

i.e.,

$$\frac{1}{T} \geq \alpha\left(\frac{18.3}{B + \tau/T + 1}\right).$$

Using $B = D/T$ we can rewrite the previous inequality as follows:

$$\frac{1}{T} \geq \alpha\left(\frac{18.3}{(D + \tau)/T + 1}\right).$$

This last inequality gives us the trade-off between parameters T and τ of the GCRA(T, τ) that we should request from the network. The best choice corresponds to the cheapest tariff. Note that the best choice depends on the statistics of the traffic through its effective bandwidth. Remember that our analysis ignores the factor of the exponential in calculating the loss probability.

Traffic Shaping

The sources can use a simple procedure to reduce the resources that they require from the network. This procedure is called traffic shaping. We explain how the source can implement traffic shaping and the effect of this procedure on the resource utilization.

Consider a real-time stream **v**. We propose the following traffic-shaping procedure. The source selects some duration T that is a fraction of the acceptable delay of the traffic in the network. The source is equipped with a buffer whose input is **v**. The source reads out its buffer at a rate $r(t)$ that is equal to the average rate of **v** during the interval $[t - T, t]$. One can show that by implementing this procedure the source delays the stream by at most T. Moreover, the rate $r(t)$ tends to be more regular (less bursty) than **v**.

The traffic shaping delays the cells at the source instead of having them buffered by the network. The rationale is that the network should be used to transport cells and not to store them. An advantage of traffic shaping is that it prevents one stream from perturbing other streams excessively.

6.4.4 Deterministic or Statistical?

One important debate among network researchers is whether networks should use deterministic or statistical procedures. We examine various aspects of that debate in this section. We start with a numerical evaluation based on what we learned in the previous sections.

A Comparison

We try to compare the number of connections that can be accepted by using the statistical and the deterministic approaches. The comparison is made on a hypothetical model of a video source.

As a numerical example, let us imagine video connections with a mean rate of 1.5 Mbps that can produce bits at 9 Mbps for random periods with durations of up to 10 s corresponding to active movie scenes. We model such a source as being produced by a line rate of 9 Mbps with a maximum burst of 10 s. Thus, for such a source, $R = 9$ Mbps, $R/SCR = 6$ and the maximum number of back-to-back cells M is such that $M/R = 10$ s, i.e., $M \approx 2.1 \times 10^5$

cells. We assume an acceptable delay of 100 ms, i.e., $9 \times 10^5/424 = 2100$ cell transmission times at the line rate R. From (6.7) we find that the effective bandwidth of this source is very close to 1 and that the source must be treated by the switch as a constant rate source with rate 9 Mbps. Thus, if the output line rate of the switch is 155 Mbps, the number of video connections that can be accepted with a delay of 100 ms is approximately $155/9 = 17$.

This example points to the conservatism of this admission strategy. If the source could be viewed as having a rate equal to their mean rate 1.5 Mbps, more than one hundred connections could be accepted. If we want to make sure that no cell gets delayed by more than 100 ms, then we cannot accept more sources than if they had their peak rate of 9 Mbps. Indeed, it is possible, although very unlikely, that the sources that we accept keep their peak rate for 10 s. If we accept N sources, then the peak rate is $9 \times 10^6 \times N$ for 10 s and during that time, the buffer accumulates $10(9N - 155) \times 10^6$, which exceeds 155×10^5, i.e., 100 ms of buffering, as soon as N exceeds 17.

We now examine the multiplexing approach. To evaluate the number of sources that the network can accept, we assume that the sources have two rates: 9 Mbps with probability 0.12 and 0.5 Mbps with probability 0.88. This model is approximate but may not be unreasonable. In an actual implementation, the network would measure its spare capacity. With this model we can use the Bahadur-Rao approximation or the binomial distribution as follows. Let us define the random variables Y_n for $n \geq 1$ as being iid with $P(Y_n) = 1 = 0.12 = 1 - P(Y_n) = 0$. We choose some number c and we find the largest value $N(c)$ of N such that

$$P\left(Y_1 + \cdots + Y_N > \frac{Nc}{8.5}\right) = 10^{-8}.$$

In addition, we want $N(c + 0.5) \leq 155$. The justification is that the rate of source n is modeled by $8.5Y_n + 0.5$. We then maximize $N(c)$ over c. The solution to the exercise is $N = 34$.

Thus, it is plausible that a statistical method that accepts a small level of risk (a few bad seconds every few days or weeks) can double the number of connections carried by the switch.

Pros and Cons of Deterministic Approaches

The deterministic approaches based on leaky buckets have the following advantages:

- the source can easily ensure that its traffic meets the specifications,
- the network can easily verify that the traffic meets the specifications,

- the network can guarantee hard bounds on delays and avoid all losses because of buffer overflows, and
- since the quality of service is specified in terms of hard bounds, the users can verify that the network provides the requested quality of service.

The main disadvantage of deterministic approaches is that they are based on worst cases: worst T seconds of a stream. The method assumes that all the connections that go through any one node will exhibit their worst-case behavior during the *same* T-second period.

Difficulties with Statistical Approach

The enforcement and policing of statistical traffic descriptions poses unresolved issues. The network uses some quantities for admission and routing of real-time traffic and others for interactive traffic. For real-time traffic, the network would need to know the parameters of the Bahadur-Rao bound (6.13). The network also needs to know the effective bandwidth and the decoupling bandwidth of the traffic. For interactive traffic, the network needs to know the effective and decoupling bandwidths of the traffic. These quantities are not readily available.

Although we can develop procedures to measure the necessary statistics, it is not clear that we should ask the users to perform these measurements, although protocols for doing so are conceivable. It is likely that repeated videoconferences with the same equipment will have similar statistics. Also, movies that are distributed for video-on-demand applications could remember their statistics. However, all this bookkeeping appears rather cumbersome, and it is tempting to look for solutions that make it unnecessary.

A Proposal

We propose an approach that has the advantages of both the deterministic and statistical approaches. This approach makes simple enforcement and policing possible, and it also permits efficient resource utilization.

We propose that users specify deterministic bounds, such as *PCR*, *SCR*, *BT*, and *CDVT*, for their traffic and statistical descriptions of the quality of service.

The network admits and routes a new connection by assuming worst-case behavior. The network then measures the statistics of the traffic: effective and decoupling bandwidths and the Bahadur-Rao parameters if it is a real-time connection. The network then uses these statistics to keep track of the resources that are now occupied by the ongoing connections.

To prevent the users from always behaving in the worst possible way, we propose that the billing be based on the actual resources that they use. In this way, the resource utilization is charged fairly to the different users.

There is one question that this proposal does not answer: how do users verify that they are getting the statistical quality of service that they requested?

6.5 SUMMARY

Virtual circuit networks like ATM seek to combine the gains from statistical multiplexing that packet switching creates with the guaranteed performance that circuit switching offers. When they succeed in this goal, virtual circuit networks will be able to reap the benefits of economies of scale, service integration, and network externalities. In order to succeed, however, network engineers must solve a number of control problems of admission, routing, flow and congestion control, and resource allocation.

Problems of routing and admission are similar to those that occur in circuit-switched networks; flow- and congestion-control problems arise in packet-switched networks. And we reviewed the approaches developed in the context of circuit and packet switching.

Problems of resource allocation are peculiar to virtual circuit networks. (In circuit-switched networks each connection gets a fixed resource, whereas in datagram networks no resources are allocated to a connection.) The major difficulty in designing resource-allocation mechanisms is that they depend on: (1) characteristics of the user traffic, (2) the available resources, and (3) the quality of service guaranteed to the user. Of these three items, the first is the most difficult to characterize, and we presented two approaches.

The deterministic approach is being pursued by the ATM Forum. It is easy to implement, but its worst-case assumptions will give very conservative answers, leading to significant underutilitzation of the network. The statistical approach at the present time is still being developed in research laboratories. It is far from standardization, but progress is rapid. We have described the most important accomplishments of both approaches.

The discussion of the statistical approach does make use of the concepts and framework of probability theory. Without an introduction to probability theory, the reader will find it difficult to follow the discussion. However, we have tried to minimize technical concepts so that the material is accessible. The full range of technical detail is concentrated in Chapter 7. That chapter is meant only for the reader with a strong background in probability.

6.6 NOTES

For a more detailed discussion of routing in circuit-switched networks, see [K94]. The gradient projection algorithm of section 6.3.3 is studied in [BG92]. For an analysis of the window flow-control scheme described in section 6.3.4, see [FMM91]. For details of the ATM Forum's recommendations summarized in section 6.4.2, see [A93]. A more detailed analysis of the pricing model of section 6.4.2 can be found in [CWW96]. The formulas (6.11) and (6.12) are derived in [AMS82]. The Bahadur-Rao theorem appears in [BR60]. Our discussion is based on [H95]. There is by now a significant literature using large deviations theory to study buffer overflow probabilities and to calculate effective bandwidth; see [W86, Hu88, B90, K91, GH91, CWe94, DZ93, DV93, KWC93]. The small buffer analysis is developed in [SW95]. Theorem 6.4.1 is obtained independently in [CWe94] and [BD94]. Admission control procedures based on effective bandwidth and decoupling bandwidth are presented in [HW94]. An introduction with an extensive bibliography to this material is given in [W95].

6.7 PROBLEMS

1. What are reasonable delays for interactive database queries, for remote control of a video server, for videoconferences, for telephone conversations, for transferring X rays?
2. Consider the transmission of a video program with a rate of 1.5 Mbps. Assume that the cell loss rate is equal to 10^{-8}. What is the average time between losses? What is the probability that the transmission of a one-hour video will not have any error?
3. What is the transmission rate required to transmit in 1 s a 4-inch by 6-inch photograph with a resolution of 1,200 dots per inch and 8 bits per pixel?
4. Assume a typical Web surfer makes 1-MB requests at random times at an average rate of 20 requests per day. Assume also that there are 10 million Web surfers and that they access 10,000 Web servers. To simplify, we make the gross assumptions that these servers are equally likely to be consulted. What is the average rate with which one server serves 1-MB requests? What is the minimum connection bandwidth and throughput of a server that is 100 times more popular than average? How would caching the information in distributed servers help run this application?

5. Consider the exchange of important data over a network with a typical end-to-end delay of 5 s and an average transmission time of 30 Kbps. Discuss the effects of a cell error rate of 10^{-4} on such an application. What is the probability that a 1-MB file will be corrupted by transmission errors? Assume that errors are checked for each block of 64 KB in this file. How many blocks are likely to be corrupted? What is the average time until the file is successfully received? How would this time change if the error checking were performed only for the complete file? *Hint:* To solve this problem, you need to know that if a coin has probability p of landing on heads in any one flip, then it takes on the average $1/p$ coin flips until the coin first lands on heads.

6. Consider a large office building with 1,000 telephone sets. Assume that an employee uses a telephone about 30 minutes during a typical 8-hour business day. What is the average number of telephone calls ongoing at a typical time during the business day? Assume that 15% of these calls are outside calls. Give a rough estimate of the number of outside lines that are required.

7. Consider the transmission of messages over a transmission line equipped with a buffer. The messages have lengths that are exponentially distributed with mean L bits. The transmitter has rate C bps. The messages arrive as a Poisson process with rate λ messages per second.
   (a) Using formula (6.1), find the average delay per message. For $\lambda = 10/s$ and $L = 8 \times 10^6$, find the minimum value of C so that the average delay does not exceed 0.1 s.
   (b) How does the average delay per packet change if C and L are multiplied by the same constant?
   (c) Let us pretend that our queue models a Web server that answers requests. We decide to divide up the files in the server into subfiles that are K times smaller. As a result, the rate of requests for the subfiles becomes $K\lambda$. What is the new average delay per subfile?

8. Consider the network of Figure 6.11. Assume that we can choose the parameters (p, μ_1, μ_2) subject to $\mu_1 + 4\mu_2 \leq 2\lambda$. Find the values of the parameters that minimize the average delay per packet.

9. As in the previous problem, consider the network of Figure 6.11. Describe a possible implementation of the distributed-gradient routing algorithm for this network. Explain the estimates that the nodes must perform and the information they must exchange.

10. Window flow control is claimed to be inefficient for connections with a large bandwidth × delay product. We discuss a simple model that traces the inefficiency to the long feedback delay. Imagine an M/M/1 queue fed by two streams with respective rates λ and α. The service rate is μ. The average delay through the queue is $1/r$ where $r = \mu - \lambda - \alpha$. We assume that λ is controlled after a feedback of T time units. We also assume that the delay through the queue stabilizes after one unit of time. The feedback algorithm is as follows. At time t, the measured delay is $r(t) = \mu - \lambda(t-1) - \alpha(t-1)$. The target delay is $1/r$, which corresponds to a target value of $\mu - \lambda - \alpha$. If $r(t) \neq r$, then the feedback informs the source to decrease λ by $r - r(t)$. This message reaches the source at time $t + T$. If the source is aware of the delay, it replaces λ by $\lambda(t) - r + r(t)$. Thus,

$$\lambda(t+T) = \max\{0, \lambda(t) - r + \mu - \lambda(t-1) - \alpha(t-1)\}.$$

Try this algorithm with $r = 1$, $T = 10$, $\mu = 3$, $\lambda(t) = \alpha(t) = 1$ for $t \leq 20$, $\alpha(21) = 2$, and $\alpha(t) = 1$ for $t \geq 22$. Propose methods to stabilize the algorithm.

11. Consider the following analogy to a call admission problem. An elevator can carry 2,000 kg. We want to decide an admission mechanism for people to get into the elevator. The rule of the game is that once somebody is in we cannot ask her to leave. Assume that everybody weighs less than 125 kg. The first procedure is to admit 16 people, assuming the worst case. The second procedure is based on statistics that give us the distribution of the weight of an arbitrary person. For simplicity, assume that the weights are Gaussian with mean 70 and standard deviation 15.
   (a) Find the number of persons that can be admitted so that the probability that the elevator is overloaded does not exceed 10^{-5}.
   (b) The third method consists in measuring the total load of the elevator as people get in and to close the doors when that load exceeds 1,875 kg. Estimate the expected number of people that can get into the elevator using the same assumptions as in part a.

12. Write a program to simulate the GCRA(T, τ) algorithm and verify the graphs in Figures 6.19 and 6.20.

7 CHAPTER Control of Networks: Mathematical Background

In this chapter you will learn the mathematical analyses that underlie the control techniques used and the resulting network performance measures described in Chapter 6. Some sections of this chapter demand from the reader a sophisticated background in stochastic processes. A basic knowledge of multivariate random variables and Markov chains is essential for these sections. You can skip those sections and still find accessible the discussion on deterministic models. If you are able to follow the more mathematical material, you will yourself be able to participate in the mathematical discussions on networking published in the research journals. But even if you are unable to follow the argument, Chapter 6 makes accessible the conclusions of those discussions.

We start by reviewing some key results on Markov chains in section 7.1. We apply these results to the study of circuit-switched networks in section 7.2 and of datagram networks in section 7.3. Section 7.4 explains the analysis of virtual circuit networks.

7.1 MARKOV CHAINS

In this section, we review Markov chains and discuss some key results.

7.1.1 Overview

A Markov chain is a model of the random motion of an object in a discrete set of possible locations. Two versions of this model are of interest to us: discrete time and continuous time. In discrete time, the position of the object—called the *state* of the Markov chain—is recorded every unit of time, i.e., at times 0, 1, 2, and so on. In continuous time, the state is observed at all times $t \geq 0$. One can think of the continuous-time model as being a discrete time model where the time unit is infinitesimally small. The state of the Markov chain changes randomly. In discrete time, there is a die at every location. Every time unit, the Markov chain tosses the die at its current location to decide where to jump next. In that way, the law of the future motion of the state depends only on the present location and not on previous locations. This key property that the Markov chain has of "forgetting" its past locations greatly simplifies the analysis.

Engineers use Markov chains to model the progression of the calls that a telephone network carries and of packets that a datagram or virtual circuit network transports. The randomness in these models reflects the uncertainty about when users place calls or send packets and about the length of packets and their destination. The randomness also captures the transmission errors and failures of devices.

The theory of Markov chains tells us how to calculate the fraction of time that the state of the Markov chain spends in the different locations. Network engineers use that theory to estimate the delays and losses of packets in networks or the fraction of time that telephone calls are blocked because all the circuits are busy. The engineers then use these estimates to design and control networks, as we explained in Chapter 6.

In this section, we review the main results of the theory of Markov chains, and we illustrate these results with examples.

7.1.2 Discrete Time

One is given a set $\mathbf{X}$, called the *state space*. We call the elements of $\mathbf{X}$ *states*. The set $\mathbf{X}$ is countable. That is, $\mathbf{X}$ is either a finite set $\mathbf{X} = \{i_1, \ldots, i_N\}$ (for some finite number N), or $\mathbf{X}$ is infinite but we can enumerate its elements exhaustively as $\mathbf{X} = \{i_1, i_2, i_3, \ldots\}$. For instance, the set of nonnegative integers $\mathbf{Z}_+ = \{0, 1, 2, 3, \ldots\}$ is countable, but the set of real numbers between 0 and 1, i.e., the interval $[0, 1]$, is not countable. For our purpose, the relevance of a set being countable is that if the elements in a collection $\mathbf{A}$ of positive real numbers add up to one, then the collection must be countable.

We denote typical elements of $\mathbf{X}$ as i, j, k. For $i \in \mathbf{X}$ we are given a list of nonnegative numbers $\{P(i,j), j \in \mathbf{X}\}$ that add up to one. That is,

$$0 \leq P(i,j) \leq 1 \text{ for all } i,j \in \mathbf{X} \text{ and } \sum_{j \in \mathbf{X}} P(i,j) = 1 \text{ for all } i \in \mathbf{X}.$$

We think of $P = \{P(i,j), i,j \in \mathbf{X}\}$ as a matrix. This matrix is N by N if $\mathbf{X}$ is finite and has N elements. If $\mathbf{X}$ is infinite, then the matrix P is infinite. The matrix P is called a *transition probability matrix*.

We are also given a probability distribution π_0 on $\mathbf{X}$, i.e., a collection of nonnegative numbers $\{\pi_0(i), i \in \mathbf{X}\}$ that add up to one. That is,

$$0 \leq \pi_0(i) \leq 1 \text{ for all } i \in \mathbf{X} \text{ and } \sum_{i \in \mathbf{X}} \pi_0(i) = 1.$$

We now define a random sequence $\mathbf{x} = \{x_0, x_1, x_2, \ldots\}$ that takes values in $\mathbf{X}$ as follows:

$$P(x_0 = i_0, x_1 = i_1, \ldots, x_n = i_n) = \pi_0(i_0)P(i_0, i_1)P(i_1, i_2) \times \cdots \times P(i_{n-1}, i_n) \quad (7.1)$$

for all $n \geq 0$ and all $i_0, i_1, \ldots, i_n \in \mathbf{X}$.

The random sequence $\mathbf{x}$ is called a Markov chain with transition probability matrix P and initial distribution π_0.

The definition of $\mathbf{x}$ specifies that x_0 is selected in $\mathbf{X}$ according to the probability distribution π_0 and that if $x_n = i$, then $x_{n+1} = j$ with probability $P(i,j)$, independently of the values x_m for $m < n$. The interpretation is that x_n is the position at time n of some object that moves randomly in the set $\mathbf{X}$. If $x_n = i$, then the object picks up a die located at state i and rolls the die. With probability $P(i,j)$, the outcome of the roll is j and the object moves to position j at time $n+1$.

Figure 7.1 illustrates three Markov chains. The leftmost diagram shows two states 0 and 1. The arrow from 0 to 1 is marked with the letter a, which represents some number in $(0, 1)$. The meaning of that arrow is that $P(0, 1) = a$. Similarly, the other arrows mean that $P(0, 0) = 1 - a$, $P(1, 0) = b = 1 - P(1, 1) \in (0, 1)$. Such a diagram is called a *state transition diagram* and it specifies the transition probability matrix of some Markov chain on $\mathbf{X} = \{0, 1\}$.

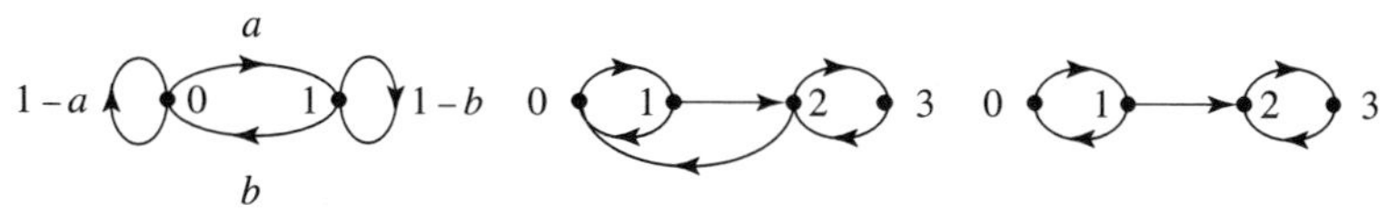

FIGURE 7.1 State transition diagrams of three discrete-time Markov chains.

The two other diagrams in Figure 7.1 are other state transition diagrams. By convention, if there is an arrow from i to j, then $P(i, j) > 0$. The state transition diagram is a directed graph. A *path* in such a graph is a succession of arrows such that the end of one arrow is the start of the next arrow. A path corresponds to a possible trajectory of the Markov chain in the state space.

We can calculate the probability distribution of x_n as follows. For $n = 1$ we have

$$\pi_1(i) := P(x_1 = i) = \sum_{j \in \mathbf{X}} P(x_0 = j, x_1 = i) = \sum_{j \in \mathbf{X}} \pi_0(j) P(j, i). \tag{7.2}$$

Define $\pi_n(i) := P(x_n = i)$ for $i \in \mathbf{X}$ and let π_n be the row vector with elements $\{\pi_n(i), i \in \mathbf{X}\}$. We can rewrite (7.2) as

$$\pi_1 = \pi_0 P$$

where the right-hand side denotes the product of the row vector π_0 by the matrix P. When $\mathbf{X}$ is infinite, the matrix multiplication involves infinite sums.

By repeating the argument that led us to (7.2) you can verify that

$$\pi_{n+1} = \pi_n P, \qquad n \geq 0. \tag{7.3}$$

From this identity we conclude that

$$\pi_n = \pi_0 P^n \tag{7.4}$$

where P^n is the nth power of the matrix P defined as the product of P by itself n times.

For instance, with the transition matrix of the leftmost diagram of Figure 7.1,

$$P = \begin{bmatrix} 1-a & a \\ b & 1-b \end{bmatrix}, \tag{7.5}$$

we find

$$P^n = \begin{bmatrix} 1-a_n & a_n \\ b_n & 1-b_n \end{bmatrix},$$

where

$$a_n = \frac{a + b(1-a-b)^n}{a+b} \text{ and } b_n = \frac{b + a(1-a-b)^n}{a+b}.$$

We say that a probability distribution π is *invariant* for the transition probability matrix P if

$$\pi = \pi P. \tag{7.6}$$

The equations (7.6) are the *balance equations* for the transition probability matrix P. Note that if π is invariant for P and if $\pi_0 = \pi$, then (7.4) implies that $\pi_n = \pi$ for all $n \geq 0$.

For the transition probability matrix P given in (7.5) we find that the only solution $\pi = [\pi(0), \pi(1)]$ of the balance equations (7.6) such that $\pi(0) + \pi(1) = 1$ is

$$\pi = \left[\frac{b}{a+b}, \frac{a}{a+b}\right]. \tag{7.7}$$

Before describing the main results about discrete-time Markov chains we must introduce a few definitions.

Definition 7.1.1 (Irreducibility) Let P be a transition probability matrix on the state space $\mathbf{X}$. The transition matrix P is *irreducible* if it is possible for a Markov chain with transition matrix P to move from any state i to any other state j in finite time. In other words, P is irreducible if there is a path from every i to every other j in the state transition diagram that corresponds to P.

A Markov chain with transition probability matrix P is said to be irreducible if P is irreducible. ■

For instance, looking at the state transition diagrams of Figure 7.1, we find that the two leftmost Markov chains are irreducible, whereas the rightmost one is not irreducible.

The following theorem tells us about the possible invariant distributions of an irreducible Markov chain.

Theorem 7.1.1 An irreducible Markov chain has at most one invariant distribution. It certainly has one if it is finite. ■

We noted earlier that the leftmost Markov chain of Figure 7.1 has a unique invariant distribution that is given by (7.7). The theorem tells us that the Markov chain in the center of Figure 7.1 also has a unique invariant distribution.

The invariant distribution, when it exists, measures the fraction of time that the Markov chain spends in the various states. This relationship is expressed in the following theorem.

Theorem 7.1.2 Let $\mathbf{x} = \{x_n, n \geq 0\}$ be an irreducible Markov chain on $\mathbf{X}$. Then

$$\lim_{N\to\infty} \frac{1}{N} \sum_{n=0}^{N-1} 1\{x_n = i\} = \pi(i), i \in \mathbf{X} \tag{7.8}$$

where π is the unique invariant distribution of the Markov chain if it exists and $\pi(i) := 0$ for $i \in \mathbf{X}$ if the Markov chain has no invariant distribution. ■

In the identity (7.8), the notation $1\{x_n = i\}$ has the value 1 if $x_n = i$ and the value 0 otherwise. Thus, the quantity

$$\frac{1}{N} \sum_{n=0}^{N-1} 1\{x_n = i\}$$

is the fraction of time that the Markov chain spends in state i during the first N units of time. This fraction of time is random because it depends on the particular realization of the sequence $\mathbf{x}$ that happens to occur. The theorem says that, in the long term, this fraction of time approaches a nonrandom quantity $\pi(i)$.

Thus, if the Markov chain has an invariant distribution π, then $\pi(i)$ is the long-term fraction of time that the Markov chain spends in state i, for $i \in \mathbf{X}$. If the Markov chain does not have an invariant distribution, then the fraction of time that the Markov chain spends in any one state is negligible. What happens in that case is either that the Markov chain is wandering off to infinity, if one enumerates the states in any arbitrary way, or that it visits all the states infinitely often but makes such large excursions in the state space that it spends a negligible fraction of time in any finite set of states.

The following theorem gives another important interpretation of the invariant distribution. However, we need one more definition to state that result.

Let P be the transition probability matrix of an irreducible Markov chain. Define

$$d(i) = \gcd\{n \geq 1 | P^n(i,i) > 0\}, i \in \mathbf{X}. \tag{7.9}$$

In this definition, gcd $\mathbf{A}$ denotes the greatest common divisor of the elements of the set $\mathbf{A}$. For instance, $\gcd\{6, 9, 12, 15, \ldots\} = 3$ and $\gcd\{1, 4, 6, 8, 10, 12, \ldots\} = 1$. The set in (7.9) is the collection of numbers of steps n such that the Markov chain can go from the state i back to itself in n steps.

For instance, consider the leftmost Markov chain of Figure 7.1 and assume that $0 < a < 1$ and $0 < b < 1$. One finds that $\{n \geq 1 | P^n(0,0) > 0\} = \{1, 2, 3, 4, \ldots\}$ so that $d(0) = \gcd\{1, 2, 3, 4, \ldots\} = 1$. Similarly, one can verify that $d(1) = 1$. Now assume that $a = b = 1$. Then one finds $d(0) = \gcd\{2, 4, 6, 8, \ldots\} = 2$ and $d(1) = \gcd\{2, 4, 6, 8, \ldots\} = 2$.

It can be proved that, for any irreducible Markov chain, $d(i) = d$ for all $i \in \mathbf{X}$. This leads us to the following definition.

Definition 7.1.2 (Aperiodicity) Let P be an irreducible probability transition matrix on $\mathbf{X}$. Define

$$d = \gcd\{n \geq 1 | P^n(i, i) > 0\}, i \in \mathbf{X}.$$

If $d > 1$, then P is said to be *periodic with period d*. If $d = 1$, then P is said to be *aperiodic*.

A Markov chain with transition probability matrix P is also said to be aperiodic or periodic with period d. ∎

Thus, the leftmost Markov chain in Figure 7.1 is aperiodic if $0 < a < 1$ and $0 < b < 1$; it is periodic with period 2 if $a = 1 = b$. Note that in the latter case, the values x_n alternate between 0 and 1 for $n \geq 0$. Thus, in that case, if $\pi_0(0) = \alpha$ with $0 < \alpha < 1$, then $\pi_{2n}(0) = \alpha$ for $n \geq 0$ and $\pi_{2n+1}(0) = 1 - \alpha$ for $n \geq 0$. This example shows that the distribution at time n, π_n, alternates between two values and therefore does not converge. The periodicity 2 of the Markov chain is reflected in the periodicity of its distribution as a function of time.

The Markov chain in the center of Figure 7.1 is aperiodic. Indeed, $\{n \geq 1 | P^n(0, 0) > 0\} = \{2, 3, 4, 5, \ldots\}$ so that $d = 1$.

One may hope that if the Markov chain is aperiodic, the distributions π_n may converge. This is indeed the case, as the following theorem makes precise.

Theorem 7.1.3 Let $\mathbf{x}$ be an irreducible and aperiodic Markov chain with invariant distribution π. Then, for any initial distribution π_0,

$$\pi_n(i) \to \pi(i) \text{ as } n \to \infty, \text{ for all } i \in \mathbf{X}. \quad ∎$$

The above theorem tells us that if we start a Markov chain $\mathbf{x}$ that is irreducible and aperiodic and has an invariant distribution π and if we wait long enough, then the probability of finding the Markov chain in state i is close to $\pi(i)$. The interpretation is that the Markov chain *approaches steady state*.

It is often comforting to be able to show that a Markov chain is positive recurrent even if one is not able to calculate its invariant distribution. For instance, if the Markov chain is the model of a buffer occupancy, then showing that it is positive recurrent tells us that the buffer empties relatively frequently and gives us a sense of stability of the system.

The following result is one of a number of useful sufficient conditions for positive recurrence. This result has the advantage of being intuitive and of being applicable to many queuing systems.

Theorem 7.1.4 Let **x** be an irreducible Markov chain on **X** and $V : \mathbf{X} \to [0, \infty)$ some function. Define the *drift* $\Delta(x)$ of $f(.)$ at x by

$$\Delta(x) := E[V(x_{n+1}) - V(x_n) | x_n = x], \text{ for } x \in \mathbf{X}.$$

Assume that there is some *finite* subset S of **X** and some constants $D > 0$ and $A < \infty$ such that

$$\Delta(x) \leq -D < 0, \text{ for } x \notin \mathrm{S}, \tag{7.10}$$

and

$$\Delta(x) \leq A < \infty, \text{ for all } x. \tag{7.11}$$

Then the Markov chain **x** is positive recurrent. ■

We leave the proof of this result as an exercise. Roughly speaking, if the fraction of time that **x** spends in S were negligible, then the expected value of $V(x_n)$ would keep decreasing at rate $-D < 0$ forever. This cannot be since $V(x) \geq 0$. Thus, **x** must spend a positive fraction of time in the finite set S. In view of Theorem 7.1.2, it follows that **x** must be positive recurrent.

7.1.3 Continuous Time

Engineers often use continuous-time models of networks. As we stated in the overview of this section, one can view a continuous-time Markov chain as a discrete-time Markov chain with an infinitesimally small time unit. However, it is easier to work with a more direct definition. Before introducing that definition we need to recall a few facts the about the exponential distribution.

Definition 7.1.3 (Exponential Distribution) The random variable τ is *exponentially distributed with rate* $\lambda > 0$ if

$$P(\tau > t) = e^{-\lambda t}, \text{ for } t \geq 0. \tag{7.12}$$

For convenience, when $\lambda = 0$, we define τ to be an infinite random variable (i.e., $\tau = +\infty$ with probability 1). ■

The following properties follow from this definition.

Theorem 7.1.5 (Properties of Exponential Distribution) Let τ be exponentially distributed with rate λ. Then

(a) The mean value $E(\tau)$ of τ is given by

$$E(\tau) = \frac{1}{\lambda};$$

(b) The random variable is *memoryless*. That is,

$$P[\tau > t + s | \tau > s] = P(\tau > t), \text{ for all } s, t \geq 0. \quad \blacksquare$$

The memoryless property can be interpreted as follows. Assume that a light bulb has an exponentially distributed lifetime. Then knowing how old the bulb is does not help predict how long it will still live. In other words, an old bulb is exactly as good as a new one: the bulb does not age. (That is, until it suddenly dies.)

Using exponential distribution we can construct a continuous-time Markov chain. We first define a rate matrix Q.

Definition 7.1.4 (Rate Matrix) Let $\mathbf{X}$ be a countable set. A *rate matrix* Q on $\mathbf{X}$ is a collection $Q = \{q(i,j), i, j \in \mathbf{X}\}$ of real numbers such that

$$0 \leq q(i,j) < \infty, \text{ for all } i \neq j \in \mathbf{X}, \text{ and}$$

$$-q(i,i) = q(i) := \sum_{j \neq i} q(i,j) < \infty, \text{ for all } i \in \mathbf{X}. \quad \blacksquare$$

We are now ready to define a continuous-time Markov chain.

Definition 7.1.5 (Continuous-Time Markov Chain) Let $\mathbf{X}$ be a countable set and Q a rate matrix on $\mathbf{X}$. Let also π be a probability distribution on $\mathbf{X}$. We define a continuous-time Markov chain $\mathbf{x} := \{x_t, t \geq 0\}$ on $\mathbf{X}$ with rate matrix Q and initial distribution π as follows.

First one chooses x_0 with distribution π in $\mathbf{X}$. That is, $P(x_0 = i) = \pi(i)$ for $i \in \mathbf{X}$.

Second, if $x_0 = i$, one selects a random time τ that is exponentially distributed with rate $q(i)$. The process $\mathbf{x}$ is defined so that

$$x_t = i \text{ for } 0 \leq t < \tau.$$

Third, at time $t = \tau$, the process $\mathbf{x}$ makes a jump from its initial value i to a new value j that is selected independently of τ and so that

$$P[x_\tau = j | x_0 = i, \tau] = \Gamma(i,j) := \frac{q(i,j)}{q(i)}, j \neq i.$$

The construction then resumes from $x_\tau = j$ at time τ, independently of the process before time τ. $\blacksquare$

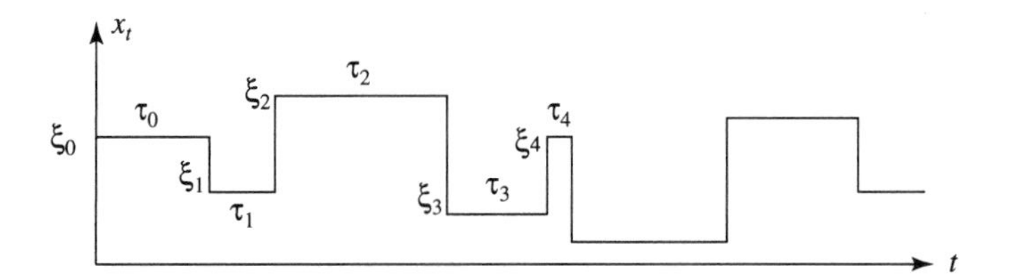

7.2 FIGURE

Trajectory of the continuous-time Markov chain x_t. The successive states are $\xi_0, \xi_1, \xi_2, \ldots$ and the successive holding times are $\tau_0, \tau_1, \tau_2, \ldots$.

Figure 7.2 shows a typical realization of the process $\mathbf{x}$. Thus, the process starts in some state ξ_0 and keeps that value for τ_0, then visits a sequence of states $\xi_1, \xi_2, \xi_3, \ldots$ where it stays for $\tau_1, \tau_2, \tau_3, \ldots$, respectively. The sequence of successive states is such that

$$P(\xi_0 = i_0, \xi_1 = i_1, \ldots, \xi_n = i_n) = \pi(i_0)\Gamma(i_0, i_1)\Gamma(i_1, i_2)\cdots\Gamma(i_{n-1}, i_n).$$

Moreover, given this sequence of states, the successive *holding times* $\tau_0, \tau_1, \ldots, \tau_n$ are independent and exponentially distributed with rates $q(i_0), q(i_1), \ldots, q(i_n)$, respectively.

We assume in this text that the rate matrix Q is such that the jump times do not accumulate. That is, we assume that

$$\sum_{n=0}^{\infty} \tau_n = \infty, \text{ with probability } 1.$$

With this assumption, the construction that we described defines a process $\mathbf{x}$ over $[0, \infty)$. The rate matrix Q is said to be *regular* if it has that property. We always assume that the rate matrices are regular.

The memoryless property of the exponential distribution implies that the Markov chain $\mathbf{x}$ starts afresh from x_t at time t, for any $t \geq 0$. That is, we have the following property.

Theorem 7.1.6 (Markov Property) Let $\mathbf{x}$ be a continuous-time Markov chain with rate matrix Q on $\mathbf{X}$. Then, for any set $\mathbf{A}$ of trajectories in $\mathbf{X}$,

$$P[(x_s, s \geq t) \in \mathbf{A} | x_t = i; x_u, u < t] = P[(x_s, s \geq 0) \in \mathbf{A} | x_0 = i]. \quad \blacksquare$$

The set $\mathbf{A}$ in the theorem is any set of trajectories of $\mathbf{x}$ in $\mathbf{X}$ of the form

$$\mathbf{A} = \{\mathbf{x} | x_{t_k} \in S_k \text{ for } k = 1, \ldots, K\}$$

where $K \leq \infty, 0 \leq t_1 < t_2 < \cdots < t_K$, and S_k is a subset of $\mathbf{X}$ for $k = 1, \ldots, K$.

This result says that the only information about the trajectory of $\mathbf{x}$ up to time t that is useful for predicting the trajectory after time t is the current value x_t. The intuitive justification of that theorem is that the past values of $\mathbf{x}$ are irrelevant because of the way successive values are selected. Moreover, the past holding times are irrelevant because the future ones depended only on the future states. Finally, the elapsed value of the current holding time is irrelevant for predicting when the next jump will occur because that holding time is memoryless.

Next we study the invariant distribution of a continuous-time Markov chain. We first need to define irreducibility.

Definition 7.1.6 (Irreducibility) A rate matrix Q on $\mathbf{X}$ is *irreducible* if $q(i) > 0$ for all $i \in \mathbf{X}$ and if the transition probability matrix Γ defined by

$$\Gamma(i,j) = \begin{cases} \frac{q(i,j)}{q(i)}, & i \neq j \in \mathbf{X} \\ 0, & i = j \in \mathbf{X} \end{cases}$$

is irreducible.

A continuous-time Markov chain with rate matrix Q is said to be *irreducible* if its rate matrix Q is irreducible. ■

Thus, a continuous-time Markov chain is irreducible if it can go from any state to any other state in finite time. With this definition we can state the main result about continuous-time Markov chains.

Theorem 7.1.7 (Invariant Distribution) Let $\mathbf{x}$ be an irreducible Markov chain on $\mathbf{X}$ with rate matrix Q and initial distribution π.
(a) The distribution π is invariant, i.e.,

$$P(x_t = i) = \pi(i), \text{ for all } i \in \mathbf{X} \text{ and } t \geq 0,$$

if and only if π solves the following *balance equations*:

$$\sum_{i \in \mathbf{X}} \pi(i) q(i,j) = 0, \text{ for all } j \in \mathbf{X}.$$

(b) The Markov chain is stationary if and only if its initial distribution is invariant.
(c) The Markov chain has either no or one invariant distribution. It certainly has one if $\mathbf{X}$ is finite.
(d) If the Markov chain has one invariant distribution π, then

$$\lim_{t \to \infty} P(x_t = i) = \pi(i), \text{ for all } i \in \mathbf{X}, \text{ and}$$

$$\lim_{T \to \infty} \frac{1}{T} \int_0^T 1\{x_s = i\} ds = \pi(i), \text{ for all } i \in \mathbf{X}.$$

(e) If the Markov chain has no invariant distribution, then

$$\lim_{t\to\infty} P(x_t = i) = 0, \text{ for all } i \in \mathbf{X}, \text{ and}$$

$$\lim_{T\to\infty} \frac{1}{T}\int_0^T 1\{x_s = i\}ds = 0, \text{ for all } i \in \mathbf{X}. \quad \blacksquare$$

It is often useful to consider a Markov chain reversed in time. The following theorem summarizes the key features of that process.

Theorem 7.1.8 (Reversing Time) (a) Let $\mathbf{x}$ be a stationary continuous-time Markov chain on $\mathbf{X}$ with rate matrix Q and invariant distribution π. Then,

$$\tilde{\mathbf{x}} := \{x_{T-t}, 0 \le t \le T\}$$

is a stationary continuous-time Markov chain on $\mathbf{X}$ with invariant distribution π and with rate matrix $\tilde{Q}$ given by

$$\tilde{q}(i,j) = \frac{\pi(j)q(j,i)}{\pi(i)}, i, j \in \mathbf{X}.$$

The process $\tilde{\mathbf{x}}$ is said to be $\mathbf{x}$ reversed in time.
(b) Let $\mathbf{x}$ be a continuous-time Markov chain on $\mathbf{X}$ with rate matrix Q. If π is a distribution on $\mathbf{X}$ and Q' is a rate matrix on $\mathbf{X}$ such that

$$\pi(i)q(i,j) = \pi(j)q'(j,i), \text{ for all } i, j \in \mathbf{X},$$

then π is invariant for $\mathbf{x}$ and Q' is the rate matrix of $\mathbf{x}$ reversed in time. $\blacksquare$

The expression for $\tilde{Q}$ given in part (a) of the theorem can be understood as follows. Assume that $\mathbf{x}$ is stationary with invariant distribution π. Then

$$P(x_0 = i, x_t = j) = \pi(i)q(i,j)t + o(t), \text{ for all } i \ne j \in \mathbf{X}.$$

But if $\tilde{\mathbf{x}}$ is $\mathbf{x}$ reversed in time, then

$$\begin{aligned} P(\tilde{x}_0 = i, \tilde{x}_t = j) &= \pi(i)\tilde{q}(i,j)t + o(t) \\ &= P(x_t = i, x_0 = j) = P(x_0 = j, x_t = i) \\ &= \pi(j)q(j,i)t + o(t), \end{aligned}$$

which shows that $\pi(i)\tilde{q}(i,j) = \pi(j)q(j,i)$.

With this rapid review of the main results on Markov chains, we are ready to analyze models of communication networks. We start with circuit-switched networks before moving on to packet-switched networks.

7.2 CIRCUIT-SWITCHED NETWORKS

In this section we explain the basic theory of circuit-switched networks. We start by discussing the case of a single switch in section 7.2.1. In section 7.2.2 we examine the case of a network.

7.2.1 Single Switch

Consider the situation shown in Figure 7.3. Phone calls arrive at a single switch as a Poisson process with rate ρ. That is, the times between successive calls are independent and exponentially distributed with rate ρ. There are N circuits to carry the calls. If a call arrives when all the lines are busy, then the call is rejected (blocked). The durations of the calls are independent and exponentially distributed with rate 1.

Markov Chain Model

We want to calculate the fraction of calls that are blocked. To perform this calculation, we observe that the number x_t of calls in progress at time $t \geq 0$ is a Markov chain with the transition diagram shown in Figure 7.3. In this diagram, an arrow from state i to state j is labeled with the value $q(i,j)$ of the entry of the rate matrix that corresponds to that pair of states.

To explain the transition diagram, assume that there are n calls in progress at time t, so that $x_t = n$. The next transition of $\mathbf{x}$ will occur when either a new call is placed or when one of the n ongoing calls is completed. The time σ until the next call arrives is exponentially distributed with rate ρ. The time τ until the first call in progress terminates is the minimum of n independent

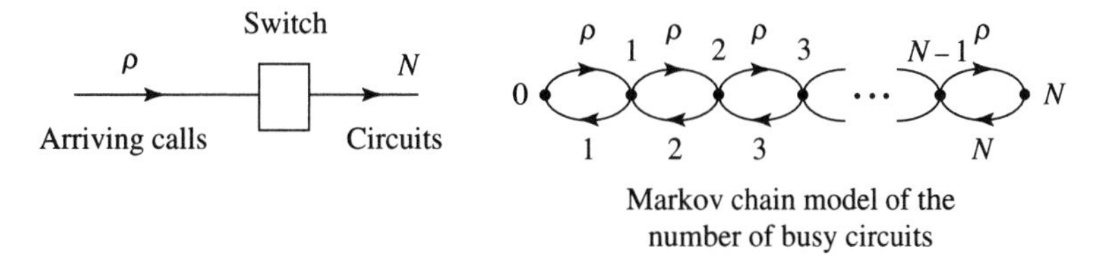

FIGURE 7.3 Model of a circuit switch. Calls arrive at rate ρ and are carried by N circuits.

exponentially distributed random variables $\tau_1, \ldots, \tau_n$ that are the residual values of the durations of the calls in progress at time t. These residual values are exponentially distributed with rate 1 because the original call durations are exponentially distributed and are therefore memoryless, so the residual durations are distributed exactly as the original durations. Now, the minimum τ of the random variables $\tau_1, \ldots, \tau_n$ is exponentially distributed with rate n because

$$\begin{aligned} P(\tau > t) &= P(\tau_1 > t, \tau_2 > t, \ldots, \tau_n > t) \\ &= P(\tau_1 > t)P(\tau_2 > t) \cdots P(\tau_n > t) = e^{-t}e^{-t} \cdots e^{-t} = e^{-nt}. \end{aligned}$$

The next transition of $\mathbf{x}$ after time t occurs after the minimum of τ and σ. This time is exponentially distributed with rate $\rho + n$ because

$$P(\min\{\tau, \sigma\} > t) = P(\tau > t, \sigma > t) = P(\tau > t)P(\sigma > t) = e^{-nt}e^{-\rho t} = e^{-(n+\rho)t}.$$

Now, when this transition occurs, the probability that it is due to a new call instead of a call termination is given by

$$\begin{aligned} P[\sigma < \tau \mid \min\{\sigma, \tau\} \in (s, s+\epsilon)] &= \frac{P(\tau > s, \sigma \in (s, s+\epsilon))}{(n+\rho)\epsilon \exp\{-(n+\rho)s\}} \\ &= \frac{[\exp\{-ns\}] \times [\rho\epsilon \exp\{-\rho s\}]}{(n+\rho)\epsilon \exp\{-(n+\rho)s\}} \\ &= \frac{\rho}{n+\rho}. \end{aligned}$$

Thus, the Markov chain $\mathbf{x}$ is such that

$$q(n) = \rho + n \text{ and } \Gamma(n, n+1) = \frac{\rho}{n+\rho}.$$

Since $\Gamma(n, n+1) = q(n, n+1)/q(n)$, we conclude that $q(n, n+1) = \rho$ and consequently that $q(n, n-1) = n$.

One can understand the diagram of Figure 7.3 more directly than by doing the above calculations. The diagram shows that when $x_t = n$, i.e., when there are n calls in progress, a new call arrives with rate ρ, so that $\mathbf{x}$ jumps from n to $n+1$ with rate ρ. Also, a call terminates, and $\mathbf{x}$ jumps from n to $n-1$ with rate n. This rate n is the sum of the n rates of completion of the calls in progress.

Invariant Distribution

The diagram shows that $\mathbf{x}$ is irreducible. By Theorem 7.1.7, we calculate the invariant distribution of $\mathbf{x}$ by solving the balance equations

$$\sum_{i \in X} \pi(i) q(i,j) = 0, \quad \text{for all } j \in X.$$

By using the definition $q(i) = \sum_{j \neq i} q(i,j) = -q(i,i)$, we can rewrite these equations as

$$\pi(i) q(i) = \sum_{j \neq i} \pi(j) q(j,i), \quad \text{for all } i \in X. \tag{7.13}$$

The interpretation of the equation (7.13) for a particular value of i is that the rate of transitions out of state i should equal the rate of transitions into state i.

Thus, for the Markov chain of Figure 7.3, the balance equations are as follows:

$$\begin{aligned} \pi(0)\rho &= \pi(1) \\ \pi(1)(\rho+1) &= \pi(0)\rho + \pi(2)2 \\ &\cdots \\ \pi(N)N &= \pi(N-1)\rho. \end{aligned}$$

Remembering that the distribution π must sum to one, we find that the solution of these balance equations is given by

$$\pi(n) = \frac{\rho^n/n!}{\sum_{m=0}^{N} \rho^m/m!}, \quad \text{for } n = 0, 1, \ldots, N.$$

In particular, we find that

$$P(x_t = N) = \pi(N) = E(\rho, N) := \frac{\rho^N/N!}{\sum_{m=0}^{N} \rho^m/m!}.$$

Erlang Loss Formula

This formula $E(\rho, N)$ is called the *Erlang loss formula*. That formula is the fraction of time that the N circuits are busy. This fraction of time is also the fraction of the calls that arrive when all the circuits are busy and are therefore blocked. To see why this is the case, we calculate the probability $\alpha(n)$ that a call that arrives finds n circuits busy: We find

$$\begin{aligned} \alpha(n) &= P[x_t = n | \text{a call arrives in } (t, t+\epsilon)] \\ &= \frac{P(x_t = n) P[\text{a call arrives in } (t, t+\epsilon) | x_t = n]}{P(\text{a call arrives in } (t, t+\epsilon))} \\ &= \frac{\pi(n)\rho\epsilon}{\rho\epsilon} = \pi(n). \end{aligned}$$

In the last equation, we used the fact that an arrival occurs in the next ϵ time units with probability $\rho\epsilon$, independently of x_t, by definition of the arrival process.

This calculation shows that the fraction of calls that find n circuits busy is equal to the fraction of time that n circuits are busy. In particular, for $n = N$, the fraction of calls that are blocked is the fraction of time that all the circuits are busy. Consequently, the blocking probability is given by the Erlang loss formula.

Insensitivity

We have calculated the blocking probability at a switch by assuming that the call durations are exponentially distributed. It turns out that the blocking probability does not change if the call durations have another distribution with the same mean value.

In other words, the blocking probability is *insensitive* to the actual distribution of the call durations. We explain a proof of that important result in section 7.3.5.

7.2.2 Network

Consider the network of Figure 7.4. The network has K switches that are connected by L links. Link i has n_i circuits. A route is an acyclic set of links. When a call is routed along a route, it uses one circuit in each link along the route.

We assume that calls are placed along route r as a Poisson process with rate λ_r for $r = 1, \ldots, R$. The call durations are independent and exponentially distributed with rate 1. We want to analyze the blocking probability of calls.

Denote by x_t^r the number of calls in progress along route r for $r = 1, \ldots, R$ and for $t \geq 0$. Let $x_t = (x_t^r, r = 1, \ldots, R)$. Because of the memoryless property of the exponential distribution, the process $\mathbf{x} = \{x_t, t \geq 0\}$ is a Markov chain. The state space X of that Markov chain is

$$\mathrm{X} = \{x \in \mathbf{Z}_+^R \mid \sum_{r=1}^{R} x^r A(r, i) \leq n_i, \text{ for } i = 1, \ldots, L\}.$$

In this expression, $A(r, i)$ takes the value 1 if route r goes through link i and the value 0 otherwise. Thus, $\sum_{r=1}^{R} x^r A(r, i)$ is the number of calls that go through link i. That number must be at most n_i since link i has n_i circuits and can carry at most n_i calls.

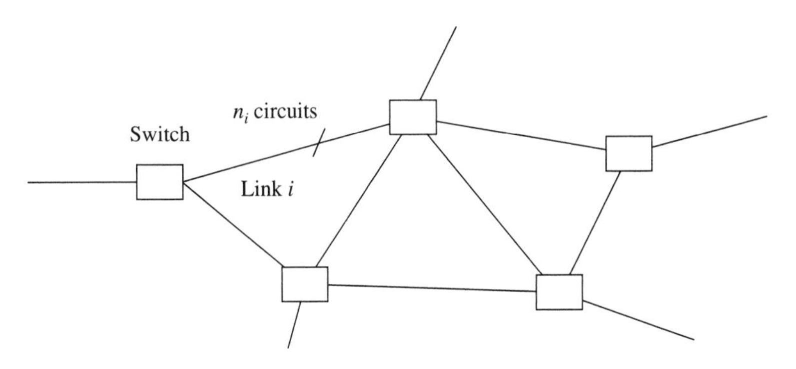

7.4 FIGURE

Circuit-switched network. The network has K switches that are interconnected by L links. Link i has n_i circuits.

If the links had an infinite number of circuits, then the numbers of calls in progress along different routes would be independent because they would not interfere with each other. In that case, the number of calls in progress along route r would be modeled by a Markov chain on $\{0, 1, 2, \ldots\}$ with a rate matrix Q such that $q(n, n+1) = \lambda_r$ and $q(n, n-1) = n$. By solving the balance equations of this Markov chain we would find that its invariant distribution is given by

$$\pi_r(n) = \frac{\lambda_r^n}{n!} e^{-\lambda_r}, \text{ for } n \geq 0.$$

Thus, if the links had infinitely many circuits, the invariant distribution π of $\mathbf{x}$ would be, by independence,

$$\pi(x^1, \ldots, x^R) = \pi_1(x^1) \ldots \pi_R(x^R).$$

However, since the links have a finite number of circuits, calls along different routes interfere with each other. Remarkably, the invariant distribution $\pi_{\mathbf{X}}$ in this case remains proportional to π. That is, the invariant distribution is given by

$$\pi_{\mathbf{X}}(x^1, \ldots, x^R) = \frac{\pi(x^1, \ldots, x^R)}{\pi(\mathbf{X})}, \text{ for } (x^1, \ldots, x^R) \in \mathbf{X}$$

where $\pi(\mathbf{X}) := \sum_{x \in \mathbf{X}} \pi(x)$. Note that the denominator normalizes the distribution so that it adds up to one over $\mathbf{X}$.

To prove this result we denote by Q the rate matrix of the Markov chain $\mathbf{x}$ on $\mathbf{X}$, and we observe that

$$\pi_X(x)q(x,y) = \pi_X(y)q(y,x), \text{ for all } x, y \in X. \tag{7.14}$$

To see this, let $x \in X$ and $y = x + e_r$ where e_r is the unit vector in direction r. That is, $y^i = x^i + 1\{i = r\}$. Then $q(x,y) = \lambda_r$ since a transition from x to y corresponds to an arrival of a call along route r and $q(y,x) = x^r + 1$ since the transition from y to x is the completion of a call along route r when $x^r + 1$ calls are in progress along that route. Thus, equation (7.14) reads

$$\frac{1}{\pi(X)}\pi_1(x^1)\cdots\pi_r(x^r)\cdots\pi_R(x^R)\lambda_r = \frac{1}{\pi(X)}\pi_1(x^1)\cdots\pi_r(x^r+1)\cdots\pi_R(x^R)(x^r+1),$$

and we see that this equation is satisfied because of the form of π_r. Other instances of pairs of states x and y can be checked along similar lines.

We can use the invariant distribution π_X to calculate the fraction of calls that are blocked. To do this, we consider a call that is placed along route r. Such a call is blocked if it is placed when the network is in a state $x \in X$ such that $x + e_r$ is no longer in X. Let us denote the set of such states by X_r. The fraction of time that the network state is in X_r is $\pi_X(X_r)$. Arguing as we did for a single switch, we can show that the fraction of calls placed along route r that find the network state in X_r is equal to the fraction of time $\pi_X(X_r)$ that the network state is in that set. Thus, the blocking probability B_r for a call placed along route r is given by

$$B_r = \pi_X(X_r) = \frac{\pi(X_r)}{\pi(X)}. \tag{7.15}$$

Complexity Considerations

In principle, the network engineers can use these formulas to calculate blocking probabilities and rates of revenues of the network, as we discussed in Chapter 6. However, the calculation of the numerator and denominator in (7.15) is very time-consuming because of the large number of elements in the sets X and X_r. That number of elements is of the order of

$$\Pi_{i=1}^{L} n_i.$$

To reduce the complexity of the calculations, it is possible to develop recursions in the n_i. It is also possible to evaluate the numerator and denominator of (7.15) by Monte Carlo simulations. The idea is to generate at random a vector x according to the distribution π. This generation is made particularly simple by the product form of π. By repeating the experiment we then estimate $\pi(X)$ as the fraction of the samples x that happen to fall in X and similarly for $\pi(X_r)$. A simple calculation can be used to estimate the number of random samples that should be generated in order to estimate the blocking probability within a few percent with a high degree of confidence (say 95%). This simulation can also be speeded up by exploiting importance sampling methods.

Erlang Fixed Point

To circumvent the numerical complexity of evaluating (7.15), researchers have developed an approximation called the *Erlang fixed-point approximation*. This approximation is based on assuming that a call is blocked independently by the different links along its route. With this assumption, a call along route r will not be blocked and appear on link i along route r with probability

$$\Pi_{i \neq j \in r}(1 - B_j)$$

where B_j is the probability that link j blocks the call and where the product is over all the links j other than i along route r. Thus, the rate λ_r of calls along route r contributes a rate

$$\lambda_r \times \Pi_{i \neq j \in r}(1 - B_j)$$

of calls on link i since this rate is the rate of calls that are not blocked by other links along route r. Thus, by summing over the different routes r we find that the rate ρ_i of calls that are placed on link i is given by

$$\rho_i = \sum_{r=1}^{R} \lambda_r \times \Pi_{i \neq j \in r}(1 - B_j).$$

Now, the blocking probability B_j at link j is a function of the rate ρ_j of calls on that link. In fact, $B_j = E(\rho_j, n_j)$ where $E(\rho, n)$ is the Erlang loss formula for n circuits faced by a Poisson process of calls with rate ρ. Thus, we find that

$$\rho_i = \sum_{r=1}^{R} \lambda_r \times \Pi_{i \neq j \in r}(1 - E(\rho_j, n_j)), \text{ for } i = 1, \ldots, L. \tag{7.16}$$

These equations form a set of fixed-point equations that define the rates ρ_i in terms of themselves. These equations are the *Erlang fixed-point equations*. It has been shown that these equations provide a good approximation of the blocking probabilities when the network is large.

7.3 DATAGRAM NETWORKS

In this section we apply the theory of continuous-time Markov chains to the analysis of datagram networks. In the process, we derive the key results on product-form queuing networks.

7.3.1 M/M/1 Queue

An M/M/1 queue is a waiting room equipped with a service facility where customers arrive as a Poisson process with rate λ; the customers are served by a single server, and their service times are independent and exponentially distributed with rate $\mu > \lambda$. The first M in the notation M/M/1 means that the arrival process is memoryless (Poisson), the second M means that the service times are memoryless (exponentially distributed). The 1 indicates that there is a single server.

Because of the memoryless property of the exponential distribution, the number x_t of customers in the queue at time t, including the one in service, is a Markov chain with transition diagram shown in Figure 7.5. The balance equations for the Markov chain are as follows:

$$\pi(0)\lambda = \pi(1)\mu$$

$$\pi(n)(\lambda + \mu) = \pi(n-1)\lambda + \pi(n+1)\mu \text{ for } n \geq 1$$

The solution of these equations is easily verified to be

$$\pi(n) = (1-\rho)\rho^n, n \geq 0, \text{ with } \rho := \frac{\lambda}{\mu}. \tag{7.17}$$

In particular, it follows that this invariant distribution is such that

$$\pi(i)q(i,j) = \pi(j)q(j,i), \text{ for all } i,j \geq 0.$$

For instance, $\pi(n)q(n, n+1) = (1-\rho)\rho^n\lambda = (1-\rho)\rho^{n+1}\mu = \pi(n+1)q(n+1, n)$. It follows from Theorem 7.1.8 that the rate matrix Q is also the rate matrix of the Markov chain $\mathbf{x}$ reversed in time. Thus, the Markov chain has the same rate matrix when it is reversed in time, and it is therefore statistically the same after time reversal. We say that the Markov chain is *time reversible*.

Note that the departures from the M/M/1 queue before time t become the arrivals after time t when the time is reversed. Since the queue remains an M/M/1 queue after time reversal, the arrivals after time t are a Poisson process with rate λ that is independent of the queue length x_t at time t. We then reach the following conclusion.

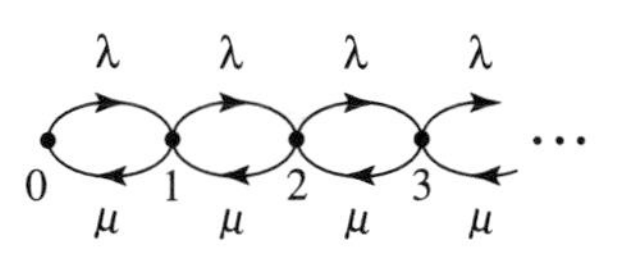

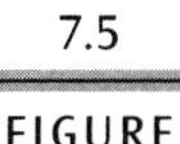

Markov chain model of an M/M/1 queue with arrival rate λ and service rate μ. The state of the Markov chain is the number of customers in the queue or in service.

Theorem 7.3.1 (Quasi Reversibility of the M/M/1 Queue) Consider a stationary M/M/1 queue. The departures before time t from that queue are a Poisson process with rate λ that is independent of the state of the queue at time t.

A queue with that property is said to be *quasi reversible.* ■

From the invariant distribution of the queue length, we can derive the average queue length and the average delay through the queue. For the average queue length we find

$$E(x_t) = \sum_{n=0}^{\infty} n\pi(n) = \sum_{n=0}^{\infty} n(1-\rho)\rho^n = \frac{\rho}{1-\rho} = \frac{\lambda}{\mu-\lambda}.$$

For the average delay through the queue, we first perform a direct calculation. The probability that a customer arrives when there are n customers in the queue is $\pi(n)$. To see this, note that

$$P[x_t = n | \text{a customer arrives in } (t, t+\epsilon)]$$
$$= \frac{P[\text{a customer arrives in } (t, t+\epsilon) | x_t = n] P(x_t = n)}{P(\text{a customer arrives in } (t, t+\epsilon))} = \frac{\lambda\epsilon\pi(n)}{\lambda\epsilon} = \pi(n),$$

as claimed. Now, the average delay of a customer who arrives when there are n customers in the queue is the average of $n+1$ service times, i.e., $(n+1)/\mu$. Indeed, such a customer must wait for the n customers who were ahead of him to be served and then for his own service time before he can leave the queue. Consequently, the average delay T of a customer in the M/M/1 queue is given by

$$T = \sum_{n=0}^{\infty} \frac{n+1}{\mu}\pi(n) = \sum_{n=0}^{\infty} \frac{n+1}{\mu}(1-\rho)\rho^n = \frac{1}{\mu-\lambda}.$$

Note that the average queue length $L := E(x_t)$ and the average delay T are related by the relationship

$$L = \lambda T.$$

This relationship, called *Little's result*, holds for very general queuing systems. One intuitive justification for Little's result is as follows. Assume that, on average, λ customers go through some service system per unit of time, each customer spends T units of time in the system, and L customers are in the system at any time. We want to argue that $L = \lambda T$. If each customer pays the system at a unit rate while in the system, the system gets paid at an average rate equal to the average number L of customers in the system. On the other hand, each customer pays the system an average amount T equal to the average time spent in the system. Since λ customers go through the system per unit of time and each pays an average of T, we conclude that the system gets paid at an average rate equal to λT. Hence, $L = \lambda T$, as we claimed.

The M/M/1 queue models a transmitter with rate c bps where packets arrive as a Poisson process with rate λ and have independent identically distributed (iid) lengths exponentially distributed with rate μc bits. Indeed, the successive transmission times of the packets are then exponentially distributed with rate μ. The results of this section enable us to calculate the average delay of the packets going through the transmitter. Note that if the utilization ρ of the transmitter does not exceed 80%, then the average delay T per packet does not exceed $5/\mu$. That is, the average delay does not exceed five packet transmission times. (The average packet transmission time is equal to μ^{-1}.)

In many communication systems, the interarrival times of packets and the service times are not exponentially distributed. We introduce a model for deterministic service times (such as in ATM networks) with batch arrivals.

7.3.2 Discrete-Time Queue

We model an ATM transmitter with a queue that has constant service times (equal to 1) and where A_n customers arrive at time n, for $n \geq 0$. We assume that the random variables $\{A_n, n \geq 0\}$ are iid, with mean λ and variance σ^2. To analyze the queue we can observe the number of customers in the system at the successive times. Let Y_n be the number of customers in the queue at time n. We assume that the queue operates as follows. At the beginning of the nth time epoch, there are Y_n customers in the queue. The server then serves one of these customers (if $Y_n > 0$), so that there are $(Y_n - 1)^+$ customers just after the service completion. The next batch of A_n customers then enters the queue. Consequently,

$$Y_{n+1} = A_n + (Y_n - 1)^+ = A_n + Y_n - 1\{Y_n > 0\}.$$

Because the A_n are iid, the process $\{Y_n, n \geq 1\}$ is a discrete-time Markov chain. If we assume that Y_n and Y_{n+1} have the same distribution, i.e., that the queue is in steady state, then we can use the above equation to calculate the mean queue length as follows. We first calculate the mean value of both sides of the identity above. We find

$$EY_{n+1} = EA_n + EY_n - P(Y_n > 0).$$

Since $EY_{n+1} = EY_n$, we conclude that $P(Y_n > 0) = EA_n = \lambda$. Next we take the mean value of the squares of both sides of the identity. When we use the independence of A_n and Y_n and the identity $Y_n 1(Y_n > 0) = Y_n$, we find

$$\begin{aligned} E(Y_{n+1})^2 &= E(A_n + Y_n - 1\{Y_n > 0\})^2 \\ &= E(A_n)^2 + E(Y_n)^2 + P(Y_n > 0) \\ &\quad + 2EA_n EY_n - 2EA_n P(Y_n > 0) - 2EY_n \\ &= \sigma^2 + \lambda^2 + E(Y_n)^2 + \lambda \\ &\quad + 2\lambda EY_n - 2\lambda^2 - 2EY_n. \end{aligned}$$

In steady state, $E(Y_{n+1})^2 = E(Y_n)^2$, so that we can simplify the last equality and solve for EY_n. We find

$$EY_n = L = \frac{\sigma^2 - \lambda^2 + \lambda}{2(1-\lambda)} = \frac{\sigma^2}{2(1-\lambda)} + \frac{\lambda}{2}. \tag{7.18}$$

The maximum delay experienced by a customer is equal to the maximum backlog of the queue. Let us consider a bound D on the average delay. The above equality shows that the constraint $L \leq D$ is satisfied if

$$\frac{\sigma^2}{2(1-\lambda)} \leq D - 0.5,$$

or

$$\lambda + \frac{\sigma^2}{2D - 1} \leq 1. \tag{7.19}$$

As an illustration of this rule, assume that the sequence $A_n = A_n^1 + \ldots + A_n^K$ for $n \geq 0$, where for each $k = 1, \ldots, K$ the random variables $\{A_n^k, n \geq 0\}$ are iid with mean λ_k and variance σ_k^2. Then we find that $\lambda = \lambda_1 + \cdots + \lambda_K$ and $\sigma^2 = \sigma_1^2 + \cdots + \sigma_K^2$. Consequently, the inequality (7.19) becomes

$$\sum_{k=1}^{K} \left[\lambda_k + \frac{\sigma_k^2}{2D - 1} \right] \leq 1.$$

We can write this equation as

$$\sum_{k=1}^{K} \alpha_k(D) \leq 1 \text{ where } \alpha_k(D) := \lambda_k + \frac{\sigma_k^2}{2D-1}. \tag{7.20}$$

We can think of $\alpha_k(D)$ as being the *equivalent bandwidth* of source k and of inequality (7.20) as stating that the sum of the equivalent bandwidths of the sources should not exceed the bandwidth of the transmitter (1 cell per unit of time). The inequality (7.20) indicates that the equivalent bandwidth of a source increases with its variance and decreases with the acceptable average delay through the queue.

M/GI/∞ Queue

The M/GI/∞ queue models independent delays. The customers arrive as a Poisson process with rate λ, and there are infinitely many servers waiting for customers to arrive. As soon as a customer arrives, a server starts serving her. The service times $\{\sigma_n, n \in \mathbf{Z}\}$ are iid. Thus, if the arrival time of customer n is τ_n, then her departure time is $\tau_n + \sigma_n$.

Consider the set of points $\mathbf{X} := \{x_n := (\tau_n, \sigma_n), n \in \mathbf{Z}\}$ in the plane, as shown in Figure 7.6. The points x_n in the infinite triangle $\mathbf{L}(t)$ defined as

$$\mathbf{L}(t) := \{x = (\tau, \sigma) | \tau \leq t \text{ and } \tau + \sigma > t\}$$

correspond to customers that are in the queue at time t. Indeed, if $\tau_n \leq t$, then customer n arrived before time t. Also, if $\tau_n + \sigma_n > t$, then the customer leaves at time $\tau_n + \sigma_n > t$.

The points x_n in the infinite parallelogram $\mathbf{D}(s,t)$ defined as

$$\mathbf{D}(s,t) := \{x = (\tau, \sigma) | s < \tau + \sigma \leq t\}$$

correspond to customers that leave the queue during $(s, t]$. Indeed, $\tau_n + \sigma_n$ is the departure time of customer n.

We claim that the random set of points $\mathbf{X}$ defines a Poisson measure in the plane. That means that the numbers $N(\mathbf{A}_1), \ldots, N(\mathbf{A}_M)$ of points of $\mathbf{X}$ in disjoint sets $\mathbf{A}_1, \ldots, \mathbf{A}_M$ are independent random variables that are Poisson distributed with means $\lambda_1, \ldots, \lambda_M$ where

$$\lambda_m := \lambda \iint_{\mathbf{A}_m} f(\sigma) d\tau d\sigma, m = 1, \ldots, M$$

where $f(.)$ is the probability density of the service times.

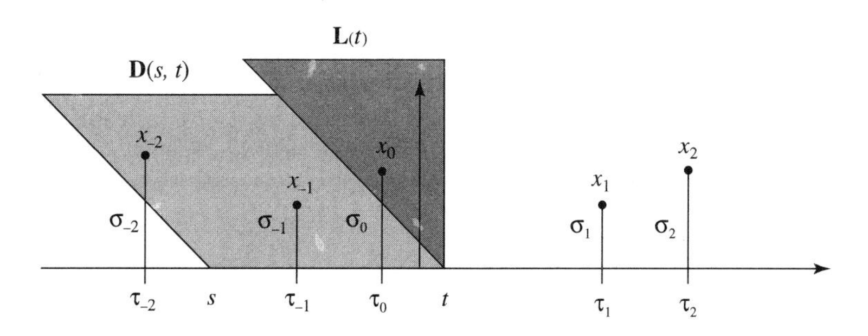

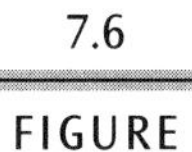
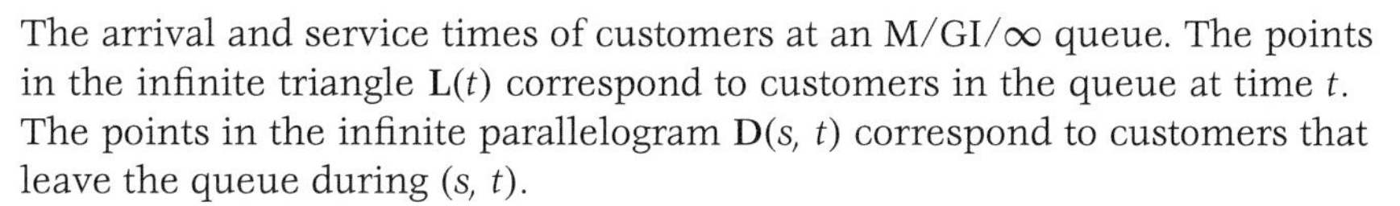
7.6 FIGURE The arrival and service times of customers at an M/GI/∞ queue. The points in the infinite triangle $\mathbf{L}(t)$ correspond to customers in the queue at time t. The points in the infinite parallelogram $\mathbf{D}(s, t)$ correspond to customers that leave the queue during (s, t).

To verify the claim, choose $\epsilon \ll 1$ and consider $\mathbf{L}_m = (m\epsilon, (m+1)\epsilon] \times [0, \infty)$ for $m \in \mathbf{Z}$. The random variables $N(\mathbf{L}_m)$ for $m \in \mathbf{Z}$ are independent random variables that are Poisson with mean $\lambda\epsilon$ because they are the increments of the Poisson arrival process with rate λ over intervals of duration ϵ. Now consider the sets $\mathbf{S}_{mn} = (m\epsilon, (m+1)\epsilon] \times [n\epsilon, (n+1)\epsilon]$ for $m \in \mathbf{Z}$ and $n \geq 0$. The random variables $N(\mathbf{S}_{mn})$ for $n \geq 0$ are obtained by sampling the Poisson random variable $N(\mathbf{L}_m)$ with probabilities $p_n := f(n\epsilon)\epsilon$. That is, each of the $N(\mathbf{L}_m)$ points of $\mathbf{X}$ in the set $\mathbf{L}_m$ is a point of $\mathbf{S}_{mn}$ with probability p_n. Indeed, each customer who arrives during $[m\epsilon, (m+1)\epsilon)$ has a service time in the interval $[n\epsilon, (n+1)\epsilon)$ with probability p_n.

It is a simple exercise to verify (see problem 10) that such a sampling of a Poisson random variable produces independent Poisson random variables with mean values $\lambda\epsilon p_n$, respectively. Also, by adding independent Poisson random variables one obtains a Poisson random variable. Thus, by approximating the sets $\mathbf{A}_m$ by unions of squares $\mathbf{S}_{mn}$ and by using the independence and Poisson distribution of the random variables $N(\mathbf{S}_{mn})$, one concludes that the random variables $N(\mathbf{A}_m)$ are indeed independent Poisson random variables with the mean values indicated in the claim.

In particular, if we choose a collection of times $t_1 < t_2 < \cdots < t_m < t$, we see by considering the disjoint sets $\mathbf{D}(t_1, t_2), \ldots, \mathbf{D}(t_m, t)$, and $\mathbf{L}(t)$ that the departures from the queue during the intervals $[t_1, t_2), \ldots, [t_m, t)$ are independent Poisson random variables with mean values $\lambda(t_2 - t_1), \ldots, \lambda(t - t_m)$, respectively, and that they are independent of the number $N(\mathbf{L}(t))$ of customers in the queue at time t.

The mean value of the queue length $N(\mathbf{L}(t))$ is equal to

$$L = \lambda \iint_{\mathbf{L}(t)} f(\sigma) d\tau d\sigma$$

$$= \lambda \int_0^\infty P(\sigma_1 > t) dt = \lambda E\sigma_1 =: \rho.$$

We have proved the following result.

Theorem 7.3.2 (Quasi Reversibility of the M/GI/∞ Queue) The M/GI/∞ queue is quasi reversible. That is, the departure process from the queue before time t is a Poisson process that is independent of the queue length at time t. Moreover, the queue length is a Poisson random variable with mean $\rho := \lambda E\sigma_1$. ■

7.3.3 Jackson Network

We now examine a network of J M/M/1 queues as shown in Figure 7.7. Customers arrive from outside as independent Poisson processes with rate γ_i into queue i. When a customer leaves queue i, he joins queue j with probability $r(i,j)$, independently of the past evolution of the network. On leaving queue i, a customer leaves the network with probability

$$r(i,0) := 1 - \sum_{j=1}^{J} r(i,j).$$

The service times of the customers are independent, and they are exponentially distributed with rate μ_i in queue i.

We assume that the probabilities $r(i,j)$ are such that every customer in the network eventually leaves it. Under this assumption, there is a unique solution to the following flow-conservation equations

$$\lambda_i = \gamma_i + \sum_{j=1}^{J} \lambda_j r(j,i), \text{ for } i = 1, \dots, J. \tag{7.21}$$

For $t \geq 0$ define

$$x_t = (x_t^1, \dots, x_t^J),$$

where x_t^j is the length of queue j at time t.

Because of the memoryless property of the exponential distribution and of the way the routing decisions are taken, the process $\mathbf{x} = \{x_t, t \geq 0\}$ is a

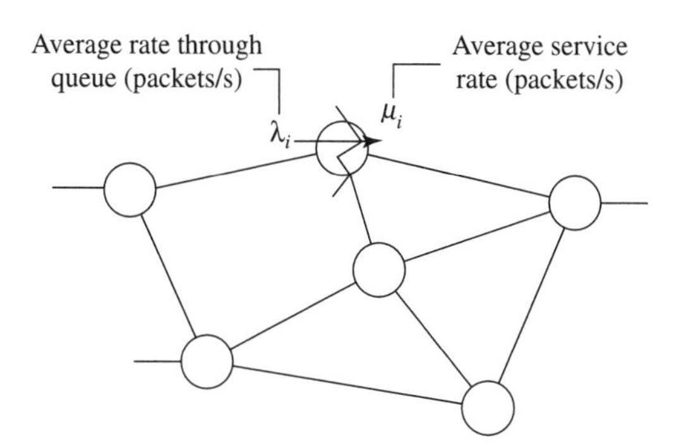

7.7 FIGURE Jackson network. This is a network of single-server queues. The service times in the various queues are independent and are exponentially distributed with rate μ_i in queue i. Customers arrive from outside as independent Poisson processes with rate γ_i into queue i. The routing is Markov.

Markov chain. The following theorem gives the invariant distribution of that Markov chain.

Theorem 7.3.3 (Product Form) Assume that the solution $(\lambda_1, \ldots, \lambda_J)$ of (7.21) is such that $\lambda_i < \mu_i$ for $i = 1, \ldots, J$. Then the Markov chain $\mathbf{x}$ admits the following invariant distribution:

$$\pi(x^1, \ldots, x^J) = \pi_1(x^1) \cdots \pi_J(x^J), \tag{7.22}$$

where, for $j = 1, \ldots, J$,

$$\pi_j(n) = (1 - \rho_j)\rho_j^n, \text{ for } n \geq 0 \text{ where } \rho_j := \frac{\lambda_j}{\mu_j}. \quad \blacksquare \tag{7.23}$$

We give a simple proof of this theorem that uses time reversal. Define a new network with the same M/M/1 queues but with different arrival rates γ_j' and different routing probabilities $r'(i,j)$. These values are selected so that

$$\lambda_i r(i,j) = \lambda_j r'(j,i), \text{ for } i,j \in \{1, \ldots, J\}$$

and

$$\gamma_i' = \lambda_i r(i,0).$$

These rates are calculated by *reversing the arrow* in Figure 7.7. Denote by Q' the rate matrix of this new network. A direct verification shows that

$$\pi(x)q(x,y) = \pi(y)q'(y,x), \text{ for all } x \text{ and } y. \tag{7.24}$$

For instance, let $y = x + e_j - e_i$ so that a transition from x to y occurs when a customer moves from queue i to queue j. You then find that

$$q(x, y) = \mu_i r(i, j) \text{ and } q'(y, x) = \mu_j r'(j, i).$$

Equation (7.24) then reads

$$\pi(x)\mu_i r(i, j) = \pi(x + e_j - e_i)\mu_j r'(j, i),$$

and this equation is satisfied in view of the form of π and of the definition of $r'(j, i)$. Theorem 7.1.8 then implies that π is indeed the invariant distribution for the Markov chain $\mathbf{x}$ as we wanted to prove.

It follows from the above result that the distribution of the vector of queue lengths in the network is identically the same as it would be if the arrival processes at all the queues were independent with rate λ_i at queue i. (It can be shown that the processes are not Poisson in general and that they are certainly not independent.)

In particular, we conclude that the average queue length in queue i is the same as the average queue length of an M/M/1 queue with arrival rate λ_i and with service rate μ_i. This average queue length is

$$L_i = \frac{\lambda_i}{\mu_i - \lambda_i}.$$

Consequently, the average number L of customers in the network is given by

$$L = \sum_{i=1}^{J} L_i = \sum_{i=1}^{J} \frac{\lambda_i}{\mu_i - \lambda_i}.$$

Using Little's result, we conclude that the average delay T of a customer in the network is given by

$$T = \frac{1}{\gamma} \sum_{i=1}^{J} \frac{\lambda_i}{\mu_i - \lambda_i}, \tag{7.25}$$

where $\gamma = \sum_{j=1}^{J} \gamma_j$ is the total rate at which customers enter the network.

7.3.4 Buffer Occupancy for an MMF Source

In section 6.4.3 we described a stochastic source as a Markov-modulated fluid (MMF). We considered the situation in which this source feeds into a buffer

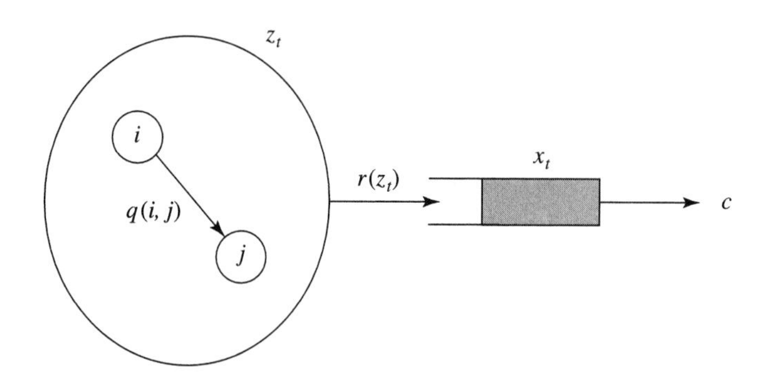

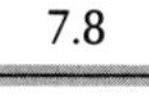

7.8 FIGURE When the source is in state z_t, it emits fluid at rate $r(z_t)$. The buffer is drained at rate c.

that is drained at a constant rate. We derive the stationary distribution of the buffer occupancy. The state of the Markov source is z_t, the occupancy of the buffer is x_t. When $z_t = j$, the source emits fluid at rate $r(j)$. The transmission rate is c. See Figure 7.8.

Define the steady-state distribution of the Markov chain (z_t, x_t) by

$$\pi(i, x) = P(z_t = i \text{ and } x_t \leq x).$$

Consider the evolution of the buffer occupancy and of the source between time t and time $t + dt$. Note that, for $x > 0$, $z_{t+dt} = i$ and $x_{t+dt} \leq x$ if and only if for some j one has $z_t = j$, $x_t \leq x + (c - r(j))dt$, and if the source then jumps from state j to state i. Indeed, under these conditions, during most of the interval $[t, t + dt]$ the source produces fluid at rate $r(j)$ so that the buffer occupancy increases by $(r(j) - c)dt$.

Consequently, for $x > 0$,

$$\begin{aligned}\pi(i, x) &= \sum_{j \neq i} \pi(j, x + (c - r(j))dt)q(j, i)dt + \pi(i, x + (c - r(i))dt)(1 + q(i, i)dt) \\ &= \sum_{j \neq i} [\pi(j, x)q(j, i)dt] + \pi(i, x)(1 + q(i, i)dt) + \frac{\partial}{\partial x}\pi(i, x)(c - r(i))dt.\end{aligned}$$

Hence,

$$\frac{\partial}{\partial x}\pi(i, x) = -(c - r(i))^{-1} \sum_j \pi(j, x)q(j, i).$$

Denote by $\pi(x)$ the row vector

$$\pi(x) = [\pi(1, x), \ldots, \pi(M, x)]$$

and by $\frac{d}{dx}\pi(x)$ the row vector

$$\frac{d}{dx}\pi(x) = \left[\frac{\partial}{\partial x}\pi(i, x), \ldots, \frac{\partial}{\partial x}\pi(M, x)\right].$$

We can then rewrite the differential equations as

$$\frac{d}{dx}\pi(x) = \pi(x)A$$

where A is the matrix $A = [a(i, j), 1 \leq i, j \leq M]$ with

$$a(i, j) = \frac{q(i, j)}{r(j) - c}.$$

To solve these equations we must take into account the boundary conditions that we derive by considering the case $x = 0$. Define the set S of possible values j of z such that $r(j) \leq c$. By reproducing the derivation above for $x = 0$ we find that for $i \in S$ one has

$$\frac{\partial}{\partial x}\pi(i, 0)\{r(i) - c\} = \sum_{j \in S} \pi(j, 0)q(j, i).$$

Similarly, we find that for i not in S, $\pi(i, 0) = 0$.

These boundary equations enable us to solve the linear equations. We can write the solution as follows:

$$\pi(x) = \sum_{k=0}^{M-1} a(k)e^{\beta_k x}, x \geq 0$$

where the β_k are the eigenvalues of A—that we assume distinct for simplicity—and the $a(k)$ are proportional to the corresponding eigenvectors. One eigenvalue, say β_0, is zero and corresponds to the eigenvector ϕ equal to the stationary distribution of z_t; the other eigenvalues are negative if the system is positive recurrent. Since $\pi(i, \infty) = \phi(i)$ we conclude that $a(0) = \phi$. That is,

$$\pi(x) = \phi + \sum_{k=1}^{M-1} a(k)e^{\beta_k x}, x \geq 0.$$

To clarify the above calculations, let us solve explicitely for $\pi(i, x)$ when the MMF is an on-off Markov fluid with on-rate equal to 1. In that case, the process

z_t takes the values 0 and 1 and its rate matrix is such that $q(0,1) = \lambda$ and $q(1,0) = \mu$. We have $r(0) = 0$ and $r(1) = 1$, and we assume that $0 < c < r(1) = 1$; otherwise there is no queuing possible. Moreover, we must assume that the average input rate is less than c, i.e., $\lambda/(\lambda+\mu) < c$, otherwise the queue fills up and has no invariant distribution. We find that

$$A = \begin{bmatrix} \frac{\lambda}{c} & \frac{\lambda}{1-c} \\ -\frac{\mu}{c} & \frac{-\mu}{1-c} \end{bmatrix}.$$

The eigenvalues of A are 0 and $\beta := \lambda/c - \mu/(1-c) < 0$ with the corresponding eigenvectors $\phi = [\mu/(\lambda+\mu), \lambda/(\lambda+\mu)]$ and $v := [1-c, c]$.

Thus we know that $\pi(x) = \phi + ave^{\beta x}$, and we find the constant a by using the boundary condition $\pi(1,0) = 0$. This gives $a = -(\lambda/c)/(\lambda+\mu)$. Putting all this together yields

$$\pi(0,x) = \phi(0) - \phi(1)\frac{1-c}{c}\exp\{\beta x\}$$

$$\pi(1,x) = \phi(1) - \phi(1)\exp\{\beta x\}$$

with $\phi(0) = \mu/(\lambda+\mu) = 1 - \phi(1)$ and $\beta := \lambda/c - \mu/(1-c)$. From these expressions we can calculate the steady-state probability that the buffer occupancy exceeds x. We find

$$P(x_t > x) = 1 - \pi(0,x) - \pi(1,x) = \frac{\phi(1)}{c}\exp\{\beta x\}.$$

Figure 7.9 shows a few examples.

7.3.5 Insensitivity of Blocking Probability

We mentioned at the end of section 7.2.1 that the blocking probability for a loss network does not depend on the distribution of the holding times of telephone calls. We now prove that result.

Consider the network in the left side of Figure 7.10. The figure shows a closed network with one $M/GI/\infty$ queue and one $M/M/1$ queue with service rate λ. Assume that there are N customers in the network. The outputs of the $M/M/1$ queue are calls being placed and the outputs of the $M/GI/\infty$ queue are calls being terminated. Calls are placed with rate λ as long as there are fewer than N calls in progress (represented by all N customers in the $M/GI/\infty$ queue). When there are N calls in progress, new calls are blocked. Thus, the call blocking probability is the probability that there are

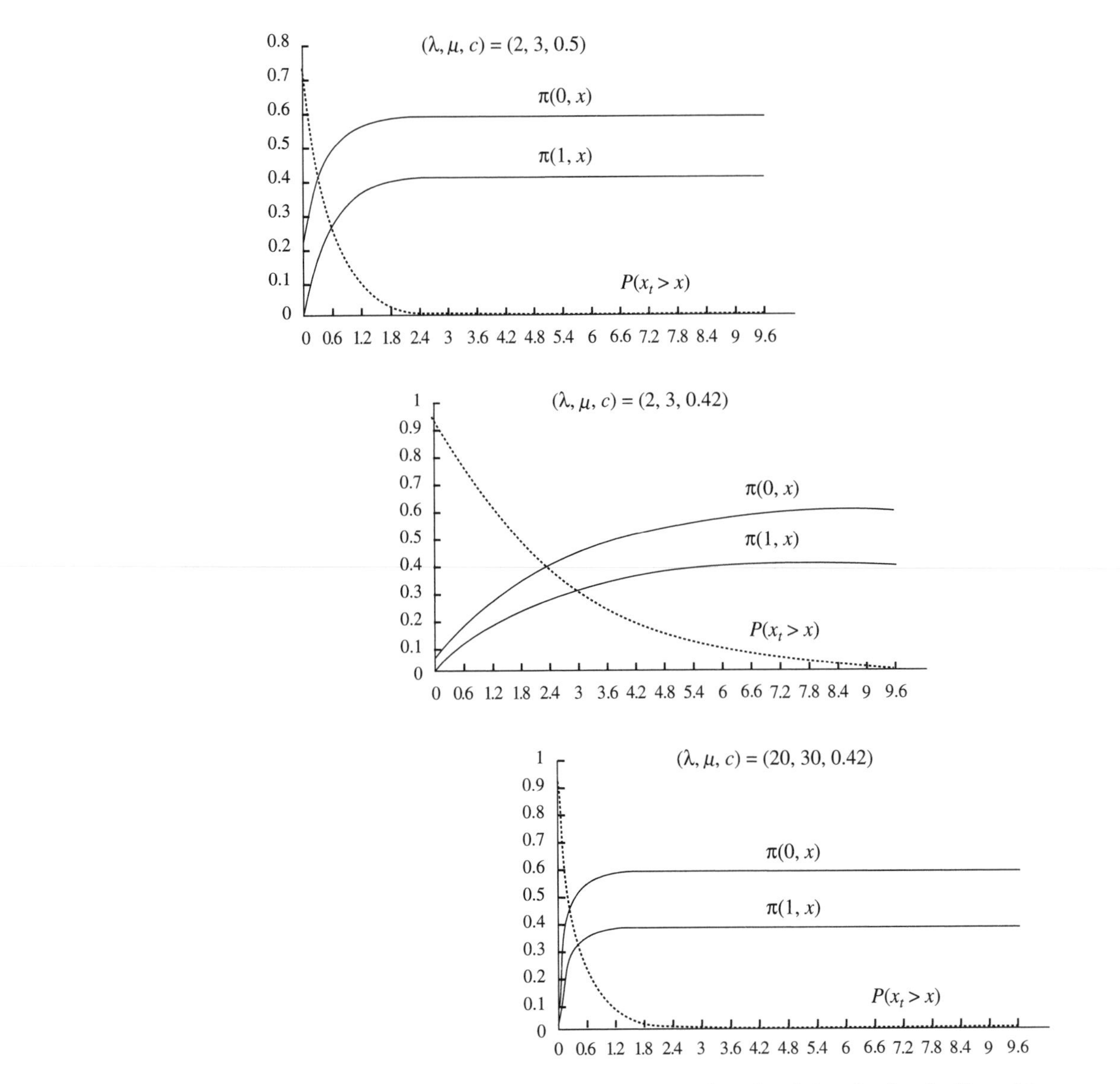

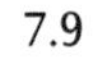

FIGURE 7.9

The invariant distribution of an on-off Markov fluid that feeds a buffer with a constant service rate. Three sets of parameter values are illustrated.

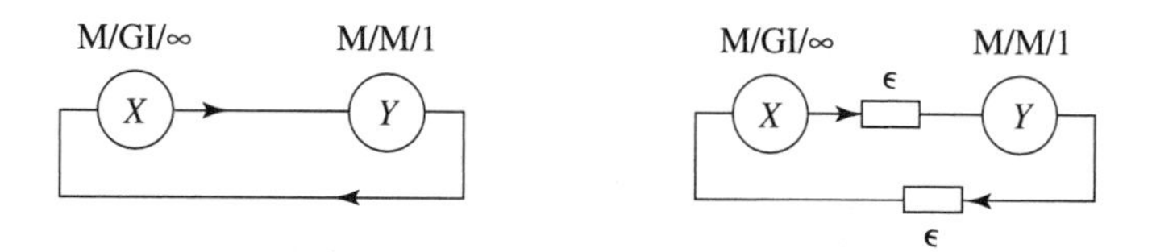

7.10 FIGURE

Network model of loss system. The network in the left-hand side of the figure models a loss system. A customer who enters the $M/GI/\infty$ queue represents a call being placed. A customer who leaves that queue represents a call being terminated. Thus, calls are placed with rate λ and have iid holding times. The network on the right-hand side is modified by inserting two small delays to simplify the analysis.

N customers in the $M/GI/\infty$ queue. We show that this probability depends on the distribution of the call holding times—the service times in the $M/GI/\infty$ queue—only through their mean value.

To analyze the network, we introduce arbitrarily small delays of duration $\epsilon \ll 1$ as shown in the right-hand side of Figure 7.10. Assume that at time 0, the states of the $M/GI/\infty$ queue and the $M/M/1$ queue are independent and have the invariant distributions that correspond to Poisson inputs with rate $\alpha < \lambda$. Moreover, assume that the states of the little delay lines at time 0 are such that their outputs during $[0, \epsilon)$ are independent Poisson processes with rate $\alpha > 0$ and that these outputs are independent of the states of the queues. Consequently, during $[0, \epsilon)$ the *inputs* of the two queues are independent Poisson processes with rate α.

Moreover, because the queues have the invariant distribution for those input processes and are quasi reversible, the *outputs* of these queues are independent Poisson processes during $[0, \epsilon)$. In addition, these outputs are independent of the states of the queues at time ϵ. Thus, at time ϵ, the state of the network has the same distribution as at time 0 and we can repeat the argument during $[\epsilon, 2\epsilon)$, and so on. Consequently, during any time interval $[n\epsilon, (n+1)\epsilon)$ the outputs of the queues are independent Poisson processes with rate α.

Moreover, at any time t the queue lengths X in the $M/GI/\infty$ queue and Y in the $M/M/1$ queue are independent and are Poisson with mean $\alpha E\sigma$ and geometric with parameter α/λ, respectively. That is,

$$P(X = m, Y = n) = \phi(m, n) := \frac{(\alpha E\sigma)^m}{m!} e^{-\alpha E\sigma} \left(\frac{\alpha}{\lambda}\right)^n \left(1 - \frac{\alpha}{\lambda}\right), m, n \geq 0. \quad (7.26)$$

We conclude that the above distribution is invariant for the network with the small delays ϵ.

Now consider what happens when $\epsilon \downarrow 0$. Take an arbitrary realization of the initial state of the network. For ϵ small enough, the delay lines are empty. By decreasing ϵ further, we modify the arrival times into the queue by ϵ. Eventually, for ϵ small enough, the lengths of the two queues $(X_t^\epsilon, Y_t^\epsilon)$ at time t stop changing as ϵ keeps on decreasing. Moreover, these queue lengths are equal to the value they would have for $\epsilon = 0$ whenever time t is not an arrival time into one of the queues when $\epsilon = 0$. Hence,

$$P((X_t^\epsilon, Y_t^\epsilon) = (m, n)) \to P((X_t^0, Y_t^0) = (m, n)) \text{ as } \epsilon \downarrow 0.$$

It follows that the distribution (7.26) is invariant for the network without the delay lines.

Note that for this invariant distribution the total number of customers in the network is random. If we know that the total number is N, then we can derive the invariant distribution of the network as follows. Assume that we have some Markov chain approximation $\mathbf{x} = \{x_t, t \geq 0\}$ of the state of the network. Such an approximation can be constructed by approximating the holding times by first passage times of finite Markov chains. Denote by π the invariant distribution of the Markov chain $\mathbf{x}$ on its state space $\mathbf{X}$. Assume that there is a subset $\mathbf{Y}$ of $\mathbf{X}$ such that the Markov chain $\mathbf{x}$ is irreducible on $\mathbf{Y}$ and cannot leave or enter $\mathbf{Y}$. That is, the rate matrix Q of $\mathbf{x}$ is such that

$$q(x, y) = 0 \text{ if } (x, y) \in \mathbf{Y}^c \times \mathbf{Y} \text{ or } (x, y) \in \mathbf{Y} \times \mathbf{Y}^c$$

where $\mathbf{Y}^c := \mathbf{X} - \mathbf{Y}$. The claim is then that the distribution $\pi_{\mathbf{Y}}$ is invariant for $\mathbf{x}$ where

$$\pi_{\mathbf{Y}}(x) := \frac{\pi(x) 1\{x \in \mathbf{Y}\}}{\pi(\mathbf{Y})}. \tag{7.27}$$

To verify this claim, note that

$$\sum_{x \in \mathbf{Y}} \pi(x) q(x, y) = \sum_{x \in \mathbf{X}} \pi(x) q(x, y) = 0.$$

The first equality follows from (7.27) and the second from the invariance of π.

Thus, the invariant distribution of the network of the left-hand side of Figure 7.10 is the distribution (7.26) normalized on the set of states that correspond to N customers in the network. In particular, we find that the invariant probability that there are m customers in the M/GI/∞ queue and $N - m$ in the M/M/1 queue is given by

$$P[X=m, Y=N-m|X+Y=N] = \frac{\phi(m, N-m)}{\sum_{n=0}^{N} \phi(n, N-n)}$$
$$= \frac{\rho^m/m!}{\sum_{n=0}^{N} \rho^n/n!}, \text{ for } m=0,\ldots,N$$

where $\rho := \lambda E\sigma$.

In particular, for $m = N$, this formula gives the blocking probability that is thus seen to depend on the distribution of the service times only through their mean value.

7.4 ATM NETWORKS

The analysis of ATM networks reflects one important characteristic of these networks: their large bandwidth-delay product. As we explained in Chapter 2, this large bandwidth-delay product makes feedback control largely ineffective. Consequently, ATM networks use open-loop control strategies. In this section we discuss some basic results that network researchers use to analyze the performance of ATM networks.

In section 7.4.1 we discuss deterministic models of networks. These models enable researchers to gain some insight in the behavior of buffers and leaky buckets. In section 7.4.2 we explain large deviations of iid random variables. We explore a generalization in section 7.4.3. We then apply the results to the analysis of loss probabilities in a buffer in section 7.4.4.

7.4.1 Deterministic Approaches

In this section we develop performance bounds for buffers and traffic rates using deterministic rather than stochastic analysis.

Linear Bounds

Consider a source that transmits at most $B + Rt$ bits in any interval of t seconds, for *any* possible value of t. We say that such a source produces a (B, R)-traffic to recall these constraints.

Assume that a (B, R)-traffic goes through a first-come, first-served buffer (initially empty) with a capacity to store B bits and equipped with a transmitter with rate C bps, with $C \geq R$. We claim that the buffer never loses any bit and that it delays the input stream by at most B/C seconds.

To verify the claim, assume that the buffer loses bits at some time T and that it was empty for the last time before T at time $T - S$. Then, during

$[T - S, T]$, the buffer transmits exactly CS bits and at most $B + RS$ bits enter the buffer. Since $B + RS - CS \leq B$, it is not possible for the buffer occupancy at time T to exceed B. This contradicts the assumption that the buffer loses bits at time T.

Consider a node that transmits bits from its buffer at rate C whenever the buffer is nonempty. Assume that one (B, R)-traffic and another (B', R')-traffic share that buffer. Assume that $C > R + R'$. The input stream is a $(B + B', R + R')$-traffic. If the node transmits the bits in their order of arrival, then the node delays the (B, R)-traffic by at most $(B + B')/C$. However, if we do not assume that the node sends the bits in their order of arrival but only that the node keeps transmitting whenever it is nonempty, then we find that the node delays its input traffic streams by at most $T = (B + B')/(C - R - R')$.

To see this, note that the buffer cannot remain nonempty for an interval with a duration longer than T, since in that interval the node transmits CT bits, and at most $B + B' + (R + R')T = CT$ bits enter the buffer. Moreover, if the node transmits all the bits of the (B, R)-stream in their order of arrival, then the node delays these bits by at most $S = (B + B')/(C - R')$. Indeed, the worst backlog facing a bit of the (B, R)-traffic is $B + B'$ and the node runs out of that backlog and of the subsequent arrivals of the (B', R')-traffic after S seconds.

Let us summarize the above observations in the form of a theorem.

Theorem 7.4.1 ((*B*, *R*)-traffic) By definition, a (B, R)-traffic carries at most $B + Rt$ bits in any time interval of duration t seconds. A network node can avoid losing bits of a (B, R)-traffic by reserving a buffer capacity of B bits and a bandwidth of R bps for that traffic.

Moreover, a network node that transmits at rate C whenever it is not empty and that is shared by a (B, R)-traffic and a (B', R')-traffic delays its inputs by at most $(B + B')/(C - R - R')$ s. If the node transmits the bits of the (B, R)-traffic in their order of arrival, then it delays those bits by at most $(B + B')/(C - R')$ s. ∎

This theorem can be used to derive bounds on delays in network nodes and therefore the end-to-end network delay.

Leaky Bucket

To ensure that the source satisfies the above conditions, the user can control that source with a (B, R)–leaky bucket. This leaky bucket accumulates credits (tokens) at the continuous rate R and can accumulate B units of credit. To transmit a bit, the token must contain at least one unit of credit.

Let us verify that the output of the (B, R)–leaky-bucket controller is a (B, R)-traffic. To do this, we consider an interval of time $[S, S + t]$ of duration t $(t \geq 0)$. We must show that the output of the leaky-bucket controller produces at most $B + Rt$ bits in that time interval. Let A be the number of tokens in the counter at time S. Note that $A \leq B$. During the time interval $[S, S + t]$ the counter accumulates Rt tokens. Thus, $A + Rt$ tokens are available to the source to transmit bits. Consequently, the output produces at most $A + Rt \leq B + Rt$ bits in the time interval of duration t. Hence the output is indeed a (B, R)-traffic, as claimed.

We have seen that it is a simple matter for the user to guarantee that the traffic is a (B, R)-traffic. What is more difficult to evaluate is the effect of the leaky-bucket controller on the source traffic. The leaky-bucket controller delays the bit stream that the source produces. The statistics of the delay depend on the statistics of the source bit stream.

Thus, by using a leaky-bucket controller and by reserving buffer capacity and bandwidth, the network can guarantee that it will not lose user bits and that it will not introduce a delay larger than B/R. However, it is up to the user to select the parameters (B, R) so that the controller does not delay the source traffic excessively.

The network can verify that the traffic is a (B, R)-traffic by checking whether the traffic goes through a (B, R)–leaky-bucket controller without any delay. Indeed, only a (B, R)-traffic can go without delay through a (B, R)–leaky-bucket controller.

Assume that a user sets up a connection and specifies that the traffic will be a (B, R)-traffic. What should the network do if the traffic does not go through the (B, R)-controller? The ATM Forum recommendation is that the network should mark the cells that the leaky bucket delays as low-priority cells by setting their CLP header bit. These cells become candidates for discarding by network nodes that experience congestion.

We now show that the (B, R)–leaky-bucket controller—we call it *LB*—delays the traffic the least among all the first in, first out controllers that output a (B, R)-traffic. Consider the leaky-bucket controller *LB* and another controller *LB*′ whose output is also a (B, R)-traffic. The two controllers have the same input. We claim that every bit leaves *LB* before *LB*′. To prove the claim we argue by contradiction. Assume that at least one bit leaves *LB*′ before *LB* and that this occurs for the first time at time t. Thus, at time t the token counter of *LB* must be empty. Denote by $t - T$ the last time before time t that the token counter of *LB* was full. It follows that during $[t - T, t]$ the output of *LB* carries $B + R \times T$ bits. Moreover, the output of *LB*′ during $[t - T, t]$ must carry at least the same bits as the output of *LB* plus one more. Indeed, before time t, all the bits leave LB' after LB, and the output of LB' catches up with and exceeds that of LB at time t. But this fact implies that LB' outputs at least $B + R \times T + 1$ bits in T seconds. Consequently, the output of LB' is not a (B, R)-traffic.

As we explained earlier, the GCRA differs slightly from the leaky bucket described in the previous section. The difference is the quantization: the GCRA(T, τ) is a leaky bucket that removes T units of fluid per cell and not one infinitesimal unit per bit. We analyzed in section 6.4.2 the effect of a GCRA controller on the traffic.

Burstiness

We can use deterministic models to clarify the notion of burstiness. Roughly speaking, a traffic stream is more bursty than another if it requires more buffering at a transmitter. After defining this notion more precisely we show that a leaky-bucket controller reduces the burstiness.

For the purpose of this discussion, it is easier to view traffic as a fluid than as a stream of discrete packets. We define a message as a time-varying bit rate. That is, a message $\mathbf{m}$ is a nonnegative function of time $\mathbf{m} = \{m(t), t \geq 0\}$. We assume that the function $\mathbf{m}$ is integrable and that $\int_0^\infty m(t)dt = M$. The interpretation is that $m(t)$ is the bit rate of the message at time t and that M is the total number of bits of the message.

Assume that $\mathbf{m}$ goes through a buffer with service rate c bps. We denote the maximum number of bits that the buffer must store by $b_m(c)$. To calculate $b_m(c)$, we note that the buffer occupancy $x(t)$ satisfies, with $x(0) = 0$,

$$\frac{d}{dt}x(t) = \begin{cases} m(t) - c, & \text{if } x(t) > 0 \\ (m(t) - c)^+, & \text{if } x(t) = 0. \end{cases}$$

These equations express that the buffer occupancy grows at a rate equal to the input rate $m(t)$ minus the service rate c and that the buffer occupancy cannot become negative. By solving the equations we can find the maximum value $b_m(c)$ of $x(t)$ for $t \geq 0$.

We can define a (B, R)–leaky-bucket controller as a device that accumulates a token fluid at a constant rate R in a token buffer that can store up to B units of token fluid. To transmit ϵ units of (traffic) fluid, the transmitter must remove ϵ units of token fluid.

Consider some message $\mathbf{m}$. Assume that this message goes through a (B, R)–leaky-bucket controller. The output is a new message $\mathbf{n}$. The claim is that $b_n(c) \leq b_m(c)$ for all $c \geq 0$. That is, the output of the leaky bucket requires less buffering at a transmitter of any fixed rate c. Note that this buffering at the transmitter does not include the buffering in the leaky-bucket controller. We say that the message **n** is *less bursty* than message **m**.

To verify the claim, consider the situation where the message **n** is served by a buffer with a transmitter with rate c. Assume that the buffer accumulates b units of fluid. We will show that the buffer also accumulates at least b units of fluid when its input is **m**. To show this, we denote by T the first time that the buffer occupancy reaches the value b with the input **n** and by S the last time before T that the buffer was empty. Thus, during $[S, T]$ the input **n** must carry $b + c(T - S)$ units of fluid.

Some researchers have proposed the following variation on the linear bounds. Instead of defining a (B, R)-traffic, they define a $\{(B_1, R_1), \ldots, (B_K, R_K)\}$-traffic as a stream that carries at most $B_k + R_k t$ cells for all $k = 1, \ldots, K$ and for all $t \geq 0$. To enforce that condition, the source traffic should go through each of the (B_k, R_k)-regulators for $k = 1, \ldots, K$. For instance, we may recall that the VBR specification calls for two leaky buckets; see section 6.4.2.

7.4.2 Large Deviations of iid Random Variables

In this section we explore statistical methods for call admission. We want to justify the effective bandwidth results. We first explain the case of iid random variables. In the next section we extend the results to nonindependent random variables.

Cramer's Theorem

Consider a collection $\{X_n, n \geq 1\}$ of independent and identically distributed random variables with common distribution $F(.)$ and with finite mean value m. Define partial sum S_n as

$$S_n = \sum_{k=1}^{n} X_k.$$

We know from the strong law of large numbers that

$$\frac{S_n}{n} \to m \text{ as } n \to \infty \text{ with probability 1.}$$

Thus, the probability that S_n/n is away from m goes to 0 as n increases. It can be shown that this convergence to 0 occurs exponentially fast in n. More precisely, for $a \geq m$,

$$\lim_{n\to\infty} \frac{1}{n} \log P(S_n \geq na) = -\Lambda^*(a) \tag{7.28}$$

where

$$\Lambda^*(a) = \sup_{\theta}[\theta a - \Lambda(\theta)] \text{ with } \Lambda(\theta) = \log E(e^{\theta X_1}). \tag{7.29}$$

For this result to be valid we need $\Lambda(\theta)$ to be differentiable and finite in a neighborhood of 0. Algebra shows that $\Lambda^*(m) = 0$.

Roughly, this result states that

$$P(S_n \approx na) \approx e^{-n\Lambda^*(a)}.$$

This result is called Cramer's Theorem.

That is, it is unlikely that S_n/n is away from m, and this is exponentially unlikely in n. The value of $\Lambda^*(a)$ indicates how difficult it is for S_n/n to be close to a. If $\Lambda^*(a)$ is large, then it is very difficult for S_n/n to be close to a.

Comments and Sharpening

Why should we expect this probability to decay exponentially in n? Assume that

$$P(S_n \approx na) \approx \epsilon.$$

Then, $P(X_1 + \cdots + X_n \approx na) \approx \epsilon$. Moreover, the most likely way for $X_1 + \cdots + X_{2n}$ to be close to $2na$ is for $X_1 + \cdots + X_n$ to be close to na and for $X_{n+1} + \cdots + X_{2n}$ also to be close to na. If we believe this last sentence, then

$$\begin{aligned} P(S_{2n} \approx 2na) &= P(X_1 + \cdots + X_{2n} \approx 2na) \\ &\approx P(X_1 + \cdots + X_n \approx na)P(X_{n+1} + \cdots + X_{2n} \approx na) \\ &\approx \epsilon^2, \end{aligned}$$

which shows that $P(S_n \approx na)$ is approximately exponential in n. One could argue that there are many ways for S_{2n} to be close to $2na$ other than for both $X_1 + \cdots + X_n$ and $X_{n+1} + \cdots + X_{2n}$ to be close to na. For instance, $X_1 + \cdots + X_n$ could be approximately $n(a + q)$ for some small q, and $X_{n+1} + \cdots + X_{2n}$ could be approximately $n(a - q)$. The probability of this event is approximately

$$e^{-n\Lambda^*(a+q)} \times e^{-n\Lambda^*(a-q)}.$$

However,

$$\Lambda^*(a + q) + \Lambda^*(a - q) > 2\Lambda^*(a),$$

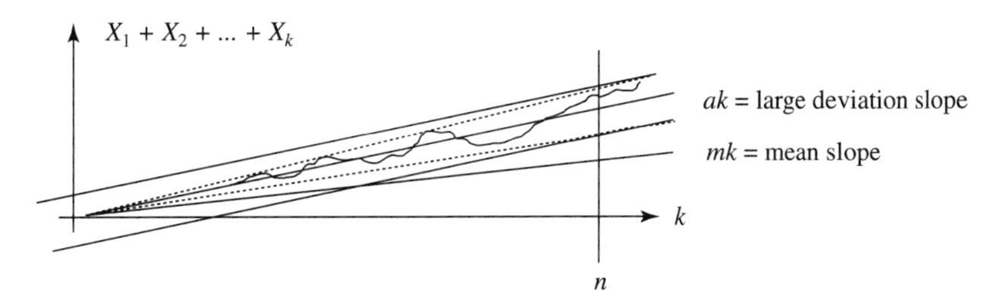

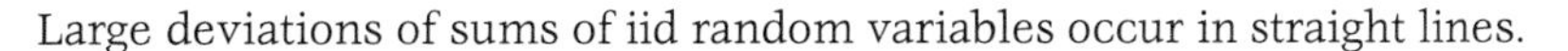

7.11 FIGURE Large deviations of sums of iid random variables occur in straight lines.

by strict convexity of the function $\Lambda^*(.)$. Consequently, the probability of that event is much smaller than ϵ^2.

It can be shown formally that, given that $S_n \approx na$, the most likely realization of that event is if $S_k \approx ak$ for $k \leq n$. Thus, the rare event $\{S_n \approx na\}$ occurs when the partial sums $X_1 + \cdots + X_k$ are approximately equal to ak for $k \leq n$. We will remember that fact by saying that *the large deviation occurs in a straight line*, as illustrated in Figure 7.11. Thus, the most likely way that the arrival sequence has an empirical rate a over some long time interval is for the arrivals to occur with an approximately constant rate a during that interval.

Proof of Cramer's Theorem

The proof of this theorem illustrates a general approach used in the theory of large deviations. To show a limit, we prove an upper bound and a lower bound. That is, we show

- for all $n \geq 1$,

$$P(S_n \geq na) \leq \exp\{-n\Lambda^*(a)\}. \tag{7.30}$$

- for all $\epsilon, \delta > 0$ there is some N such that, for all $n \geq N$,

$$P(S_n \geq na - n\epsilon) \geq (1 - \delta)\exp\{-n\Lambda^*(a)\}. \tag{7.31}$$

Let us first show the upper bound (7.30). We use Markov's inequality. For all $\theta > 0$ one has

$$P(S_n \geq na) \leq E\exp\{\theta(S_n - na)\} = \exp\{-n\theta a\}[E\exp\{\theta X_1\}]^n$$
$$= \exp\{-n(\theta a - \Lambda(\theta))\}.$$

We obtain (7.30) by minimizing the right-hand side of the above inequality over $\theta > 0$. (Recall the definition (7.29).)

The lower bound (7.31) is slightly more difficult to prove. The trick is to change the distribution of the X_n's to make it easy for them to have a sample mean close to a. To do this, we define random variables Y_n that have a distribution such that

$$P(Y_n \in (x, x+dx)) = \exp\{-\theta x\} P(X_n \in (x, x+dx)) E\exp\{\theta X\} \quad (7.32)$$

$$= \exp\{\Lambda(\theta) - \theta x\} P(X_n \in (x, x+dx)). \quad (7.33)$$

The last term in the right-hand side of (7.32) is there to normalize the distribution of Y_n. The random variables Y_n for $n \geq 1$ are iid and we choose θ so that $E(Y_n) = a$. Note that

$$P\left(\frac{Y_1 + \ldots + Y_n}{n} \in [a-\epsilon, a+\epsilon]\right) \to 1, \text{ as } n \to \infty, \quad (7.34)$$

by the weak law of large numbers (because $E(Y_n) = a$). Now we claim that for $0 < \epsilon \ll 1$

$$P\left(\frac{Y_1 + \ldots + Y_n}{n} \in [a-\epsilon, a+\epsilon]\right)$$
$$= \exp\{-\theta na + n\Lambda(\theta)\} P\left(\frac{X_1 + \ldots + X_n}{n} \in [a-\epsilon, a+\epsilon]\right). \quad (7.35)$$

To see this equality, note that with $\mathbf{X} = (X_1, \ldots, X_n)$ and $\mathbf{Y} = (Y_1, \ldots, Y_n)$ and any function $f(.)$ one has

$$E\left\{f(\mathbf{X})\frac{P_{\mathbf{Y}}(\mathbf{X})}{P_{\mathbf{X}}(\mathbf{X})}\right\} = \int f(\mathbf{x})\frac{P_{\mathbf{Y}}(\mathbf{x})}{P_{\mathbf{X}}(\mathbf{x})} P_{\mathbf{X}}(\mathbf{x}) d\mathbf{x}$$

$$= \int f(\mathbf{x}) P_{\mathbf{Y}}(\mathbf{x}) d\mathbf{x} = E\{f(\mathbf{Y})\}$$

where $P_{\mathbf{Y}}(.)$ is the probability density of the vector $\mathbf{Y}$ and similarly for $P_{\mathbf{X}}(.)$. We obtain (7.35) by choosing $f(\mathbf{x}) = 1\{x_1 + \cdots + x_n \in [na - n\epsilon, na + n\epsilon])$ and by substituting the value of the ratio of densities as specified by (7.32).

Using (7.34) we conclude that for n large enough

$$\exp\{-na\theta + n\Lambda(\theta)\} P\left(\frac{X_1 + \ldots + X_n}{n} \geq a - \epsilon\right) \geq (1-\delta). \quad (7.36)$$

We now use the fact that $E(Y) = a$. That is, since Λ is differentiable,

$$a = E(Y) = \exp\{-\Lambda(\theta)\} \int x \exp\{x\theta\} P(x \in (x, x+dx)) = \Lambda'(\theta).$$

This equality shows that θ achieves the maximum over σ of $g(\sigma) := a\sigma - \Lambda(\sigma)$ since $g'(\theta) = a - \Lambda'(\theta) = 0$. This maximum can be shown to be unique because Λ is convex. Thus,

$$a\theta - \Lambda(\theta) = \sup_{\sigma}[a\sigma - \Lambda(\sigma)] = \Lambda^*(a).$$

We use the result of this calculation in (7.36) to obtain, for large enough n,

$$P\left(\frac{X_1 + \ldots + X_n}{n} \geq a - \epsilon\right) \geq (1 - \delta)\exp\{-n\Lambda^*(a)\},$$

which is (7.31).

7.4.3 Straight-Line Large Deviations

The results on iid random variables are not sufficient to study communication networks. We need results on sequences of random variables that are not independent. In this section, we define a property of sequences of random variables, and we comment on when that property is satisfied.

Sequences with Straight-Line Large Deviations

Let $\{X_n, n \geq 1\}$ be a sequence of random variables. Define

$$S_n := X_1 + \cdots + X_n, \quad \text{for } n \geq 1.$$

Let also Λ^* be a function defined on the real line and taking values in $[0, \infty]$. We say that the random variables are of class $S(\Lambda^*)$ if

$$\lim_{\epsilon \downarrow 0} \lim_{n \to \infty} \frac{1}{n} \log P(\max_{1 \leq k \leq n} |S_k - ka| \leq n\epsilon) = -\Lambda^*(a) \tag{7.37}$$

for all $a \in (-\infty, +\infty)$.

This property means that the probability that the partial sums $\{S_k, k \geq 1\}$ follow a straight line with slope a for n time units is approximately $\exp\{-n\Lambda^*(a)\}$.

For a sequence of class $S(\Lambda^*)$, we define

$$\Lambda(\delta) := \sup_{a}[a\delta - \Lambda^*(a)]. \tag{7.38}$$

Sequences of iid random variables are of class $S(\Lambda^*)$. We will see shortly that, under specific assumptions, output processes of queues are also of such a class.

7.4.4 Large Deviation of a Queue

Our objective is to evaluate the loss rate at a queue. That is, the queue has a finite buffer capacity B and we want to estimate the fraction of arrivals that occur when the queue is full. For many arrival processes, this fraction is comparable to the fraction of times that the queue with an infinite buffer capacity has a buffer occupancy that exceeds B. That relationship holds for the processes that we consider in this section. Thus, we assume that the queue has an infinite buffer capacity, and we estimate $P(W > B)$, the invariant probability that the buffer occupancy W exceeds B. Fix $\delta > 0$. We show that this probability is such that

$$P(W > B) \approx \exp\{-\delta B\}$$

provided that the service rate of the queue is large enough. For a background on this formulation, see the discussion on buffering in section 6.4.3.

Consider the following discrete-time queuing system. For $n \geq 1$, X_n customers arrive at the queue at time n, and up to c customers that are in the queue are served at that time. We try to estimate the probability that, starting empty, the queue occupancy reaches a large value B before becoming empty again. We argue that this event can happen if the arrivals have an empirical rate $a > c$ for at least $B/(a - c)$ time units. According to (7.28), the probability that the arrivals behave in that way is approximately equal to

$$\exp\left\{-\frac{B}{a-c}\Lambda^*(a)\right\}.$$

Since the rate a can be any value larger than c, we find that the probability that the queue occupancy reaches a value B before becoming empty again is approximately equal to

$$\sum_{a>c} \exp\left\{-\frac{B}{a-c}\Lambda^*(a)\right\}.$$

We further approximate this sum of exponentials in B by the exponential with the largest exponent, i.e., by

$$\exp\left\{-\frac{B}{a^*-c}\Lambda^*(a^*)\right\}$$

where

$$\frac{\Lambda^*(a^*)}{a^* - c} = \min_{a>c} \frac{\Lambda^*(a)}{a - c}. \tag{7.39}$$

The above argument shows that the probability that the buffer occupancy reaches a large value B in a busy cycle decays exponentially in B, at least in a first approximation. We should note the many approximations that we made along the way. We were only trying to get the rate of decay of the exponential, and we neglected all possible estimates of the coefficient in front of the exponential. For specific models we could calculate the coefficient of the exponential rather precisely. However, since actual traffic models in networks are largely unknown, there is little practical interest in pursuing that line of development.

Effective Bandwidth

A natural question is to ask what the value of c should be for the decay rate to be equal to a specified value, say δ. That is, we want to find the smallest possible value of c such that

$$\min_{a>c} \frac{\Lambda^*(a)}{a - c} = \delta. \tag{7.40}$$

We denote that value of c by $\alpha(\delta)$, and we call it the *effective bandwidth* of the arrival stream. The interpretation is that the effective bandwith $\alpha(\delta)$ is the rate at which that stream must be served so that the buffer occupancy decays as an exponential with rate δ.

The claim is that

$$\alpha(\delta) = \frac{\Lambda(\delta)}{\delta}. \tag{7.41}$$

To prove the claim, we note that (7.40) implies that, for all $c \geq \Lambda(\delta)/\delta$, one has

$$\delta c \geq \Lambda(\delta) = \sup_a[\delta a - \Lambda^*(a)] = \delta a^* - \Lambda^*(a^*), \text{ for some } a^* > c,$$

so that, for all a,

$$\delta c \geq \delta a - \Lambda^*(a)$$

and therefore, for all $a > c$,

$$\frac{\Lambda^*(a)}{a - c} \geq \delta.$$

On the other hand, if $c < \Lambda(\delta)/\delta$, then

$$\delta c < \Lambda(\delta) = \delta a^* - \Lambda^*(a^*) \text{ and } \frac{\Lambda^*(a^*)}{a^* - c} < \delta.$$

Consequently, $\Lambda(\delta)/\delta$ is indeed the smallest value of c so that $\min_{a>c}[\Lambda^*(a)/(a-c)] = \delta$.

If we recall the definition (7.29) of $\Lambda(\delta)$, we note the important result that *the effective bandwidth is additive.* That is, if $X_n = X_n^1 + X_n^2$ where the sequences of random variables $\{X_n^1\}$ and $\{X_n^2\}$ are independent, then we find that

$$\begin{aligned}\alpha(\delta) &= \frac{1}{\delta}\log Ee^{\delta(X_n^1+X_n^2)} = \frac{1}{\delta}\log[Ee^{\delta X_n^1} \times Ee^{\delta X_n^2}] \\ &= \frac{1}{\delta}\log Ee^{\delta X_n^1} + \frac{1}{\delta}\log Ee^{\delta X_n^2} \\ &= \alpha_1(\delta) + \alpha_2(\delta),\end{aligned}$$

where $\alpha_i(\delta)$ is the effective bandwidth of the sequence $\{X_n^i, n \geq 1\}$ $(i = 1, 2)$.

This result is very appealing because it suggests that we can treat the various streams as if they had a fixed rate equal to their effective bandwidth. However, we should keep in mind the crude approximations made along the way. We will sharpen these results later.

Summarizing, we have motivated (not really proved) the following result.

Theorem 7.4.2 Consider a discrete-time queue with X_n arrivals at time n $(n \geq 0)$ that serves c customers per unit of time in the queue. If the sequence of arrivals is of class $S(\Lambda^*)$, then the invariant probability $P(W > B)$ that the queue length W exceeds B is such that, for large B,

$$\frac{1}{B}\log P(W > B) \approx -\min_{a>c}\frac{\Lambda^*(a)}{a-c}.$$

Moreover, the minimum value of c such that the right-hand side of the above inequality is at most $-\delta < 0$ is called the *effective bandwidth* of the arrival sequence and is equal to $\alpha(\delta)$ where

$$\alpha(\delta) = \frac{\Lambda(\delta)}{\delta}$$

with $\Lambda(\delta)$ defined by (7.38).

In particular, the effective bandwidth of the sum of two independent arrival sequences is the sum of their effective bandwidths. ■

Loosely speaking, we can think of the random arrival streams as having constant arrival rates given by their effective bandwidths. This effective bandwidth has a value between the mean and the peak rate of the stream, and it measures in a precise way the burstiness of that stream.

Decoupling Bandwidth

We consider once again the same queuing system that serves c customers per unit of time, but we now explore the properties of the departure process from the queue. Specifically, we want to analyze the effective bandwidth of the departure process. Our motivation is to be able to analyze a network of buffers instead of a single buffer.

The effective bandwidth $\alpha_c(\delta)$ of the departure process from the queue is the rate at which a subsequent buffered link should serve that departure process so that the occupancy at the buffer of that link has an exponential tail with rate δ.

To calculate the effective bandwidth $\alpha_c(\delta)$, we argue as follows. Since large deviations of the arrivals occur along straight lines, the most likely way for the departures to produce customers at a rate $a < c$ over a long time interval is for the arrivals to produce customers at the same rate. In other words, it is much less likely that the arrivals will be more busy for a while to fill up the queue and then less busy to average out the departure rate to a. Also, it is not possible for the departure rate to exceed c customers per unit of time. Hence, the probability that the departure produces customers with rate a for $n \gg 1$ time units is approximately equal to

$$\exp\{-n\Lambda_c^*(a)\} \text{ where } \Lambda_c^*(a) = \begin{cases} \Lambda^*(a), & \text{for } a \leq c \\ +\infty, & \text{for } a > c. \end{cases} \tag{7.42}$$

Using the effective bandwidth result that we explained for the arrival process, we conclude that

$$\alpha_c(\delta) = \frac{\Lambda_c(\delta)}{\delta} \tag{7.43}$$

where

$$\Lambda_c(\delta) := \sup_a [a\delta - \Lambda_c^*(a)]. \tag{7.44}$$

At this point, a figure helps understand the above formulas. Figure 7.12 shows $\Lambda^*(.)$, $\alpha(\delta)$, and $\alpha_c(\delta)$.

For a given value of $\delta > 0$, $\alpha^*(\delta)$ is such that

$$\frac{d}{dc}\Lambda^*(\alpha^*(\delta)) = \delta. \tag{7.45}$$

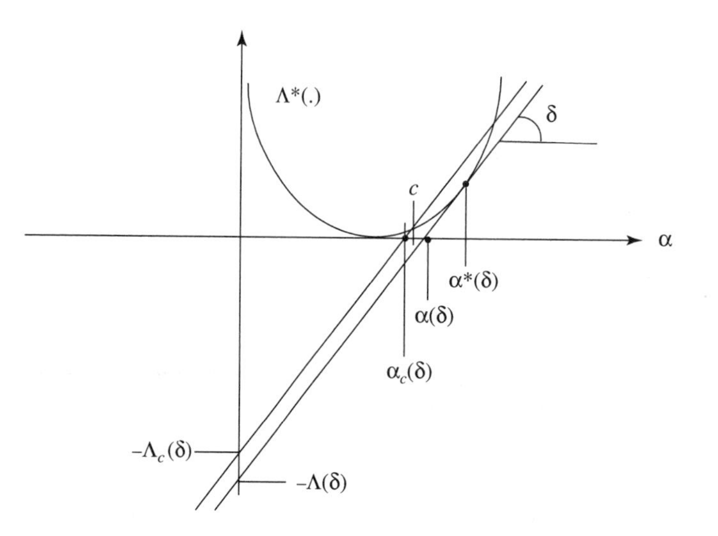

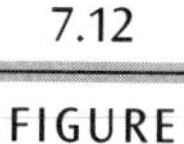
7.12 FIGURE

Effect of a buffer on the effective bandwidth. The figure shows the main quantities involved in the calculation of the effective bandwidth of the output of a buffer. In particular, the figure shows how the service rate c reduces the effective bandwidth from $\alpha(\delta)$ to $\alpha_c(\delta)$.

Observe that $\alpha^*(\delta)$ is the rate of the arrival process that is most likely to make the buffer overflow when the service rate c is equal to $\alpha(\delta)$. Indeed, $\alpha^*(\delta)$ is the value of a that maximizes

$$\frac{\Lambda^*(a)}{a - \alpha(\delta)},$$

as you can see from the figure.

Note that

$$\alpha_c(\delta) = \begin{cases} \alpha(\delta), & \text{if } c \geq \alpha^*(\delta) \\ c - \frac{\Lambda^*(c)}{\delta}, & \text{if } c < \alpha^*(\delta). \end{cases} \tag{7.46}$$

We are interested in the first case of the above equality. That case states that if the service rate c of the buffer is large enough, then the effective bandwidth of the output is equal to that of the input. In other words, as far as large deviations are concerned, the buffer is transparent: it does not perturb the stream that goes through it. We call this result a *decoupling* property. What is surprising about the result is that it is exact, not an approximation. We will see

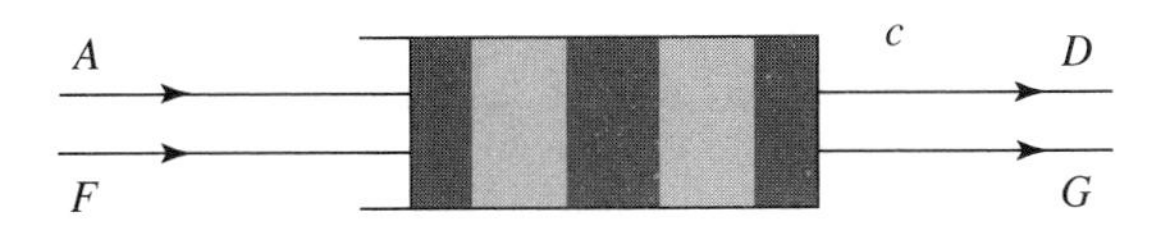

7.13 FIGURE

Two streams A and F share a transmitter that transmits at rate c whenever the queue is not empty. Stream A comes out as stream D. The effective bandwidth of stream D depends on the statistics of A and F and on the actual service discipline. However, if c is large enough, then the effective bandwidth of D is the same as that of A.

shortly that this result has interesting implications for the design of networks. But first, we must extend the decoupling property to the case of interfering traffic.

7.4.5 Large Deviation of a Multiclass Queue

Consider the situation shown in Figure 7.13. Two streams A and F are sharing a buffer with service rate c. Stream A goes out as stream D and stream F goes out as stream G. We are interested in stream D. Specifically, we would like to calculate the effective bandwidth $\alpha_D(\delta)$ of stream D.

To be able to characterize the stream D we need to assume that the queuing discipline is not very strange but instead possesses some reasonable continuity property with respect to the workload. We define that property next.

Fluid Continuity

Intuition suggests that if we neglect the discreteness of the arrivals of cells in a fast network, then the streams look like fluids. The rates of the fluids are random, and, on rare occasions, the rate of a particular stream may deviate from its mean rate.

Consider one queue with a number of input and output streams. It is reasonable to expect that if the input streams are fluids with constant rates over a long period of time, then so are the output streams. Moreover, if the input streams are close to being fluids with constant rates, then the same should be true of the output streams.

To make these ideas precise, we need to define what we mean by two streams being close to each other. That is, we must define a metric.

Let $\{A_k, k \geq 1\}$ be a sequence of real numbers. We define $A(0,0] = 0$, $A(0,k] = A_1 + \cdots + A_k$ for $k \geq 1$, and

$$||A|| := \limsup_{n \to \infty} \frac{1}{n} \max_{1 \leq k \leq n} |A(0,k]|. \tag{7.47}$$

We define the constant process λ such that $\lambda(0,k] = \lambda k$ for all $k \geq 1$ and the constant process μ such that $\mu(0,k] = \mu k$ for all $k \geq 1$.

We now come back to the queue of Figure 7.13. We say that the queue is *fluid continuous at* (λ, μ) if, when initially empty, it is continuous with respect to the metric (7.47) at the point (λ, μ), i.e., if for all $\epsilon > 0$ there is some $\delta > 0$ such that

$$||A - \lambda|| \leq \delta \text{ and } ||F - \mu|| \leq \delta \tag{7.48}$$

implies that

$$||D - \lambda|| \leq \epsilon \text{ and } ||G - \mu|| \leq \epsilon. \tag{7.49}$$

The interpretation is that if the input streams look like constant processes in the fluid limit, then so do the output streams.

Fluid Continuity of FCFS Queues

Consider the queue of Figure 7.13. Assume that the queue is FCFS (first come, first served). We show that this queue is fluid continuous at (λ, μ) whenever $\lambda + \mu < c$.

Let λ and μ be two positive numbers such that $\lambda + \mu < c$. Consider two arrival processes A and F such that

$$\lambda t - \epsilon T \leq A(0,t] \leq \lambda t + \epsilon T, \text{ for } 0 \leq t \leq T \tag{7.50}$$

and

$$\mu t - \epsilon T \leq F(0,t] \leq \mu t + \epsilon T, \text{ for } 0 \leq t \leq T. \tag{7.51}$$

Assume that the queue is empty at time 0. The work to be done in the queue at time t, W_t, satisfies

$$W_t = \max_{0 \leq u \leq t} \{A(u,t] + F(u,t] - c(t-u)\}$$

where $A(u,t] = A(0,t] - A(0,u]$ and similarly for $F(u,t]$.

Consequently, we find using (7.50) and (7.51) that

$$W_t \leq 4\epsilon T, \text{ for } 0 \leq t \leq T.$$

As a result, the delay faced by any customer who arrives in [0, T] in the queue is less than δ where

$$\delta := \frac{4\epsilon T}{c}.$$

(This is where we use the FCFS assumption.)

This delay bound implies that

$$A(0, t - \delta] \leq D(0, t] \leq A(0, t].$$

Combining these inequalities with (7.50), we conclude that

$$D(0, t] \leq A(0, t] \leq \lambda t + \epsilon T, \text{ for } 0 \leq t \leq T$$

and

$$D(0, t] \geq A(0, t - \delta] \geq \lambda(t - \delta) - \epsilon T = \lambda t - \lambda \frac{4\epsilon T}{c} - \epsilon T, \text{ for } 0 \leq t \leq T.$$

Consequently,

$$|D(0, t] - \lambda t| \leq \epsilon \left(\frac{4}{c} + 1\right) T, \text{ for } 0 \leq t \leq T.$$

A similar inequality can be proved for the process G. These inequalities prove the fluid continuity of the queue.

Decoupling of Fluid-Continuous Queues

We want to show that the output of the multiclass queue has the same effective bandwidth as the input and that, moreover, that output is still sufficiently well-behaved to be able to analyze the possible output of a subsequent queue. Specifically, we explain that if the inputs A and F have straight-line large deviations and if the transmitter rate c is large enough, then the effective bandwidth of D is the same as that of A, and D once again has straight-line large deviations.

Theorem 7.4.3 Assume that the queue of Figure 7.13 is fluid continuous at (λ, μ) whenever $\lambda + \mu < c$. Assume also that the processes A and D are independent and of class $S(\Lambda_A^*)$ and $S(\Lambda_D^*)$, respectively. Then

$$\alpha_D(\delta) = \alpha_A(\delta), \text{ if } c \geq \lambda_F + \alpha_A^*(\delta) \text{ and } c \geq \alpha_A(\delta) + \alpha_F(\delta) \tag{7.52}$$

where λ_F is the average rate of F and $\alpha_A^*(\delta)$ is the decoupling bandwidth of A. Moreover, the processes D and G are of class $S(\Lambda_1^*)$ and $S(\Lambda_2^*)$, respectively, for some functions Λ_1^* and Λ_2^* that we describe later. ■

The interpretation of this result is that if the transmitter is fast enough, then it does not disturb the stream A, even though this stream shares the buffer with another stream F. Note that in general $\alpha_D(\delta)$ may be larger or smaller than $\alpha_A(\delta)$. For instance, if $F = 0$ as in the previous result, then $\alpha_D(\delta) \leq \alpha_A(\delta)$. On the other hand, if the stream A has constant rate λ_A and if F is bursty, then one may expect that D is bursty and has average rate λ_A, so that $\alpha_D(\delta) > \alpha_A(\delta)$, at least as long as c is not very large.

Mathematically, the result (7.52) states that the probability that D generates *at* bits over a long period of time with duration t is approximately equal to the probability that A produces *at* bits in t time units, as long as $a + \lambda_F < c$. That is, other possibilities, such as when F first backlogs the queue to make it later output mostly A bits, are much less likely.

The queue is stable and empties in finite time so that when we calculate the average rate of the output we can assume that the queue is initially empty.

Assume that the queue is in steady state at time 0. We can get a lower bound on the probability that it empties in some arbitrarily small fraction β of t time units. That probability goes to one as t increases. We show that the average rate from the time when the queue is empty until time t is at least a with probability at least $\exp\{-t\Lambda_A^*(a)\}$. To show this, assume that A produces *at* bits in t time units and that F produces bits at its average rate λ_F also for t time units. We assume that these rates are essentially constant over the time interval (since large deviations occur along straight lines). The probability of this event is approximately $\exp\{-t\Lambda_A^*(a)\}$. When this event occurs, the process D carries bits at an essentially constant rate a because of the fluid continuity of the queue. Consequently,

$$P(D \text{ has an essentially constant rate } a \text{ over } [0, t]) \geq \exp\{-t\Lambda_A^*(a)\}. \quad (7.53)$$

Second, we want to show the reverse inequality. That is, we want to show that

$$P(D \text{ has an essentially constant rate } a \text{ over } [0, t]) \leq \exp\{-t\Lambda_A^*(a)\}. \quad (7.54)$$

Intuitively, inequality (7.54) follows because the output D cannot be faster than A. That is, the most likely way for D to produce *at* bits in t time units is for A to also produce *at* bits in the same t time units. The difficulty in making this intuition precise is that it is conceivable that the queue might be unusually full of A bits at the start of the time interval and that the output could produce the at bits even if the input had a smaller rate, say $a' < a$ during the t time units. That possibility has a negligible probability under the assumption $c > \alpha_A(\delta) + \alpha_F(\delta)$, as we show next. At time 0, the queue contains a random number W_0 of bits. This random number is distributed as the steady-state queue length of a queue with input process $S := A + F$ and with service rate c. That random variable W_0 can be written as

$$W_0 = \max_{u \geq 0}\{S(-u, 0] - cu\} = \max_{u \geq 0}\{A(-u, 0] + F(-u, 0] - cu\} \tag{7.55}$$

where $S(-u, 0]$ is the number of bits in $S = A + F$ during the time interval $(-u, 0]$. To see why (7.55) holds, note that

$$W_0 = W_{-u} + S(-u, 0] - cu + cE(-u, 0]$$

where W_{-u} is the number of bits in the queue at time $-u$ and $E(-u, 0]$ is the time during $(-u, 0]$ when the queue is empty. Consequently, $W_0 \geq S(-u, 0] - cu$, and the equality holds when $-u$ is the last time before time 0 when the queue is empty.

Now, W_0 is an upper bound on the number of A bits in the queue at time 0 since some of the bits may be F bits. Therefore, for $t > 0$,

$$D(0, t] \leq W_0 + A(0, t],$$

so that for $0 < \theta < \delta$, $\epsilon > 0$, and t large enough,

$$\begin{aligned}
E\exp\{\theta D(0, t]\} &\leq E\exp\{\theta(W_0 + A(0, t])\} && (7.56)\\
&\leq E\max_{u \geq 0}\exp\{\theta(A(-u, 0] + F(-u, 0] - cu + A(0, t])\} && (7.57)\\
&\leq \sum_{u \geq 0} E\exp\{\theta(A(-u, 0] + F(-u, 0] - cu + A(0, t])\} && (7.58)\\
&\leq \sum_{u \geq 0} \exp\{(\Lambda_A(\theta) + \epsilon)t + (\Lambda_A(\theta) + \Lambda_F(\theta) + 2\epsilon - \theta c)u\} && (7.59)\\
&\leq C\exp\{(\Lambda_A(\theta) + 3\epsilon)t\}. && (7.60)
\end{aligned}$$

In the above derivation, the inequality (7.59) follows from the definitions of Λ_A and Λ_F. Indeed, we assume that

$$\Lambda_A(\theta) = \lim_{t \to \infty} \frac{1}{t} \log E\exp\{\theta A(0, t]\}.$$

Consequently, for $\epsilon > 0$ and $t > 0$ large enough,

$$E\exp\{\theta A(0, t]\} \leq \exp\{(\Lambda_A(\theta) + \epsilon)t\}. \tag{7.61}$$

Similarly, by stationarity of F, for $\epsilon > 0$ and $u > u(\epsilon, \theta)$ for $u(\epsilon, \theta)$ large enough,

$$E \exp\{\theta F(-u, 0]\} = E \exp\{\theta F(0, u]\} \leq \exp\{(\Lambda_F(\theta) + \epsilon)u\}. \tag{7.62}$$

We get the corresponding inequality for A:

$$E \exp\{\theta A(-u, 0]\} = E \exp\{\theta A(0, u]\} \leq \exp\{(\Lambda_A(\theta) + \epsilon)u\}. \tag{7.63}$$

The inequality (7.59) follows from (7.61)–(7.63). To derive the inequality (7.60) we first note the inequality (for $\theta > 0$)

$$\Lambda_A(\theta) + \Lambda_F(\theta) < c\theta,$$

which follows from the assumption

$$\alpha_A(\delta) + \alpha_F(\delta) = \frac{\Lambda_A(\delta)}{\delta} + \frac{\Lambda_F(\delta)}{\delta} < c,$$

so that the same inequality holds for $0 < \theta \leq \delta$, since $\Lambda(\theta)/\theta$ is increasing in $\theta > 0$. To get (7.60) we then define C by

$$C = \sum_{u \geq 0} \exp\{(\Lambda_A(\theta) + \Lambda_F(\theta) + \epsilon - \theta c)u\}.$$

We now use (7.60) to prove (7.54). For $a \geq \lambda_A$ and $0 < \theta \leq \delta$, we find, using Markov's inequality,

$$\begin{aligned} P(D \text{ has rate } a \text{ during } [0, t]) \leq P(D(0, t] \approx at) &\leq E \exp\{\theta(D(0, t] - at)\} \\ &\leq \exp\{(\Lambda_A(\theta) + \epsilon - \theta a)t\} \end{aligned}$$

for $\epsilon > 0$ and t large enough.

Thus,

$$\lim_{t \to \infty} \frac{1}{t} \log P(D \text{ has rate } a \text{ during } [0, t]) \leq \Lambda_A(\theta) - \theta a.$$

Since this inequality holds for all $0 < \theta \leq \delta$, we conclude that

$$\lim_{t \to \infty} \frac{1}{t} \log P(D \text{ has rate } a \text{ during } [0, t]) \leq -\sup_{0 < \theta \leq \delta} (\theta a - \Lambda_A(\theta)).$$

Now we observe that for $a \leq \alpha_A^*(\delta)$ one has

$$\sup_{0 < \theta \leq \delta} (\theta a - \Lambda_A(\theta)) = \Lambda_A^*(a).$$

Therefore,

$$\lim_{t\to\infty} \frac{1}{t} \log P(D \text{ has rate } a \text{ during } [0,t]) \leq -\Lambda_A^*(a),$$

which is the precise statement of (7.54).

This argument also shows that D has straight-line large deviations and that it is of class $S(\Lambda_1^*)$ with $\Lambda_1^*(a) = \Lambda_A^*(a)$ for $a < \min[c - \lambda_F, \alpha^*(\delta)]$. By symmetry, the argument shows that F is of class $S(\Lambda_2^*)$ with $\Lambda_2^*(a) = \Lambda_F^*(a)$ for $a < \min[c - \lambda_F, \alpha^*(\delta)]$.

7.4.6 Bahadur-Rao Theorem

We used this theorem in Chapter 6 to estimate the statistical multiplexing gain. We sketch a derivation of the result.

Theorem 7.4.4 (Bahadur-Rao) Let $\{Y_n, n \geq 1\}$ be iid random variables with $\Lambda(\theta) := \log M(\theta)$ where $M(\theta) := E[\exp(\theta Y_1)]$. Then

$$P(Y_1 + \cdots + Y_N > Nc) \approx \frac{1}{\sqrt{2\pi}\sigma_c\theta_c\sqrt{N}} \exp\{-N\Lambda^*(c)\}. \tag{7.64}$$

In the above expression, θ_c achieves the maximum in

$$\Lambda^*(c) = \sup_\theta [\theta c - \Lambda(\theta)]$$

where

$$\sigma_c^2 = \Lambda''(\theta_c).$$

Proof Define the iid random variables Z_n so that

$$P(Z_n \in (x, x+dx)) = \frac{\exp\{\theta_c x\} P(Y_n \in (x, x+dx))}{M(\theta_c)}.$$

Note that $EZ_n = c$ and $\text{var}(Z_n) = \Lambda''(\theta_c) = \sigma_c^2$. With $Z^N = Z_1 + \cdots Z_N$, $Y^N = Y_1 + \cdots + Y_N$, $\mathbf{Y} = (Y_1, \ldots, Y_N)$, and $\mathbf{Z} = (Z_1, \ldots, Z_N)$, we find

$$\begin{aligned} P(Y^N > Nc) &= E_{\mathbf{Y}}[1\{Y^N > Nc\}] = E_{\mathbf{Z}}\left[1\{Z^N > Nc\}\frac{P_{\mathbf{Y}}}{P_{\mathbf{Z}}}\right] \\ &= M(\theta_c)^N E[\exp\{-\theta_c Z^N\} 1\{Z^N \geq Nc\}] \\ &= \exp\{-N\Lambda^*(c)\} E[\exp\{-\theta_c W^N\} 1\{W^N \geq 0\}] \end{aligned}$$

where $W_n = Z_n - c$ and $W^N = Z^N - Nc$. Let $V = W^N/(\sigma_c\sqrt{N})$. Then

$$P(Y^N > Nc) = \exp\{-N\Lambda^*(c)\}E[\exp\{-\theta_c\sigma_c\sqrt{N}V\}1\{V \geq 0\}].$$

By the central limit theorem, we know that V is approximately Gaussian with zero mean and unit variance. We then write

$$\begin{aligned}
P(Y^N > Nc) &\approx \exp\{-N\Lambda^*(c)\}\frac{1}{\sqrt{2\pi}}\int_0^\infty \exp\{-\theta_c\sigma_c\sqrt{N}x\}\exp\{-x^2/2\}dx \\
&= \exp\{-N\Lambda^*(c)\}\frac{1}{\sqrt{2\pi}}\exp\left\{\frac{1}{2}(\theta_c\sigma_c\sqrt{N})^2\right\}\int_0^\infty \exp\left\{-\frac{1}{2}(\theta_c\sigma_c\sqrt{N}+x)^2\right\}dx \\
&= \exp\{-N\Lambda^*(c)\}\frac{1}{\sqrt{2\pi}}\exp\left\{\frac{1}{2}(\theta_c\sigma_c\sqrt{N})^2\right\}\int_{\theta_c\sigma_c\sqrt{N}}^\infty \exp\left\{-\frac{x^2}{2}\right\}dx \\
&\approx \exp\{-N\Lambda^*(c)\}\frac{1}{\sqrt{2\pi}}\exp\left\{\frac{1}{2}(\theta_c\sigma\sqrt{N})^2\right\}\frac{1}{\theta_c\sigma_c\sqrt{N}}\exp\left\{-\frac{1}{2}(\theta_c\sigma_c\sqrt{N})^2\right\} \\
&= \frac{1}{\sqrt{2\pi N}\sigma_c\theta_c}\exp\{-N\Lambda^*(c)\},
\end{aligned}$$

as we wanted to prove. In the next to last equality we used the approximation

$$\int_y^\infty \exp\left\{-\frac{x^2}{2}\right\}dx \approx \frac{1}{u}\exp\left\{-\frac{u^2}{2}\right\}, \quad \text{for } u \gg 1. \qquad \blacksquare$$

7.5 SUMMARY

The performance evaluation of circuit-switched networks requires computation of the blocking probabilities; the evaluation of network-congestion and flow-control strategies in packet-switched networks requires computing queue length distributions; and the evaluation of (small) buffer overflow probabilities and multiplexing gain in virtual circuit networks requires computing "tail" probabilities. This chapter provided a rapid but reasonably complete summary of the mathematical models and results underlying all these computations.

The results on blocking probabilities first appeared 70 years ago; those on queue length distributions started in the 1960s; and the first results on tail length distributions were published in the late 1980s. These last results are available only in journals and conference proceedings.

These results are difficult to comprehend fully without a strong background in probability theory. However, it is worth the effort to understand the results, because they provide ways of analyzing resource allocation and admission control for high-performance virtual circuit networks. An indirect method of control is through pricing of network services. That approach is considered in Chapter 8.

7.6 NOTES

For general discussions of queuing theory see [K75, K79, W88]. Also see the references cited in section 6.6. More extensive treatments of the deterministic approaches of section 7.4.1 are given in [C91, LV95]. For elaboration of the intuition presented at the end of section 7.4.2, consult [DZ93]. For a discussion of the Bahadur-Rao theorem see Theorem 1.3 of [D91].

7.7 PROBLEMS

1. Let $\mathbf{x} = \{x_n, n \geq 0\}$ be a discrete-time Markov chain with transition probability matrix P and invariant distribution π. Show that $\{z_n = (x_n, x_{n+1}), n \geq 0\}$ is a Markov chain. Find its invariant distribution.

2. Consider a continuous-time Markov chain on $\{1, 2, 3\}$ with $q(1) = q(1, 2) = a > 0, q(2) = q(2, 3) = b > 0, q(3) = q(3, 1) = c > 0$. Find the rate matrix of the Markov chain reversed in time.

3. Let $\mathbf{A} = \{A_t, t \geq 0\}$ and $\mathbf{B} = \{B_t, t \geq 0\}$ be two independent Poisson processes with rate λ. Define $x_t = x_0 + A_t - B_t$ where x_0 is a random variable independent of $\mathbf{A}$ and $\mathbf{B}$. Show that $\mathbf{x} = \{x_t, t \geq 0\}$ is a Markov chain. Is it irreducible? Does it have an invariant distribution?

4. Consider a continuous-time Markov chain $\mathbf{x} = \{x_t, t \geq 0\}$ with rate matrix $Q = \{q(i, j), i, j \in \mathbf{X}\}$. Assume that ν is the invariant distribution of $\{\xi_n, n \geq 0\}$, the Markov chain $\mathbf{x}$ observed at its jump times. Relate ν to π, the invariant distribution of $\mathbf{x} = \{x_t, t \geq 0\}$.

5. Consider a simple telephone network where calls that arrive as a Poisson process with rate λ are sent with probability p to a group of M lines and with probability $1 - p$ to another group of N lines. Find the value of p that minimizes the blocking probability of a call. Compare the resulting blocking probability to that corresponding to the calls being served by $M + N$ lines.

6. Consider K telephone sets that share $N < K$ outgoing lines of a PBX. Assume that each telephone generates a new call, when it is not busy, as a Poisson process with rate ρ. Calls have independent holding times that are exponentially distributed with rate 1. Model the number of calls in progress by a Markov chain and calculate the probability that the N lines are busy. Relate that probability to the probability that a call is blocked. Are these probabilities the same?

7. Show that the blocking probability in problem 6 is insensitive with respect to the distribution of the call holding times. *Hint:* Consider a closed network with an $M/GI/\infty$ and an $M/M/\infty$ queue.

8. Consider K telephone sets that are independently busy or idle with probabilities p and $1-p$, respectively. Use the central limit theorem to estimate a number N such that the probability that more than N phones are busy is about 1%. Relate this result to that of problem 6.

9. Consider the following model of a fixed wireless network. There are N radio transmitters that share M radio channels. Each radio station can serve K telephone sets. Each telephone set generates calls with rate ρ when it is not busy. Find a Markov description of the network when the call holding times are independent and exponentially distributed with rate μ. What is the blocking probability of this network? Design a Monte Carlo procedure for estimating the call blocking probability.

10. Let $\{A_t, t \geq 0\}$ be a Poisson process with rate λ. Define the processes $\{B_t, t \geq 0\}$ and $\{C_t, t \geq 0\}$ as follows. At each arrival time of $\{A_t, t \geq 0\}$, one flips a coin. With probability p, the outcome is heads and the arrival time is defined to be an arrival time of $\{B_t, t \geq 0\}$. Otherwise, that arrival time is defined to be an arrival time of $\{C_t, t \geq 0\}$. Show that the processes $\{B_t, t \geq 0\}$ and $\{C_t, t \geq 0\}$ are independent Poisson processes with rate λp and $\lambda(1-p)$, respectively.

11. Consider a network of queues with Poisson external arrivals and Markov routing as in Figure 7.7. Assume that some queues are $M/M/1$ and others are $M/GI/\infty$. Find the invariant distribution of the network.

12. ATM cells arrive at a transmitter equipped with a buffer according to a Poisson process with rate λ (in cells per cell transmission times). What is the maximum value of λ if the average delay must be less than five cell transmission times?

13. An ATM source produces A_n cells during the nth cell transmission time. Assume that $P(A_n = m/p) = p = 1 - P(A_n = 0)$. Calculate how many such

sources can go through a transmitter equipped with a buffer if the average delay per cell must be less than five cell transmission times.

14. Let $\mathbf{x} = \{x_t, t \geq 0\}$ be a Markov chain on a finite-state space $\mathbf{X} = \{1, 2, \ldots, m, m+1\}$ with rate matrix Q. Assume that $P(x_0 = i) = \pi(i), i \in \mathbf{X}$. Define $\tau = \inf\{t > 0 | x_t = m+1\}$. Calculate $E(\exp\{-s\tau\})$. *Hint:* Let $\alpha(i, s) = E[\exp\{-s\tau\}|x_0 = i]$. Argue that if $x_0 = i$, then τ is equal to the holding time of state i plus the time to go from the next state to $m+1$. Thus, $\alpha(i, s) = [1/(q(i) + s)] \sum_j \alpha(j, s) \times q(i, j)$ for $i \neq m+1$. Derive a set of equations and solve them to obtain $E(\exp\{-s\tau\})$. Explain how to use these ideas to construct a Markov chain model of the M/GI/∞ queue.

15. A simple network consists of two nodes in tandem with respective service rates C_1 and C_2. A (B_1, R_1)-traffic (see section 7.4.1) enters node 1 and continues on through node 2. A (B_2, R_2)-traffic also enters node 2. Find bounds on the delay of stream 1 through the network.

16. Let $\{X_n, n \geq 1\}$ be a sequence of iid Poisson random variables with mean λ. Calculate the effective bandwidth of that sequence. Compare that effective bandwidth to that of an iid sequence of $\{0, A\}$ random variables with the same mean rate.

17. Let $\{X_n, n \geq 1\}$ be iid random variables with values in $\{0, 1\}$. Think of these as being on-off sources. Find how many sources can go through a transmitter with rate C using the Bahadur-Rao approximation. Compare your answer with the result that you would obtain if the sources were Gaussian with the same mean and variance. Assume $C = 20$, $p = 0.2$, and $p_e = 10^{-8}$.

8 CHAPTER Economics

In this chapter you will be introduced to the basic economic concepts and normative principles of pricing of network services. By the end of this chapter you will be able to judge whether the economic aspects of networking decisions made in an organization are sound. Most books on networking do not discuss economics, and most networking engineers and computer scientists (and, of course, students) are not well prepared to formulate and assess economic arguments. All too often as a result, they are virtually excluded from technical decisions in which they have expertise, because they are unable to draw out the economic implications of those decisions. Important technical decisions are thereby left in the hands of business school graduates even when they have a limited understanding of networking technology.

Like the study of other economic commodities, our discussion of communication services focuses on characteristics of factors of supply and demand and on their interaction in the market.

The most important supply factors are the technology of network elements (communication links and switches) and the management rules that can be used to control these elements so that the network can supply services that users want. The technology of network elements is described in Chapters 9 and 10, while Chapters 6 and 7 were devoted to a study of network control. A more complete study of supply should include an assessment of costs: the capital costs of network elements (hardware and software) and the network operating costs (maintenance, depreciation, administration). Because much of the network technology we have discussed is not yet deployed, it is not possible to get realistic estimates of costs, and we will be unable to discuss network cost in any depth.

Demand factors determine the kinds and amounts of services users want, the trade-offs users make between different dimensions of service quality, and their willingness to pay. The demand for network services in large part is *derived demand*. That is, users demand network services not for themselves, but because those services provide a means to an end that users really want. For example, users may access an on-line database in order to obtain news; or they may use a video-on-demand service to watch a movie; or they may telecommute. In each of these examples, users have alternative means to achieve their ends: news may be obtained from a newspaper; videos may be rented from a store; and one may travel to work. The price and convenience of these alternatives will affect the derived demand.

Network services providers and consumers interact in the market. This interaction is mediated through a system of prices or charges. Producers compare their costs with what they can charge to determine how large a network they should build, how it should be configured, and what sorts of services to offer for sale. Consumers compare their willingness to pay with the charges and determine the kinds and amounts of service to purchase. The setting of prices and charges depends in part on technology that limits the range of service quality that can be offered. Prices are also determined in part by the industrial organization of network providers (whether the industry is competitive or monopolistic) and the extent of governmental regulation.

There is a substantial theoretical and empirical literature on the economics of long-established networks that provide a narrow range of services: the telephone and cable and broadcast TV. However, few studies discuss the economics of high-performance networks that can meet virtually any communication need. This is because few such networks exist and because the most important of these, the Internet, was until recently subsidized and noncommercial. Lacking significant prior work, our discussion will largely be based on economic "first principles" and by analogy with markets for services that resemble networking services.

We begin in section 8.1 by considering four standard types of charges for communication services. Implementing these charges requires appropriate protocols and network control capabilities. These may not be present as is the case for the Internet, which was not designed to accommodate a system of charges.

In section 8.2 we will study a billing scheme for the Internet. In section 8.3 we present data about Internet demand at the University of California-Berkeley in order to study the usefulness of the billing scheme. We will find that a limitation of the current Internet protocols is its inability to provide any guarantee of service quality—an observation that we have made several times before.

In sections 8.4 and 8.5 we introduce two models to study pricing of communication services. The first model is useful for a data network like Internet, with a single quality of service and no admission control, making the network susceptible to congestion. The second model is for ATM networks, where service quality guarantees are possible and new user requests can be rejected if network resources to serve those requests are not available. We use the second model to calculate how many requests of different types of service can simultaneously be served. We also develop a variant of this model to suggest that a good way to charge for those services is in terms of the resources they consume.

A summary is provided in section 8.6.

8.1 NETWORK CHARGES: THEORY AND PRACTICE

In this section we introduce the economic principles underlying network charges. The principles apply to all networks, but our main examples will be drawn from the familiar Internet.

At first fully subsidized by the United States government, Internet subsidy has diminished over time, disappearing altogether in 1995. Internet cost is estimated at $200 million annually. It is now run by commercial carriers and network access providers, who recover costs (and profits) through charges.

A word on terminology: we will use the terms *charge* and *price* interchangeably, although a useful distinction can be made. A price is normally associated with one unit of service: if you buy n units of service, you pay n times the unit price. A charge is a more general form of price. For example, a charge may consist of a fixed component plus a unit price. Mathematically, we may think of a charge as a nonlinear price and a price as a linear charge.

8.1.1 Economic Principles

Although one may imagine many kinds of charges, it is important to distinguish only four types: a fixed or access charge, a usage charge, a congestion charge, and a service quality charge.

A *fixed* charge is a monthly subscription fee for access to the network. The access may be limited to a certain period of time each day, or it may be unlimited. The fee is paid independently of how many connections the subscriber actually makes, or how much data is transferred, i.e., it is independent of the subscriber's use of the network. Ideally, the level of the fixed charge should equate network cost and user benefit. The fixed charge should cover just the additional cost incurred to provide network access. Thus the fixed monthly fee charged by the telephone company should pay for the local loop connecting the subscriber to the central office. Similarly, a company providing Internet access to a customer over a phone line incurs a fixed cost for hardware (modem, host memory, etc.), software, and administration (setting up and managing the customer account) that should be recovered through the access fee.

On the demand side, a user would subscribe to the network only if her willingness to pay exceeds the benefit of access. Because access confers the right to use the network whether or not one actually uses it, users may subscribe to preserve their options. (Common examples of access charges are the membership dues one pays to join a club or a museum; membership gives the right to use the club or visit the museum.)

A *usage* charge depends on the amount of use. It is, therefore, based on each connection or call the subscriber makes. (We use the terms *connection* and *call* interchangeably to refer to a TCP connection, an ATM virtual circuit connection, or a telephone call.) The usage charge can be calculated in many ways. It may be a function of the duration of the connection, the amount of data transferred, or the end-to-end distance. Telephone usage charges, for example, depend on the call duration and the distance between the calling parties. Usage charges, too, should equate cost and benefit. More network resources (transmission links, buffers, routers or switches, system operations and maintenance, etc.) are devoted to subscribers or connections with greater usage, and so they should pay a greater share of the network cost. Ideally, the usage charge should equal the cost of the additional resources that a connection or call needs. Thus, the usage charge for a 64-Kbps telephone call of a certain duration should equal the cost over the duration of the call of increasing by 64 Kbps the capacity of the links and switches along the route of the call.

Fixed and usage charges of network services should reflect economies of scale, i.e., the underlying costs grow more slowly than in proportion to the number of subscribers or the amount of use.

A *congestion* charge depends on the amount of traffic or load that the network is carrying at the time of the subscriber's connection. Congestion charges are responsive to the state of the network: the charge is higher when the network is congested, and there is no congestion charge when it is uncongested. The rationale for a congestion charge is as follows. Service quality deteriorates rapidly as the network approaches congestion. In any network that uses statistical multiplexing, such as the Internet or an ATM network, congestion increases queuing delays or packet loss due to buffer overflow. In the telephone network, which uses time-division multiplexing to reserve a fixed bandwidth for each call, congestion increases the blocking probability.

In order to prevent congestion and the resulting quality degradation, the number of connections must be limited. There are many ways to do this, but it would socially be better to permit connections network users deem more valuable and prevent connections users think are less valuable. One way to do this is by imposing a congestion charge: users would initiate only those connections that they value more than the charge and postpone making less-valued connections to a time when the congestion charge is low or zero. In this way congestion charges can be used to discourage less-valued connections. The level of congestion charge should be just sufficient to prevent congestion: too low a charge will not prevent congestion, and too high a charge will reduce the number of calls to a level below what the network can accommodate.

Sometimes the traffic pattern varies regularly over the day, and congestion predictably occurs during certain busy hours of the day. In these cases a congestion charge may be approximated by a "time of use" charge: there is a higher usage charge during busy periods. (The difference between the usage charge during the busy and nonbusy periods is a congestion charge.) This is what telephone companies do when they impose a lower usage charge at night and on weekends.

We should not confuse usage and congestion charges. A usage charge reflects the cost of network resources that are being used. A congestion charge is a means to give network access to more valuable connections and to deny access to less valuable connections. Congestion charges reflect overall network demand and bear no direct relation to network cost. Of course, a large congestion charge is an indication of large demand, suggesting an opportunity for profit by investing in increasing network capacity.

Finally, a network may provide different qualities of service, some of which require more network resources than others. The *quality* charge reflects this difference in resource use. Telephone networks and data networks today typically provide only one service quality, so quality charges are uncommon. However, with the growth of high bandwidth applications that require guaranteed service quality (e.g., guaranteed delay bounds), services will be differentiated by quality, and quality charges will become commonplace. ATM networks are expected to provide different service quality. (The different charges for overnight, first-class, and bulk-mail delivery imposed by the postal system is an example of quality charges.)

In summary, economic theory suggests that a network user should pay a four-part bill comprising a fixed charge for the additional fixed network costs, a usage charge equal to the cost of resources used, a congestion charge that limits less-valued connections, and a quality charge for additional resources needed for higher-quality service. Practice, however, does not strictly follow theory, for historical, technical, and economic reasons.

8.1.2 Charges in Practice

Telephone companies have the most elaborate pricing schemes, including fixed charges, usage charges that depend on distance and call duration, and congestion charges that are approximated by time of use charges. Often, there are quantity discounts for both fixed and usage charges, reflecting the economies of scale. In the United States the system of charges is closer to what the theory prescribes. However, it is not ideal. For example, because they are further away from the switch, the cost of the local loop is greater for subscribers in lightly populated rural areas than in densely populated urban areas. Nevertheless, the access charges are the same for rural and urban subscribers.

Data network charges are unsophisticated, by contrast. A local area network usually belongs to a single enterprise that provides free access to its members. Because the enterprise bears all of the network costs, there is little incentive to charge users, and network costs are part of the enterprise's "overhead" costs. However, as the use and cost of LANs have increased, a fixed, internal charge per host or per user or department is often imposed to recover the cost. This is especially the case in enterprises with large network costs and with a distinct department that acquires network assets and provides network services. An additional benefit from instituting such an internal charge is that it elicits information about the value members place on network services. That information can be used to make decisions about network expansion.

Companies like CompuServe and America Online provide Internet access to individual subscribers. They, too, impose only a fixed charge. The charge includes the cost of additional services, such as access to databases that provide news and allow transactions such as airplane reservations and on-line catalog shopping.

More recent network services such as SMDS and Frame Relay include a form of usage charge. In SMDS, for example, users subscribe to a service parameterized in terms of the sustained information rate, the maximum burst size, and the maximum number of interleaved messages (see section 3.7), and the charge for the service depends on these parameters. Since these parameters reflect the network resources devoted to providing this service, the charge includes a usage component. ISDN charges include a usage component.

We have already noted that quality charges are uncommon today. However, networks such as the Internet distinguish between normal and expedited packets, and the latter have higher priority and face less delay. This example suggests that one way of achieving service quality differentiation is through priority of service, which guarantees preferential treatment (but not a guaranteed delay or loss bound). In principle, ATM will permit the most sophisticated forms of charges and service definition.

8.1.3 Vulnerability of the Internet

Historical factors explain why in the early 1990s Internet users faced only fixed charges. Until 1990, the Internet was heavily subsidized, and most users belonged to universities and research institutions. The Internet was rarely congested. When congestion did occur, hosts automatically exercised flow control. Thus fixed charges were adequate.

Internet traffic grew because commercial network access providers extended the user group beyond the academic and research communities and because of the popularity of applications such as WWW. (Another spurt of growth may occur from commercial transactions on the Internet.) This growth has made the Internet vulnerable and exposed the difficulty in serving applications that need guaranteed service quality.

Internet provides a single service of uncertain quality—best-effort service. The network accepts all connections and tries to deliver the data packets. There is no admission control, and flow control is left to hosts. When a router's buffers become full, all connections through the router suffer packet loss or increased queuing delay. Hosts detect this condition because of retransmission timeouts. They are then expected to reduce their packet rate. However, there is no requirement that hosts adopt such a responsible policy. Such a requirement would be very difficult to enforce. A selfish user may deliberately decide not to exercise flow control. Indeed, if others act responsibly and reduce their packet rate, the selfish user will receive a greater share of network bandwidth and buffers. This perverse incentive further encourages abuse. The resulting congestion causes retransmission, reducing network throughput and inflicting poorer service quality on all.

An Internet service provider faces the same perverse incentive. The typical provider has access to the Internet of a certain capacity (link speed and router or gateway capacity). The provider gives users access for a fixed charge. To maximize profits, the provider has the incentive to give access to as many users as possible, since there is no service quality requirement. The service received by each user (measured by delay or loss) will of course deteriorate as the number of users increases.

In a best-effort service network everyone receives the same service quality, measured, say, in terms of delay. This quality depends on the total network load. Since there is no admission or flow control, the load is unpredictable, and so there can be no guaranteed bound on the delay. The network thus cannot meet the needs of applications that require such delay bounds. More generally, the network poorly serves connections that need a predictable service quality. In Chapter 3 we presented proposals to upgrade the Internet protocols that seek to provide QoS guarantees. If these proposals are implemented, the vulnerability of the Internet will be reduced.

Because the Internet is a datagram network, it lacks the control functions and associated protocols (like ATM networks) that are needed in order to provide different, guaranteed levels of service quality. It may be possible, nevertheless, to provide an indirect form of control by instituting a proper pricing system, including both usage and congestion charges. We now present a design for such a system. We will then assess the kinds of control that can be exercised with such a pricing system.

8.2 A BILLING SYSTEM FOR INTERNET CONNECTIONS

A billing system for usage charges must meter the traffic and collect appropriate data for each TCP connection. If there is a congestion charge that depends on real-time changes in the network state, the billing system must monitor the network state and provide real-time price feedback to users. Thus, a practical billing system must have the following features:

- *No changes to existing Internet protocols and applications.* Because the installed base of hosts, routers, and user software is huge, the billing system must work without requiring changes to existing Internet protocols and widely used applications such as FTP, e-mail, and Mosaic.

- *User involvement.* In order to bill individual users, the system must determine the user's identity, explain the charges, and obtain approval for those charges in a secure manner.
- *On-line reporting of network usage.* In order to institute a congestion charge, the system must collect and report in real time aggregate network usage data so that the appropriate congestion charge can be calculated.
- *Sharing of information and resources.* Applications like WWW encourage the sharing of information and resources between remote sites. Billing systems should be able to cooperate and identify users and bill them accordingly.

We will describe a billing system with these features. We will do so in three stages. In the first stage, the system is limited to a single administrative domain, like the Berkeley campus. The second stage extends the design to multiple cooperative sites. The third stage permits billing across noncooperative sites using a third party such as a credit card company.

8.2.1 Stage 1

This is the single-campus billing system, illustrated in Figure 8.1. In this example, a user wishes to initiate a TCP connection from the Berkeley campus to the machine *remote* at Stanford. To make the example more general, the user is sitting at machine *usermachine* but is logged on locally to the machine *local*. The user will initiate the connection from *local*. The user's application (FTP, Mosaic, etc.) will initiate the connection, and *local* will send a connection setup message (a TCP SYN message) to the machine *remote*. This will traverse the Berkeley campus, reaching the gateway that connects the campus to the outside world. The gateway, shown here as *billing gateway* or BGW, will recognize the TCP SYN message as an attempt to initiate a new connection. Before BGW will allow the TCP SYN to pass and the connection to be established, it communicates with the access controller to request authorization for this connection.

Figure 8.2 explains the communications required to set up the connection. The event diagram on the left illustrates the normal three-way handshake between *local* and *remote* required to set up a TCP connection. The diagram on the right shows the additional messages needed for billing.

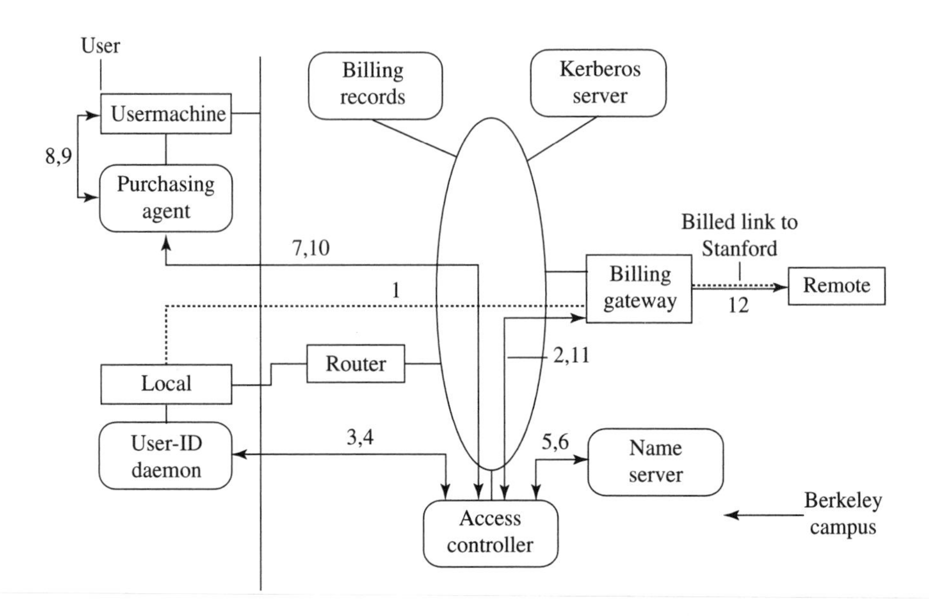

FIGURE 8.1 Single-site access control and billing system. Numbers on links refer to messages in Figure 8.2.

The access controller, or AC, is responsible for deciding whether to allow the connection to be established. (In case the network offered variable quality of service, the AC would decide the network resources to be used by the connection. Internet does not now offer such variable quality.) The AC requires an authorization to proceed from the user and an account code to debit the charge. The AC achieves this in two steps. First it identifies the user. Second, it indicates to the user the price of the connection and requests that the user provide an authenticated account code to pay for the connection and specify any desired resource or cost limits.

The AC identifies the user by communicating with machine *local* (the machine that initiated the connection) and providing cost information about the TCP connection. The user-ID daemon running on *local* determines which user initiated this connection, returning the name to AC.

For AC to request authorization from the user, it must locate the machine where the user is sitting. This is achieved by the name server. The AC then requests authorization from the user by communicating with the user's purchasing agent or PA. The PA is assumed to run on *usermachine* and is responsible for purchasing network services on the user's behalf, with or without

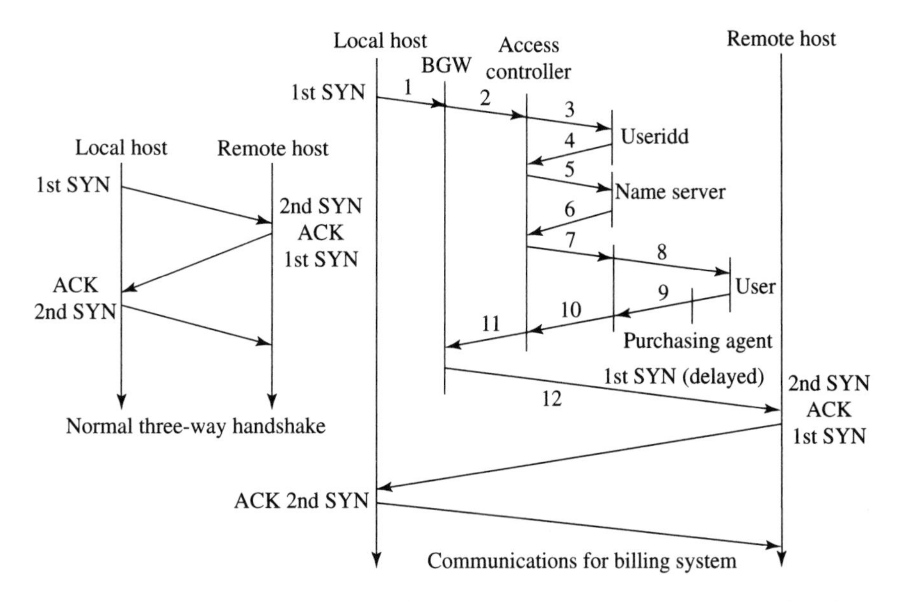

8.2 FIGURE

Communications required to establish connection without (left) and with (right) a billing system.

the user's direct involvement. The PA may be configured by the user, for example, to check with the user (perhaps with a dialog box) every time a new connection is requested. Alternatively, and more realistically, the PA may be configured to purchase connections up to a certain value, to a certain number of locations, for certain services, at certain times of the day, etc., without direct user involvement. Thus there are many possibilities for flexibility in the PA, and the PA can be made simple to use and to configure.

When the PA has decided to purchase the connection on the user's behalf, it confirms this to the AC, supplying an account code and the limits up to which it is prepared to pay. It is important that this communication be authenticated so that the AC can be sure that the authentication came from the user that initiated the connection. This may be achieved using a secure ticket mechanism such as Kerberos.

Next, the AC verifies that the user is authorized to charge to the account code provided (using its own records or by contacting billing records). The AC indicates to BGW that the connection may proceed, and the TCP SYN message is allowed to continue on its way to *remote*. The BGW creates a record in its address tables so that future messages that are part of this connection are metered and forwarded without delay by the BGW. The BGW periodically reads the metering information from the address tables and provides feedback to the user (if requested) and to the billing records system, which is responsible for preparing and issuing bills (each month, say) to the user.

We emphasize certain features of the stage 1 design. The design requires no change outside of the administrative domain (Berkeley campus). Billing software must be installed in hosts within the administrative domain, and data-collection software must be installed in the billing gateway. Lastly, the billing overhead penalty, measured by the additional communications defined in Figure 8.2, must not significantly increase the delay in setting up TCP connections.

8.2.2 Stage 2

The preceding design needs to be extended to a system with multiple cooperating sites. In exchanging network traffic with a remote site, a user accumulates communication charges in both directions. This is particularly the case in client-server applications in which the user (client) obtains information from a remote server (e.g., anonymous FTP, Mosaic). A (stage 1) billing system is used for recovering charges for traffic from its site to the outside world. The problem is that the user is only physically present, and presumably trusted to pay, at most at one of the sites: the access controller on one campus may not trust the user on a remote campus. Furthermore, the billing records system may not have the means to recover payment from a remote initiating user. This problem can be overcome in the case of billing systems on two campuses that trust each other. We will explain this by continuing with our previous example.

Assume that the connection is initiated from the machine *stanford* and that Stanford University maintains a compatible billing system. When the billing gateway (BGW) recognizes the TCP SYN message, it will contact the local (Berkeley) access controller requesting authorization for the connection. The Berkeley AC in turn requests authorization from the Stanford AC. The latter requests and obtains authorization in the same way as described above. The Stanford AC authorizes the connection by communicating with the Berkeley AC using an authenticated protocol. This authorization obligates Stanford to pay for transport costs accrued by Berkeley on behalf of the user. As costs accumulate, the Berkeley billing records function provides feedback to Stanford's billing records system, which, in turn, may provide feedback to the user at *stanford*.

Once again, the design requires no change outside of the cooperative sites. Billing software in hosts within each site is the same as in stage 1. There will be a small additional delay caused by the extra communications between the cooperating billing gateways.

8.2.3 Stage 3

The stage 1 design can be used for charging users in a single domain, and stage 2 permits users in one domain to be charged for the use of the network in another, cooperative domain. The stage 3 design permits noncooperative sites to recover charges from each other through a trusted third party. This feature can be used to assist small enterprises offering access to specialized data and other information services for sale worldwide over a network, like Internet. Examples of these small information enterprises (SIEs) may include a specialized museum, a ticket-reservation system for a small theater, an on-line library, teachers offering remotely supervised coaching, or a small consultancy group. The SIEs may be located any place in the world with Internet access.

A major obstacle to starting such an SIE is the overhead of billing and accounting for services provided to a dispersed population of customers. An SIE cannot afford to collect individual billing records and recover costs separately from all its customers. Two components of the cost must be recovered from the customer:

1. The transportation cost of sending the information to the user (the "shipping" cost). This depends on the location of the user and the transport charges in effect at the time of the transaction.
2. The cost of the information provided. This depends on the information requested by the user and is unknown to the transport billing system.

The user (SIE customer) must know in advance the total cost of obtaining the information. It would be undesirable for the user to have to try and determine these two components separately.

The operation of this design is illustrated in Figure 8.3. If the user and SIE are located within the same or cooperative billing domains, the user could be charged for the information service via the stage 1 or 2 billing system. In this case, when the connection is established for an item of premium information (e.g., the transfer of a picture from a digital library), the SIE would indicate the premium price to the transport billing system. This would simply be added to the price that the user pays for data transport. As before, the user would receive the price information before the connection is established and would receive real-time feedback as the connection proceeds.

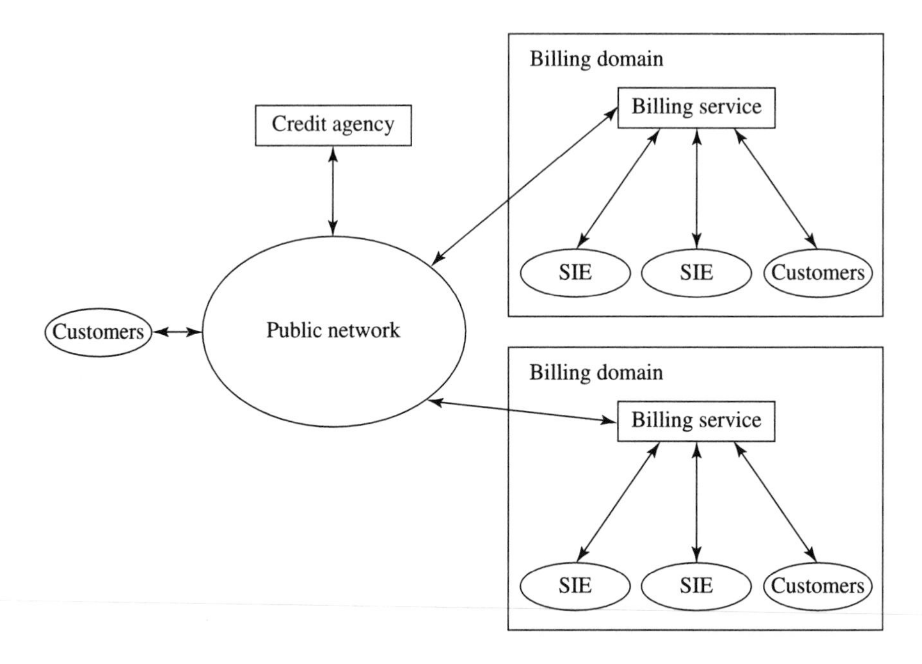

8.3 FIGURE

A system for small information enterprises (SIEs) that provides accounting and bill-collecting services.

Often, the user will not be inside a billing domain. In this case, it is assumed that the user will use a trustworthy credit agency, similar to a credit card company. If this company provides a method to authenticate, promise, and deliver payment, the billing service can recover cost on behalf of the SIE by communicating with the credit agency.

8.2.4 Functions of Internet Charges

The design sketched above gives us a good idea of the capabilities of a billing system that does not require changes to Internet protocols or existing applications. We will now discuss which of the four types of charges summarized in section 8.1 can be implemented with this billing system and the economic functions that can be supported.

To discuss these changes, let us abstract the organization of the Internet into the simplified structure of Figure 8.4. There is a high-speed backbone network. Individual users are grouped into several administrative domains.

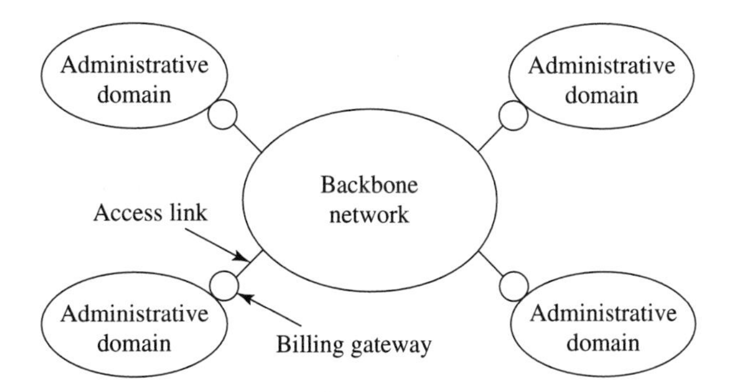

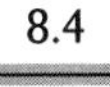

8.4 FIGURE The Internet is modeled as a set of billing domains with an access link to the backbone network. The access link is controlled by a billing gateway.

Each domain is connected to the backbone network by an access link, through a billing gateway. A user in one domain who wants to set up a connection with a host or user in another domain must accept charges imposed by one or more billing gateways involved in the connection. The simplest charge that can be implemented is of course a fixed charge, levied per user or per host. Such charges are widely used because they are easy to implement and do not require the elaborate billing scheme described above. The fixed charge can support the cost-recovery function. An administrative domain incurs the cost of gateway, access link, and maintenance. The cost will increase with the capacity of the gateway (measured, say, by the number of packets per second that can be processed) and the speed or bandwidth of the access link. The fixed charge spreads this cost over the users who benefit from Internet access.

However, the fixed-charge scheme, by itself alone, suffers from two disadvantages. Since there is no usage charge, the scheme encourages network overuse. Second, users who generate little traffic pay much more, on average, per unit of traffic than users who generate a lot of traffic. As a result, low-volume users may find the fixed charge too high and may not subscribe to the service at all. This is inefficient, since the network could easily accommodate these low-volume users.

An economically more efficient cost-recovery scheme is a two-part charge: a fixed charge and a usage charge. The design presented above can implement this scheme. If there is no need to provide real-time feedback to the user, then the design can be simplified. The usage charge could be calculated for each connection, proportional to the number of messages or bytes in the connection. The billing gateway must be capable of monitoring each connection and recording the number of bytes or packets transmitted in each connection. It may be necessary to incorporate as part of the billing system a means to authenticate user identity.

The billing system can also implement a limited form of congestion charge. The charge increases when congestion increases. The billing system must be able to communicate these charges to the users in real time so they may decide whether to pay the higher price for immediate service or postpone their use of the network to a time when this charge is low. Then traffic that is more highly valued will be transmitted sooner than less valued traffic. To implement congestion charges, however, the billing gateway must measure the level of congestion in some way. However, measurement of congestion is not easy in the current structure of the Internet.

Ideally, congestion measurement should include the queuing delay along the route of the connection. This queuing delay is equal to the sum of the queuing delays that occur at the individual routers within the backbone network and at the gateways of the two end hosts. Since measurements of these individual delays are not currently available, a surrogate measurement must be used that can be implemented at a single site or at multiple cooperating sites. One possibility is for the billing gateway of Figure 8.1 to keep track of the average delay of the connections going through it and to base a congestion charge on this average delay. (This average delay could be measured by the difference in the time that a packet is transmitted and its acknowledgment is received.) Note, however, that this delay is the result of many factors: delays in the gateway itself, queuing delays along the various routes of the various connections, the delays at the hosts in generating the acknowledgments, and the delays faced by those acknowledgments.

Congestion charges can serve one important function. As we will see in section 8.4, the revenue from a properly designed congestion charge should pay for the cost of any capacity of the access link (see Figure 8.4). If the revenue is larger than this cost, this indicates user willingness to pay for faster access; if the revenue does not cover this cost, users do not value the faster service sufficiently to pay for it, and the access link capacity should be reduced. This capacity cost is a significant part of the cost of Internet access.

Lastly, we discuss the possibility of quality charges for Internet access. Within the current structure of Internet protocols and Internet routers, it is impossible to guarantee any bounds on delay or loss, because it is impossible to reserve communication capacity or buffers for a particular connection. A certain degree of service quality differentiation is, nevertheless, possible within a single administrative domain by implementing different priority levels. We can imagine how this may be done for, say, two levels—high and low, using Figure 8.1. The billing gateway maintains two buffers. High-priority packets are stored in the high-priority buffer, low-priority packets are stored in the low-priority buffer. Low-priority packets are transmitted only when the high-priority buffer is empty. Users would face a higher price for high-priority traffic. They would divide their applications into two types with less-delay-sensitive applications (like e-mail or news) choosing cheaper, low-priority connections. The high-priority traffic sees the full bandwidth of the access length available for itself. It faces a lower delay, but the low-priority traffic faces a higher delay.

The charges discussed above deal only with transmission services. As suggested in the discussion of the small information enterprises, in practice, transmission services will be bundled with information content (databases, bulletin boards, libraries, etc.) for which users will be charged. The charge for transmission and content will similarly be bundled together. The stage 3 billing system can be used for this purpose.

In summary, the structure of the Internet protocols and routers and gateways currently installed in the Internet prevent the introduction of charges based on congestion or quality. Nevertheless, it is not difficult to introduce fixed, usage, average congestion, and priority-based charges, and to charge for content. Such charges will lead to more efficient use of the Internet. Users will then take into account the traffic their applications generate, and service providers will be more responsible for the quality of their service.

8.3 INTERNET TRAFFIC MEASUREMENTS

The results presented here are derived from a trace of network traffic on the Berkeley campus FDDI backbone network. The trace was obtained by logging all TCP/IP headers that appeared on the backbone. The trace includes connections between hosts within the campus. However, the analysis presented here considers only WAN connections (i.e., connections from Berkeley to the "rest of the world") for a 24-hour period commencing at midnight September 15, 1994. Approximately 22,000 hosts are attached to the campus network, of which 3,172 hosts participated in WAN TCP connections during the study period. The campus community is composed of 40,000 students, faculty, and staff members. The campus network is connected to the Internet by two 1.5-Mbps T-1 links and a single boundary router. (This router is the billing gateway of Figure 8.1.)

Host type	Number of hosts	Per host average			In category (%)	
		Connections	Datagrams	Bytes	Datagrams	Bytes
Multiuser	1,541	273	35,768	7,183,160	80.3	83.2
PCs	629	25	3,406	740,954	3.1	3.5
Unknown	972	103	11,719	1,814,937	16.6	13.3

8.1 TABLE

Amount of TCP WAN traffic by host type. During the trace period there were 530,000 connection attempts of which 370,000 were successful.

During September 1994, Berkeley contributed the 13th largest amount of WAN traffic to the NSFNET out of over 22,000 registered networks. (NSFNET is part of the Internet backbone.) During the 24-hour study period, there were 18 GB (gigabytes) of WAN traffic in 98 million datagrams. Of this WAN traffic, 94% of the bytes and 92% of the datagrams were for TCP traffic. Of this TCP traffic, 83% of the bytes and 80% of the datagrams were from multiuser hosts (see Table 8.1).

8.3.1 Connection Statistics

The trace data was obtained by logging all TCP/IP headers that appeared in the campus backbone network. A 50-ns time stamp was attached to each log entry. Analysis of this data shows that 80% of the connections took between 25 ms and 520 ms to establish, with a median of 107 ms and an average time of 411 ms. A billing system should not impose an extra delay in the connection time more than 50% of the average connection setup delay.

There were a maximum of 1,100 active connections at any instant and a peak rate of 52 new connection setups per second. These statistics indicate the billing gateway performance necessary to keep up with TCP WAN use.

8.3.2 Diversity of Usage

Table 8.2 shows the usage per host averaged over the trace duration for the top 10 active subdomains within Berkeley's network. These statistics show significant variation in usage by subdomain. There is even greater variation by host within each subdomain. For example, the EECS department subdomain has the largest number of hosts with an average TCP usage of 2.3 MB and a standard deviation of 17 MB. This large variation by individual and subdomains suggests that significant efficiency gains are to be made from usage-based pricing.

Subdomain	Number of hosts	Per host average			In category (%)	
		Connections	Datagrams	Bytes	Datagrams	Bytes
cs.berkeley.edu	275	315	42,248	8,792,253	22.4	31.5
eecs.berkeley.edu	365	108	14,619	2,347,930	10.3	11.2
cc.berkeley.edu	102	1,743	107,153	7,999,763	21.1	10.6
lib.berkeley.edu	229	208	15,672	1,952,972	6.9	5.8
hip.berkeley.edu	299	41	6,350	951,868	3.7	3.7
astro.berkeley.edu	69	78	18,460	3,614,468	2.5	3.2
icsi.berkeley.edu	64	87	14,518	3,608,875	1.8	3.0
biochem.berkeley.edu	56	161	15,183	3,276,475	1.6	2.4
math.berkeley.edu	63	121	15,499	2,825,546	1.9	2.3
ocf.berkeley.edu	22	397	108,559	7,378,574	4.6	2.1

8.2 TABLE

Average usage by hosts in the top 10 subdomains. Table excludes hosts that serve the entire campus. These hosts are mostly in cc (central campus) and hip (home IP) subdomains.

The average demand or usage by academic departments is about 3 MB per day per host. We may predict that as more computers are introduced in departments currently with relatively few computers (compared with CS or EECS), WAN TCP traffic will grow at this rate.

8.3.3 Potential for Congestion Pricing and Traffic Shaping

Congestion pricing would increase prices during periods of high demand. This would shift less-delay-sensitive traffic towards periods with lower prices, i.e., periods with lower demand. (A similar shift can be accomplished by administratively imposed traffic shaping.) It seems reasonable to suppose that most e-mail (SMTP) and bulletin-board (NNTP) traffic is less delay-sensitive than file transfer (FTP) traffic. In the trace, SMTP, NNTP, and FTP account for 11, 22, and 47 percent of all bytes transferred, respectively. If we take the 33% of e-mail and bulletin-board traffic and spread it out over the 24 hours of the trace duration, the total amount of traffic that lies above the average rate is reduced by about 30%, as shown in the left panel of Figure 8.5. The panel on the right shows that by buffering traffic for 1 s, the 100-ms peak traffic rate is reduced by 40% (compare top and middle graphs), and buffering for 20 min, the peak is reduced by 60% (compare middle and lowest graphs).

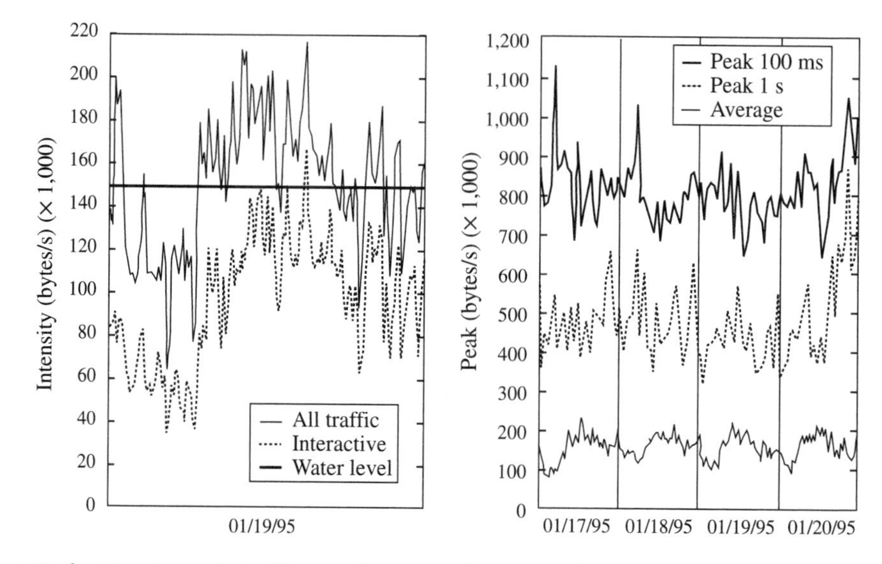

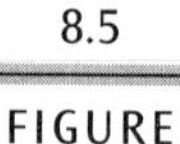

8.5 FIGURE

Reduction in peak traffic rate by spreading e-mail and bulletin-board traffic uniformly over the day (left); peak traffic rate measured over 0.1-s and 1-s intervals and 20-minute average (right).

8.4 PRICING A SINGLE RESOURCE

The billed link between the Berkeley campus and the Internet is an expensive resource. Its cost increases as the link capacity is increased, but the service to users will improve, too. Thus there is a need to balance service benefits against capacity cost.

Assume for now that the capacity is fixed. How can one maximize user benefits? The data show that if we divide the traffic between delay-insensitive (e-mail and bulletin-board services) and delay-sensitive (the rest) traffic, there are two ways to reduce peak capacity and accommodate more traffic. First, by shifting delay-insensitive traffic, perhaps by several hours, to low utilization periods during the day, the peak rate can be decreased by 20%. Second, by buffering delay-sensitive traffic for a very short time on the order of a second, one can reduce the peak rate by 40%.

The shifting over time of these two types of traffic can be accomplished by administrative procedures, or by a system of time of use and congestion charges.

The administrative procedures would classify some traffic types as delay-insensitive and restrict their transmission to low utilization periods of the day. The remaining, delay-sensitive traffic would be buffered for short periods of time (on the order of 1 s). This approach is easy to implement but it has two defects. It makes the determination of which traffic is delay-insensitive a matter of bureaucratic choice. Such choice is likely to be crude. For example, although e-mail typically is delay-insensitive, some e-mail connections may be urgent, but an administrative procedure cannot recognize these urgent messages. Second, the administrative classification will have to be extended over time to include new applications as they develop. This extension is bound to be arbitrary, since there is no accurate way to predict the purposes for which the new applications will be used.

A much better way to shift traffic over time is to allow users themselves to make the choice. In an ideal world, users would be altruistic and voluntarily transmit their delay-insensitive traffic during periods of low demand. But altruism is unreliable. A more reliable mechanism is through a system of prices. If users are charged a lower (or zero) rate at night, then they will have an incentive to shift their delay-insensitive traffic to those periods. For delay-sensitive traffic, a congestion charge could be imposed to keep down the peak demand. We will next study these two price mechanisms.

It is useful to keep in mind that our discussion applies to any situation in which a single resource is shared by many users. This resource could be an access link, a disk system, a pool of computers, or a highway.

8.4.1 Usage-Based Prices

We develop an economic model that will focus our discussion on usage-based pricing. The model is built by specifying three elements: the users' demand for service; the network capacity, i.e., the amount of service that the network can supply; and the interaction between demand and supply through prices.

We consider a single service (say, best-effort service) that is differentiated by time of day. We divide the day into periods denoted by $t = 1, \ldots, T$. For notational simplicity, consider only two periods: $t = 1$ is the "peak" period, and $t = 2$ is the "off-peak" period. We consider a population of potential users indexed by $i = 1, \ldots, I$.

Consider a user's decision regarding one connection, say, to send e-mail or to browse through a WWW site. The user must choose both the period ($t = 1$ or 2) and the amount of traffic to transmit (measured in bytes, say). We model the user's preferences by the *utility function*

$$u_t(x) = u(x) - d_t x, \qquad x \geq 0, \ t = 1, 2.$$

Here x is the amount of traffic, $u(x)$ is the benefit (measured in dollars) that the user derives from sending x, and $d_t x$ is the loss or benefit reduction (also measured in dollars) suffered from sending x in period t. Typically, $d_1 < d_2$, i.e., most users prefer sending the traffic during the peak period.

Suppose the price for sending 1 byte in period t is p_t. If the user selects period t, she will decide to transmit a message of the size that will maximize her net benefit, i.e., she will solve the problem

$$\max u(x) - d_t x - p_t x.$$

The optimum message size, x_t, is given by setting to zero the derivative with respect to x:

$$\frac{\partial}{\partial x}[u(x) - d_t x - p_t x] = 0 \text{ or } u'(x_t) = p_t + d_t. \tag{8.1}$$

We denote the net benefit she will derive from this transaction by

$$V(p_t + d_t) = u(x_t) - (p_t + d_t)x. \tag{8.2}$$

Here, $u'(x) := \partial u/\partial x$ is the downward sloping curve in Figure 8.6; x_t is the size where this curve intersects the line through $p_t + d_t$; and $V(p_t + d_t)$ is the shaded area between the curve and the line. We can solve (8.1) to obtain her demand as a function of $(p_t + d_t)$. We write this as $x_t = D(p_t + d_t)$. In terms of Figure 8.6, $D(p_t + d_t)$ is simply the downward sloping curve $u'(x)$ expressed as a function of the ordinate or y coordinate. Note that D decreases as p_t increases.

Thus, if the user decides to transmit in period 1, her benefit is $V(p_1 + d_1)$, and if she decides to transmit in period 2, her benefit is $V(p_2 + d_2)$. Clearly, she will choose the period that yields a larger benefit, i.e., she will choose period 1 if $p_1 + d_1 < p_2 + d_2$, and she will choose period 2 otherwise.

Let us now consider an arbitrary user i, whose utility function is given by $u_t^i(x) = u^i(x) - d_t^i x$. This user will select period 1 if $p_1 + d_1^i < p_2 + d_2^i$ and transmit $D^i(p_1 + d_1^i)$; otherwise he will select period 2 and transmit $D^i(p_2 + d_2^i)$. Thus, depending on the relative price difference $p_1 - p_2$ and the relative urgency of their connection $d_2^i - d_1^i$, users will segment themselves into two subsets, I_1 and I_2:

$$I_1 = \{i|p_1 - p_2 < d_2^i - d_1^i\}, \qquad I_2 = \{i|p_1 - p_2 \geq d_2^i - d_1^i\}. \tag{8.3}$$

Users in group I_1 will transmit during period 1; those in period 2 will transmit during period 2. The resulting *aggregate* traffic demand in each period will be

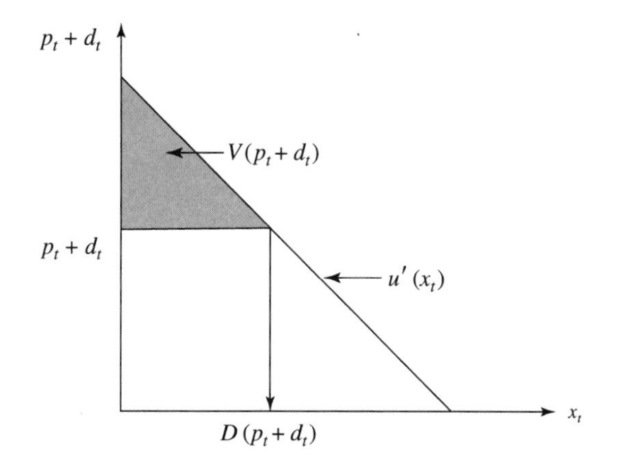

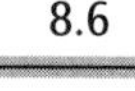

8.6

FIGURE

A user's demand curve $D(p_t + d_t)$ is obtained from the marginal utility $u'(x_t)$. The shaded area is the user's surplus $V(p_t + d_t)$.

$$D_1(p_1, p_2) = \sum_{i \in I_1} D^i(p_1 + d_1^i), \qquad D_2(p_1, p_2) = \sum_{i \in I_2} D^i(p_2 + d_2^i).$$

The results are intuitive: demand will shift from one period to another as the first period becomes relatively more expensive; moreover, if both prices, p_1 and p_2, are increased, keeping $p_1 - p_2$ fixed, then both D_1 and D_2 will decrease.

Note that once the network sets the two prices, users will themselves decide which period to use and how much traffic to send in each connection. Thus a user will send urgent e-mail during the expensive period and nonurgent e-mail during the cheaper period.

Now that we understand how demand is affected by prices, we turn to a more difficult question: how should the network set the prices p_1, p_2? The first thing to observe is that the prices should be such that the traffic demand in each period does not exceed the "supply," i.e., the amount of traffic that can be transmitted over the billed link during each period. Let C_t be the total amount of traffic (in bytes) that can be transmitted in period $t = 1, 2$. Then, one requirement on prices is

$$D_t(p_1, p_2) \leq C_t, \qquad t = 1, 2.$$

However, this still leaves a large range of choice for the prices.

In order to determine the optimum prices, we shall temporarily adopt the role of an omniscient planner who knows the utility function of each user and who will act on the user's behalf to determine for each connection which period to use and how much traffic to generate. The planner will make these choices in a way that maximizes total benefit, i.e., the planner will solve the maximization problem

$$\max \sum_{i \in I_1} [u^i(x_1^i) - d_1^i x_1^i] + \sum_{i \in I_2} [u^i(x_2^i) - d_2^i x_1^i] \tag{8.4}$$

$$\text{subject to} \sum_{i \in I_1} x_1^i \leq C_1, \qquad \sum_{i \in I_2} x_2^i \leq C_2. \tag{8.5}$$

The planner determines which connections i to assign to the two periods, i.e., the sets I_1, I_2, and how much data x_t^i to transmit, subject to the requirement (8.5) that the total traffic in each period be less than the available capacity. The planner will do this in a way that maximizes the benefits summed over all users (8.4). The solution to this is called the *social welfare optimum*. The following important result of economics characterizes this optimum.

Theorem 8.4.1 The optimum is characterized by two prices p_1, p_2 such that equations (8.1) and (8.3) hold and, moreover, for each period t

$$\frac{\partial u^i}{\partial x_t^i} = p_t + d_t^i, \quad \text{for all } i \in I_t \tag{8.6}$$

$$I_1 = \{i | p_1 - p_2 < d_2^i - d_1^i\}, I_2 = \{i | p_1 - p_2 \geq d_2^i - d_1^i\} \tag{8.7}$$

$$\sum_{i \in I_t} x_t^i = C_t. \quad \blacksquare \tag{8.8}$$

The result says that the social welfare optimum can be achieved through a market mechanism in which the network optimally sets two prices p_1 and p_2. Each user then selects the period to transmit and how much traffic to generate so as to maximize her own benefit (equations (8.6), (8.7)). Lastly, the optimal price for a period is such that the demand in each period equals that period's capacity (equation (8.8)).

The result suggests a practical implementation of an adaptive price-setting rule to find the optimum prices. The network begins with an arbitrary pair of prices (p_1, p_2), and measures the aggregate demand in response to these prices. If the demand in a period t is lower than its capacity C_t, the network lowers the price p_t; if it exceeds capacity, p_t is increased.

For future reference, let us note that the network revenue generated by this usage-based pricing equals

$$R_{usage} = \sum_{t=1}^{2} \sum_{i \in I_t} p_t x_t^i = \sum_{t=1}^{2} p_t C_t. \quad (8.9)$$

The second equality follows from (8.8).

8.4.2 Congestion Prices

We now consider delay-sensitive traffic. By definition, users suffer significant reduction in benefit if this traffic is delayed even by a fraction of a second. This delay is queuing delay. It occurs when the *rate* Λ of total user traffic (measured, say, in bytes/s) approaches the link capacity M (bytes/s).

Queuing delay is very different from the situation considered previously, where we compared the total traffic demand D_t (bytes) over period t with the number of bytes (C_t) that can be transmitted during that period. (C_t is the product of M and the duration of period t.)

As before, we consider a population of users, indexed by $i = 1, \ldots, I$. The benefit that user i derives by transmitting (delay-sensitive) traffic at rate λ^i bytes/s is given by the utility function

$$u^i(\lambda^i) - \gamma^i \times d \times \lambda^i.$$

Here $u^i(\lambda^i)$ is the dollar value of transmitting λ^i bytes/s; d is the delay faced by each byte that is transmitted, and γ^i converts the delay into user i's perceived dollar cost. Suppose that the user is charged a congestion price of p_c per unit rate. (The price p_c is the price per unit of bandwidth, its unit is dollars per byte/s. By contrast, usage price discussed previously is dollars per byte.) Then user i will choose to transmit at rate λ^i, which solves the following problem

$$\max_{\lambda^i} u^i(\lambda^i) - \gamma^i d \lambda^i - p_c \lambda^i.$$

The optimum rate λ^i is obtained by solving the equation

$$\frac{\partial u^i}{\partial \lambda^i} = \gamma^i d + p_c. \quad (8.10)$$

Knowing u^i, γ^i we can solve (8.10) and obtain user i's demand for bandwidth $\lambda^i = D^i(p_c)$ as a function of the congestion price p_c. Let us also note the *aggregate* bandwidth demand

$$\Lambda = \sum_i \lambda^i = D(p_c) = \sum_i D^i(p_c).$$

We now consider the problem of the omniscient planner who chooses λ^i on behalf of user i so as to maximize the total benefit

$$\sum_i [u^i(\lambda^i) - \gamma^i d\lambda^i].$$

The queuing delay d is a function of the total traffic rate $\Lambda = \sum \lambda^i$ and the link capacity M, which we write as $d = f(\Lambda, M)$. So the planner's problem is

$$\max \sum_i u^i(\lambda^i) - f(\Lambda, M) \sum_i \gamma^i \lambda^i.$$

The optimum values of λ^i are obtained by solving the equations

$$\begin{aligned} \frac{\partial u^i}{\partial \lambda^i} &= \gamma^i f(\Lambda, M) + \frac{\partial f}{\partial \Lambda}(\Lambda, M) \sum_j \gamma^j \lambda^j \\ &= \gamma^i d + \frac{\partial f}{\partial \Lambda}(\Lambda, M) \sum_j \gamma^j \lambda^j, \qquad i = 1, \ldots, I. \end{aligned} \tag{8.11}$$

The right-hand side of (8.11) is the sum of two terms. The first term is the cost of delay directly suffered by user i. The second term is the increase in the delay cost suffered by all users due to a unit increase in user i's traffic rate. This term is therefore called the *congestion cost*. Comparing (8.10) and (8.11) we see that the welfare optimum is achieved if the congestion price charged to each user equals the congestion cost.

Theorem 8.4.2 The optimum is achieved by charging each user the congestion price of

$$p_c = \frac{\partial f}{\partial \Lambda}(\Lambda, M) \sum_j \gamma^j \lambda^j \tag{8.12}$$

per bytes/s for one unit of time, say one hour. ■

As an example, we compute the congestion price for an M/M/1 queuing model. The delay is given by

$$f(\Lambda, M) = \frac{1}{M} \frac{\Lambda}{M - \Lambda},$$

from which the congestion price can be calculated to be

$$\frac{\partial f}{\partial \Lambda} \sum_i \gamma^i \lambda^i = \frac{\sum_i \gamma^i \lambda^i}{M^2} \left[\frac{\rho}{1 - \rho} + \frac{\rho^2}{(1 - \rho)^2} \right],$$

where $\rho := \Lambda/M$ is the utilization. We see that as ρ approaches one, the congestion price increases rapidly. Facing such a large price, users will reduce their traffic rate.

The Berkeley campus data showed that the total traffic rate varies widely from one second to the next. The congestion price will vary equally rapidly. It is not practical for the network to communicate such a price variation on a second by second basis and for users to react to such rapid price variation. Thus schemes to implement congestion pricing must proceed indirectly. We describe one such scheme, based on the notion of a *reservation price*.

In this scheme, whenever user i sets up a connection, she provides the network her "reservation price" p_i. (p_i is i's maximum "willingness to pay" for traffic in that connection.) The understanding is that the network will buffer her packets until the congestion price falls below p_i, at which time her traffic will be forwarded. Thus the network maintains a list of users, ordered by their reservation prices. The network continually computes the congestion price and transmits those packets whose reservation price exceeds the congestion price. Users are charged according to the prevailing congestion price. This scheme guarantees that users would pay less than their reservation price.

The revenue (per hour) from the congestion price is

$$R_{cong} = p_c \sum_i \lambda^i = p_c \Lambda. \tag{8.13}$$

Most users would find it bewildering to face a complex pricing scheme such as congestion prices, which can shift unpredictably over time. These schemes may be proper for large corporate users who are purchasing transmission services on behalf of many individuals in the corporation. The scheme may also be sensible for network access providers. These providers may purchase services in the congestion "market" but make them available to their customers at a fixed price, larger than the average congestion price that the access provider faces. From a social viewpoint, this is a good arrangement: the access provider is bearing the full cost of congestion and absorbing the risk, whereas the individual users face no risk but pay a "risk premium" above the average cost.

8.4.3 Cost Recovery and Optimum Link Capacity

The revenue collected by the usage-based and congestion pricing schemes may exceed or fail to cover the cost of the billed link. To simplify the comparison between revenue and cost, we will ignore congestion prices. So total revenue is

$$R = R_{usage} = \sum_{t=1}^{2} p_t C_t = \sum_{t=1}^{2} p_t M L_t,$$

where M is the link capacity (bytes/s) and L_t is the duration of period t. Since the two periods $t = 1$ and 2 add up to one day, this is the revenue per day.

So far we have assumed that the link capacity M is fixed. We will now suppose that M can be varied and that the daily cost of renting a link of capacity M is $r(M) = r_{fix} + r_{var} \times M$. Thus the cost has a fixed component r_{fix} and a variable component $r_{var}M$ that is proportional to the capacity. (A cost structure for a link comprising a fixed term and a variable term is very common. The variable cost may not be proportional to the capacity, as we have supposed here for simplicity.) Hence the net benefit $B(M)$ of a link of capacity M is the revenue minus cost, i.e.,

$$B(M) = \sum_{t=1}^{2} p_t M L_t - r_{fix} - r_{var} M.$$

The optimal capacity is obtained by maximizing $B(M)$, i.e., by setting $\partial B / \partial M = 0$. This gives

$$\sum_{t=1}^{2} p_t L_t = r_{var}. \tag{8.14}$$

The left-hand side is the increase in revenue resulting from a unit increase in the link capacity and the right-hand side is the cost of that incremental capacity.

If we select the optimal capacity, then the daily revenue will be $r_{var}M$, which is the variable part of the capacity cost. The fixed cost r_{fix} will *not* be covered by the usage-based price scheme. If it is not possible to cover this through a subsidy, then an alternative is to impose a fixed charge on users. That is, any user who wants access to the billed link will have to pay a daily subscription charge, regardless of how much traffic the user will transmit.

Suppose this fixed charge is p_{fix}. If there are I users in all, then this fixed charge should be $p_{fix} = r_{fix}/I$. Let us study the impact of this fixed charge on a user who decides to transmit x_t bytes during period t. This user will now pay $p_{fix} + p_t x_t$. The net benefit she now derives is given by (compare (8.2))

$$V = u(x_t) - (p_t + d_t)x_t - p_{fix}.$$

There are two cases to consider.

If this is a user with a large demand x_t, then $V > 0$, despite the fixed cost. This user will pay the fixed charge and generate the same traffic as before. On the other hand, if this is a user with a small demand x_t, then $V < 0$, i.e., this user finds the fixed cost to be so large that the net benefit is negative. Thus low-demand users will refuse to subscribe. This is undesirable, since the link capacity is large enough to accommodate them. We conclude that, ideally, the fixed charge should be levied only on the high-demand users who would continue to subscribe. However, it may not be possible in practice to discriminate in this way between high-demand and low-demand users.

We can summarize the discussion in this section. We considered a single resource, such as the billed link for the Berkeley campus. We studied how access to this link should be charged by a three-part pricing scheme: a usage-based price that varies by time of day and encourages users to shift their delay-insensitive traffic to periods with a low demand; a congestion price that encourages users to reduce their traffic rate when the resource is congested; and a fixed charge to recover fixed costs, ideally imposed only on high-demand users. In the next section we consider multiple resources.

8.5 PRICING FOR ATM SERVICES

In the preceding section we discussed the pricing of a single service that was provided using a single resource. The service demand was measured either by the number of bytes or the rate (in bytes/s) of the user's traffic. The capacity of the resource was correspondingly measured either by the maximum number of bytes that the link can transmit in a given period or the maximum transmission rate. Because of this direct correspondence between the service that users demand and the capacity of the resource, we can think of the price either as a price per unit of service or as the rent for a portion of the resource capacity sufficient to produce that unit of service. Thus, for example, the congestion price p_c is the price to transmit traffic at a rate of 1 byte/s for one hour. It can also be regarded as the hourly rent of $1/M$th of the link capacity of M bytes/s.

It becomes essential to distinguish between prices for services and rents for resources in ATM networks, because different resources are used to provide many different transmission services. We will focus on two sets of resources: the capacities of the different links and the buffers associated with each link. We will be concerned with services that transmit traffic with certain burstiness characteristics within a certain delay. Thus different services offered by the network will be distinguished by burstiness parameters and delay bounds.

Users and the network service provider enter into a contract. The contract specifies the burstiness parameters, the delay, and the price. It obligates the user to make sure that her traffic will conform to the burstiness parameters. It obligates the network to transfer conforming traffic within the specified delay, in exchange for the specified price. (We saw an example of such a contract in the form of GCRA in section 6.4.2.)

The network (service provider) meets its obligation by (1) selecting a route and (2) by *reserving* bandwidth and buffers in each link along the route in amounts sufficient to meet the delay requirement. The network can choose different routes, and it can dedicate different amounts of resources to meet the requirement. The network will select those combinations of routes and resources that will maximize its revenue. We will first describe a model that allows us to formulate the question of revenue maximization. A variant of this formulation will address the question of optimum prices for services. This will be the counterpart of Theorem 8.4.1. Finally, we will suggest an alternative formulation in which users directly rent resources—buffers and bandwidth—from the network and decide how to satisfy their own service requirements.

8.5.1 A Model of ATM Resources and Services

We consider ATM networks comprising a set of links, L, interconnected by switches. Suppose that these are output buffered switches, with one buffer per link. Then each link $l \in L$ is characterized by its transmission capacity of C_l (ATM) cells per second (or cps), and its buffer size of B_l cells. The resources of the network as a whole are given by the set of pairs

$$\{(C_l, B_l) \mid l \in L\}. \tag{8.15}$$

Figure 8.7 gives a model of a network with four links.

We suppose that the network provider has selected a set of routes R. Each route $r \in R$ designates a set of links that comprise this route. (For example, one possible route in the network of Figure 8.7 is $r = \{1, 2\}$ comprising links 1 and 2.) A route could be defined by a virtual path. Lastly, we suppose that the network provider has designed a set of service provision activities A. To each activity of type $a \in A$ there corresponds a route and bandwidth and buffer

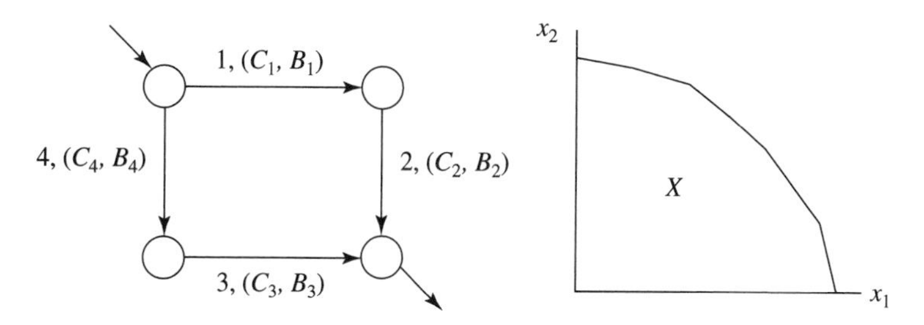

8.7 FIGURE

The network has four links; link i has bandwidth C_i and buffers of size B_i. The feasible set of activities is X.

reservations on the links along the path. Formally, an *activity* a is defined by its route $r_a \in R$ and a list of resources along the route,

$$a \to \{(c_{la}, b_{la}), \qquad l \in r_a\}. \tag{8.16}$$

The interpretation is that in order to carry out the activity a the network must reserve c_{la} cps of bandwidth and b_{la} cells of buffer in each link l along the route r_a.

When a user selects a service (to be described later), the network will choose an activity type a such that the resources (8.16) reserved by it can meet the delay requirements of the service that the user has selected. Of course the provider can do this only if free resources are available. Thus one wants to know how many activities of different types can be accommodated by the available network resources, given by (8.15). This leads to the following definition.

It is feasible to simultaneously accommodate x_a of type $a \in A$ provided that

$$\sum_a \epsilon_{la} c_{la} \times x_a \le C_l, \qquad l \in L, \tag{8.17}$$

$$\sum_a \epsilon_{la} b_{la} \times x_a \le B_l, \qquad l \in L. \tag{8.18}$$

Here, we define the numbers $\epsilon_{la} = 1$ or 0 according as link l belongs or does not belong to route r_a. Thus the inequalities (8.17), (8.18) imply that there are enough bandwidth and buffers in each link l to meet the requirements simultaneously placed by x_a activities of type a.

Consider again the network of Figure 8.7. Suppose the resources of links 1 and 2 are, respectively,

$$C_1 = 10^6 \text{ cps}, \quad B_1 = 10^6 \text{ cells}, \quad C_2 = 0.8 \times 10^6 \text{ cps}, \quad B_2 = 2 \times 10^6 \text{ cells}.$$

(We ignore the other links.) Consider two activities $a = 1$ and $a = 2$, both designating the same route $\{1, 2\}$. The resources required by activities 1 and 2 are

$$\{c_{11} = 10^3 \text{ cps}, b_{11} = 10^3 \text{ cells}, c_{21} = 10^3 \text{ cps}, b_{21} = 10^3 \text{ cells}\},$$
$$\{c_{12} = 10^2 \text{ cps}, b_{12} = 10^2 \text{ cells}, c_{22} = 0.5 \times 10^2 \text{ cps}, b_{22} = 2.5 \times 10^2 \text{ cells}\}.$$

Then it is feasible to accommodate x_1 activities of type 1 and x_2 activities of type 2 if

$$10^3 x_1 + 10^2 x_2 \leq 10^6,$$
$$10^3 x_1 + 0.5 \times 10^2 x_2 \leq 0.8 \times 10^6,$$
$$10^3 x_1 + 10^2 x_2 \leq 10^6,$$
$$10^3 x_1 + 2.5 \times 10^2 x_2 \leq 2 \times 10^6.$$

These inequalities represent the constraints imposed by C_1, C_2, B_1, B_2, respectively. They define the feasible set X of all activities that satisfy these four inequalities. X is a convex set, because the inequalities (8.17), (8.18) are linear. (The set X in Figure 8.7 is not drawn to scale.)

Note how the activity definition (8.16) transforms the set of available resources (8.15) into the feasible set X. We will now see how each activity is transformed into services that the provider can offer to users.

We consider a population of users who wish to transmit a stream of cells, whose instantaneous rate is $m(t), 0 \leq t \leq T$; $m(t)$ is measured in cps and T is the duration of the stream in seconds. It will be convenient to call such a stream a *message*. Figure 8.8 displays an example where the user's message is a sequence of video frames. A frame contains 512 × 512 8-bit pixels, and is generated every 33 ms. The user requires a service that will deliver a frame every 33 ms to the display.

Suppose the message rate is as shown in the figure. If the network reserves a bandwidth of 150 Mbps (the peak rate) in every link along the route, then the received signal will be the same as $m(t)$ (except for a constant propagation and switch processing delay). An alternative is to allocate 65 Mbps and some buffers (given by the shaded area, 85 × 14 Kb) in each link; the received signal then will be $m'(t)$ as shown. Note that both allocations meet the user's requirements. Observe that the allocations have the form of activities, i.e., they reserve bandwidth and buffers in the links along a route. Clearly, if a much smaller bandwidth or fewer buffers are allocated, then the user's requirement of delivering frames cannot be met. Thus the user's message must

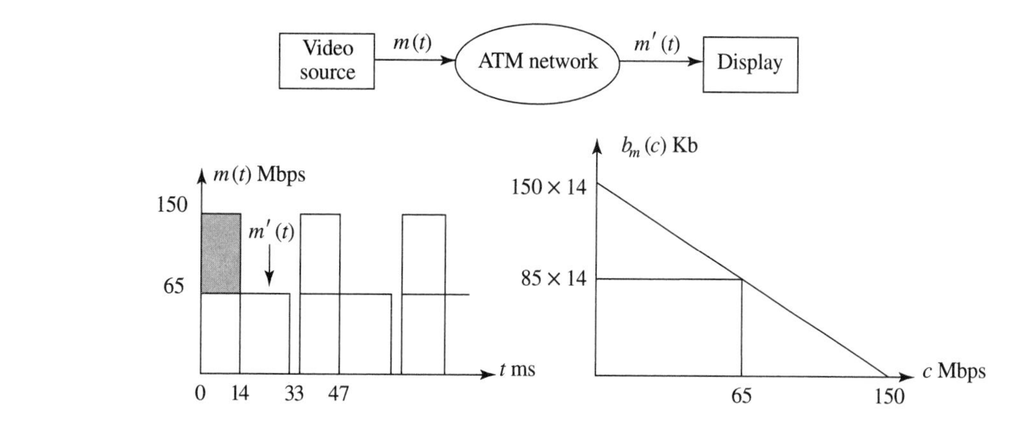

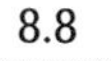

8.8 FIGURE The message $m(t)$ is a sequence of video frames once every 33 ms. The received message is $m'(t)$. $b_m(c)$ is the burstiness curve of $m(t), 0 \leq t \leq 33$ ms.

conform to the resources allocated for it. We now specify what a conforming message is, using the idea of a burstiness curve. (We have already introduced the notion of burstiness in Chapter 6. We repeat it here only to make the discussion self-contained.)

The burstiness curve of a message $m(t), 0 \leq t \leq T$, is the function $b_m(c)$ that gives the size of the buffer needed to transmit message m without loss over a link of capacity c cps. Clearly, the larger the rate c, the smaller is the required buffer size $b_m(c)$. At one extreme, if $c = 0$, the entire message must be buffered, so

$$b_m(0) = \int_0^T m(t)dt.$$

At the other extreme, if c is larger than the peak rate of m, no buffer is needed, so

$$b_m(c) = 0, \qquad c \geq \max_t m(t).$$

A useful property of the burstiness curve is that it is convex. Figure 8.8 shows the burstiness curve of one frame for the video source.

Suppose the message m is transmitted over a route r comprising links $l \in r$, and suppose c_l cps of bandwidth and b_l cells of buffers are reserved for this message in link l. Then this message will be transmitted without loss if in every link the reserved buffer size exceeds the burstiness curve at the reserved rate, i.e., if

$$b_l \geq b_m(c_l), \qquad l \in r.$$

Moreover, if $c_{min} = \min\{c_l | l \in r\}$ is the minimum reserved bandwidth, then the total end-to-end delay suffered by the message is

$$\text{delay} = \frac{b_m(c_{min})}{c_{min}} + \text{propagation and processing delay.} \tag{8.19}$$

We can now describe the services that the network provider may offer to users. A *service* is a four-tuple $s = (r_s, b_s(c), c_s, \delta_s)$, where r_s is a route, b_s is a burstiness curve, c_s is the minimum transmission bandwidth,

$$\delta_s = \frac{b_s(c_s)}{c_s} + \text{propagation and processing delay}$$

is the guaranteed delay. (Any nonnegative, convex, decreasing function $b(c), c \geq 0$ is a burstiness curve.) Let S denote the set of services offered by the network. Each service s is sold at a price p_s per unit of time. (The reader will note that this definition of service is a generalization of the ATM Forum's GCRA proposal.)

If a user wishes to transmit a message $m(t), 0 \leq t \leq T$, she can purchase a contract for service $s = (r_s, b_s, c_s, \delta_s)$ for time T. She must pay $p_s \times T$, and the contract requires that her message be *compliant*. This means that her message must be less bursty than the service burstiness curve, i.e.,

$$b_m(c) \leq b_s(c), \qquad c \geq c_s. \tag{8.20}$$

In return, the contract requires the network to transmit her message without loss, over route r_s, and with a delay not exceeding δ_s. To fulfill its side of the contract, the network provider selects an activity a that can meet the requirements of s. This means the two routes are the same, and the resources reserved by a are sufficient, i.e.,

$$r_a = r_s; \text{ and } b_{la} \geq b_s(c_{la}), c_{la} \geq c_s, \qquad l \in r_a. \tag{8.21}$$

Let us check that this selection is proper. Let $c_{min} = \min\{c_l | l \in r_s\}$. Then

$$\frac{b_m(c_{min})}{c_{min}} \leq \frac{b_m(c_s)}{c_s} \leq \frac{b_s(c_s)}{c_s}.$$

The first inequality follows since $c_{min} \geq c_s$ by (8.21), and the second inequality follows from (8.20). Finally, using this inequality in (8.19) implies that the delay suffered by the message is less than δ_s, as required by the contract.

8.5.2 Revenue Maximization

We have seen above that in order to fulfill a contract for service s, the network provider must undertake an activity a that satisfies (8.21). There may be several activities that meet this requirement. For example, we saw in Figure 8.8 that the service requirement could be met by providing peak bandwidth and no buffers or a lower bandwidth and some buffers. The network provider must consider which activity to assign to each service so as to maximize the network revenue. We formulate this revenue maximization problem.

For each service s let A_s be the subset of activities that satisfies (8.21). Let $x_{sa}, s \in S, a \in A_s$ be the number of units of service s that are assigned to activities of type a. Then the number of units of service s that are sold with this assignment is

$$n_s = \sum_{a \in A_s} x_{sa}, \qquad s \in S.$$

The revenue (per unit of time) from this assignment will be

$$\text{Revenue} = \sum_s p_s n_s = \sum_s \sum_{A_s} p_s x_{sa}.$$

On the other hand, the number of simultaneous activities of type a resulting from this assignment is

$$x_a = \sum_s x_{sa}, \qquad a \in A,$$

where we adopt the convention that $x_{sa} = 0$ if $a \notin A_s$. This assignment of services to activities must be feasible, i.e., the activities $\{x_a, a \in A\}$ must meet the constraints (8.17), (8.18).

Combining these observations, we can see that the revenue is maximized by the assignment $\{x_{sa}, s \in S, a \in A\}$, which solves the following problem

$$\max \sum_{s \in S} \sum_{a \in A} p_s x_{sa}$$

$$\text{subject to} \sum_s \sum_a \epsilon_{la} c_{la} \times x_{sa} \leq C_l, \qquad l \in L,$$

$$\sum_s \sum_a \epsilon_{la} b_{la} \times x_{sa} \leq B_l, \qquad l \in L$$

$$x_{sa} = 0, \qquad a \notin A_s.$$

This is a linear programming problem, which can be solved using standard algorithms.

There is a useful geometric picture that relates the set X of feasible simultaneous activities and the set N of feasible service units. Let us say that it is feasible to sell simultaneously service units n_s of type s, if there is an assignment x_{sa} such that

$$n_s = \sum_{a \in A_s} x_{sa}, \qquad s \in S,$$

and the corresponding activities

$$x_a = \sum_{s} x_{sa}, \qquad a \in A,$$

are feasible, i.e., these activites are in X. Let N be the set of feasible service units. Then, like X, the set N is a convex set. The maximum revenue is then given by maximizing $\sum_s p_s n_s$ over the set N.

We summarize the two steps developed in this section. We start by modeling the ATM network as a collection of links L, with each link l being characterized by its bandwidth and buffers, (c_l, b_l). Next, the network provider designs a set of activities A, with each activity a associated with a route and bandwidth-buffer reservation along the route. This determines the set X of feasible activities. The network provider offers for sale a set S of services and finds which activities A_s can fulfill each service s. This determines the set N of services that can be supplied for sale. Figure 8.9 illustrates the two steps: from resources to activities, and from activities to services.

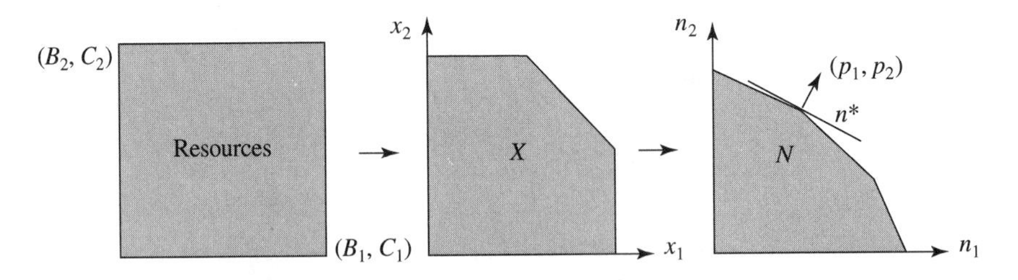

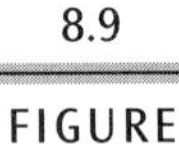

8.9 FIGURE

Network resources are transformed into the set X of feasible activities, which in turn is transformed into the set N of feasible services. For prices as shown, revenue is maximized at n^*.

8.6 SUMMARY

Technical discussions on networking carried out in textbooks, in journals, or in conferences normally do not discuss economic issues, either in terms of costs or in terms of the use of prices and other market mechanisms as a form of control and resource allocation. There are obvious reasons for this. The most important networks—the telephone and CATV networks—until recently were subject to regulation that effectively set prices for their services. The most important data networks—LANs—were privately owned, and the Internet was fully subsidized and free. Equally importantly, each network offered a single, undifferentiated service, so economic issues generally reduced to questions of cost and scale economics and, of course, regulation.

With the rapid commercialization of the Internet with its potential to provide services that compete with the existing networks, the remarkably rapid developments in ATM, and, more generally, the movement within the entire communications industry to compete to reap the economies of service integration and network externalities, economic issues are coming to the foreground. Network engineers and computer scientists who previously were engaged in purely technological concerns are finding that economic "variables" can be used to manage the network just as much as admission and flow control.

This chapter provided a very brief introduction to a variety of microeconomic models that can be used to formulate questions of network control and performance. In several important ways the economic models build on top of the models developed in Chapters 6 and 7. We made particular use of the deterministic models of traffic. We believe that network engineers are in a particularly good position to address economic issues because of their familiarity with the technology. We hope that this introduction will encourage them to turn their attention to those economic issues.

As we mentioned in the introduction, a more complete consideration of supply factors must include the costs of networking technology. Those estimates are not available, at least in the published literature, but we can provide an appreciation of the advances that are bringing such dramatic changes in networking. The next two chapters are devoted to those advances.

8.7 NOTES

For a general introduction to microeconomics, see [V93]. The material in sections 8.2 and 8.3 is adapted from [EMV95]. Section 8.4 is based on [MV95]. Section 8.5 is based on [JV91, JV94]. For Internet accounting requirements see RFC 1272.

8.8 PROBLEMS

1. Suppose that it costs more to provide (telephone or CATV or Internet) access in rural areas than in urban areas. Would it be more efficient to impose larger access charges on rural experiences? On what public policy grounds should the government require the same access charge, which, in effect, implies that urban subscribers subsidize rural subscribers?

2. Suppose that wages and salaries are lower in rural than in urban areas but that network access charges are the same. Consider a business that serves its customers through the network. (For example, a mail-order business takes most of its orders over the telephone.) Would you expect such a business more likely to be located in rural areas? Can you find any empirical evidence that might support or deny such an expectation?

3. Suppose a company sells Internet access through dial-up modems. The company has two modem pools: the fast modems have a speed of 28.8 Kbps and the slow modems have a speed of 2.4 Kbps. Subscribers can dial either modem pool. Suppose that from noon to 6 P.M., the fast pool is congested, i.e., more subscribers wish to gain network access than the number of modems in the fast pool. What is your prediction about subscriber behavior during the busy period in terms of (1) their willingness to keep on dialing the fast modem pool until they get through, (2) their willingness to dial into the slow modem pool, and (3) their unwillingness to attempt access during the busy period?

4. If a company creates a unique product (software or hardware) that is considered valuable by users, the company can charge much more for this product than its cost. After some time, however, other companies will develop the same or a substitute product, and competition will drive the price to the cost of production. Can you build a model of consumer behavior and market organization that supports this? A market for a particular product is said to be contestable if the period for which the first company enjoys a monopoly is relatively short. Do you think that software product markets are likely to be more contestable or less contestable than hardware products? Why?

5. Companies A and B both make computers. A's computer has an open architecture, B's has a proprietary architecture. How would you compare the prospects of these two companies?

6. This problem seeks to model the vulnerability of the Internet. Suppose n users have TCP connections with traffic rates $x_1, \ldots, x_n$ going through the

same router. The queuing delay they all experience is some function $d(x)$ where $x = x_1 + \cdots + x_n$.

(a) Since queuing delay increases with the traffic rate, $d(x)$ must be an increasing function of x. What more can you say about the behavior of d? Suggest a specific form of d based on the discussion of section 6.3.

(b) Suppose the value to user i of the connection is given by $V_i(x_i, d)$ where the function V_i is increasing in x_i and decreasing in d. Is this form of the value function plausible? Suppose the n users cooperate with each other and decide to set their traffic rates x_i so as to maximize their total benefit,

$$max_{x_1,\ldots,x_n} \sum_i V_i(x_i, d).$$

Derive equations that can be solved to find the optimum rates $\{x_i^*\}$.

(c) Suppose the users are noncooperative. We model noncooperative behavior by saying that user i selects x_i to maximize $V_i(x_i, d)$, taking $d = \sum_{j \neq i} x_j$ as fixed. Let the resulting traffic rates be x_i^u. Do you expect x_i^u to be larger or smaller than x_i^*? Can you support your guess by a mathematical argument?

(d) One way of achieving the cooperative solution is through congestion pricing. Another way is through some sort of social regulation in which users who do not reduce their traffic rate during congestion are subject to some kind of social sanction. Can you think of such a regulation mechanism? Compare such social regulation with market regulation through pricing. What are the pros and cons of the two schemes? (The discussion in section 8.4 should help in answering this question.)

7. Proposals to upgrade Internet protocols were discussed in Chapter 3. The argument for an upgrade is that it will permit the Internet to support applications that require QoS guarantees. The argument against the upgrade is that it will render obsolete the large installed base of IP software around the world. (The situation is more complicated than this, but we ignore the complications.) Assess the two arguments. Clearly, users who want QoS guarantees will favor the upgrade and would be willing to bear the upgrade cost, and those who do not want such guarantees will be unwilling to bear the cost since they derive no additional benefit. Can you propose an upgrade strategy whereby the former can compensate the latter for the additional cost?

8. Analyze the additional TCP connection setup time resulting from implementing the billing system of section 8.2.

9. Propose a billing scheme for ATM connections.
10. Talk to your network manager and figure out the cost of Internet access as a function of bandwidth, number of users, total traffic, and any other parameters you find important.
11. Find out the charges for Internet access offered by three network access providers. Why are the charges different? Are they appealing to different segments of users? What costs are incurred by the access provider?
12. A small business office seeking Internet access has two options (among others). The office can install a number of modems to an access provider. Alternatively, the office can connect its computers via Ethernet and rent or lease a high-speed link (e.g., an ISDN or Frame Relay or SMDS service). Compare the two options.
13. Video stores offer to rent or sell a video of the *same* movie. Since the rental price for the video is much less than the sales price, it seems surprising that anyone would buy the video rather than rent it. Can you explain why both markets coexist? (Saying that consumers are irrational is not an explanation. Part of the answer may be discovered by finding out which movies are offered for sale.)
14. Software and books are both easy to copy, and copyright laws do not provide much protection. The pricing strategies for these two products seem quite different. The price of software has gone down steadily as manufacturers have gone for the mass market. The price of books has gone up steadily as publishers have restricted their market to libraries and businesses that are more likely to abide by copyright laws. Thus the software producers have accommodated to the ease of illegal copying by lowering prices, thereby reducing the incentive to make illegal copies, whereas publishers have limited their markets. Compare the two strategies.
15. We saw the diversity of Internet traffic on the Berkeley campus. Propose a pricing scheme that takes this diversity into account and that is simple to implement, understandable to users, and recovers costs.
16. Suppose the telephone switch on the University of California-Berkeley campus costs $300,000 per year. Suppose 10,000 phones are served by this switch. This cost can be recovered in at least two different ways. First, one may charge a fixed fee of $30 per month for each phone, or one may charge a usage price of $\$p$ per phone call, with p adjusted to recover the switch cost. Discuss the relative merits of these two forms of charges.

9 CHAPTER Optical Links

Until now we have regarded a transmission link as a bit way—a device that transports a stream of bits from one end of the link to the other at a fixed rate, with some error. We will now study the optical links that implement such bit ways with very large rates over long distances and at a cost that is much lower than that of copper links. In addition, optical links are less susceptible to interference, have a longer lifetime, and are less expensive to maintain than copper links.

For the foreseeable future, networks will use optical links for new links and for replacing existing copper links. The operations of multiplexing, demultiplexing, and switching of optical signals are carried out today after the signals are converted into electrical signals. But it is possible to carry out these operations directly in the optical domain, and experimental systems have demonstrated the potential of this approach. When this approach is fully developed, there will be further reduction in cost.

In this chapter, we explain the operating principles and the characteristics of optical links. We discuss the limitations of different types of links. We briefly introduce the novel optical multiplexing and demultiplexing techniques made possible by optical communication technology.

In section 9.1 we examine the characteristics of a link seen from end to end and show that the bandwidth-distance product of a communication link is an important figure of merit. In section 9.2 we study the use of fibers as a communication medium and the two phenomena that limit their performance: attenuation and dispersion. In section 9.3 we explain the basic characteristics and operating principles of light sources and detectors and the mechanisms that limit their performance. In section 9.4 we study coherent optical demodulation, subcarrier multiplexing, and wave-division multiplexing. In section 9.5 we derive the quantum limits of on-off keying (OOK) with homodyne detection. Lastly, in section 9.6 we introduce some of the other important components of optical links.

The reader who is interested only in understanding optical links as a bit way need read only section 9.1.

Optical links offer the only practical means of high-speed transmission over distances longer than a few kilometers, but there are alternatives for shorter distances and for broadcast applications. The three major alternatives are copper links in the form of wire pairs or coaxial cable, broadcast radio over a small area as in today's cellular radio or future "personal communication systems," and broadcast satellite. Copper links can have a cost advantage for short distances and also because the large installed copper base may make it advantageous to use copper for replacement and expansion. Radio has the overwhelming advantage of the capability to support user mobility as in cellular phone and related data services. Lastly, the large area covered by satellite broadcast may make it very economical for certain applications, as we see today with Direct Broadcast Satellite video. Despite the importance of these alternatives, we will consider optical links only.

9.1 OPTICAL LINK

The optical link is one of the two most important network elements (the other is the switch). In order to model it as a communication system, we regard an optical link as consisting of a transmitter, a fiber, and a receiver, as in the top of Figure 9.1. The input bit stream is represented by an electrical input signal (Data IN). The transmitter converts this input signal into an optical signal using a particular modulation scheme. For example, in on-off keying, or OOK, a light source is turned on for a 1 bit and turned off for a 0 bit. Thus a bit stream of 1s and 0s is converted into a sequence of light and dark (no light) pulses. The modulated optical signal propagates over the fiber and reaches the receiver where it is demodulated into an electric signal (Data OUT) from which the input bit stream is recovered, possibly with error.

As a communication system, a link is characterized by a pair of numbers (B, L). Here B bps is the bit rate and L km is the maximum distance of the fiber for which the error rate is below a specified amount. For optical links, this bit error rate or BER is on the order of 10^{-9}. We may transmit B bps over a distance of $2L$ km with two (B, L) links in series: at the receiver of the

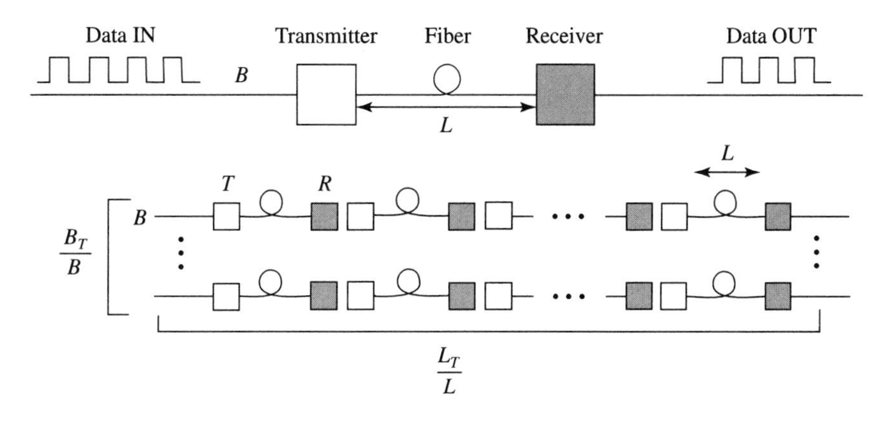

9.1

FIGURE

A link of rate B_T and length L_T can be built by a series-parallel connection of $B_T/B \times L_T/L$ links each of rate B and length L.

first link, the original bit stream is regenerated (with some error) and used to modulate the transmitter of the second link. (If each link's BER is 10^{-9}, the series connection BER is of the same order.)

Suppose we want to build a communication system that can transmit B_T bps over a distance of L_T km using (B, L) optical links. We can achieve our aim with B_T/B parallel systems, each system consisting of L_T/L links in series, as shown in the bottom of the figure. Thus, we need $(B_T \times L_T)/(B \times L)$ optical links. Hence, as a communication system, the value of a link that can transmit B bps over L km is proportional to the bit way–distance product $B \times L$. If a link has a $B \times L$ product twice as large as another link, then one should be willing to pay twice as much for it since only half as many are needed. In addition to being an economically significant figure of merit, we will see in section 9.2.2 that $B \times L$ is an important parameter in the limitations of optical links caused by dispersion.

We summarize the ways in which link performance is affected by the limitations of its three components: transmitter, fiber, and receiver.

The transmitter is a light source that is turned on or off accordingly as the bit to be transmitted is 1 or 0. The transmitter's limitations are determined by the power P_T of the light source and its modulation bandwidth, i.e., the maximum rate at which the light source can be turned off or on. There are two light sources: light-emitting diodes or LEDs and laser diodes or LDs. LEDs are very cheap with an output power of 1 mW (milliwatt) and with a modulation bandwidth of 100 MHz. LDs have an output power of 10 mW and a modulation bandwidth of 3 GHz.

The receiver includes a detector that converts incident light into electrical current. Thermal noise in the receiver and randomness of the incident signal determine receiver sensitivity, P_R, defined as the minimum received power per bit needed to detect signals of a specified bit rate with a specified bit error rate or BER. BER is on the order of 10^{-12}.

The difference $P_T - P_R$ between the power injected by the transmitter and the receiver sensitivity is dissipated by the fiber. This dissipation, AL, is proportional to the length L of the fiber. So the maximum length of the fiber is given by $(P_T - P_R)/A$.

Thus, the transmitter limits the maximum value of P_T, the receiver places a lower bound on P_R, and the fiber determines A. Together, these limitations determine the maximum bandwidth-distance product of the optical link.

We now examine these limitations in more detail.

9.2 FIBER

As the optical signal propagates over the fiber it gets distorted due to attenuation and dispersion. Attenuation is the reduction in power in the optical signal and dispersion is the spreading of a pulse of light. At any given bit rate, the distortion, and hence the error rate, increases with the length of the fiber.

9.2.1 Attenuation

The attenuation of a fiber is expressed in decibels per km (dB/km). To explain why these units are appropriate, we first show that attenuation is exponential in the fiber length. Consider an optical fiber propagating a beam of light. Suppose the power of the beam launched into the fiber is P_T. As the beam travels along the fiber, some of its power is dissipated. Suppose that after traveling l km of fiber, the power in the beam is $P(l)$. $P(l)$ is proportional to P_T. We denote by $a(l)$ the attenuation factor, i.e., $P(l) = a(l)P_T$. The power in the beam after $l_1 + l_2$ is $P(l_1 + l_2)$, which may be expressed in different ways,

$$P(l_1 + l_2) = a(l_1 + l_2)P_T = a(l_1)P(l_2) = a(l_1)a(l_2)P_T.$$

The first equality follows directly from the definition of $a(l)$. The second expression is obtained by writing $P(l_1 + l_2)$ as the power $P(l_2)$ attenuated by l_1 km of fiber. We conclude that

$$a(l_1 + l_2) = a(l_1) \times a(l_2),$$

from which it follows that $a(l)$ must be of the form

$$a(l) = e^{-\alpha l}, \qquad l \geq 0.$$

Since $a(l) < 1$, we must have $\alpha > 0$. By modifying the expression for α, the function $a(l)$ can be rewritten as

$$a(l) = 10^{-\frac{Al}{10}}.$$

The attenuation after L km is such that

$$10 \log \frac{P_T}{P(L)} = A \times L,$$

so that the attenuation in decibels is equal to A multiplied by the distance L in km. Thus, A is the attenuation of the fiber in decibels per km.

The attenuation coefficient A of the fiber depends on the fiber material and also on the wavelength λ of the light. Figure 9.2 shows A for an all-glass fiber as a function of λ, measured in μm or microns.

The figure indicates two different physical causes of attenuation. Rayleigh scattering is due to the crystal structure of the fiber. This scattering results in

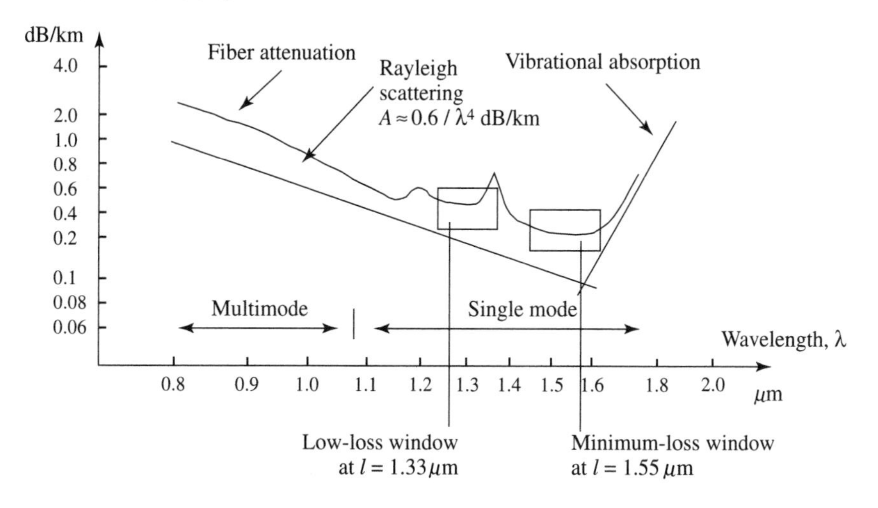

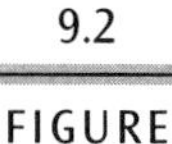

9.2 FIGURE

Attenuation in all-glass fiber is measured in dB/km. There are two low-loss windows near 1.3 and 1.55 μm.

an attenuation inversely proportional to the fourth power of the wavelength, i.e., $A = 0.6/\lambda^4$. (Because the scale in the two axes of Figure 9.2 is logarithmic, the attenuation due to scattering is a straight line.) At wavelengths larger than 1.7 μm, the dominating source of attenuation is the vibrations of the crystal induced by the light. The two peaks in the attenuation curve, at 1.2 μm and at 1.4 μm, are caused by vibrations of OH^- ions.

In the figure we see two "windows" of wavelengths where the attenuation is at a minimum. One of these windows is at 1.33 μm and its attenuation is 0.4 dB/km. The other window is at 1.55 μm and its attenuation is 0.25 dB/km.

The width of each of these windows translates into an enormous bandwidth. For instance, the window around 1.33 μm has a width of 100 nm (nanometer) (1 nm = 10^{-9} m). The range of frequencies of light carried in this window goes from $c/(\lambda + 100)$ nm to c/λ where $\lambda \approx 1.33$ μm, i.e., from $(3 \times 10^8)/(1.43 \times 10^{-6}) = 2.098 \times 10^{14}$ Hz to $(3 \times 10^8)/(1.33 \times 10^{-6}) = 2.256 \times 10^{14}$ Hz. Therefore, this window covers a range of frequencies of about 16×10^{12} Hz. This range can be used to transmit information at very high rates.

To appreciate the size of this bandwidth, recall that a regular phone call requires 64 Kbps. Therefore, a bit rate of 10^{13} bps can carry $\frac{10^{13}}{64 \times 10^3} \approx 1.6 \times 10^9$ simultaneous phone calls, which is several orders of magnitude more than the number of telephone calls in progress at any time in the world. Thus, an optical fiber is capable of transmitting a huge bit rate with a small attenuation per km. (In order to utilize this bandwidth one must modulate an optical signal at that rate—a challenge that engineers have not yet fully met.)

A receiver is characterized by its *sensitivity*, P_R. This is the power it needs in order to detect with a specified BER the bits transmitted at a certain rate. We can determine the maximum usable length of an optical fiber from its attenuation coefficient A if we know the transmitted power P_T and the receiver sensitivity P_R. To determine that maximum length, we use the formula expressing the received power $P(L)$ after L km in which we set $P(L) = P_R$, and we solve for L. This gives $L = \frac{10}{A} \log_{10} \frac{P_T}{P_R}$.

It is convenient to express P_T and P_R in dBm. By definition, a power p in watts is equal to $P(\text{dBm})$ where

$$P(\text{dBm}) := 10 \log_{10} \frac{p}{1 \text{ mW}}.$$

With this definition, we can rewrite the formula for the maximum usable length L as

$$L = \frac{1}{A}\{P_T(\text{dBm}) - P_R(\text{dBm})\}. \tag{9.1}$$

We illustrate the use of formula (9.1) by means of an example. Suppose $P_R = -45$ dBm (about 3×10^{-9} watts) at a rate of 1 Gbps and a BER of 10^{-9}. Suppose next that the attenuation is $A = 0.2$ dB/km. Finally, suppose the transmitter power P_T is 1 mW (0 dBm). Then the maximum fiber length is

$$L = \frac{1}{0.2}\{0 - (-45)\} = 225 \text{ km},$$

so that the $B \times L$ product of this link is 225 Gbps×km.

Another way to use the formula (9.1) is to express the power loss as

$$P_T(\text{dBm}) - P_R(\text{dBm}) = A(\text{dB/km}) \times L(\text{km}).$$

Besides attenuation in the fiber, the main causes of power loss between transmitter and receiver are the coupler between the source and the fiber, the splices between sections of fiber, and the coupler between the fiber and the receiver. Thus, if the light between the transmitter and the receiver goes through two couplers, N splices, and L km of fiber, then the power loss is given by

$$A(\text{dB/km}) \times L(\text{km}) + 2 \times C(\text{dB}) + N \times S(\text{dB})$$

where C is the power loss at a coupler (in dB) and S is the power loss at a splice. The *power budget* analysis of the communication link is the comparison of this power loss with the total acceptable loss $P_T(\text{dBm}) - P_R(\text{dBm})$.

The formula (9.1) for the maximum usable length of a fiber also applies to the maximum usable length of a coaxial transmission line. The values for the attenuation coefficient, transmitted power, and the receiver sensitivity for coaxial cable are of course different from those for an optical link. Typically, a microwave transmitter with a bit rate of 100 Mbps can inject a power of 1 watt into the coaxial cable, so that $P_T = 30$ dBm. This transmitter power is significantly larger than that transmitted in an optical fiber. A microwave receiver can be made very sensitive and requires only $P_R = -75$ dBm for a BER of 10^{-9}. Thus a microwave receiver requires much less signal power than an optical receiver to achieve the same BER. The attenuation coefficient of a coaxial cable around 100 MHz is about 30 dB/km, much larger than that for optical fiber. Using the formula then shows that the maximum usable length for the coaxial cable is 3.5 km. This results in a $B \times L$ product of 0.35 Gbps×km, about three orders of magnitude less than the 225 Gbps×km product for the optical fiber of the example. Recalling our earlier discussion, we can conclude that the economic value of the optical link is three orders of magnitude larger than that of the copper link.

This comparison shows that the dominant advantage of fibers over copper is due to their much lower attenuation over a large range of frequencies. Copper has a large attenuation at high frequencies because of the skin effect, which causes the electric current to localize in a thin region at the periphery of the conductors. As a result of this skin effect, only a small cross-section of the copper carries the current so that the effective resistivity of the cable is large at high frequencies.

Formula (9.1) also explains that reducing the attenuation coefficient by a factor of 10 increases the maximum length by the same factor, whereas increasing the transmitted power or decreasing the power required at the receiver by a factor of 10 has a much smaller impact on the maximum length.

When the received power drops to the value of the receiver sensitivity, the input bit stream is regenerated and used to modulate the optical signal of the next transmitter in series, as shown in Figure 9.1. Thus regeneration serves to amplify the optical signal power, with a gain $P_T - P_R$, which is about 30 to 45 dB. Regeneration requires conversion from the optical to the electric domain. The bit rate of an optical signal that can be amplified by regeneration is limited by the maximum bandwidth of electronic amplifiers, which is a few GHz.

Optical amplifiers can increase the power of the optical signal without converting first into an electric signal. When optical amplifiers become economical, the need for (electronic) signal regeneration will be reduced considerably, and there will be an increase in the bandwidth-delay product of optical links. However, the gain of power amplifiers is not greater than regeneration. The decisive advantage of optical amplifiers is their enormous bandwidth, which is comparable to the low-loss windows of Figure 9.2, on the order of 10^{12} Hz. Wave-division multiplexing is the only modulation scheme that makes use of this bandwidth.

Three generations of optical links have been used to date. We summarize the characteristics of the transmitter, receiver, and fiber used in each generation. The first generation used an AlGaAs (aluminum gallium arsenide) laser or LED as the optical power source providing $P_T = 1$ mW at a wavelength of 0.85 μm; multimode fibers (with a core diameter of 50 μm compared with 8 μm for single-mode fibers) with an attenuation coefficient $A = 2.5$ dB/km; and silicon PIN or avalanche photodiode (APD) diodes as detectors with a sensitivity of $\bar{N} = 300$ photons per bit for BER $= 10^{-9}$.

Receiver sensitivity expressed as $\bar{N}$, the average number of photons received per bit, can be converted into required receiver power P_R by the formula

$$P_R = \bar{N}Bh\nu = 2 \times 10^{-7} \times \bar{N} \times \frac{B \text{ Gb}}{\lambda \ \mu\text{m}} \text{ mW} \approx 7 \times 10^{-14} B \times \text{mW}.$$

Here B is the bit rate in bps. P_R is obtained by multiplying ($\bar{N} \times B$), the average number of photons per second, by the energy ($h\nu$) of each photon at the frequency $\nu = c/\lambda$. For instance, if the receiver needs 300 photons per bit, then its sensitivity is equal to $7 \times 10^{-14}B$ mW, when the transmission rate is B Gbps.

The second generation optical link used lasers with $P_T = 1$ mW in the low-loss window of 1.3 μm, single-mode fiber with attenuation coefficient $A = 0.4$ dB/km, and InGaAs (indium gallium arsenide) PIN or APD diodes as detectors with a sensitivity of $\bar{N} = 1{,}000$ photons per bit for a BER of 10^{-9}. The third generation uses lasers with $P_T = 1$ mW in the minimum-loss window at 1.55 μm, single-mode fiber with attenuation coefficient $A = 0.25$ dB/km, and with a receiver similar to that of the second generation.

Thus the principal advance from one generation to the next is the reduction in attenuation coefficient. First-generation links are used where the distance between transmitter and receiver is short, so that the small distance-bandwidth product is not a limitation. (For very short distances, copper coaxial cable may be sufficient.) Third-generation links are used for long distances.

9.2.2 Dispersion

Dispersion is the spreading of pulses of light as they propagate along a fiber. To understand how dispersion limits the bandwidth-distance product, suppose the transmitter turns the source of light on for T seconds to represent a 1 and turns it off for T seconds to represent a 0. ($T = 1/B$ is the bit time, where B is the bit rate in bps.) Thus, the transmitter represents the input bit string by a succession of light (and dark) pulses. To recover the bits, the receiver must distinguish the periods when the light is on from those when it is off. Consequently, as the pulses representing 1s spread, they overlap epochs that represent 0s; the dispersion will not confuse the receiver if the pulse spread is less than $T/4$. In that case, as seen in the bottom part of Figure 9.3, the receiver will see a 0 between two 1s as a small period when the light is off. However, if the pulse spread is larger than $T/4$, then the receiver may not be able to distinguish the 0s and the 1s.

The conclusion that the pulse spread should be limited to $T/4$ can also be reached by viewing the fiber as a low-pass filter with transfer function $H(\omega)$. We explain that alternative viewpoint. The top of Figure 9.3 shows

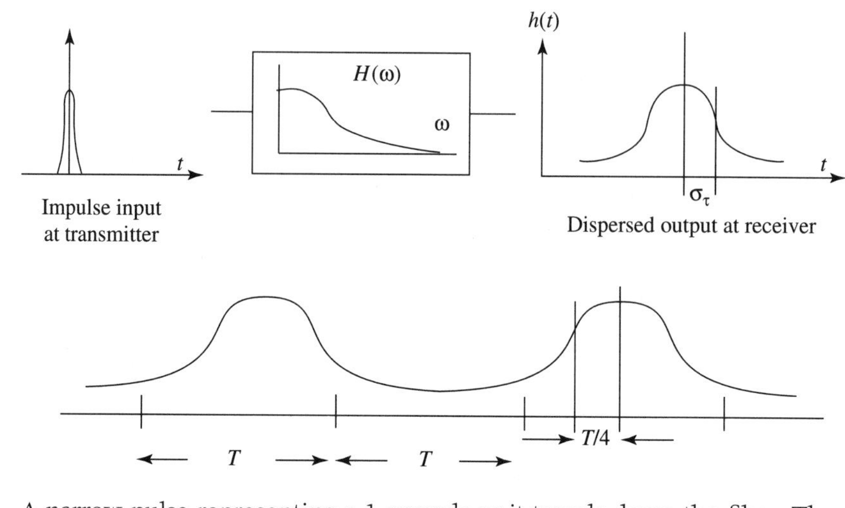

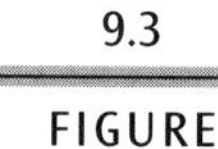

FIGURE 9.3 A narrow pulse representing a 1 spreads as it travels down the fiber. The dispersion should be less than one-quarter of the bit time, $T/4$, to prevent errors.

the impulse response of the fiber as a bell-shaped pulse of light, $h(t)$, with a spread designated by σ_τ. (That is, an impulse of light originating at the source spreads into the shape $h(t)$ when it reaches the receiver.) By approximating the exponential function by the first two terms of its Taylor expansion, i.e., $e^{j\omega t} \approx 1 + j\omega t - 1/2\omega^2 t^2$, we can approximate the transfer function $H(\omega)$ by the quadratic function

$$H(\omega) = \int h(t)e^{j\omega t}\,dt \approx H(0)\left(1 - \frac{1}{2}\omega^2\sigma_\tau^2\right).$$

In terms of the impulse response $h(t)$,

$$H(0) = \int h(t)dt, \qquad \sigma_\tau^2 = \frac{\int t^2 h(t)dt}{H(0)}.$$

Consider the string of bits 1010101 . . . being transmitted with OOK at $B = 1/T$ bits per second. The OOK signal that represents this string is a periodic square wave with period $2T$. The energy in this periodic signal is concentrated around its fundamental frequency $1/2T$ Hz, or $2\pi/2T = \pi B$ radians per second. The gain of the fiber at that frequency is $|H(\pi B)|$. Using the expression for $H(\omega)$ derived above, we can calculate the power loss, P_D, due to pulse spreading. This power loss, expressed in dB, is

$$P_D = -10 \log \frac{H(\pi B)}{H(0)} \approx -10 \log \left[1 - \frac{1}{2} \pi^2 B^2 \sigma_\tau^2 \right] \approx 21 (\sigma_\tau B)^2.$$

Communication engineers require the power loss P_D to be less than 1 dB (which is comparable to losses incurred from other elements such as couplers), from which we conclude that the pulse spread σ_τ must be smaller than $1/4B = T/4$, a quarter of the bit period.

We will soon see that the pulse spread is proportional to the length L of the fiber, i.e., it is equal to αL where α is a constant that depends on the fiber. Consequently, requiring that the pulse spread be less than $1/4B$ amounts to requiring that $\alpha L \leq 1/4B$, i.e.,

$$B \times L \leq \frac{1}{4\alpha}.$$

Thus, dispersion limits the bandwidth-distance product. This dispersion limit, together with the attenuation limit, determines the maximum usable length of the fiber at a given bit rate. The dispersion limit depends on the fiber (through the coefficient α). We will explain the physical cause of this limit for the following types of fibers: step index, graded index, and single mode.

Figure 9.4 shows a step-index fiber. It consists of a cylindrical core made of a material with refractive index η surrounded by a cladding made of a material with refractive index $(1 - \Delta)\eta$. For all-glass fiber, $\eta \approx 1.46$ and $\Delta \approx 0.01$. The speed of light in a material with refractive index η is equal to c/η, where $c = 3 \times 10^5$ km/s is the speed of light in a vacuum. Thus, the speed of light in glass is about 2×10^5 km/s.

A light ray in the core that hits the cladding with an angle less than the critical angle θ_c is subject to total reflection and comes back into the core with

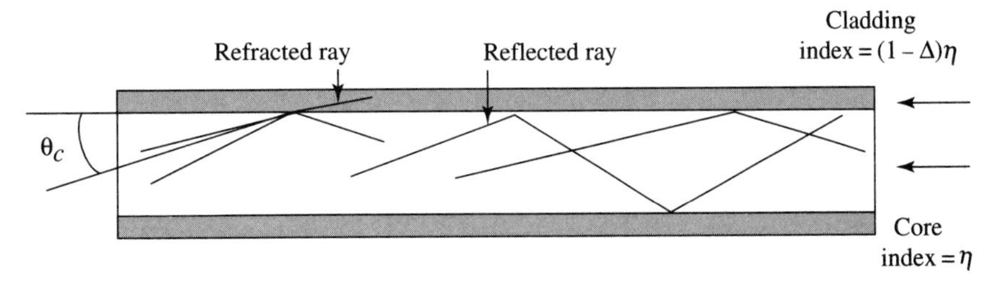

FIGURE 9.4 In a step-index fiber, light rays propagating at angles less than θ_c get reflected into the fiber.

the same angle. The angle θ_c is given by $\cos\theta_c = (1 - \Delta)$, so that $\theta_c \approx \sqrt{2\Delta}$. (For $|\theta| \ll 1$ one has $\cos\theta \approx 1 - \theta^2/2$.) A light ray in the core that hits the cladding with an angle larger than θ_c is refracted into the cladding and is eventually absorbed. It follows that rays that make an angle less than θ_c with the axis of the fiber propagate by undergoing a succession of total reflections at the boundary between the core and the cladding.

Different angles of propagation are called *propagation modes*. Observe that different modes travel different distances to go through L km of fiber. Hence they travel at different speeds or *group velocities* along the fiber. The distance traveled by a propagation mode with angle θ is equal to $L/\cos\theta$. The time taken by such a mode to go through the L km of fiber is therefore equal to

$$\frac{L/\cos\theta}{c/\eta} = \frac{L\eta}{c\,\cos\theta}.$$

The mode parallel to the fiber axis, $\theta = 0$, is the fastest. The slowest mode is the one making an angle θ_c with the fiber axis. The difference in propagation times of these two modes is the full width of the pulse spread,

$$2\sigma_\tau = \frac{L\eta}{c}\left[\frac{1}{\cos\theta_c} - 1\right] \approx \frac{L\eta}{c}\left[\frac{1}{2}\theta_c^2\right] \approx \frac{L\eta}{c}\Delta.$$

The pulse spreading due to the different propagation modes is called *modal dispersion*. We saw earlier that the pulse spread σ_τ should be less than $1/4B$. Using the expression for σ_τ above, we conclude that for a step-index fiber, modal dispersion places the limit

$$B \times L < \frac{c}{2\eta\Delta} = \frac{3 \times 10^5}{2 \times 1.46 \times 0.01} = 10 \text{ Mb}\times\text{km}.$$

This stringent limit, which is worse even than that of coaxial cable, led researchers to design a different type of fiber with a much higher limit. These fibers have a graded-index (GRIN) profile illustrated in Figure 9.5. In such a fiber, the refractive index decreases continuously away from the fiber center, as shown by the parabolic profile on the right in the figure.

Rays propagate in a GRIN fiber by being continuously refracted. The resulting rays follow sine-wave-like trajectories as shown in the figure. Different shapes of rays are also called *modes of propagation*. Note that the longest trajectories go through regions of the fiber that have a smaller refractive index and, hence, a faster propagation speed. Consequently, the difference in propagation times of different modes is smaller in a GRIN fiber than in a step-index

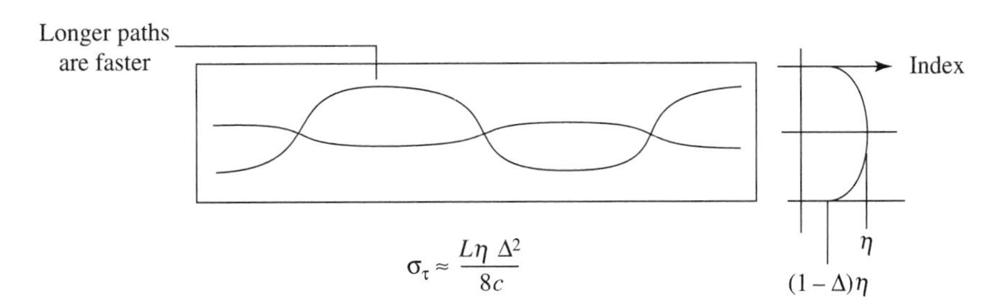

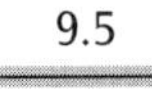
9.5 FIGURE

In a graded-index (GRIN) fiber, modes that have longer paths travel faster resulting in lower dispersion than in step-index fibers.

fiber. It follows that the pulse spread is smaller in a GRIN fiber. The pulse spread can be computed to be

$$\sigma_\tau = \frac{L\eta}{8c}\Delta^2.$$

In order to keep $\sigma_\tau < 1/4B$, we must then have

$$B \times L < \frac{2c}{\eta\Delta^2} \approx 4 \text{ Gbps}\times\text{km},$$

which is two orders of magnitude better than the 10 Mbps×km limit for step-index fiber.

Step-index and graded-index fibers are called *multimode* because light travels in several different modes in these fibers. As a result, light is subject to modal dispersion when it propagates through these fibers. It is possible to design fibers with no modal dispersion. Maxwell's equations imply that when the core diameter of a step-index fiber is less than 8 μm only one mode can propagate through the fiber. Such fibers are called *single-mode* fibers. There is no modal dispersion in a single-mode fiber. However, there is some dispersion in these fibers because of the nonlinear dependency of the refractive index on the wavelength ($d^2\eta/d\lambda^2 \neq 0$). This nonlinear dependency results in different group velocities for pulses of light of different wavelengths. Since a light pulse generated by an optical source is composed of different wavelengths, the pulse spreads as it propagates. This effect is called *material dispersion*. This is illustrated in the top left part of Figure 9.6, which shows two pulses of wavelengths λ_1 and λ_2 starting at the transmitter at the same time but arriving at the receiver at two different times. The dependence of the travel time on wavelength is shown in the top right of the figure.

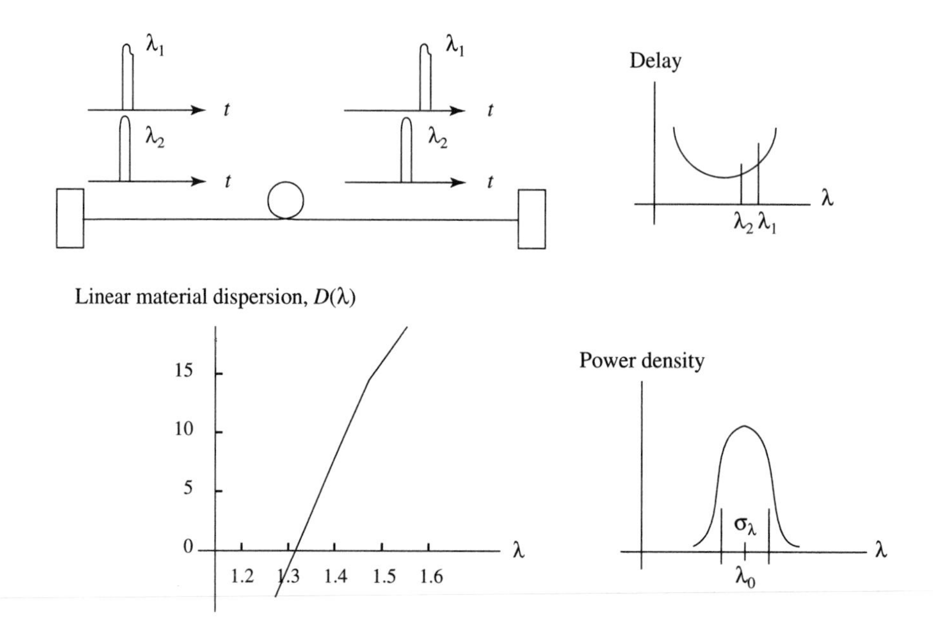

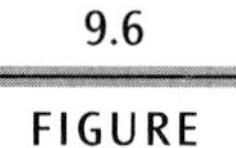
9.6 FIGURE

In a single-mode fiber a light pulse spreads because it is composed of components of different wavelengths that travel at different speeds. This is called material dispersion.

Material dispersion is computed as follows. Let $D(\lambda)$ be the linear material dispersion, expressed in ps/km.nm. $D(\lambda)$ is the difference in travel times in picoseconds ($1 \text{ ps} = 10^{-12}$ s) for light rays with wavelengths that differ by 1 nm, per km of fiber. Suppose the spectrum of the optical source has a width of σ_λ, centered at λ_0 as in the bottom right of Figure 9.6. Then the pulse spread

$$\sigma_\tau = L \times D(\lambda_0) \times \sigma_\lambda.$$

In order to achieve $\sigma_\tau < 1/4B$, we must have

$$B \times L < \frac{1}{4D(\lambda_0) \times \sigma_\lambda} \approx \frac{1}{4 \times 1(\text{ps/km.nm}) \times 1(\text{nm})} = 250 \text{ Gbps}\times\text{km}.$$

The material dispersion limit of 250 Gbps×km assumes some typical values for $\lambda_0 = 1.33\ \mu$m. This limit is much larger at 1.33 μm than at 1.55 μm because D is much smaller at that wavelength. New materials have been developed for which the linear material dispersion is very small around 1.55 μm, the wavelength where the attenuation is minimal. The single-mode fibers made with these new materials have a large attenuation limit and a large material

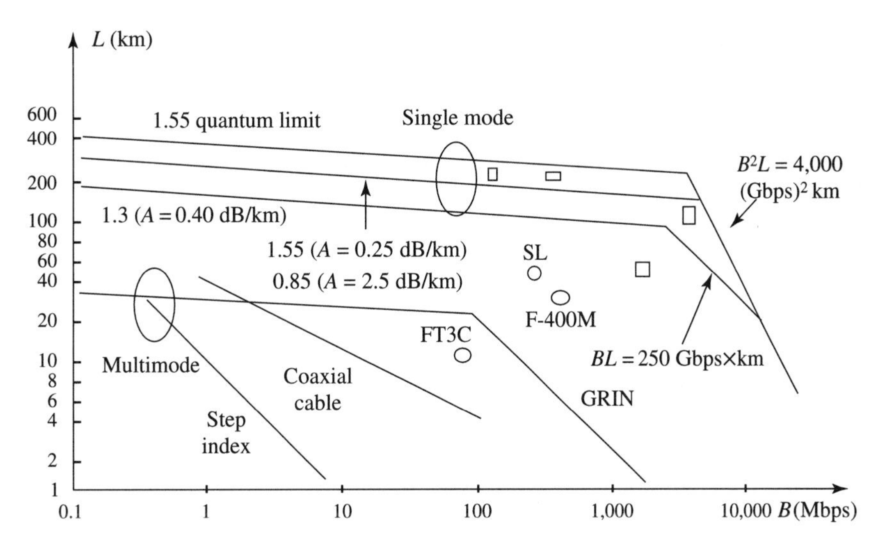

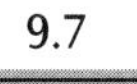
9.7 FIGURE

The maximum repeaterless distance as a function of bit rate, for three generations of optical fiber.

dispersion limit. As a result, these fibers can be used for more than 100 km at a bit rate of 10 Gbps.

Figure 9.7 summarizes our discussion of optical fibers. It shows the maximum repeaterless distance for fibers as a function of the bit rate. The maximum distance is shown for three different types of fiber: step index, graded index, and single mode. The limits on the distance are determined by the attenuation limit and by the dispersion limit. (In practice, for single-mode fibers the attentuation limit is more important; for multimode fibers the dispersion limit is more important.) Also shown is the theoretical maximum distance assuming a typical transmitted power and the minimum theoretical receiver sensitivity for on-off keying. This minimum sensitivity is called the quantum limit for OOK, and it is explained in section 9.5. In addition, the figure shows some values (denoted by small circles or boxes) of the length achieved by experimental or commercial systems.

9.3 SOURCES AND DETECTORS

An optical link consists of a light source connected to a fiber that terminates at a detector. In the preceding section we studied the limitations on the fiber medium due to attenuation and dispersion.

The usefulness of a light source for communication depends on its power; coherence (the smaller the spectral width of the source the more coherent it is said to be); the ease with which the optical signal can be modulated; and the modulation bandwidth, reliability, and cost. Scientists and engineers have developed two types of semiconductor sources: light-emitting diodes or LEDs and laser diodes or LDs.

In a semiconductor an electron occupies an energy level in one of two bands. The band at the lower energy level is called the *valence* band. The band at the higher level is the *conduction* band. When an electron is bonded to its parent atom, it is in the valence band. When an electron has broken free from its parent atom, it is in the conduction band. It is said to leave behind a *hole*. A free electron and hole pair is called a *carrier.*

At room temperature, a small fraction of the electrons are in the conduction band. By introducing certain "impurities" (atoms of different structure) into the semiconductor, one can increase the number of free electrons (negatively charged carriers) or holes (positively charged carriers). The result is an *n-type* or *p-type* semiconductor. An *intrinsic* semiconductor is one with no impurities.

The difference in energy between the conduction and valence bands is the *bandgap energy*, W_g, expressed in electron-volts and illustrated in the top right of Figure 9.8. W_g is a characteristic of the semiconductor material. LEDs, LDs, and photodetectors are usually made from a semiconductor alloy of InP (indium phosphide), InAs (indium arsenide), GaAs (gallium arsenide), and GaP (gallium phosphide). By varying the proportions of these components one can vary the bandgap energy. The table in Figure 9.8 shows that for these alloys, W_g ranges over 0.73 to 1.55 electron-volts.

9.3.1 Light-Emitting Diodes

Free electrons in the conduction band live there for a random amount of time and then combine with holes in the valence band. (In the figure, this is represented by h^+, e^-.) The excess energy is sometimes emitted as a photon of light. This process is called *spontaneous radiative emission*. (The excess energy may also be dissipated in other ways.) The wavelength of the emitted photon is inversely proportional to the energy,

$$\lambda = \frac{hc}{W_g} = \frac{1.24}{W_g(eV)} \mu\text{m},$$

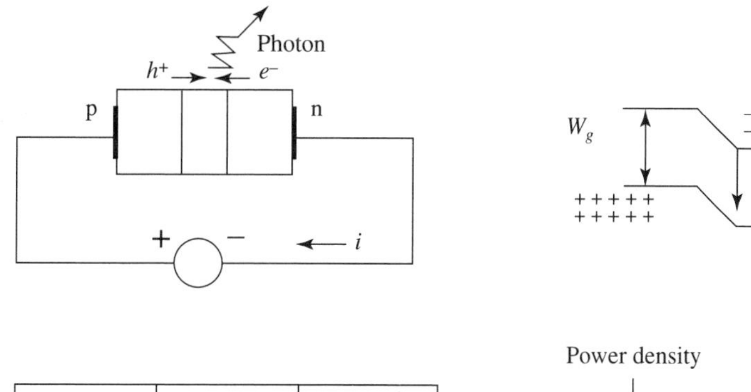

Material	λ(μm)	W_g(eV)
GaAs	0.9	1.4
AlGaAs	0.8–0.9	1.4–1.55
InGaAs	1.0–1.3	0.95–1.24
InGaAsP	0.9–1.7	0.73–1.35

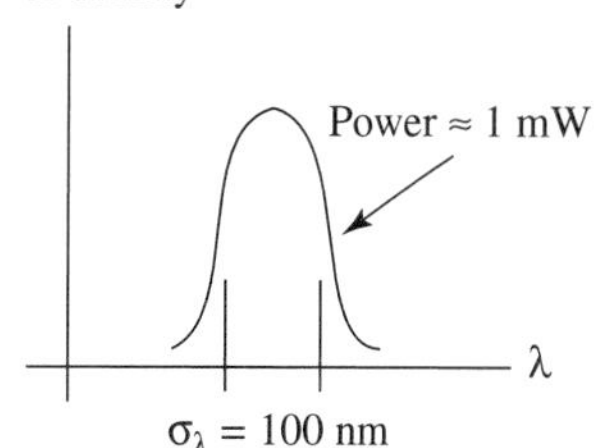

9.8 FIGURE

In a semiconductor, photons are emitted by spontaneous combination of free electron-hole pairs. The wavelength of the light depends on the semiconductor material. The light emitted by an LED has a large spectral width.

where h is Planck's constant, and c is the speed of light. As the table shows, the wavelengths for the alloys used in LEDs and LDs cover the range 0.8–1.7 μm suitable for transmission over optical fibers (see Figure 9.2).

The spontaneous emission of photons will increase with the number of free electron-hole pairs. This is accomplished by applying a forward bias current i to a p–n junction semiconductor as shown in the top left of Figure 9.8. The current injects a stream of electrons which travel from the n-type region (on the right) across the depletion region into the p-type region (on the left), where they combine with the holes that are there due to the impurities. (By convention, the arrow designating the current flow is drawn in the direction opposite to the electron flow.) Many electron-hole combinations will produce photons. Some of those photons will be absorbed in a reverse process that generates electron-hole pairs. As the current i is increased, more photons are generated than are absorbed. The intensity or power P of the light generated by this spontaneous emission increases in proportion to the injected current,

$$P = \eta i W_g,$$

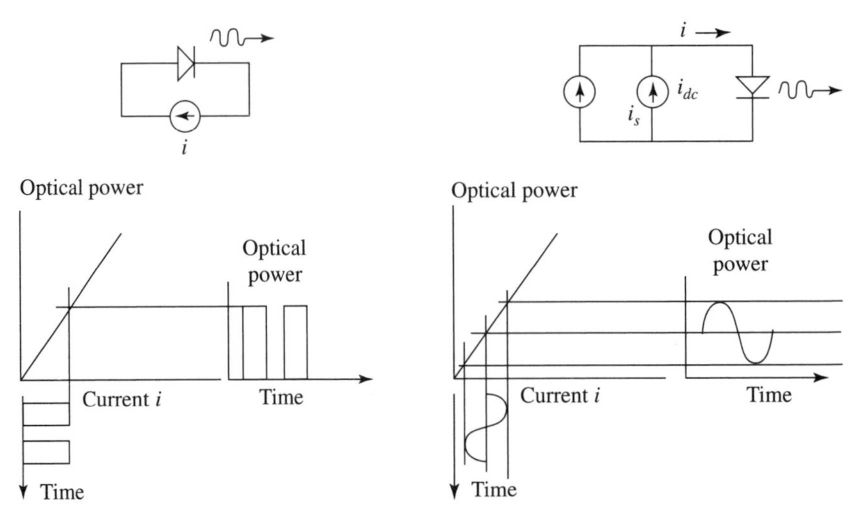

9.9 FIGURE LEDs are modulated by varying the injected current. The modulation may be digital or analog.

where η is the efficiency of conversion of injected electrons into photons. 1 mW is the typical output power of an LED and of an LD before threshold.

The spectrum of the light source also limits the maximum distance before regeneration, since, as we saw in the previous section, the narrower the spectrum, the smaller the dispersion. The light generated by the process of spontaneous radiative emission is incoherent, i.e., the photons have random phase, frequency, and polarization. The resulting power spectrum has a large width, on the order of 100 nm, indicated in the bottom right of Figure 9.8.

A transmitter consists of a light source that is modulated by the input data. LEDs and LDs both can be modulated directly by varying the injected current as suggested in Figure 9.9. This is a distinct advantage since additional circuitry for modulation is not needed. In direct digital modulation (i.e., OOK), the LED injection current is turned on or off depending on whether the input signal—the input data bit—is 1 or 0, as in the left half of the figure. In direct analog modulation, as in the right half of the figure, the LED current consists of a bias current i_{dc} plus a term proportional to the input signal. (The bias current i_{dc} does not carry any information, and so a fraction of the LED power is wasted.)

There is a limit on the bandwidth of the modulating signal, determined by how rapidly the intensity of the light emitted by the LED can be changed.

That depends on how quickly the electron and hole combine—the spontaneous recombination lifetime, τ. To understand this relationship we consider a sinusoidal input signal $I_s \sin \omega t$ of frequency ω radians per second. The output signal is also sinusoidal, $I_{ac} \sin \omega t$. The ratio of the two magnitudes or the gain is

$$\frac{I_{ac}(\omega)}{I_s(\omega)} = \frac{1}{[1+\omega^2\tau^2]^{1/2}}.$$

When $\omega = 1/\tau$, the gain is $1/\sqrt{2}$ or 3 dB. It is customary to call this the 3 dB bandwidth. (For larger values of ω, the gain is even smaller.) Since $1/\tau \approx 10^8$, an LED can be modulated by an analog signal of bandwidth 100 MHz or by direct digital modulation at a bit rate of about 200 Mbps.

9.3.2 Laser Diodes

The chance that an atom emits a photon is enhanced if another photon of the right state is present. Because the emitted photon is of the same frequency, phase, propagation direction, and polarization as the first, the intensity of light (the optical power) is enhanced. This *stimulated* emission is the basis of light generation in LDs. *Laser* is the acronym for light amplification by stimulated emission of radiation.

An LD is a p–n junction semiconductor placed between a pair of mirrors. As the forward bias current applied to the LD is increased, incoherent light is generated by spontaneous emission, just as in the case of LEDs. The power output increases in proportion to the input current. Because of the mirrors, photons get reflected and stay in the semiconductor for a longer time, increasing the chance of stimulated emission. (The mirrors are semitransparent, so the photons leave after a few reflections.) As the current is increased, it eventually crosses a threshold, and stimulated emission dominates. Because the photons are now coherent, the output power rises rapidly with input current as shown in the upper part of Figure 9.10.

The mirrors provide feedback. Depending on the distance between the mirrors, this leads to sustained oscillations at certain wavelengths, while light at different wavelengths is attenuated. Thus the output spectrum changes from its large width characteristic of spontaneous emission to a very narrow spectrum about 3 nm wide, and consisting of a few lines or discrete wavelengths, after threshold is reached.

LD output power values of up to 10 mW are typical. The modulation bandwidth of an LD is about 3 GHz, limited by mode partition noise (MPN),

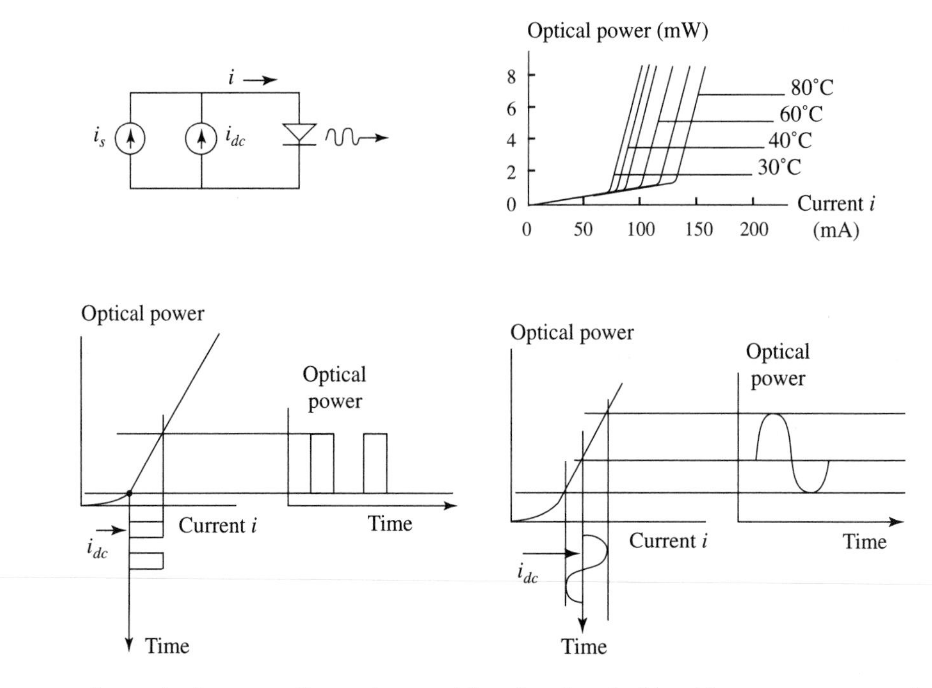

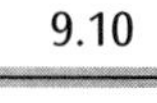

9.10 FIGURE

The optical power of LDs rises rapidly after threshold, with a narrow spectral width. Modulation may be analog or digital.

chirp, and delay. We briefly describe these phenomena. MPN occurs when the relative power of the various lines in the laser spectrum varies considerably from one pulse to the next. As a result, the pulses are dispersed as they travel down the fiber. MPN can be eliminated by "single-frequency lasers," but these suffer chirp (a shift in the laser frequency). Delay occurs as follows. Suppose the laser has initially 0 bias current and the current is rapidly increased to correspond to 1. There is a random delay between the leading edge of the current pulse and the light output caused by the random amount of time it takes for stimulated emission to build up from spontaneous emission. Delay can be reduced by biasing near to threshold.

Figure 9.10 also illustrates direct modulation of LDs. A bias current, i_{dc}, is necessary even for digital modulation so that the LD operates above threshold. Analog modulation requires a larger bias current. As shown in the figure, the power output is sensitive to temperature; thus accurate control of the power requires temperature control.

9.3.3 Detectors

We now study the receiver. Its function is to convert the received optical signal into an electrical signal and demodulate it to recover the modulating signal—the input data at the transmitter.

In order to determine whether a 1 or 0 is transmitted during a specific bit time requires several operations: photo detection, amplification, filtering, and decision. Figure 9.11 indicates the principal receiver components. The photodetector converts the received optical signal into electric photocurrent, $i_{ph}(t)$. The transimpedance preamplifier converts the photocurrent into a voltage signal at a usable level. The low-pass filter reduces the noise introduced by the preamplifier by cutting off frequencies beyond the bandwidth B of the input data signal. The decision circuitry includes an equalizer to restore the data pulse shape and a timing extractor, and it compares the processed signal with a threshold to declare whether a 1 or 0 bit is received.

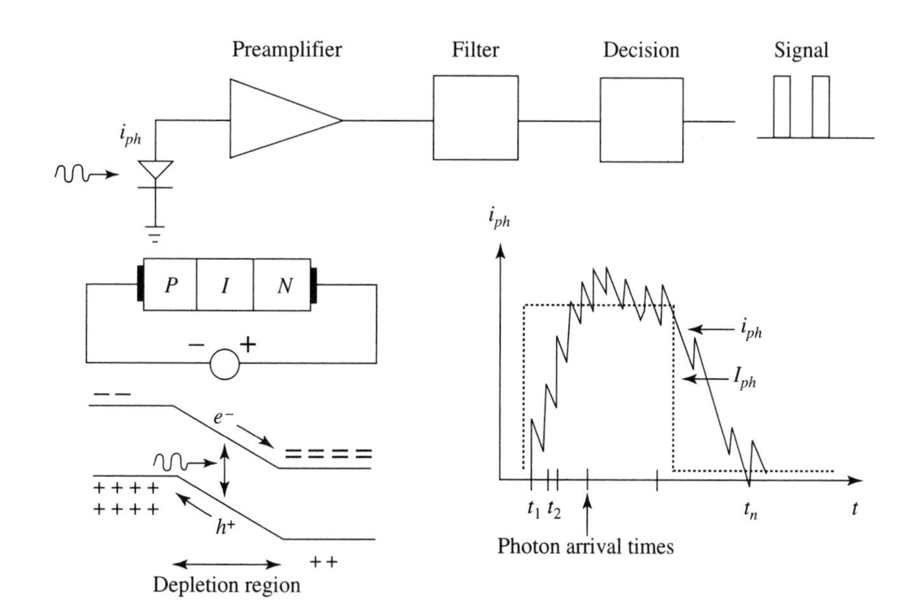

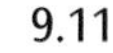

FIGURE 9.11 The receiver consists of a photodetector, preamplifier, filter, and decision circuitry. The current i_{ph} produced by the photodetector is a shot noise process. The two other noise sources are the dark current and the thermal noise in the preamplifier.

The performance of this receiver is measured by its *sensitivity*. By convention, this is the minimum received optical power P needed to achieve a BER of 10^{-9}, at a specified bit rate B. The BER is a function of the signal to noise ratio,

$$\text{SNR} = \frac{\text{Average signal energy per bit}}{\text{Receiver noise power}}. \tag{9.2}$$

In general, the larger is the SNR the smaller is the BER. As a rule of thumb, which we justify in section 9.5, in order to achieve a BER of 10^{-9}, we need SNR ≥ 6. (For analog data, of course, there is no notion of bit error rate, and receiver performance is measured by SNR defined as the ratio of average signal power to noise power.) We now consider the factors that affect the numerator and denominator of the SNR (9.2).

As we will explain below, the signal power is given by the average photocurrent, I_{ph}, which is proportional to the received power, $I_{ph} = R \times P$. Here R is the responsivity of the photodetector. Thus the average energy per bit is

$$I_{ph} \times T = \frac{I_{ph}}{B} = \frac{R \times P}{B},$$

where $T = 1/B$ is the bit time. For example, if $B = 100$ Mbps, then $T = 10 \times 10^{-9} = 10$ ns (nanoseconds). We can immediately see that SNR is proportional to P and inversely proportional to B.

We now consider the denominator in (9.2). There are three independent sources of noise: the photodetector shot noise, the photodetector dark current, and the preamplifier thermal noise. These are independent sources of noise, and so their effect is additive:

$$\langle i^2 \rangle_{total} = \langle i^2 \rangle_{shot} + \langle i^2 \rangle_{dark} + \langle i^2 \rangle_{thermal},$$

where $\langle i^2 \rangle_{total}$ is the variance of the total noise, and the terms on the right are the variances of the three individual noise components. In order to achieve a BER of 10^{-9}, we must have SNR ≥ 6,

$$\frac{R \times P}{B} \geq 6\sqrt{\langle i^2 \rangle_{total}}\,.$$

In practical receivers, the thermal noise dominates, and we may neglect the two other noise terms. However, as we will see in section 9.5, it is possible to design a receiver that "amplifies" the photocurrent so much as to swamp out the effect of both the dark current and the thermal noise, and only the shot noise remains. It is therefore of interest to study all three noise components.

To understand shot noise we need to describe the operation of the basic photodetector, the PIN photodiode. It converts the photons in the received optical signal into electrons following a process that is the reverse of the spontaneous radiative emission in LEDs. The PIN photodiode is a p–i–n junction in which the p- and n-type regions are separated by an intrinsic region in order to widen the depletion region. The photodiode is operated under reverse bias, so there are virtually no free electron-hole pairs in the depletion region. When a photon with appropriate energy (wavelength) impinges on the depletion region, it may excite an electron in the valence band into a free electron in the conduction band. In this way, a photon is absorbed and produces an electron-hole pair (e^-, h^+), shown in the bottom left of Figure 9.11. The free electron flows into the external circuit of the detector producing the photocurrent i_{ph}. The photodiode is made from the same semiconductor material as the LED or LD source, so it is sensitive to light of wavelengths produced by these sources.

The photocurrent is a *shot noise* process. It is the sum of a sequence of impulses that coincide with the random arrival times, t_k, of the photons that produce free electrons,

$$i_{ph}(t) = \sum_k h(t - t_k).$$

The average value of this photocurrent, $I_{ph}(t)$, is proportional to the modulating signal. We can therefore express the photocurrent as the sum of a term that is proportional to the modulating signal and a zero mean shot noise term.

$$\begin{aligned} i_{ph}(t) &= I_{ph}(t) + [i_{ph}(t) - I_{ph}(t)] \\ &= [\text{signal}] + [\text{shot noise}]. \end{aligned}$$

Suppose the modulating signal has two values, a high value corresponding to 1 and a low value corresponding to 0. Then $I_{ph}(t)$ has a higher value when a 1 is transmitted and a lower value when a 0 is transmitted. The actual photocurrent, $i_{ph}(t)$, fluctuates around the average value, I_{ph}, as shown in the bottom right of the figure.

We note a characteristic property of the shot noise process. Consider a bit time when a 1 is transmitted and suppose that N photons are received at the photodetector during this time. Then N is a random number with Poisson probability distribution, i.e.,

$$\text{Prob}\{N = n\} = \frac{\bar{N}^n}{n!} e^{-\bar{N}}.$$

It can be verified by direct calculation that the mean value of N and its variance are both equal to $\bar{N}$.

Finally, we note the relation between the optical power P received by the photodiode and the average photocurrent I_{ph}. The quantum efficiency of a

PIN diode is measured either as η, the ratio of the number of electrons produced to the number of incident photons or, more usefully, as the *responsivity* R, the ratio of the average photocurrent I_{ph} to the incident power P,

$$R = \frac{I_{ph}}{P} = \eta \frac{q}{hf} \text{ amp/W}.$$

Here q is the electron charge, and hf is the energy of a photon with frequency f. Typical parameter values for an AR (antireflection) coated InGaAs PIN diode are $\eta = 0.9$ and $R = 1$ amp/W for wavelengths $\lambda \sim 0.8 - 1.5\ \mu$m. Thus, for example, if $P = 1\ \mu$W (-30 dBm), then the average photocurrent is about 1 μA (microamp).

The second source of noise is the *dark current*. This is the photocurrent produced even when no external light is impinging on the photodiode. Dark current is caused by the spontaneous thermal excitation of an electron into the conductance band in the depletion region. Dark current is added to the photocurrent due to the received optical signal. It is also a shot noise process. The larger the dark current, the larger has to be the received power when a 1 is transmitted (so that the 1 and 0 can be distinguished with sufficiently low error), and so the lower is the sensitivity. Typical values of dark current range between 1 and 5 nA (nanoamps).

The third noise source is the *thermal noise* current in the preamplifier. This is a Gaussian white noise, measured by its variance $\langle i^2 \rangle_{thermal}$. In order to minimize this variance, the low-pass filter shown in the figure is designed to cut off frequencies beyond the bandwidth of the modulating signal. The variance of the thermal noise is then proportional to this modulation bandwidth. For binary signals with a bit rate of B bps, the modulation bandwidth is approximately $B/2$ Hz.

In conclusion we note that the photocurrent can be significantly increased by the use of an avalanche photodiode or APD. In an APD an incident photon triggers an "avalanche" of electrons, so that $I_{ph} = M \times R \times P$, where R is the responsivity of the ordinary PIN photodiode. The APD's "internal gain" M improves sensitivity, since the same amount of received optical power produces a larger photocurrent. However, the gain M is random, with an average value near 50. Because M enters multiplicatively, the variance in the photocurrent grows with the magnitude of the received signal. As a result the photocurrent has a larger variance when a 1 is received than when a 0 is received. Detection of 1 compared to detection of 0 in this case requires a greater average value of the photocurrent than in the case of a PIN photodiode (for the same BER), which reduces sensitivity. On balance, use of an APD improves sensitivity, but not by as much as the average value of the APD gain.

9.4 NONDIRECT MODULATION

Until now our discussion on sources and detection focused on direct modulation in which the intensity of the transmitted light is proportional to the modulating signal. Direct modulation is the simplest and the most common modulation scheme for transmission of digital data. We now discuss three nondirect modulation schemes: coherent detection, subcarrier multiplexing, and wave-division multiplexing.

9.4.1 Coherent Detection

In RF (radio frequency) modulation, *coherent detection* refers to any scheme in which the receiver has a local oscillator that is phase-locked to the transmitted carrier wave. (In optical communication this is the lightwave carrier with wavelength of 0.8 to 1.5 μm or frequency of 1.3 to 2.5 $\times 10^{14}$ Hz.) In optical communication, the term *coherent detection* is employed more broadly, whenever the receiver uses a local oscillator, whether or not the phase of the transmitted carrier is recovered.

The low power (about -80 to -50 dBm) of the received signal limits sensitivity of direct detection schemes. As we will explain, coherent detection overcomes this low power limit by a kind of amplification. (Optical amplifiers can have a similar effect.) In coherent detection, the received lightwave, $E_S \cos(\omega_S t + \phi_S)$, is combined with a much stronger local oscillator (LO) lightwave, $E_L \cos(\omega_L t + \phi_L)$. The combined signal is

$$E = E_S \cos(\omega_S t + \phi_S) + E_L \cos(\omega_L t + \phi_L),$$

and its power is

$$E^2 = E_S^2 + E_L^2 + 2E_S E_L \cos[(\omega_S - \omega_L)t + (\phi_S - \phi_L)],$$

after components at frequencies larger than ω_S or ω_L are filtered out. The combined signal illuminates a photodetector. The resulting photocurrent, I_{ph}, is proportional to this signal power. Writing $P_S = E_S^2$ and $P_L = E_L^2$ as the received signal and LO signal power, respectively, we get

$$I_{ph} = R\{P_S + P_L + 2\sqrt{P_S P_L} \cos[(\omega_S - \omega_L)t + (\phi_S - \phi_L)]\},$$

where R is the photodiode responsivity.

The term P_L is a constant and can be filtered out. The term P_S is the received power in the case of direct detection—as mentioned above its magnitude is on the order of -50 dBm. The magnitude of the third term $\sqrt{P_S P_L}$

is much larger than P_S (since P_L, the local oscillator power, is on the order of 1 mW or 0 dBm). This provides amplification. (For $P_S = -30$ dBm and $P_L = 0$ dBm, this is a power gain of 15 dB.) The magnitude $\sqrt{P_S P_L}$ can be made so large as to swamp out the dark current and preamplifier thermal noise terms, leaving only the shot noise.

There are two variants of coherent detection. The case $\omega_L = \omega_S$ (and $\phi_L = \phi_S$) is called *homodyne detection*. (This is illustrated later in Figure 9.14.) The case $\omega_L \neq \omega_S$ is *heterodyne detection*. Heterodyne detection can be combined with FSK (frequency-shift key) modulation (a 1 corresponds to modulating the intensity of the lightwave carrier by a sine wave of frequency f_1, a 0 corresponds to modulating it by a sine wave of frequency f_0) or with OOK modulation (a 1 corresponds to turning on the lightwave carrier, and 0 corresponds to turning it off). Homodyne detection can, in addition, be combined with PSK (phase-shift key) modulation (the 1 and 0 correspond to shifting the phase of the carrier by different amounts).

The power gain of coherent detection is achieved at the cost of increased receiver complexity. This complexity is easily understood in the case of homodyne detection, which requires a local oscillator locked to the phase of the received carrier wave. To build the phase-locked loop requires a coherent laser at the transmitter and a tunable, coherent laser at the receiver. From the discussion in section 9.5, we can conclude that homodyne detection can lead to an improvement in receiver sensitivity of about 15 dBm, which may be insufficient to justify the additional cost of more complex equipment.

9.4.2 Subcarrier Multiplexing

Subcarrier multiplexing or SCM is illustrated in Figure 9.12. It is a very practical scheme, similar to current radio or TV broadcast, the chief difference being that the transmission medium is optical fiber instead of free space.

In the system envisaged in the figure, N analog or digital baseband signals modulate different local microwave oscillators at different RF subcarrier frequencies, $f_1, \ldots, f_N$. The electrical signal obtained by adding the modulated subcarriers now modulates a single laser. (The word *subcarrier* is used to distinguish the local oscillators from the lightwave "carrier.") At the receiver, direct detection is followed by "downconverting" to IF (intermediate frequency) or to baseband.

This scheme can be used to combine the separate distribution systems of cable TV, telephone, and data networks into a single fiber. Suppose the total bandwidth available to modulate the lightwave carrier is 500 MHz. A TV signal, after digitizing and compression, may require a bit rate of 2 Mbps.

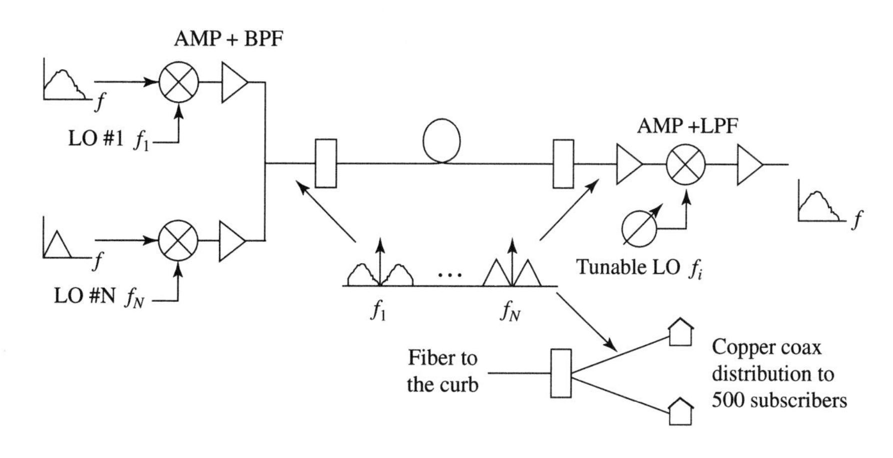

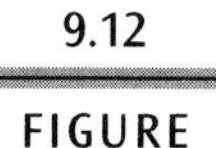
9.12 FIGURE

In subcarrier multiplexing, several signals modulate different RF subcarriers. The sum of those subcarriers modulates one laser. At the receiver, a signal is recovered by mixing with the appropriate subcarrier. Fiber to the curb uses subcarrier multiplexing.

Thus 250 TV channels would require 500 Mbps or occupy a bandwidth of 250 MHz. This would leave about 250 MHz for accommodating voice, data, and other services. At least one regional phone company is upgrading its residential copper distribution plant by fiber with subcarrier multiplexing. The phone company plan, called *fiber to the curb,* is illustrated in the bottom right of the figure. The optical fiber is brought to a residential neighborhood (curb), demodulated, and the electrical signal is amplified and distributed via short copper coaxial cable to 500 subscribers. Each subscriber must add equipment (TV set-top box) that will demodulate the subcarriers. Other subscriber receiving equipment, e.g., TV sets and phones, remains unchanged. (See the discussion on video dial tone in section 4.6.)

9.4.3 Wave-Division Multiplexing

The modulation schemes considered so far all lead to a modulation of a single lightwave carrier. Since the modulation bandwidth of a laser diode is around 3 GHz, these schemes use only a tiny fraction of the 100-nm low-loss windows (corresponding to a 10,000-GHz bandwidth) centered at 1.33 μm and

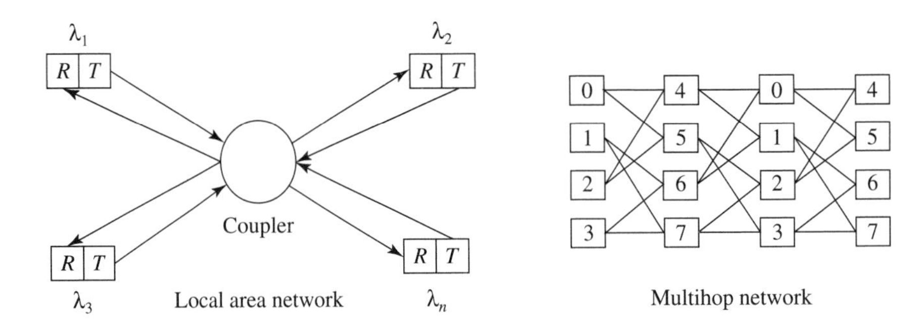

FIGURE 9.13 In wave-division multiplexing, several channels centered at different wavelengths are independently modulated. The modulated signals are combined and distributed to every receiver where a filter selects the desired channel.

1.5 μm. Wave-division multiplexing or WDM makes full use of this bandwidth. WDM divides the window into n channels centered at different wavelengths or light "colors," $\lambda_1, \ldots, \lambda_n$. The n channels are modulated independently of each other. The n modulated lightwaves are combined together and distributed among the various receivers. At each receiver, a filter selects the desired channel, the lightwave signal is demodulated, and the desired modulating signal is recovered.

The arrangement on the left in Figure 9.13 depicts two variants of an n-node local area WDM network. In transmitter-based addressing, the ith channel is permanently assigned to the transmitter at node i, which modulates a laser with fixed wavelength λ_i. The receiver at each node has a tunable filter. When a node wishes to "listen" to channel i, the receiver filter is tuned to that channel. In the second variant, receiver-based addressing, the ith channel is permanently assigned to the receiver at node i, and its fixed filter admits only that wavelength. Each transmitter modulates a tunable laser. When a node wishes to send a signal to node i, its laser is tuned to wavelength λ_i, the lightwave is modulated, and the receiver at i recovers the modulating signal. (Since tunable lasers are not yet practical, this second variant is not available.)

WDM offers several advantages. First, it makes use of the enormous bandwidth of the fiber. Second, the n channels are totally independent and may be used for entirely different purposes. Third, the scheme is ideally suited for broadcast applications. Lastly, the coupler and filter being passive elements (no conversion between electrical and optical signals is needed), they are more reliable and need less maintenance than electronic equipment.

WDM also faces several problems. First, power budget requirements limit the number of receivers. If there are n receivers, each of them receives only $1/n$th of the total power. Receiver sensitivity limits how large n can be. (Two receivers will lead to a loss of -3 dB, 2^k receivers leads to a loss of $-3k$ dB. If receiver sensitivity is -30 dBm, and the transmitted power is 0 dBm, then $k = 10$, 2^k =1,000.) This limitation can be overcome by photoamplifiers. The second technical constraint is the lack of availability of tunable filters that can be rapidly switched. The third limitation is the number of separate channels, and hence the number of independent transmitting nodes, that can be created. The number of separate channels is limited by the range of wavelengths over which the filter can be tuned and the bandwidth of the filter. Experimental results at IBM's Rainbow network show a range of 40 nm (5,000 GHz) and a 3-dB bandwidth of 0.4 nm (50 GHz), suggesting 100 channels. Rainbow has 32 channels, hence 32 transmitters.

In the local area network that we just described, every node is only one hop away from every other node. This topology requires receiver or transmitter tunability. An alternative that does not require tunability can be obtained with a multihop network shown on the right of Figure 9.13. The network has eight nodes labeled $0, \ldots, 7$ and 16 optical links. Links carry data from left to right. The link going from node i to node j is assigned a fixed wavelength λ_{ij}. (It is not necessary for the wavelengths to be distinct.) Each node has two incoming links with receivers at the link wavelengths and two outgoing links with transmitters at those wavelengths. There is a unique path from every node i to every node j with one, two, or three nodes. (In general, in these regular network structures, which we also encounter in Chapter 10, if there are 2^n nodes, each path has at most n hops.)

Suppose a packet of data is to be sent from node 0 to node 2. Just as in a packet-switched network, node 0 forwards that packet to node 5 modulating a laser with wavelength λ_{05}. At node 5, the modulated signal is received, demodulated, and converted to an electrical signal. The data packet is then forwarded to node 2, modulating a laser at wavelength λ_{52}. The optical signal is received and demodulated, and the packet is recovered by node 2.

9.5 QUANTUM LIMITS ON DETECTION

We saw in section 9.3 that receiver sensitivity is limited by shot noise, dark current, and amplifier thermal noise. In practice the thermal noise is the

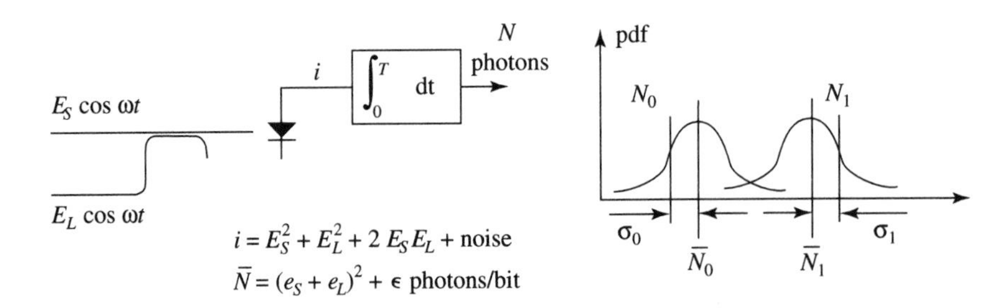

FIGURE 9.14 To calculate the quantum limits of OOK, homodyne detection is assumed, and the only sources of noise are the shot noise and dark current.

dominant noise. In principle, thermal noise can be made arbitrarily small, whereas the shot noise and dark current are fundamental quantum mechanical limitations. In this section, we calculate the limits on receiver sensitivity placed by these quantum effects. We assume homodyne detection with OOK modulation. Figure 9.14 depicts the important elements of the detector setup. When a 1 is transmitted the received signal is $E_S \cos \omega t$ for a duration equal to the bit time T. (ω is the lightwave carrier frequency.) When a 0 is transmitted, the received signal is identically zero. The received signal is combined with the local oscillator signal, $E_L \cos \omega t$, so the photocurrent i is given by

$$i = E_S^2 + E_L^2 + 2E_S E_L + \text{noise}.$$

The photocurrent is integrated and sampled at the end of the bit duration. Since i is a shot noise process, this sample is a random variable, N, which we measure in photons. The distribution of N is Poisson. So N has the same mean and variance, $\bar{N}$,

$$\bar{N} = (e_S + e_L)^2 + \epsilon \text{ photons/bit}$$

where e_S and e_L are proportional to E_S and E_L and ϵ is proportional to the magnitude of the dark current. For the range of values of $\bar{N}$ of interest, the Poisson distribution of N is well approximated by a Gaussian distribution. We will use this approximation.

First consider the simple case where there is no dark current ($\epsilon = 0$), so the only source of error arises from the fact that the photocurrent is a shot noise process. The distribution of N is as follows:

$$\text{Under 1,} \quad N = N_1 \sim \text{Gauss}\,\{(e_S + e_L)^2, (e_S + e_L)^2\};$$

$$\text{Under 0,} \quad N = N_0 \sim \text{Gauss}\,\{e_L^2, e_L^2\}.$$

Since e_L is much larger than e_S, the variance in both cases is the same:

$$\sigma_1^2 = \sigma_0^2 = e_L^2.$$

Figure 9.14 displays the probability distributions of N under the two cases. The detector will compare the sample value N with a threshold and declare that 1 or 0 is transmitted accordingly as

$$N > \frac{1}{2}(\bar{N}_1 + \bar{N}_2) \text{ or } N < \frac{1}{2}(\bar{N}_1 + \bar{N}_2).$$

This decision will yield a probability of error of at most 10^{-9} if the means under the two cases are separated by at least 12 standard deviations, i.e., if

$$(e_S + e_L)^2 - e_L^2 > 12 e_L,$$

which, since e_L is much larger than e_S, is equivalent to

$$e_S > 6 \text{ or } e_S^2 > 36,$$

so the received energy per bit when a 1 is transmitted should be 36 photons. If 1 and 0 are transmitted each with probability 0.5, then the average received energy per bit, for a BER of 10^{-9} is 18 photons/bit.

Consider now the slightly more complicated case where there is dark current, so $\epsilon = N_d$, where N_d is also Poisson distributed with mean and variance equal to $\bar{N}_d$. Using the Gaussian approximation again gives the probability distributions of N:

$$\text{Under 1,} \quad N = N_1 \sim \text{Gauss}\,\{(e_S + e_L)^2 + \bar{N}_d, (e_S + e_L)^2 + \bar{N}_d\};$$

$$\text{Under 0,} \quad N = N_0 \sim \text{Gauss}\,\{e_L^2 + \bar{N}_d, e_L^2 + \bar{N}_d\}.$$

Since $e_L \gg e_S$, the variance in both cases is the same:

$$\sigma_1^2 = \sigma_0^2 = e_L^2 + \bar{N}_d.$$

Again, a BER of 10^{-9} requires that the mean value of N under a 1 and 0 must be separated by 12 standard deviations, i.e.,

$$(e_S + e_L)^2 - e_L^2 > 12\sqrt{e_L^2 + \bar{N}_d},$$

which again leads to a sensitivity of 18 photons/bit, provided $e_L \gg \bar{N}_d$. Note that this last inequality is possible only in coherent detection.

As a final remark, we note that OOK modulation does not make use of phase information. If one uses phase-shift key (PSK) modulation combined with homodyne detection, the quantum limit sensitivity is reduced by a factor of two to 9 photons/bit for a gain of 3 dB.

While the quantum limit for OOK is 18 photons/bit, commercial receivers have sensitivities in the range 300–1,000 photons/bit.

9.6 OTHER COMPONENTS

The fiber, source (LD or LED), and detector are the key components of an optical link. An optical transmission system also requires interconnecting devices. We briefly describe the most important of these. They are all passive devices. Lastly we describe optical amplifiers.

A *splice* is required to join two optical fibers (cables) when the required span exceeds the available fiber length. Because the core of the fiber is very small, it is difficult to make a perfect joint. If the two fibers are misaligned, there will be a power loss whenever a splice is inserted. For multimode fibers this loss is about 0.1 to 0.2 dB. For single-mode fiber it is about 0.2 dB.

A *coupler* is a multiport device that can connect several incoming optical sources to several output ports. There are several types of couplers. A coupler is needed to connect a source to the fiber at the transmitter and the fiber to the detector at the receiver. A star coupler has several input and output ports, and each input signal is transmitted in equal strength on all output ports. A WDM multiplexer has several input ports and one output port. A WDM demultiplexer (splitter) has one input port and several output ports. Couplers introduce power loss (output power is less than input power) and directivity loss (input power is directed to unintended ports).

An *attenuator* is inserted to control the insertion loss. An *isolator* is used to prevent reflections along a transmission path.

In the system design of an optical transmission link, the power budget calculation accounts for all of the power from transmitter LED or LD all the way to the detector diode and the decision circuitry. The budget calculation tells us whether the optical link can operate at the required bit rate with the required BER.

We can understand the main steps in a simple power budget calculation for an optical link. Suppose P_T dBm is the power launched by the source. The source is connected to the transmission fiber by a coupler. As the signal travels down the fiber, it suffers attenuation and dispersion. At the receiver, it encounters another connector. The signal power when it reaches the detector is

$$P_R = P_T - L_{trans} - 2 \times C,$$

where C is the insertion loss of each coupler. Taking P_T as 1 mW or 0 dBm, transmission loss to be due only to attenuation, $L_{trans} = A \times l$ where $A = 1.0$ dB/km, and $l = 20$ km, and $L_C = 3$ dB, gives $P_R = -26$ dBm. Suppose the bit rate is 1 Gbps and receiver sensitivity is 1,000 photons/bit for a BER of 10^{-9}, which translates to a power requirement of $P_R = -36$ dBm. Thus this link has a power margin of 10 dBm. Some of this margin will be used up by losses due to splices.

An optical link consists of a transmitter, a fiber, and a receiver. The transmitter comprises a source, such as a laser, whose output is modulated by varying the injection current according to the input data signal. The maximum length of the fiber is limited by the receiver sensitivity. In order to transmit over a distance longer than this maximum length, the received optical signal is converted back to an electrical signal and demodulated, and the input data signal is recovered and used to modulate another laser. This process is called *regeneration*. Regeneration serves as a power amplifier, boosting the optical signal power from a level equal to the receiver sensitivity to that of the laser transmitter power. Since receiver sensitivity is about −30 dBm and laser transmitter power is about 10 dBm, we can say that the regenerator power gain is about 40 dB. The maximum bandwidth or bit rate at which a single laser can be modulated is on the order of 1 GHz or 1 Gbps. So we may say that the regenerator serves as a power amplifier with a gain of 40 dB and a bandwidth of 1 GHz.

The 1 GHz modulation bandwidth of a laser is a small fraction of the 100 nm (10,000 GHz) bandwidth of the fiber itself. Wave-division multiplexing is the only scheme that attempts to use this large fiber bandwidth. In that scheme, several lasers operating at different wavelengths are separately modulated and combined into the same fiber. If we wish to boost the power of this signal via regeneration, the different wavelengths would need to be separated, individually regenerated, and then recombined. A much more attractive alternative is offered by doped fiber *optical amplifiers*. Commercial erbium-doped fiber amplifiers provide a gain of about 22 dB over a 35 nm bandwidth at 1.5 μm. For wavelengths other than 1.5 μm, laser diode amplifiers may be used; however, these are more difficult to build.

9.7 SUMMARY

Advances in optical communication have brought about dramatic increases in transmission speed and decreases in cost of bandwidth. Together with advances in switching (studied in the next chapter) we have the foundations of high-speed communication. The performance of a link is summarized by its bandwidth-distance product—the maximum distance that the link can carry the signal of a given bandwidth before requiring regeneration. Link performance is limited by the maximum power of the light source, the attenuation and dispersion in the fiber, and the receiver sensitivity. In this chapter we studied the mechanisms that determine link performance.

The modulation bandwidth of a single laser diode is at most 5 to 10 GHz. The bandwidth of optical fiber is on the order of 10,000 GHz. That enormous bandwidth can only be used by wave-division multiplexing. When WDM becomes economical, the speed of optical links will increase enormously.

At this time the economical multiplexing, buffering, and switching of optical signals is carried out by first conversion into electrical signals. In Chapter 10 we examine switching.

9.8 NOTES

For a complete treatment of all the topics covered in this chapter, see [G93]. A broader treatment but with an emphasis on optical networks is [A94].

9.9 PROBLEMS

1. You want to build a 1-Gbps link as long as possible without regeneration. Assume a transmitter power of 1 mW and a receiver sensitivity of −30 dBm for a BER of 10^9. The dispersion at 1.5 μm is 20 ps/km.nm, and the attenuation is 0.25 dB/km. At 1.3 μm, the dispersion is zero, but the attenuation is 0.5 dB/km. What would be the maximum length if you chose (1) a wavelength of 1.5 μm and single-mode, GRIN, or step-index multimode fiber, or (2) you choose a wavelength of 1.3 μm? Assume that the bandwidth of the transmitted signal is about 1 GHz.

2. You want to interconnect a supercomputer to a device at distance 500 m and speed of 1 Gbps. How would you do it? Would you prefer to use coaxial cable or step-index multimode fiber?

3. Power transmitted through an optical fiber attenuates in proportion to distance. Suppose power is transmitted through free space in a small beam of solid angle θ for a distance L. What will be the power incident on a detector of area A, ignoring any attenuation due to absorption of light by

the atmosphere? What considerations should go into the determination of receiver sensitivity? In outdoor optical communication the ambient sunlight will have a significant impact on receiver sensitivity. How would you account for this in your receiver model?

4. How would you carry out the sensitivity calculation including thermal noise?

5. How would you carry out the quantum limit calculation with direct detection instead of coherent detection? Compare these limits with practically achievable sensitivities.

10 CHAPTER Switching

Networks use switching to achieve connectivity among users while sharing communication links. With switches networks can establish higher-capacity communication paths between users with fewer links and at lower per user cost. In this way networks can take advantage of the economies of scale in communication links.

When we view a switch from the outside as a "black box," it appears as a device with several input and output ports that terminate incoming and outgoing links. An incoming link carries multiplexed bit streams of several users. The switch guides each of these bit streams to the appropriate output port. Networks differ in the methods used to switch and to multiplex signals. The telephone network uses circuit switching and time-division multiplexing. Data networks use packet switching and statistical multiplexing. ATM uses fast packet switching and statistical multiplexing.

In this chapter we explain the operations and the design of switches, concentrating on two types of switches: circuit switches and fast packet switches. We will see that large switches are composed of small identical modules. This modular design facilitates development, construction, and testing, and it results in expandable systems.

In section 10.1 we review the tasks performed by switches, and we identify useful measures of performance of switches. In sections 10.2 and 10.3 we discuss circuit switching. The concepts introduced in these sections are also useful in the study of packet switching. In section 10.2 we introduce the principles of time-division and space-division switches. In section 10.3 we present the class of Clos networks, and we explain two key results for these networks.

One particular switch structure due to Benes is examined in detail. Our discussion then focuses on fast packet switches. In section 10.4 we present the general operations of these switches, and we introduce four designs: distributed buffer, shared buffer, output buffer, and input buffer. These are studied in sections 10.5–10.8. Section 10.9 provides a summary.

10.1 SWITCH PERFORMANCE MEASURES

The function of a switch is to provide connectivity among users while sharing resources such as links and buffers. This function is fulfilled by replacing a fully connected network with one in which switches route bit streams along shared links. Since different streams share links, in a circuit-switched network there is the possibility that a call cannot be placed when there is insufficient free capacity to set up a new connection. In a packet-switched network, packets may have to be stored (queued) in a switch before they can be transmitted. Thus, the disadvantage of sharing network resources is either blocking of calls or queuing of packets. The advantage of sharing is of course a much lower per user cost.

We first regard a switch as a "black box." The switch now appears as a device with M input links and N output links, as in the top of Figure 10.1. (We will also say that the switch has M input ports and N output ports. A link is connected to a port.) The function of the switch is to connect input links to output links so that bits or packets that arrive on one link leave on another designated link. Several performance measures can be used to compare switches: connectivity, delay, setup time, throughput, and complexity. We introduce those measures now.

Connectivity is measured by the set of pairs of input and output links that can be simultaneously connected through the switch. The larger this set, the more versatile is the switch. As it processes incoming bit streams in order to route them to the proper output ports, the switch introduces delays, and this delay is another measure of performance. For circuit switches, another component of delay is the time needed to set up a circuit. For packet switches, the switch also introduces queuing delay. The throughput of a switch is measured in terms of the number of ports and the speed of the individual input links. Finally, measures to estimate the complexity of a switch include the number of crosspoints (see below), the buffer size, and the speed of bit streams inside

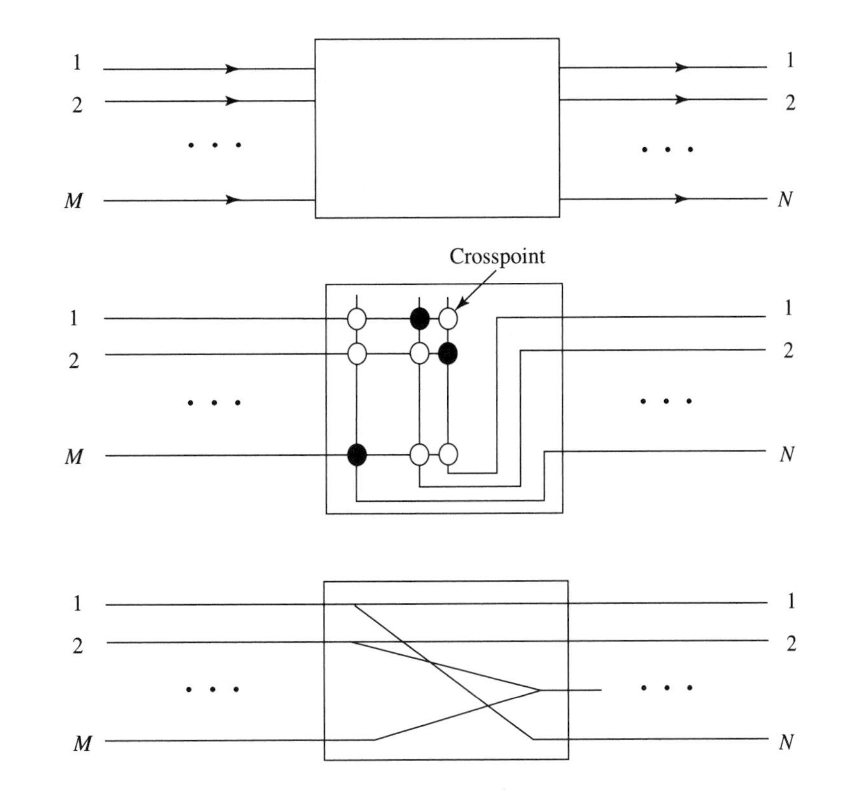

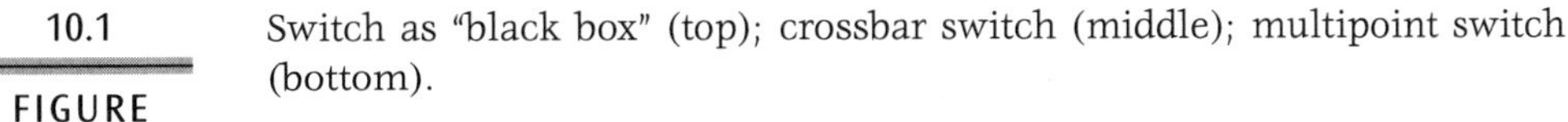

10.1 FIGURE Switch as "black box" (top); crossbar switch (middle); multipoint switch (bottom).

the switch. Switch designs that require higher speeds will need more expensive electronic components.

We now explore one measure of complexity in which a circuit switch is viewed as an organization of crosspoints. Each crosspoint is attached to two links inside the switch. A crosspoint is either closed or open. When it is closed, the crosspoint connects the two links attached to it. When it is open, the two links are not connected. The simplest organization is the *crossbar* switch in which every pair of input and output links is connected by a crosspoint, as in the middle of Figure 10.1. In the figure, the input-output pairs $(1, 2)$, $(2, 1)$, and (M, N) are connected. (A dark crosspoint means it is closed.) More complex organizations are possible where an input link is attached to an internal switch link that is in turn attached to an output link. Multistage switches can be built in this way. Any circuit switch can be viewed as such an arrangement of links connected by crosspoints, since any routing decision can be decomposed as a succession of binary decisions.

When we regard a switch as an organization of crosspoints, an obvious measure of its complexity is the number of crosspoints. For example, an M by N crossbar has $M \times N$ crosspoints. We will see that there are switches with many fewer crosspoints that can implement almost the same connections as the crossbar.

We define a *switch configuration* as the set of input-output pairs that are simultaneously connected by the switch. Thus, a switch configuration is a subset of the product $\{1, \ldots, M\} \times \{1, \ldots, N\}$, i.e., of the set of all input-output pairs. That subset specifies the input-output pairs that are connected. The switch configuration changes when a new connection is made or when an old connection is released. One measure of switch performance is the number of different configurations it can have.

If the switch has X crosspoints, then it can have at most 2^X different configurations: each crosspoint can be in any one of two states (open or closed) so that the X crosspoints admit 2^X different states. Different states of the crosspoints may or may not result in different switch configurations, but in any case different configurations must correspond to different states of the crosspoints. Thus, the logarithm (in base 2) of the number of configurations is a lower bound for the number of crosspoints required to build the switch. We state this as a proposition.

Proposition 1 If a switch is a collection of crosspoints, then

$$\text{number of crosspoints} \geq \log_2 (\text{number of different configurations}). \quad \blacksquare$$

We consider two important examples of switch complexity. The first example is the $N \times N$ point-to-point switch. By definition, a point-to-point switch can connect any input link to any output link so long as two different input links are not connected to the same output link and vice versa. Consequently, the configurations of a point-to-point switch are identified by the permutations of the numbers $\{1, \ldots, N\}$. For instance, the permutation $231 \ldots 7$ specifies that input link 1 is connected to output link 2, input link 2 to output link 3, input link 3 to output link 1, and so on. Different permutations correspond to different configurations and vice versa. Consequently, the number of configurations of an $N \times N$ point-to-point switch is equal to the number of permutations of $\{1, \ldots, N\}$, i.e., to $N!$. Proposition 1 implies that the number of crosspoints of such a switch is at least $\log_2(N!) \approx N \log_2 N$, using Stirling's approximation: $N! \approx (N/e)^N$ where $e = 2.718$. It is therefore not possible to build an $N \times N$ point-to-point switch with fewer than $N \log_2 N$ crosspoints. We will use this lower bound to evaluate some designs and also to explain the degree of blocking in switches with fewer than this number of crosspoints.

The second example is the $N \times N$ multipoint switch (see bottom of Figure 10.1). This switch can connect any input link to any set of output links. The only restriction on the configurations is that different input links must be connected to different output links. (This kind of switch has multicast capability, in which an input link is connected to several output links. Multicast is useful for videoconferencing and for video distribution.) To calculate the number of switch configurations, observe that each output link can choose arbitrarily the input link to which it is connected. Thus, this switch has N^N possible configurations. Accordingly, the switch must have at least $N \log_2 N$ crosspoints. (The $N \times N$ crossbar is a multipoint switch, too. But it has N^2 crosspoints, much larger than the lower bound of $N \log_2 N$.) In summary, an $N \times N$ point-to-point or multipoint switch requires at least $N \log_2 N$ crosspoints.

We conclude with a consideration of the other complexity measures using the model of a switch as an arrangement of crosspoints. Suppose the switch has m crosspoints and suppose it can achieve n different configurations. Whenever a particular configuration is desired, the switch controller must set the state of each crosspoint (open or closed). The time needed for this is the setup time of the switch. Whenever an existing input-output connection is terminated or a new connection is initiated, the switch configuration is changed. Clearly, satisfactory switch operation requires that the "holding time" of a configuration, the time for which the configuration is unchanged, must be much larger than the setup time. Thus, a switch with a setup time of several milliseconds can support telephone connections that last several seconds or minutes. But it would not be suitable as a packet switch with packet durations of a few microseconds.

A switch with N input lines each of which has a capacity of b bps is said to have a throughput of $N \times b$ bps. Observe that the speed of the bit stream inside the crossbar of Figure 10.1 is the same as that of the input line. Thus the internal bit rate of this switch is also b and does not increase with the number of input lines N. This is in contrast with the time-division switch studied in section 10.2, in which the internal speed equals the throughput. One may be able to reduce the internal speed by a serial-to-parallel conversion. For instance, by first converting the input stream into 16-bit words and arranging the crossbar as 16 bit-level crossbars in parallel, the speed inside the switch is reduced to $b/16$ bps. The bit stream in each output line is reconstructed by the reverse parallel-to-serial conversion. This introduces more complexity in the switch controller, and the serial-to-parallel conversion introduces some delay. But the lower switch speed may significantly reduce the switch cost.

10.2 TIME- AND SPACE-DIVISION SWITCHING

In this section we study the two important principles of time-division and space-division switching.

The telephone network uses time-division switches. Figure 10.2 illustrates the operations of such a switch. The top part of the figure shows N input signals that arrive on N different links. These signals are periodic bit streams that must go out on different output links. In the figure it is assumed that the signal arriving on link 1 must go out on link N, that arriving on link 2 must go out on link 1, . . . , and the signal arriving on link N must go out on link 3.

The time-division switch has three parts: a time-division multiplexer (MPX), a time-slot interchanger, and a time-division demultiplexer (DMX). The MPX first multiplexes the N incoming signals. That is, it divides time into slots and allocates them to the N incoming signals in a round-robin manner. The multiplexed signals arrive at the slot interchanger, which writes the

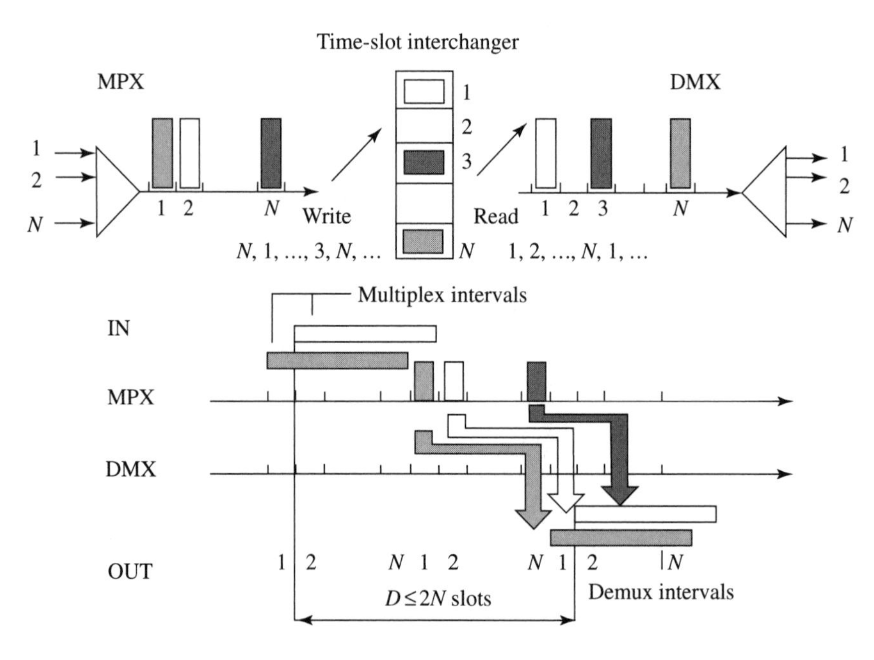

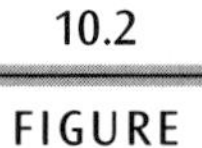

10.2 FIGURE

A time-division switch consists of a time-division multiplexer (MPX), a time-slot interchanger, and a time-division demultiplexer (DMX).

successive slots into N distinct buffers in the order $N, 1, \ldots, 3, N, 1, \ldots, 3, N, 1, \ldots$. The order is determined by the switch configuration. The output line of the time-slot interchanger then reads the N buffers in the order $1, 2, \ldots, N, 1, 2, \ldots, N, \ldots$. The output of the slot interchanger is then demultiplexed into N different signals that are sent to the output links. As is seen in the figure, this switch implements the desired connections between input and output links.

The bottom part of the figure shows the delays of the signals in a time-division switch. The timing diagram traces the evolution of signals through the three stages of the switch. From the diagram we can see that the delay from the time a signal arrives on an input link until it is placed on an output link is at most the duration of two frames of the time-division multiplexer. (A frame is a sequence of N slots.) Observe that the bit rate of the time-slot interchanger is the same as the throughput of the switch, i.e., $N \times b$ bps, where b is the bit rate of each link. Therefore the throughput of a time-division switch is limited by the maximum speed of the electronic components.

To achieve greater throughput than possible by time-division switching alone, one can combine it with space division. While a time-division switch separates signals in time, a space-division switch separates the signals in space. The simplest space-division switch is the crossbar of Figure 10.1. We may combine time-division and space-division switches as in Figure 10.3. The boxes labeled T are time-division switches. The internal speed in this combination is the same as that in the time-division switches and does not grow with the number of input ports. Such a combination also enables the switch designer to build a large switch from small modules as we study next.

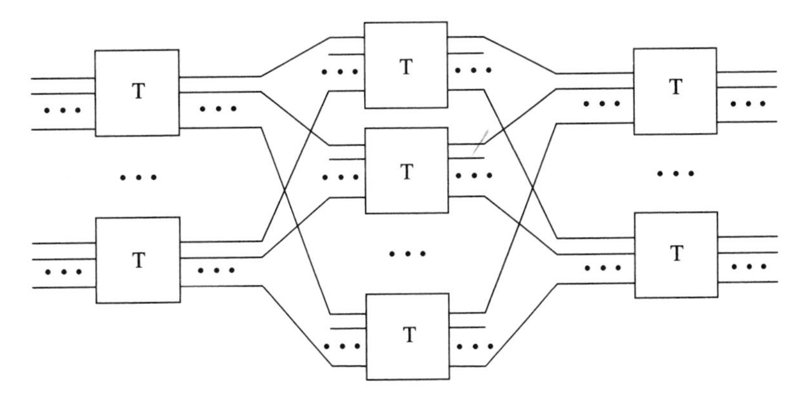

10.3 FIGURE

Space-division and time-division switching can be combined to produce large switches.

10.3 MODULAR SWITCH DESIGNS

In this section we explain how to build large switches from small modules in such a way that the switch has specified connectivity properties. We first examine one class of modular designs, called Clos networks, that will form the basic structure of large modular switches.

A Clos network is a collection of switching nodes arranged in a network as in Figure 10.4. The network is composed of three stages of switches. The first two stages are fully connected, i.e., each node in the first stage is connected to every node in the second stage. The second and third stages are similarly fully connected. (There is no direct connection between nodes in the first and third stages.) The input (leftmost stage) switches all have the same number of input lines. The output (rightmost stage) switches all have the same number of output lines. Thus, a Clos network is fully specified by five integers: $(IN,\ N_1,\ N_2,\ N_3,\ OUT)$. The network in the figure is Clos $(3, 3, 5, 4, 2)$.

We study the nonblocking properties of switches. A switch is said to be *nonblocking* if all the one-to-one connections are compatible. That is, the switch can have all the point-to-point configurations between its inputs and its outputs. A nonblocking switch can be either *strictly nonblocking* (SNB) or *rear-*

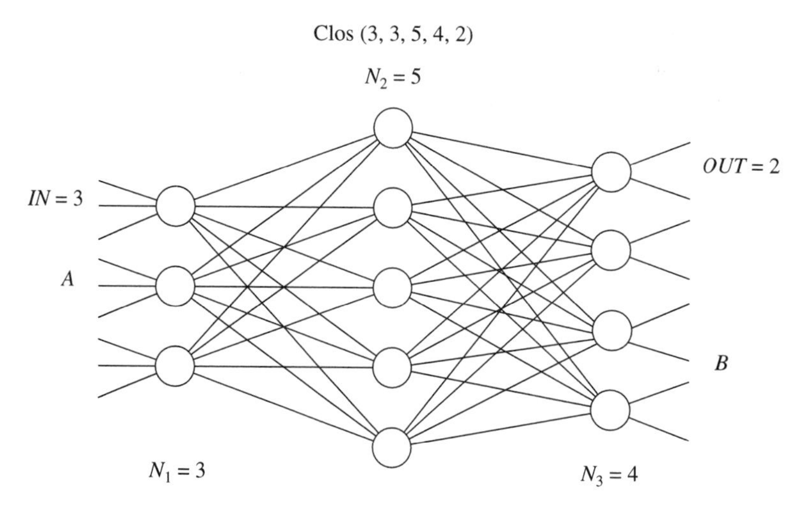

10.4 FIGURE

A Clos network is fully specified by $(IN,\ N_1,\ N_2,\ N_3, OUT)$.

rangeably nonblocking (RNB). The switch is SNB if any new connection from a free input link to a free output link can always be made without having to modify ongoing connections. Otherwise, the switch is RNB.

There is a remarkably simple characterization of SNB and RNB Clos networks, which we state as a theorem.

Theorem 10.3.1 A Clos network built from SNB modules is itself SNB if and only if

$$N_2 \geq IN + OUT - 1. \quad \blacksquare \tag{10.1}$$

The form of this result is not surprising: the switch is SNB if it has enough paths to connect input switches to output switches, i.e., if N_2 is large enough.

The proof of Theorem 10.3.1 is as follows. (See Figure 10.4.) First we suppose that inequality (10.1) is satisfied, and we show that the Clos network is SNB. To show that, assume that a new connection must be set up from input node A to output node B. We must prove that this new connection can be made without having to rearrange existing connections. When a new connection from A to B is requested, there can be at most $IN - 1$ input links busy at switch A and at most $OUT - 1$ output links busy at switch B (not including the new connection). Thus, A is connected to at most $IN - 1$ middle switches and B to at most $OUT - 1$ middle switches, for a total of at most $IN + OUT - 2$ switches. Since there are $N_2 \geq IN + OUT - 1$ middle switches, there must be one middle switch that is connected neither to A nor to B. Therefore, the new connection can be made by connecting that middle switch to A and to B.

To prove the converse, assume that the switch is SNB. We will show that inequality (10.1) must be satisfied. Consider the worst-case situation when switch A is connected to $IN - 1$ output switches excluding B and when B is connected to $OUT - 1$ input switches excluding A. In that situation, A is connected to $IN - 1$ middle switches and B is connected to $OUT - 1$ other middle switches. The switches A and B are connected to different middle switches since A and B are not connected together. Thus, $IN + OUT - 2$ middle switches are connected either to A or to B. Assume that a new connection must then be set up between the free input link of A and the free output link of B. This new connection is possible only if there is a middle switch that is not yet connected either to A or to B, which requires that $N_2 \geq IN + OUT - 1$, as had to be shown.

We now obtain the condition for a Clos network to be RNB (rearrangeably nonblocking).

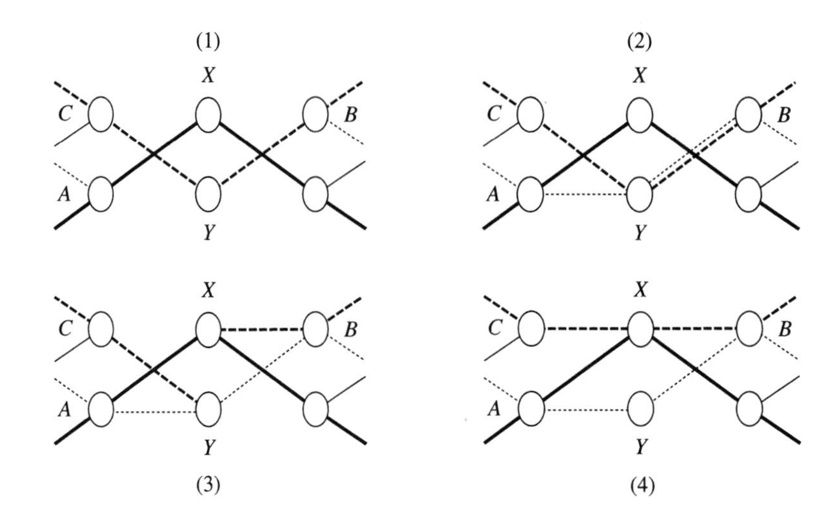

FIGURE 10.5 A new connection from A to B can be established by rerouting the connection through Y.

Theorem 10.3.2 A Clos network built from RNB modules is itself RNB if and only if

$$N_2 \geq \max\{IN, OUT\}. \tag{10.2}$$

Instead of providing a formal proof of this theorem, we illustrate, using Figure 10.5, the rearrangements that enable the switch to set up a new connection from A to B. When the connection is requested, in the top-left panel of the figure, there is one existing connection from A and one existing connection from C to B. There is no middle switch that is not already connected either to A or to B. Starting with A, we note a new connection could be made to some middle switch, Y (since $N_2 \geq IN$), as in the top-right panel. The existing connection between C and B, which goes through Y, is moved to go through X. The bottom-right panel shows the configuration after the new connection is set up.

We can use Theorem 10.3.1 to construct an $N \times N$ strictly nonblocking switch (SNB) with $N = p \times q$. The construction is shown in Figure 10.6. The input consists of q parallel planes. Each of these planes is a $p \times (2p - 1)$ SNB input switch. The input planes are attached to $2p - 1$ parallel $q \times q$ SNB intermediate switches. These $q \times q$ planes are in turn attached to q parallel $(2p - 1) \times q$ SNB output switches.

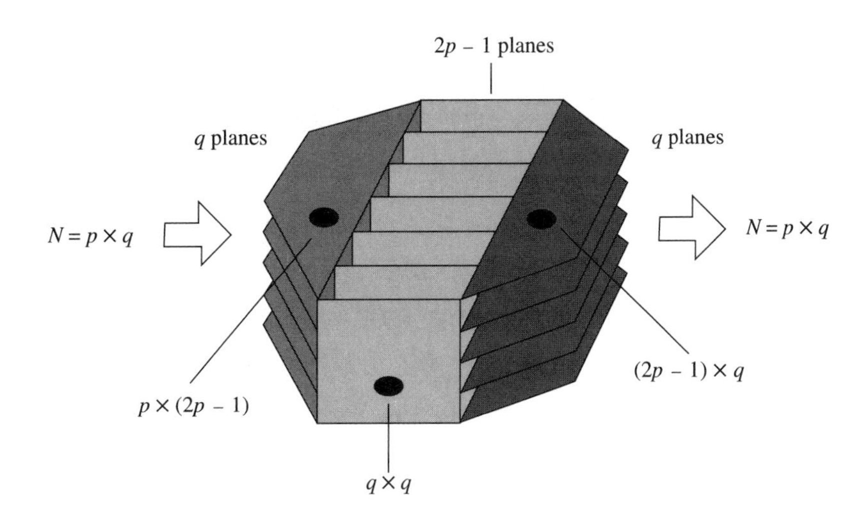

10.6 FIGURE 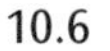

Recursive construction of a strictly nonblocking switch.

We can view each input plane as being a first stage switch in a Clos network, each middle plane as a middle stage switch, and each output plane as an output switch. Indeed, the first two stages are fully connected and so are the last two stages, so that the configuration is a Clos network. Thus, this modular switch is a $(p, q, 2p-1, q, p)$ Clos network, in the notation introduced earlier and in Figure 10.4, and by Theorem 10.3.1 this switch is strictly nonblocking.

To appreciate the advantage of this modular construction, let us assume that each of the switching modules is implemented as a crossbar. In that case, each input and output plane has $p \times (2p-1)$ crosspoints and each middle plane has $q \times q$ crosspoints. Consequently, the complete $N \times N$ switch has $2p(2p-1)q + q^2(2p-1)$ crosspoints. But if the $N \times N$ switch had been implemented as a crossbar, it would have $N^2 = p^2q^2$ crosspoints. As a numerical illustration, when $p = q = 100$, the modular design has only 4% of the crosspoints of a crossbar.

Figure 10.7 shows another modular switch design except that it is RNB instead of SNB. This design is based on Theorem 10.3.2.

Using Figure 10.8 we explain the construction of a Benes network. This is an RNB switch built in a specific recursive way. The switch is $N \times N$ where N is a power of 2. The top of the figure shows the first step of the construction, a $(2, N/2, 2, N/2, 2)$ Clos network. According to Theorem 10.3.2, the switch is RNB when its two modules are RNB. The next step is explained in the bottom of the figure: each of the two $(N/2) \times (N/2)$ modules of the Benes switch is

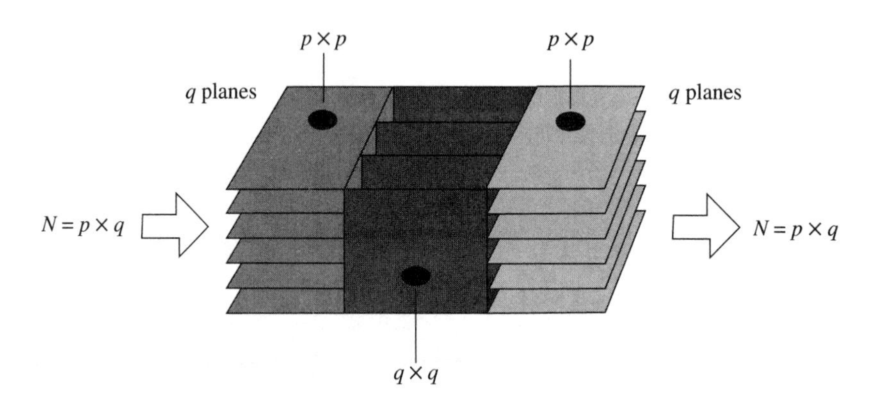

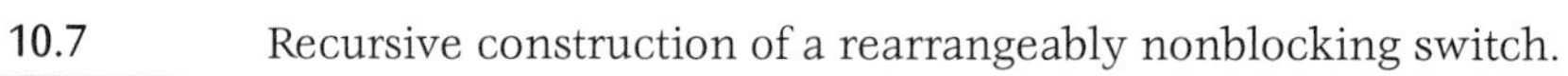

FIGURE 10.7 Recursive construction of a rearrangeably nonblocking switch.

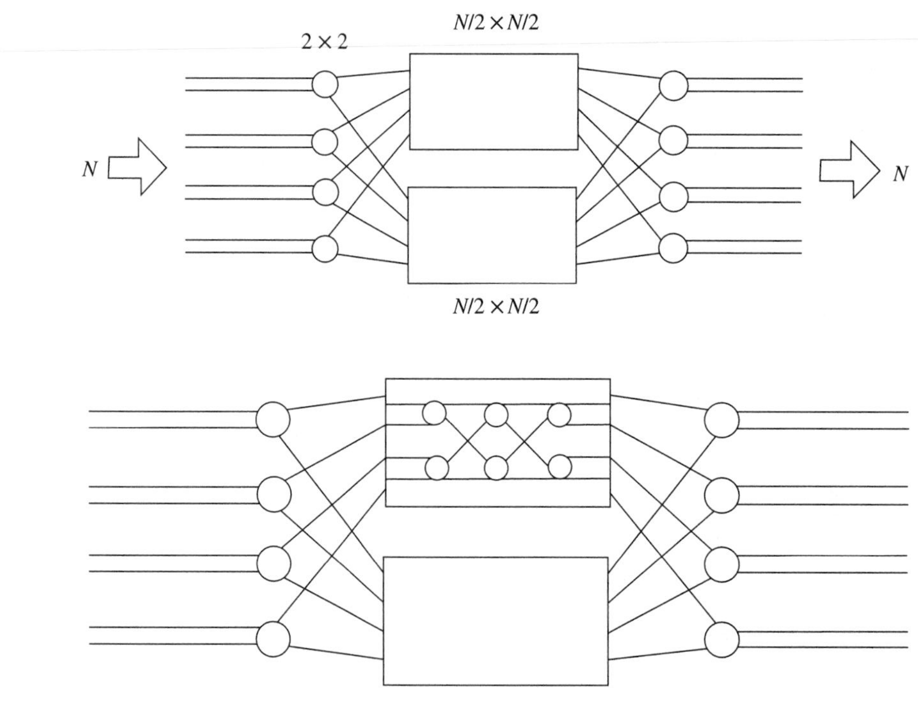

FIGURE 10.8 Recursive construction of a Benes switch.

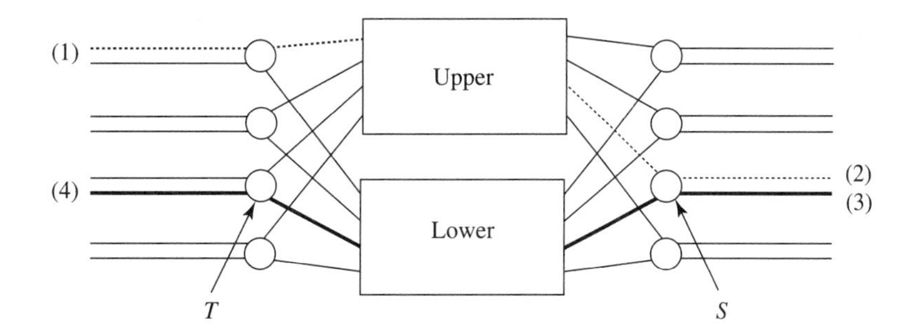

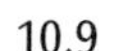

10.9 FIGURE Routing in a Benes switch can be done recursively.

again decomposed as in the first step. Continuing in this way yields a switch with $2\log_2 N - 1$ stages each of $N/2$ 2×2 switches so that the total number of crosspoints is approximately $4N \log_2 N$. So the complexity of the Benes network is four times the minimal complexity (see Proposition 1 above). Thus, the complexity of the Benes network has the optimal order.

Because the Benes switch is not SNB, the request for a new input-output connection may require rerouting existing connections. Routing in a Benes network can be performed with an algorithm that can be explained using Figure 10.9. Assume that initially no connections are made and let C be the desired set of input-output connections.

Step 1. Pick an input-output pair in C not already selected. (Terminate if none exists.)

Step 2. Connect the input (1) to the output (2) of S (say) via the upper block U.

Step 3. If the other output (3) of S desires connection to the input (4) of T (say), connect via the lower block L. Then return to Step 1.

Apply the algorithm recursively to U and L. This algorithm shows that the simplicity of the modular construction of a Benes network is matched by a simple routing algorithm.

So far we have discussed the modular construction of point-to-point switches. A similar construction can be carried out for multipoint switches. As Figure 10.10 shows, one implementation of a multipoint switch is a copy node followed by a point-to-point switch. The advantage of such an implementation is that modular designs exist for copy nodes.

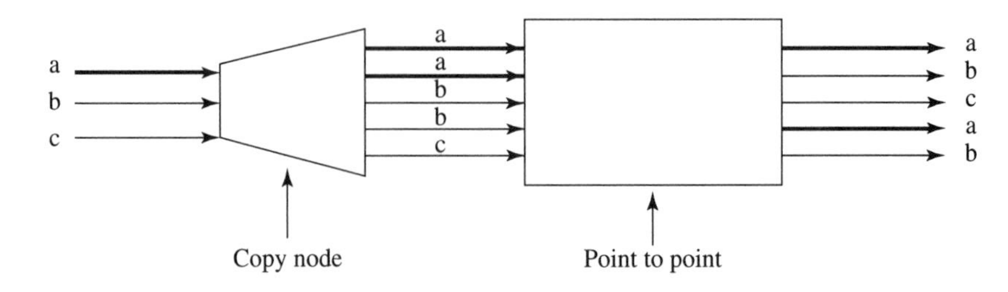

10.10 FIGURE A multipoint switch can be built by a copy node followed by a point-to-point switch.

10.4 FAST PACKET SWITCHING

The configuration of a circuit switch must be changed every time a new connection or call is to be established or an existing connection is to be torn down. The configuration of a packet switch must be changed every time a new packet arrives. Since the duration of a circuit-switched phone call lasts several minutes whereas a new packet may arrive every few μs, the frequency of changes in the switch configuration is orders of magnitude greater in the case of packet (and virtual circuit) switching than in the case of circuit switching. This difference has a major impact on switch design and performance. We first compare the functions that the two types of switches must perform.

In circuit switching, during the connection setup phase, the network assigns to the admitted call a route through the network and an idle circuit in each link along the route. Usually, a link's capacity is divided into many circuits, by time-division multiplexing. But let us suppose, for simplicity, that each link carries a single circuit. The route assigned to a call then requires each switch along the route to connect a particular idle incoming link i and a particular idle outgoing link j. Data arriving on link i must be transferred to link j for the entire duration of the connection. The switch controller must therefore maintain lists of idle incoming and outgoing links and the current configuration (i.e., a list of currently connected input-output link pairs). These lists serve two purposes: internally, to configure the switch and, externally, to notify the network's call admission and call routing procedures about the availability of idle links. These lists and the switch configurations will change whenever a new connection is set up or an existing one is torn down. The interval between changes is the call holding time, which is on the order of minutes or seconds. Note also one important additional feature of circuit switching. Once a connection between links i and j has been established, data coming in over link i is transferred out with negligible delay to link j, so no buffering is needed inside the switch and a small amount of buffering is needed at the input ports.

Packet switching differs from circuit switching in several ways. First, two packets arriving consecutively on the same incoming link may be destined for different outgoing links. In datagram networks, this can happen because successive packets are routed independently even if they have the same destination; in virtual circuits this can happen because two consecutive packets belonging to different virtual circuits and arriving on the same incoming link may leave the switch on different outgoing links. The configuration in a packet switch thus changes every time a packet arrives, so the time duration of a configuration is on the order of μs or ms. (For example, a 53-byte ATM cell takes less than 3 μs to transmit at 155 Mbps.) Second, the controller must examine each incoming packet to obtain either the destination address identifier (in the case of datagrams) or the virtual circuit identifier and then determine the outgoing links for that packet. The assignment of outgoing links to each incoming packet must be done very rapidly to maintain high throughput. Finally, unlike in circuit switching, during a short time interval it may happen that more packets are destined over the same outgoing link than can be transmitted given that link's capacity. In that event, the switch will have to buffer some packets. (A packet may be buffered also if there is contention within the switch.) This buffering function is absent in circuit switching.

Figure 10.11 illustrates the key ideas in fast packet switching schemes. The packet format must be such that the destination address or the virtual circuit identifier (called tag in the figure) is placed at the front of the packet. As the incoming packet is being read into a buffer, the tag is processed by the controller's routing algorithm, which calculates the outgoing link and, possibly, a new tag for the outgoing packet.

Suppose this is a virtual circuit-switched network. Then the routing algorithm maintains a table with entries $(VCI_{in}, Port_{in}, VCI_{out}, Port_{out})$. VCI_{in} is an entry in the incoming packet tag that gets replaced by VCI_{out} in the outgoing packet tag. VCI_{out} may be different from VCI_{in}, for reasons to be explained shortly. When a new virtual circuit is being set up, the network call admission procedure selects a route through the network, and a new entry is created in every table along the path. When the virtual circuit is torn down, the switches along the route are informed, and the corresponding entry is deleted. (The procedure for selecting a route may be similar to that used in telephone networks, or it may use a shortest-path algorithm.)

In case of a datagram network, the tag contains a destination address identifier instead of a VCI, and the routing algorithm maintains a table with

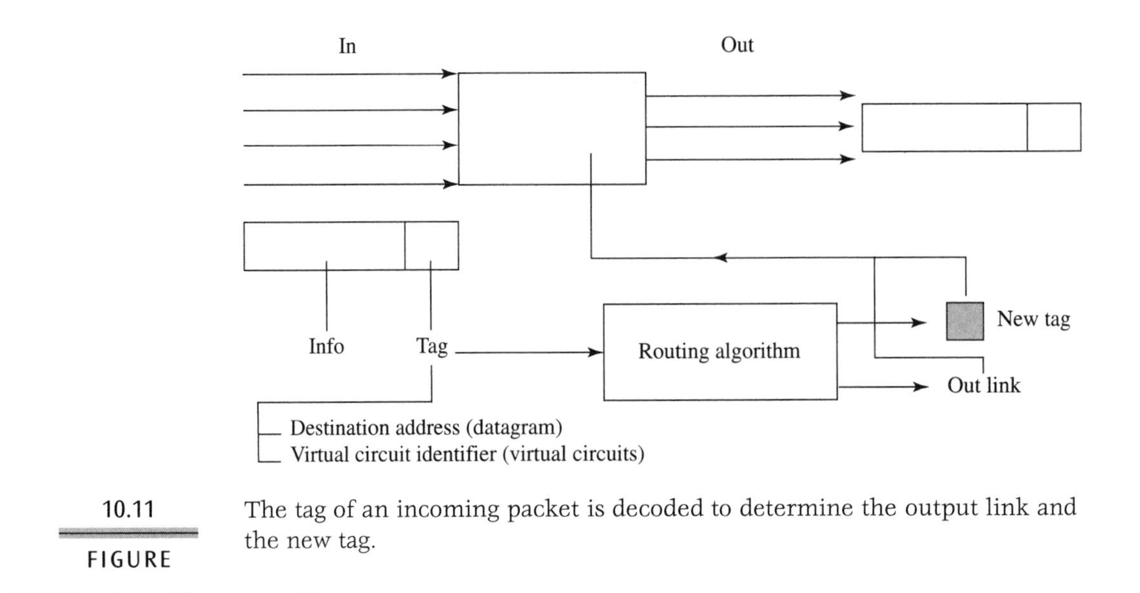

10.11 FIGURE The tag of an incoming packet is decoded to determine the output link and the new tag.

entries (destination address, $Port_{out}$). This table may be static or dynamic. In the dynamic case, the entry corresponds to the current shortest path from the switch to the destination. As the traffic pattern changes, so does the queuing delay in each link. As a result the shortest path from source to destination changes. The shortest path is periodically reestimated by the router by a separate algorithm. (A shortest-path algorithm is described in section 3.9.)

Having selected a route, Figure 10.12 illustrates how a VCI is assigned to the new connection. A new VCI is assigned by the source. A source cannot simultaneously use the same VCI for two different connections originating there. But different sources may use the same VCI. When two different sources assign the same VCI and the two virtual circuits go through the same switch, there is a potential conflict. To prevent this conflict, when the second virtual circuit is being set up, the first switch in common to the two paths will assign a new outgoing VCI to the second virtual circuit. This can be implemented by transmitting a special packet along the route of the new virtual circuit being set up. Thus, for example, in the figure, VCI #1 has already been assigned to the connection between *A* and *C*, when source *B* initiates a new virtual circuit with the same VCI. When the special connection setup packet arrives at the switch common to both routes, the conflict is noted, and the switch assigns the new (unused) VCI #2 to the new virtual circuit.

In virtual circuit switching, each packet carries a VCI, whereas in datagram switching, each packet carries a destination address. The destination ad-

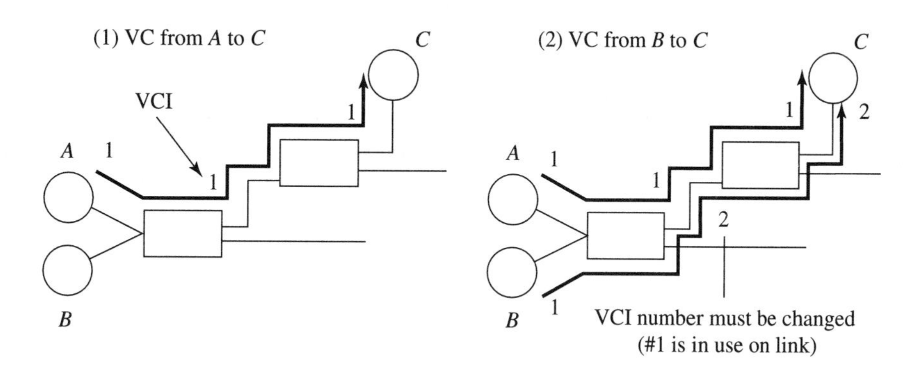

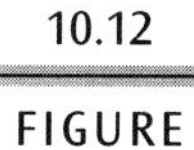
10.12 FIGURE The VCI is unique within each link. In case of conflict, a new VCI is assigned.

dress must be unique throughout the network, whereas a VCI must be unique only within each link. Therefore packets in virtual circuit networks will have a shorter header than packets in datagram networks. However, there is an extra overhead in setting up and tearing down a virtual circuit connection that is absent in datagrams.

We now introduce four fast packet switch (FPS) designs: the distributed buffer, the shared buffer, the output buffer, and the input buffer. The designs differ in the way a packet is routed through the switch and in the way it is buffered when there is contention. Accordingly, their performance, measured in throughput, delay, or buffer requirements, varies. The distributed buffer design is the only one of the four that is modular in construction. Therefore, it can be scaled to accommodate a large number of ports. The other three designs cannot be scaled without increasing the speed inside the switch; they are more suitable for local area switches.

10.5 DISTRIBUTED BUFFER

The distributed buffer switch (DBS) designs have a modular construction like, say, the Benes network. However, there is an important difference. The Benes network realizes the maximum number of possible switch configurations for a given number of crosspoints, but reconfiguring the switch is complicated. The DBS designs on the other hand can realize far fewer configurations for

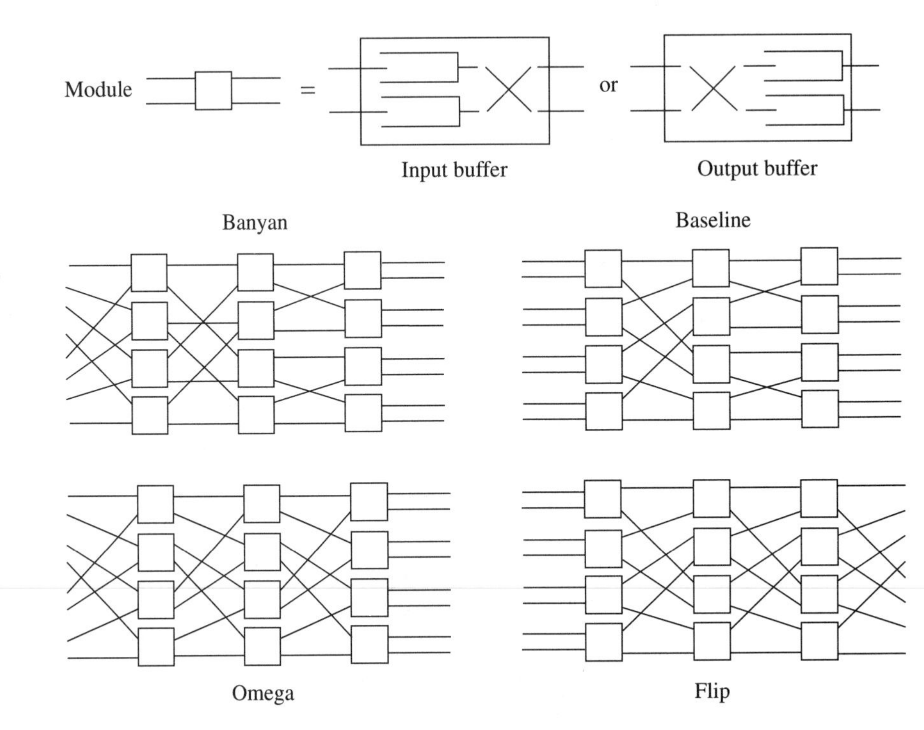

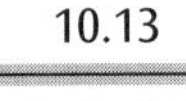

10.13 FIGURE

Four variants of distributed buffer switches. Each 2×2 module contains an input or output buffer.

the same number of crosspoints, but routing a packet through the switch—so-called tag-based or self-routing—is especially simple. The Benes network is optimized for circuit switching (where reconfiguration is relatively infrequent), the FPS designs are optimized for packet switching.

Figure 10.13 illustrates four "delta" networks. These are arrangements of 2×2 crosspoint modules. If there are $N = 2^n$ input and output ports, there are $N/2$ rows and $n = \log_2 N$ stages, for a total of $N/2 \log_2 N$ modules, hence $2N \log_2 N$ crosspoints. (Compare these arrangements with the Benes switch with twice as many stages.) Each module contains its own buffer—hence the name *distributed buffer.* (The module may have an input or output buffer, as shown, or it may have no buffer at all.) We will first discuss how all these arrangements permit rapid self-routing of a packet or cell (we use both names interchangeably) from any input port to any designated output port. The different arrangements vary in the way the route through the switch from input port to output port is selected. We will then discuss buffering.

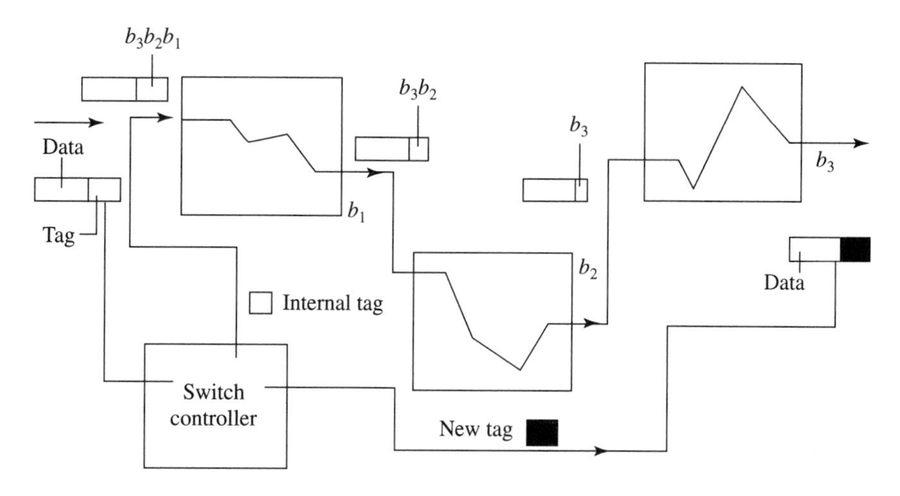

10.14
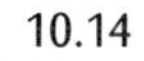
FIGURE

In self-routing, the tag is decoded bit by bit by each module to determine the route through the distributed buffer switch.

A cell arrives at a switch input port. See Figure 10.14. The cell consists of two parts: data and tag. The tag contains the destination address (if the cell is a datagram) or the VCI (if the cell is part of a virtual circuit connection). This tag is read by the switch controller, which determines the output port to which the cell must be routed.

Having determined the output port, the switch controller replaces the arriving tag by an internal tag that designates the output port. So this tag is a sequence of $n = \log_2 N$ bits. These bits are used one at a time by the $\log_2 N$ stages of 2×2 modules to determine whether the cell should go "up" or "down." In the illustration of the figure, $n = 3$, so the tag $= b_3b_2b_1$. In the first stage, bit b_1 is examined by the input port and decoded to determine whether the cell goes up or down. Bit b_1 is then stripped. At stage 2, bit b_2 is decoded to determine whether the cell goes up or down, and then stripped. Lastly, in stage 3, bit b_3 is decoded. The cell, stripped of its tag, arrives at the proper output port designated by $b_3b_2b_1$.

While the cell is being routed through the switch fabric, the switch controller calculates the new tag from the routing table. It is appended to the outgoing cell.

Figure 10.15 illustrates the omega network in which the tag is the output port number in reverse order, i.e., if the output port is ABC, then the tag is $b_3b_2b_1 = CBA$. For routing, bit 1 is interpreted as down, and bit 0 is interpreted as up. In the bottom of the figure we explain why this encoding of the output

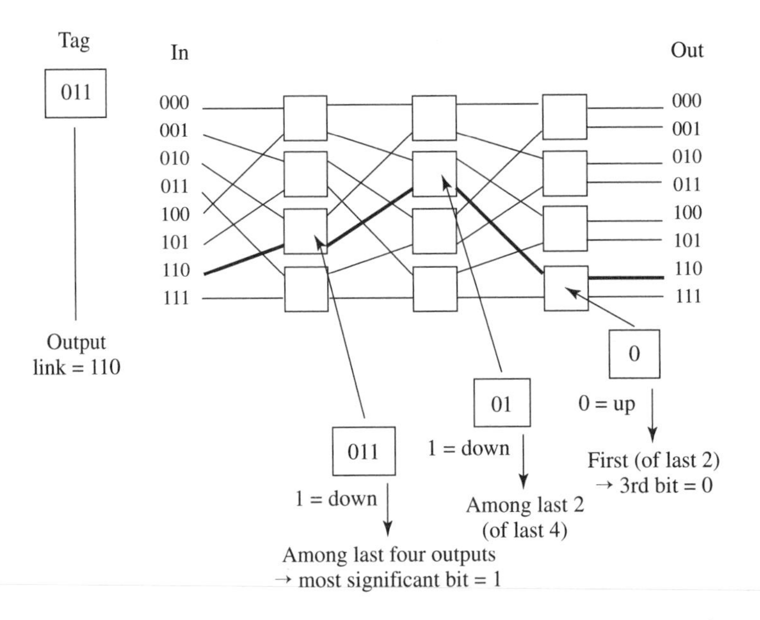

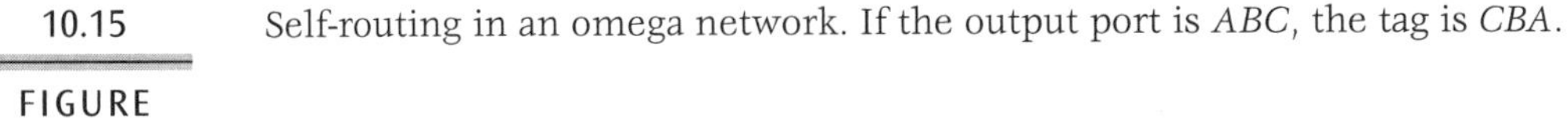

10.15 FIGURE Self-routing in an omega network. If the output port is *ABC*, the tag is *CBA*.

port leads to the correct route. Note that there is a unique route to each output port from each input port; this makes routing simple.

Figure 10.16 illustrates the banyan network. The top of the figure gives the routing. The idea is the same as for the omega network, but the tag-to-route encoding is different: if the output port is *ABC*, the tag is *CAB*, again bit 1 is down and 0 is up. The figure shows the routes from the two input ports to two output ports 100 and 001.

Having understood how packets are routed within a switch, we can evaluate the crosspoint complexity of these distributed buffer switches. There are $N/2 \log_2 N$ 2×2 modules that each admit two states (crossed or uncrossed). Hence the switch has $N^{N/2}$ states. Each state corresponds to a particular configuration. Hence these arrangements can realize only about $(N!)^{1/2}$ permutations, out of a total of $N!$ permutations. (By comparison, a Benes network with twice as many crosspoints can realize all permutations.) Thus, distributed buffer switches are not nonblocking, that is, even when the cells present at all input ports are going to different output ports (a permutation), these cells may be unable to travel simultaneously. This happens whenever they share the same path at some intermediate module, and when there is contention, one or more cells will have to be buffered. (We see in Figure 10.16 that cells destined for output ports 100 and 001 contend for the same intermediate mod-

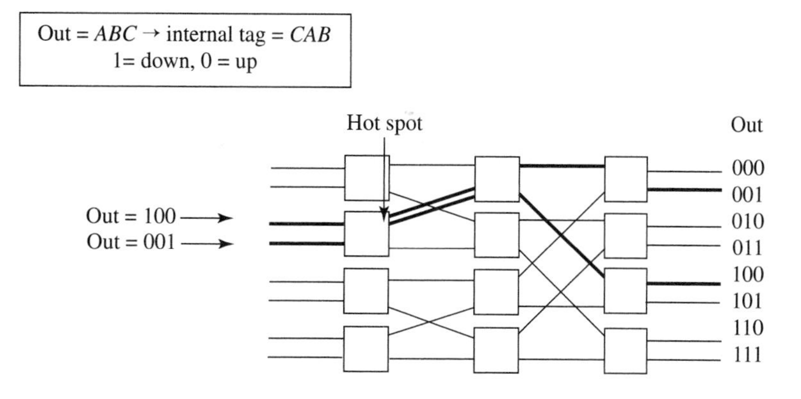

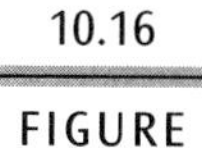
10.16 FIGURE
Queuing in a banyan network: two packets destined for different ports may contend for transfer along the same internal link causing a queuing delay.

ule and so one of these cells must be buffered.) Thus, rapid self-routing is achieved at the cost of blocking of packets at intermediate stages inside the switch with a resulting need to buffer. Intermediate modules where queues build up are called *hot spots*.

10.5.1 Impact of Hot Spots

We will calculate the reduction in throughput of a DBS due to hot spots beginning with the simplest case where the modules have no buffers. When two cells are routed through the same link, only one of them can get through and the other cell is dropped. Suppose cells arrive at each slot time, one per switch input port. The destination of each cell is random, uniformly distributed over the output ports, and different cells are independent.

Let $p(m)$ be the probability that a cell is forwarded over a link of the DBS after going through m stages, in one slot. So $p(0) = 1$, since an external cell arrives at each input link in each time slot. Consider a link in a module after $m + 1$ stages, and one slot. Suppose it is link C in Figure 10.17. By definition, the probability that a cell is not forwarded over link C is $1 - p(m + 1)$. This event occurs if and only if no cell is routed to C from either link A or B. The probability that a cell is routed from link A to C is $0.5p(m)$, so the probability that a cell is not routed from A to C is $[1 - 0.5p(m)]$. This is also the probability that a cell is not routed from B to C. Since these two events are independent, the probability that no cell is routed to C from A or B is $[1 - 0.5p(m)]^2$, and so $1 - p(m + 1) = [1 - 0.5p(m)]^2$. This gives the recursion,

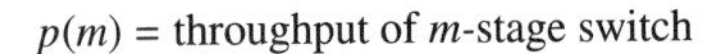

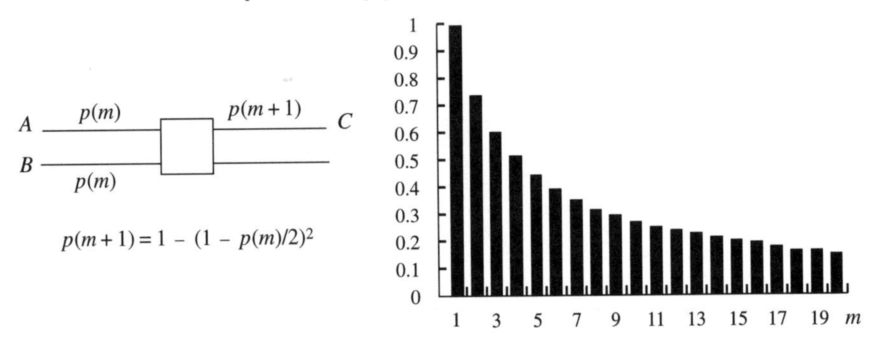

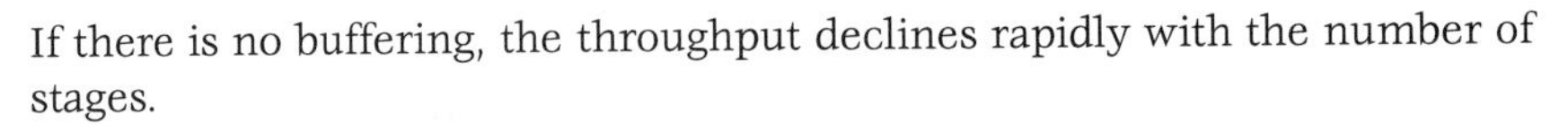

FIGURE 10.17 If there is no buffering, the throughput declines rapidly with the number of stages.

$$p(m+1) = 1 - [1 - 0.5p(m)]^2, \qquad p(0) = 1,$$

which yields $p(0) = 1$, $p(1) = 0.74$, $p(2) = 0.62$, $p(3) = 0.52$, $p(4) = 0.45$, and so on. The throughput through a link after m stages is $p(m)$. (The remainder, $1 - p(m)$ cells, are lost due to conflict.) Figure 10.17 shows that the throughput decreases rapidly as the number of stages increases.

To avoid this decline in throughput, each module must buffer cells when there is contention on its output links. As we indicated in the top of Figure 10.14, the buffer may be at the input or at the output ports of the module. The analysis of the buffer occupancy process is complicated and we shall make some simplifying assumptions. We first consider the input buffer case.

10.5.2 Input Buffers

We suppose, as before, that iid (independent identically distributed) cells arrive at the switch in each time slot and each input port with probability λ, with uniformly distributed destination. Suppose also that each buffer can hold only one cell. Then a buffered cell can move forward over a link if there is no conflict (another cell is not trying to use the same link) and if there is no blocking (the buffer at the end of the link is not full). We now make the simplifying assumption that the buffers are independent and iid per stage. (In fact this is not the case: if a buffer is occupied, it will increase the likelihood that the buffer upstream of it will also be occupied.) Consider the top-left panel in Figure 10.18.

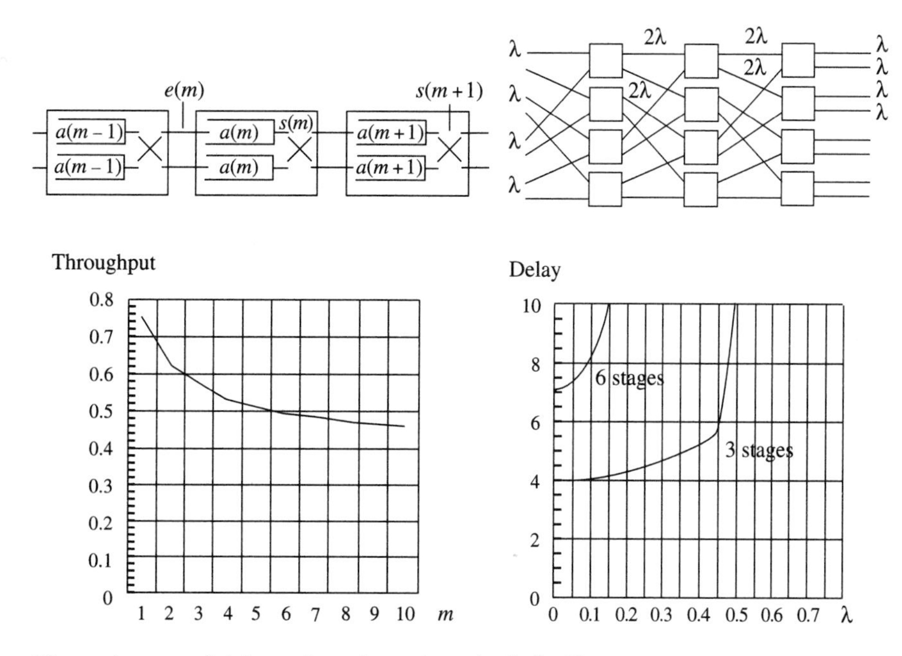

10.18
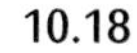
FIGURE

Throughput and delay when there is a single buffer.

For each buffer in each stage m and time slot t, define

$$a_t(m) := P(\text{queue size at end of } t = 0) =: 1 - b_t(m),$$
$$e_t(m) := P(\text{cell ready to enter queue during } t),$$
$$s_t(m) := P(\text{cell can move forward during } t \mid \text{cell is present during } t).$$

Then,

$$a_t(m) = [1 - e_t(m)][a_{t-1}(m) + b_{t-1}(m)s_t(m)],$$
$$e_t(m) = 0.75b_{t-1}(m-1)b_{t-1}(m-1) + 2 \times 0.5 \times a_{t-1}(m-1)b_{t-1}(m-1),$$
$$s_t(m) = [a_{t-1}(m) + 0.75b_{t-1}(m)][a_{t-1}(m+1) + b_{t-1}(m+1)s_{t-1}(m+1)].$$

The first equation says that the buffer is empty at the end of slot t if there is no arrival during t and either the queue was empty at the end of slot $(t-1)$ or there was a cell in the queue and it left during slot t. The second equation accounts for the likelihood of an arrival depending on whether both buffers attached to its incoming links are occupied, or exactly one of those buffers is occupied. The third equation says that a cell can be forwarded only if there is no conflict with a cell in the second buffer in the same stage m and if it cannot be blocked by a cell in the buffer in stage $m+1$.

To compute the steady-state probabilities, we can drop the index t, and get

$$a(m) = [1 - e(m)][a(m) + b(m)s(m)],$$

$$e(m) = 0.75b(m-1)b(m-1) + 0.5 \times 2 \times a(m-1)b(m-1),$$

$$s(m) = [a(m) + 0.75b(m)][a(m+1) + b(m+1)s(m+1)].$$

These equations can now be solved numerically, with appropriate boundary conditions: $e(1) \equiv \lambda$ (a cell arrives with probability λ at stage 1 in each slot) and $s(n) \equiv 1$ (a cell can leave from each output line). The maximum throughput of the DBS with n stages is $b(n)s(n)$ (for $\lambda = 1$), which is plotted in the lower left panel of Figure 10.18. Comparing this figure with Figure 10.17, we can see the improvement in throughput obtained by providing single cell buffers. As expected, the throughput declines with the number of stages, n, dropping to 40% for $n = 10$, or $2^{10} = 1{,}000$ input ports.

The delay—the expected number of slots it takes for an incoming cell to reach its output port—can be calculated from the $s(m)$. Since $s(m)$ is the probability that a cell will move forward at stage m in one time slot, the expected number of slots it will take to move forward is $1/s(m)$ and so the delay through an n stage switch is

$$\sum_{m=1}^{n} \frac{1}{s(m)}.$$

The probabilities $s(m)$ (and the other probabilities introduced earlier) depend on the distribution of the destination output port as well as the arrival rate. Figure 10.18 gives a plot of the delay versus arrival rate λ for a worst-case distribution. The top-right panel shows this worst-case distribution for the omega network. Notice how traffic arriving on four input ports and destined for four different output ports nevertheless is routed through only two internal links.

The number of conflicts (cells needing to travel over the same internal link) would decrease if cells were provided alternative paths to their destination. One way of doing this is to build parallel networks. Figure 10.19 shows the reduction in delay with two parallel baseline networks. Notice that traffic on the top four incoming ports is routed through four parallel links, two in each network.

So far we have assumed that buffers can hold only one cell. Clearly the performance will improve with larger buffers. Suppose each buffer can hold b cells. The equations that govern the evolution of the various probabilities can be derived as follows.

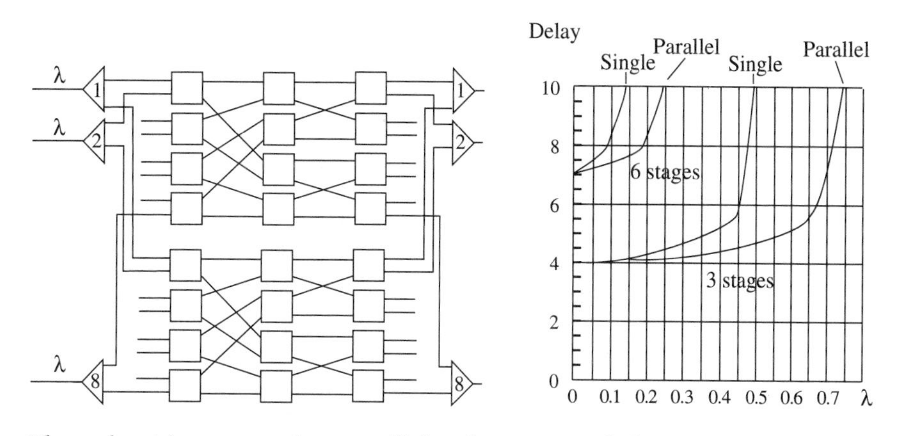

10.19
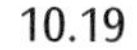
FIGURE

Throughput improves when parallel paths are provided.

For each buffer in each stage m and slot t, define

$$a_t(m, i) := P(\text{queue size at end of } t = i),$$

$$e_t(m) := P(\text{cell ready to enter queue during } t),$$

$$s_t(m) := P(\text{cell can move forward during } t \mid \text{cell is present during } t).$$

Then,

$$a_t(m, i) = a_{t-1}(m, i-1)e_t(m)(1 - s_t(m)) + a_{t-1}(m, i)[e_t(m)s_t(m)$$
$$(1 - e_t(m))(1 - s_t(m))] + a_{t-1}(m, i+1)(1 - e_t(m))s_t(m), \; 0 \le i \le b,$$

$$e_t(m) = 1 - \left[1 - \frac{1}{2}(1 - a_t(m-1, 0))\right]^2,$$

$$s_t(m) = \left[\frac{e_t(m+1)}{1 - a_t(m, 0)}\right][1 - a_t(m+1, b) + a_t(m+1, b)s_t(m+1)].$$

These equations generalize those for the case where a buffer can hold only one cell. To obtain the first equation, we begin by noting that to have i cells in the buffer at end of slot t, there must be $i-1$, i, or $i+1$ cells at the end of slot $t-1$. These possibilities account for the three terms in the right-hand side of the first equation. (If $i = 0$, the first term is absent, and if $i = b$, the third term is absent.) The second equation is based on the observation that there is no arrival into a buffer, which happens with probability $(1 - e_t(m))$, if and only if neither of the two buffers attached to its incoming links contains a cell destined for the stage m buffer being considered, which occurs with probability $[1 - 1/2(1 - a_t(m-1, 0))]$. The third equation says that a cell can be forwarded if there is at least one cell in the buffer ready to enter the queue at the next stage (the probability of this is given by the first term in square brackets) and if it cannot be blocked by a cell in the buffer at stage $m+1$ (this has probability given by the second term).

To compute the steady-state probabilities, we can drop the index t and solve the resulting recursive equations numerically, using appropriate boundary conditions as before. We can then obtain the throughput of a DBS with n stages as $s(n)[1 - a(n, 0)]$. Let $R(m)$ be the probability that a cell will move forward at stage m in one slot, so the expected number of slots it takes to move forward is $1/R(m)$ and the delay or expected number of slots it takes for an incoming cell to reach its output port through an n stage switch is

$$\sum_{m=1}^{n} \frac{1}{R(m)}.$$

To compute $R(m)$ we note that if there are i cells in the buffer, the probability that a particular cell will move forward in one slot is $s(m)/i$. Hence,

$$R(n) = s(n) \sum_{i=1}^{b} \frac{1}{i} \frac{a(n,i)}{1 - a(n,0)}.$$

Figure 10.20 displays plots of throughput as a function of the number of stages for different values of b.

10.5.3 Combating Hot Spots

We have seen that the performance of a DBS, measured by throughput or delay, is reduced by queuing or hot spots. We consider two techniques to combat hot spots. The techniques are based on insight gained from the following model of how hot spots develop. Consider a DBS with N input and output ports. Suppose that in each time slot one cell or packet arrives at each input port, destined for one of the output ports. Thus the incoming traffic is characterized by a time sequence of "destination" N-dimensional vectors of the form $(Port_{out}^1, \ldots, Port_{out}^N)$, where $Port_{out}^i$ is the output port destination of the cell at the ith input port. Associated with each destination vector are N routes through the switch, one from each input port to the corresponding output port, and one packet must be transmitted along each link of each of the N routes. The total number of packets for any link equals the number of routes through that link and may vary from 0 to N. If this number is 0 or 1, no packets will be buffered at that link; if the number is 2 or more, packets will have

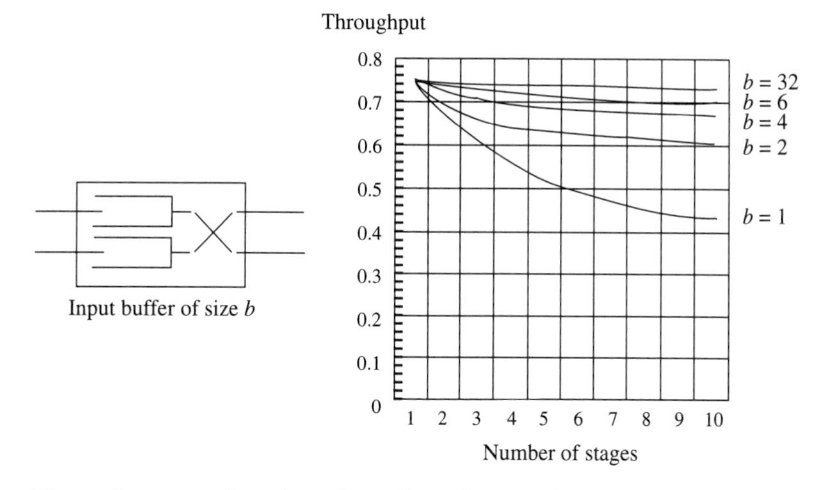

10.20 FIGURE

Throughput as a function of number of stages for buffer size $b = 2, 4, 6, 32$.

to be buffered. Some sequences of destination vectors will lead to unbalanced traffic (i.e., they cause some links to carry much more traffic than other links, creating hot spots), other sequences lead to more balanced traffic. The two techniques described below try and reduce the number of hot-spot-creating destination vectors.

The first technique is statistical in nature. It is illustrated in Figure 10.21 for the banyan switch. An additional banyan switch is added in front. It is called the randomization stage. When a cell arrives at an input port, its destination output port, Out, is replaced by another destination port, Ran. The

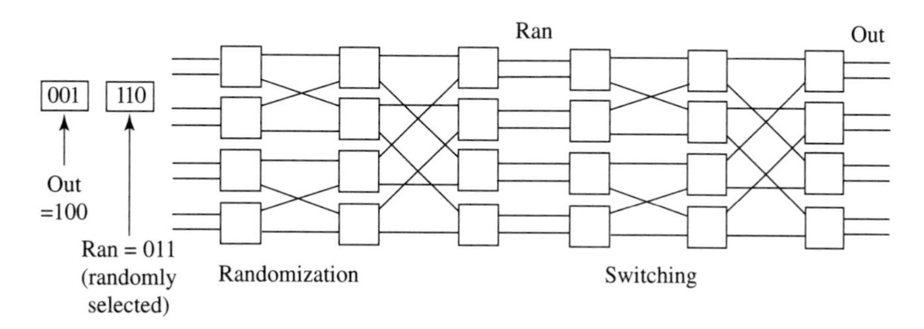

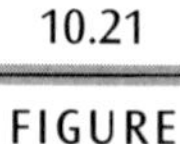

10.21 FIGURE

Introduction of a randomization stage balances the traffic entering the switching stage and reduces queuing delay.

new destination port is randomly selected with a uniform distribution over all N output ports. On leaving the randomization stage, each packet's original destination Out is restored, and it enters the second switch at input port Ran.

Thus, in the first stage, the probability distribution of the sequence of destination vectors is independent from one time slot to the next and uniformly distributed over the output ports. From the symmetry of the banyan (and other) switches, one can see that as a result the links inside the switches will be required, on average, to carry an equal number of packets. That is, on average, the traffic will be balanced, and the probability of hot spots will be low (in the first stage).

The randomization achieved in the second stage is a mirror image of the randomization in the first stage. Here, for the destination output port for any cell, the input port from which that cell originates is independent from one interval to the next and uniformly distributed over the N input ports. The symmetry of the switch ensures once again that, on average, the traffic over the links in the second stage will also be balanced, thereby reducing the occurrence of hot spots. Since randomization is statistical, this technique does not eliminate all hot spots, and so each 2×2 module must contain buffers.

The second technique is more radical: it completely eliminates contention (and hence the need for buffers) within the switch fabric. The technique is based on the observation that if the destination N-vector is sorted (in increasing order) by output port, the corresponding N routes will be disjoint. This is illustrated in Figure 10.22. Cells in the first two input ports have destinations 110 and 111, so they are sorted; the other input ports are idle (idle is highest in the ordering). We can see that the two routes have no links in common, so both cells can be transmitted simultaneously, and there will be no contention or need for buffers.

One can take advantage of this observation by a two-stage switch design in which the first stage is an automatic "sorter" in which the cells appear at the output of that stage properly sorted. Figure 10.23 illustrates one sorter called the Batcher network. If a Batcher network is followed by, say, a banyan switch, the destination vector appearing at the banyan switch will be sorted, and there will be no contention inside the switch. Thus there is no need for buffers in the banyan switch.

Of course, if two input cells with the *same* destination output port appear at the Batcher network, there will be contention. So one of those cells cannot be allowed to enter the network and must be buffered. Thus a Batcher-banyan switch must be equipped with buffers at the input. Such switches are sometimes called *input buffered*. Input buffering can reduce utilization, hence throughput. For example, suppose (referring to Figure 10.22 again) that the

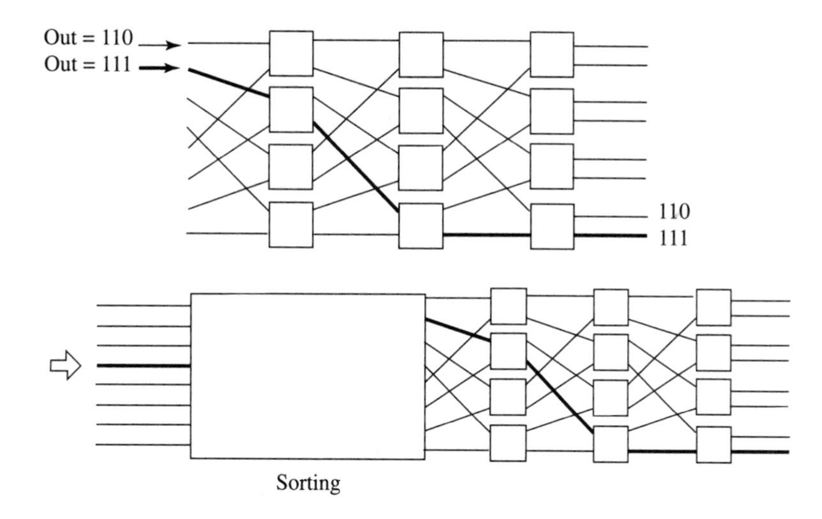

10.22 FIGURE

If the destinations of arriving cells are sorted by output port, the corresponding routes are disjoint and there is no contention.

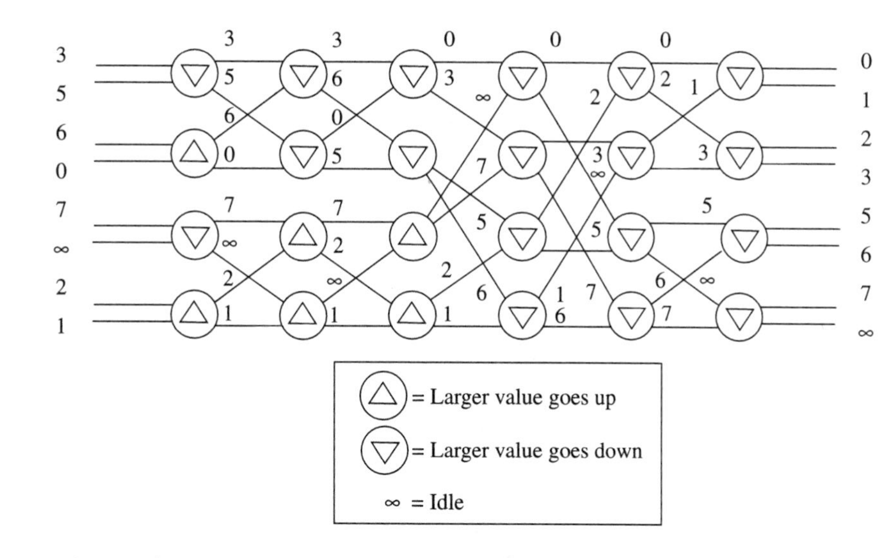

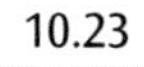

10.23 FIGURE

The Batcher network sorts incoming cells by order of output port.

first two input ports have the same destination port 110. Then one of the cells, say the one at the second port, must be buffered, and so this port (and the corresponding route) is kept idle for one cell time. We will study input buffered switches in section 10.8.

10.6 SHARED BUFFER

The modular constructions of the delta networks of Figure 10.13 are based on circuit switch designs. The other three designs are based on the design of routers for packet networks. They are not modular. In this section we study the shared buffer switch (SBS).

The name *shared buffer* is clear from Figure 10.24. There is a common pool of buffers divided into linked lists indexed by the output port name. (The figure shows three linked lists corresponding to the first three output links.) There is also one linked list of free buffers. (Thus there are $N + 1$ linked lists in all, if there are N output ports.) Each list has a begin and end pointer. The switch operates as follows. We assume that time is slotted, with one slot per cell transmission time. In each slot time the following operations take place. First, a cell from the beginning of each linked list is transmitted over the corresponding output port. The list's begin pointer is updated, and the buffer is added to the free buffer list. Second, the cells that arrived at the input ports during the slot time are examined. If a cell is destined for output port i, it is put in a free buffer (whose begin pointer is moved up) and that buffer is appended to the list for port i.

The main advantage of this scheme over the other switch designs is that a much smaller buffer is enough to achieve the same blocking probability. In a distributed buffer switch the buffers allocated to different modules are not shared, in an output buffered switch the buffers allocated to different output ports are not shared, and in an input buffer switch the buffers allocated to different input ports are not shared. As a result some buffers may be full, which may lead to blocked cells, while other buffers are not full. Another advantage is that more complex buffer-sharing schemes, involving priorities for example, may be implemented. Priorities can also be readily implemented in the input and output buffer switches, but again, at a cost of even more buffers.

The disadvantages of the shared buffer design all stem from the fact that the shared buffers must be accessed at a speed equal to the sum of the input (or output) speeds. In each time slot, one cell per input port has to be read into the buffer, and one cell per output port has to be read out. If buffers are

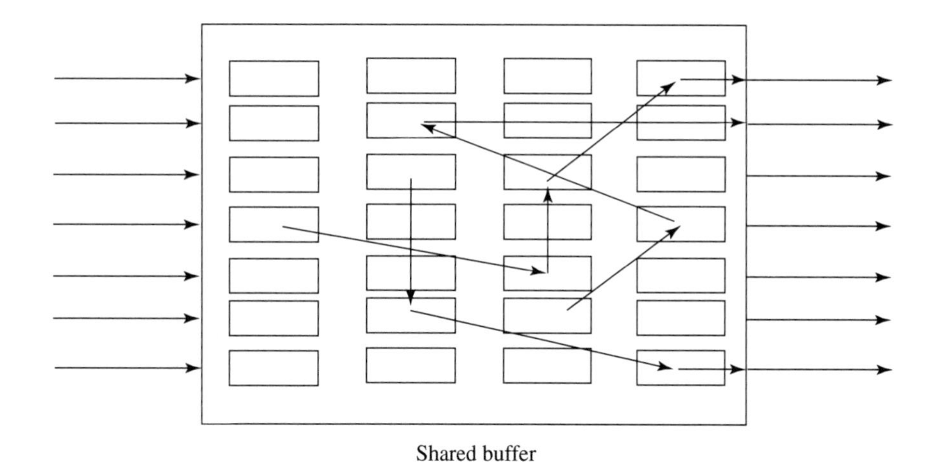

10.24
FIGURE

The shared buffer is divided into linked lists corresponding to output ports.

implemented in RAM then their speed will place a limit on the maximum throughput of an input buffered switch.

10.6.1 Queuing Analysis

We will calculate the size of a linked list for a particular output port, port 1, say. Let X_t be the size of this list at the beginning of the tth slot time. (X_t is a random variable.) Suppose that A_t cells with destination port 1 arrive during this slot. (Thus $X_t \geq 0$, and $0 \leq A_t \leq N$, if there are N input ports.) Then,

$$X_{t+1} = (X_t - 1)^+ + A_t = X_t + A_t - 1(X_t > 0), \qquad n \geq 0, \tag{10.3}$$

where we use the notation that for any number z, $z^+ = \max\{z, 0\}$; and $1(\cdot)$ is the indicator function, so $1(z > 0) = 1$ if $z > 0$, and 0 otherwise. The term $(X_t - 1)^+$ accounts for the fact that if $X_t > 0$, then one cell will be transmitted out of port 1, leaving a list of length $(X_t - 1)$. Suppose that the arrivals A_t are iid. Assume that we have reached steady state, and let $P(X)$ denote the steady-state distribution of X_t. Let $P(A)$ denote the distribution of A_t. Taking expectations on both sides of (10.3) we get

$$E(X) = E(X) + E(A) - P(X > 0),$$

so

$$P(X > 0) = E(A) =: \rho.$$

Next we square both sides of (10.3) and take expectations to get

$$E(A^2) + \rho + 2\rho E(X) - 2E(X) - 2\rho^2 = 0.$$

Suppose the variance of A_t is σ^2, so $E(A^2) = \rho^2 + \sigma^2$. Then, the average size of the list is

$$E(X) = \frac{\rho + \sigma^2 - \rho^2}{2(1-\rho)}. \tag{10.4}$$

In the special case that A_n has a Poisson distribution, $\sigma^2 = \rho$, so substituting in (10.4) we get the well-known formula,

$$E(X) = \frac{2\rho - \rho^2}{2(1-\rho)}. \tag{10.5}$$

Since on average ρ cells arrive per slot destined for output port 1, and since 1 cell can be transmitted from that port per slot, ρ is also the average utilization of the transmission bandwidth of that output link. From (10.5) we see that if the utilization is 80%, then the average number of cells in the buffer is 2.4, and if the utilization is 90%, it is 4.95.

10.7 OUTPUT BUFFER

In Figure 10.25 we exhibit an output buffer switch design. (Many routers are based on this design.) Incoming cells are multiplexed over a fast bus or any other broadcast medium. (In order to facilitate this multiplexing operation, a small buffer is needed at each input port.) Corresponding to each output link there is an address filter that reads the bus, selects the cells destined for that output link, and copies those cells into the corresponding output buffer. This design gives a better buffer utilization than the distributed buffer design but a lower buffer utilization than the shared buffer design. Of all four designs considered here, the output buffer design is best suited for multicast traffic. Indeed, cells with multiple destination ports can be handled with virtually no additional complexity. The disadvantage is that it needs a high-speed bus. The bus speed must be as large as the sum of the input link speeds.

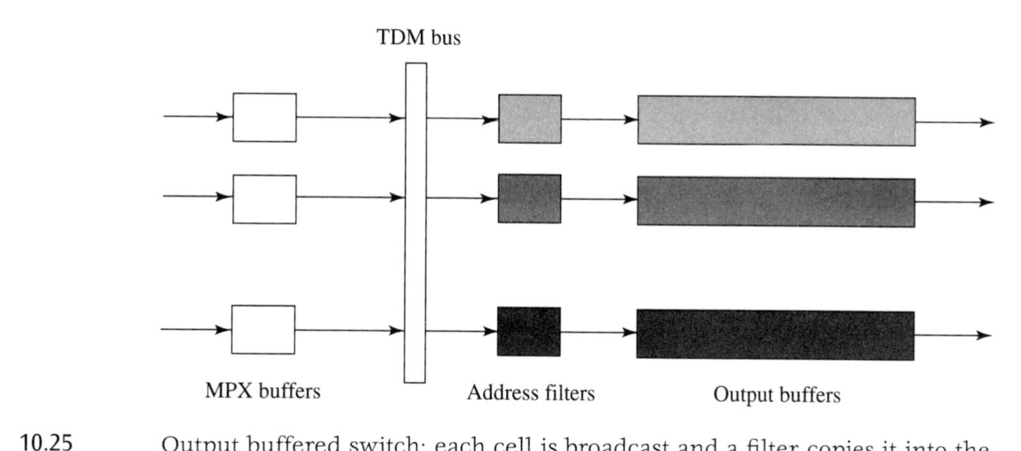

10.25 FIGURE Output buffered switch: each cell is broadcast and a filter copies it into the proper buffer.

10.7.1 Knockout

A design difficulty arises with output buffer switches. The number of cells that want to enter a given output buffer in one cell time can be as large as the number of input lines, say N. Accordingly, the number of input lines will be limited by the speed of the electronics used for the output buffers. To avoid this limitation, designers propose to limit the number of cells that can be transferred into an output buffer to some value $K < N$. If $M > K$ out of N cells are destined to the same output buffer in one cell time, then K of them are selected for transfer and the others are dropped. Various procedures exist for making sure that each of the M cells has the same probability of being selected, for fairness. One such scheme—called the knockout switch—implements a K-stage knockout tournament.

The designer must choose the value of K that keeps the fraction of dropped cells to an acceptable small value. To determine such a value, assume that the input cells have independent destinations, picked uniformly among the N output links. Assume also that there is a cell arriving at each input link with probability ρ, independently of the other input links. The probability that M cells arrive in one time slot and are destined to a specific output link is then equal to $P(M)$ where

$$P(M) = \binom{N}{M}\left(\frac{\rho}{N}\right)^{M}\left(1-\frac{\rho}{N}\right)^{N-M}$$

where the term ρ/N is the probability that a cell arrives at a given input link and is destined for that specific output link. For that specific output link, the expected number of cells dropped in one time slot is therefore equal to α where

$$\alpha = \sum_{M=K+1}^{N} (M-K)P(M).$$

Indeed, $M-K$ cells are dropped with probability $P(M)$ for $M=K+1,\dots,N$. Since the average rate of cells destined for that output link is equal to ρ, we conclude that the average fraction of cells dropped per output link is equal to $\phi := \alpha/\rho$. Using these formulas, we can determine the value of K that guarantees—under our assumptions—that the fraction ϕ of cells that are dropped is acceptable. The upshot of the analysis is that the loss rate is less than 10^{-10} for $K=12$ provided that $\rho \leq 0.9$, for any value of N.

10.8 INPUT BUFFER

The input buffer switch design is illustrated in Figure 10.26. Incoming cells are stored in buffers, one per input port. In each time slot, the crossbar (or any strictly nonblocking) switch transfers the cells at the head of each input buffer to their destination output ports, provided there is no contention. That is, if cells at the head of more than one input buffer have the same destination, the switch controller selects only one of these for transfer by the crossbar. (The design of this controller is straightforward if it knows the destinations of all the input cells. A more interesting problem is to design a switch controller that resolves contention in a decentralized manner.)

The speed of the crossbar is the same as that of the input lines. This is the major advantage over the shared buffer and output buffer designs (whose buffer and bus speeds, respectively, must equal the sum of the input line speeds). As a result, for a given very large-scale integration (VLSI) technology, the input buffer switch can have much greater throughput than these two designs. The input buffer switch, however, suffers from head of line (HOL) blocking, which can reduce its throughput considerably. We will study this next and then examine ways to overcome it.

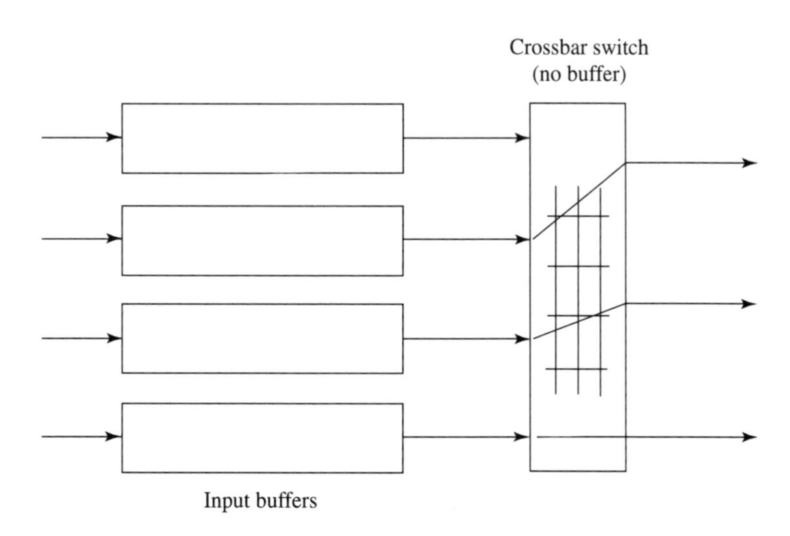

10.26 FIGURE

Input buffered switch: cells contend for transfer through the crossbar switch.

10.8.1 HOL Blocking

Figure 10.27 shows a switch with four input and three output ports. Suppose that the destination of cells at the head of the four input buffers are as shown. Then, in the next slot time, the crossbar will transfer two cells: one of the cells from buffers 1, 3, or 4 to output port 1, and the cell from buffer 2 to output port 3. The crossbar will not transfer cells in buffers 3 and 4 with destination port 2 because they encounter HOL blocking. Note that the crossbar has the capacity to transfer up to three cells per slot, so that HOL blocking reduces the utilization of this switch. We now calculate this reduction.

We consider an $N \times N$ switch with large N, and we focus attention on type 1 cells—those with destination port 1. In each slot time, a 1 cell arrives at every input port with probability λ/N. See the top of the figure. The cells in different slots and at different input ports are iid. (For example, if one cell arrives at each input port per slot, and if its destination is uniformly distributed over all N output ports, then $\lambda = 1$.) Since N is large, the total number of 1 cells that arrive at the input ports during the tth slot time has a Poisson distribution with mean λ. (The exact distribution is binomial.) Suppose that at the beginning of the tth slot, X_t cells of type 1 are at the head of their input buffers. We want to study the evolution of X_t, assuming steady state. (See the HOL queue in the bottom left of the figure.) Because the crossbar will forward only one of these cells to the output during the tth slot time,

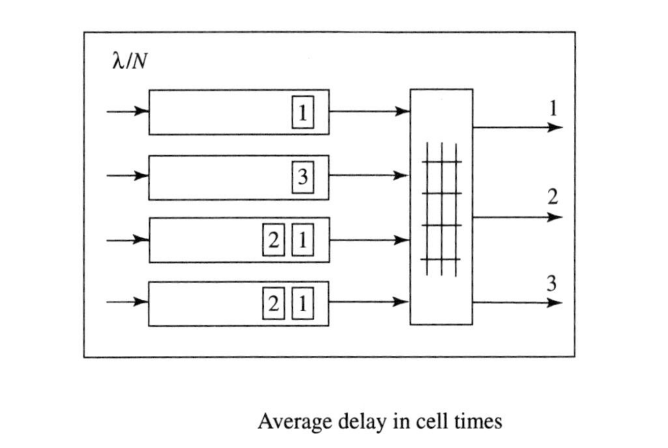

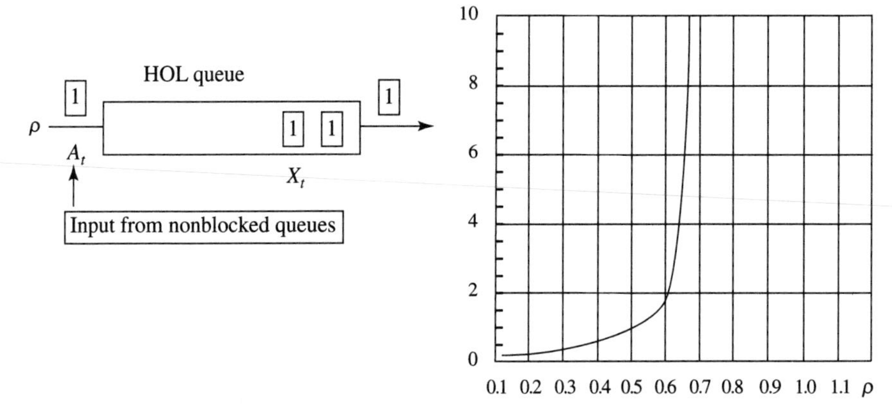

10.27 FIGURE

A queued cell prevents later cells from access to the switch fabric, causing HOL blocking.

$$X_{t+1} = (X_t - 1)^+ + A_t = X_t + A_t - 1(X_t > 0), \qquad t \geq 0,$$

where A_t is the number of new 1 cells that come to the head of the input buffers during this slot. Assume that in steady state the probability that a buffer is unblocked is Φ. Then A_t has a Poisson distribution with mean $\rho := \lambda \times \Phi$. The same calculation that led to (10.5) shows that

$$P(X_t > 0) = \rho, \text{ and } E(X_t) = \frac{2\rho - \rho^2}{2(1-\rho)} =: \beta.$$

By definition, the expected number of blocked buffers is $N(1-\Phi)$. If we assume that the destination of the cells is uniformly distributed over all N output ports, then the expected number of blocked buffers also equals $NE(X_t-1)^+ = N(\beta-\rho)$. So

$$\Phi = \frac{\rho}{\lambda} = 1 - \beta + \rho.$$

For $\lambda = 1$, this gives $\beta = 1$, i.e., $\rho = 2 - \sqrt{2} = 0.58$. Note that ρ is the throughput of 1-type cells. Thus, under the assumption of iid arrivals and uniformly distributed destinations, HOL blocking reduces throughput to 58%.

Because of this, the input rate λ must be less than 0.58, otherwise the queues in the input buffers will become unbounded. By a more detailed analysis of the HOL queue in the left of the figure, we can calculate the average delay through the switch. As expected, the figure shows that the delay becomes unbounded as the input rate approaches 0.58.

10.8.2 Overcoming HOL Blocking

We now consider two techniques to overcome the 58% throughput limit. The techniques are quite intuitive. The first technique allows the crossbar to look ahead into each input buffer for cells that could be transferred if they were not blocked by the head cell. Clearly, the throughput increases with the depth of the look ahead, as shown in Figure 10.28. An infinitely deep look ahead corresponds to input expansion, in which each input port has one buffer per output port. HOL blocking is completely overcome, and the utilization is 1, when one cell arrives at each input port per slot with a destination that is uniformly distributed over all output ports. (The plot in the figure approaches 1 as the depth of look ahead approaches infinity.) We analyze the input expansion case.

Fix an output port (port 1, say) and let X_t be the number of all type 1 cells waiting in the input buffers during slot t. Because there is no HOL blocking, one of these cells is forwarded through the switch in each slot time. So once again we have the equation,

$$X_{t+1} = (X_t - 1)^+ + A_t = X_t + A_t - 1(X_t > 0), \qquad t \geq 0,$$

where A_t is the number of new 1 cells that arrive during slot t. If the A_t are iid random variables with mean ρ and variance σ^2, then (see (10.4)) in steady state,

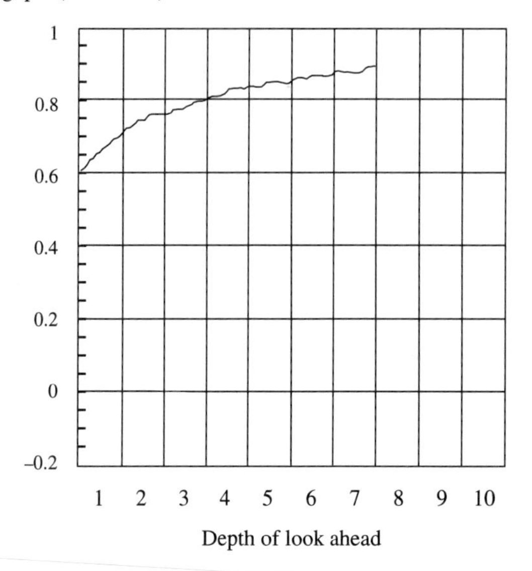

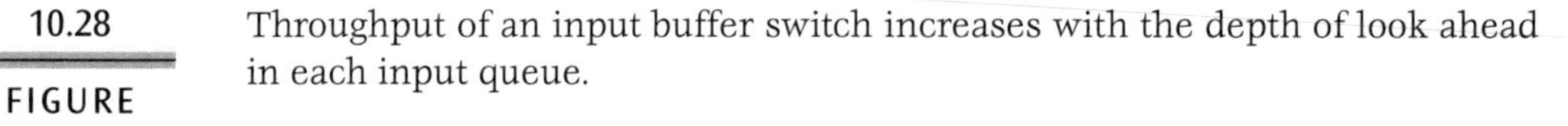

10.28 FIGURE Throughput of an input buffer switch increases with the depth of look ahead in each input queue.

$$E(X_t) = \frac{\rho + \sigma^2 - \rho^2}{2(1-\rho)}.$$

Let us observe how this analysis applies to the output buffer switch as well. Fix an output port (port 1), and let X_t be the number of cells waiting in the corresponding buffer. Since one of these cells is transmitted in each slot time, X_t evolves in the same way as the equation above and so the steady state expected value of X_t is the same. In particular, the throughput and delay of the output buffer switch and the input buffer switch with input expansion are the same. There remains an important difference however: the input buffer switch must transfer only one cell to an output port in one slot whereas the output buffer switch must be capable of transferring N cells if there are N input ports.

The second technique uses output expansion, in which each output port is replaced by, say, L output ports. The crossbar switch can transfer L cells to the same destination (in place of one cell), thereby increasing the throughput. This is shown in Figure 10.29.

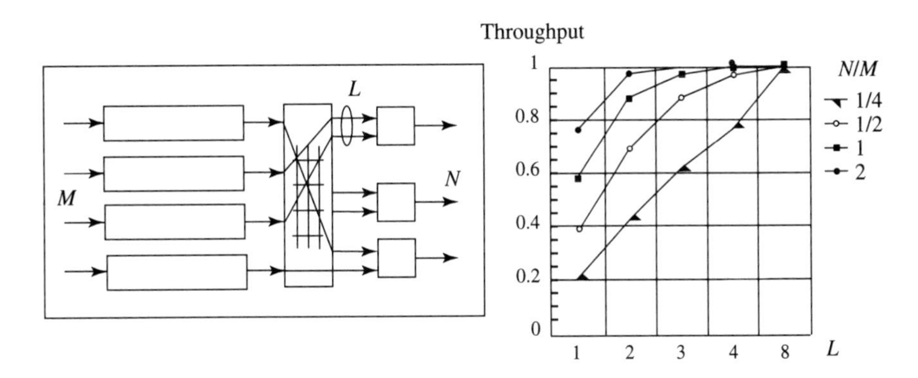

10.29 FIGURE Throughput of an input buffer switch increases with output expansion.

10.9 SUMMARY

Advances in high-speed transmission over optical fiber discussed in Chapter 9 and fast cell switches discusssed in this chapter are the two technologies that make high-performance networks affordable.

There are four basic switching architectures: input buffer, output buffer, shared buffer, and distributed buffer. Switch performance is measured in terms of throughput, delay, and complexity. Switches that implement the first three architectures are limited in their total throughput by the maximum speed of electronic circuits; hence those switches are used mostly for local ATM networks. Each architecture has its own bottlenecks that cause queuing delays. Simple models are available to analyze those queuing delays. In most cases, additional features can be added to the basic architecture to overcome bottlenecks. For example, head of line blocking of the input buffered switch can be reduced by "look ahead" schemes.

Distributed buffer switches are modular and can be scaled to arbitrarily high total throughput, and so they are used for large wide area ATM switches. The construction borrows ideas from modular telephone circuit switches. Distributed buffer switches can suffer large queuing delay that can be overcome by a randomization or sorting stage that precedes the switch.

All switches process cells within the same connection in the same order in which they arrive (first come, first served, or FCFS service), as is required for virtual circuit connections. These switches extend FCFS service to cells in different connections that share the same input and output ports, which is not required. Indeed, in order to provide different QoS to different connections,

FCFS service may be inappropriate. For instance, connections that have a guaranteed delay requirement may need to be served with a higher priority than connections that do not have such guarantees. Similarly, connections that require greater bandwidth should be served more frequently than those that require less bandwidth.

In order to provide different QoS to different connections, these switches must be enhanced. The enhancement typically consists of buffer management facilities that allow the switch to keep track of cells from different connections and to process cells from those connections in ways that meet their QoS requirements. For example, suppose there are two different priorities. Then two different buffers would be created, and higher priority buffers would be processed before the lower priority buffers. As another example, suppose delay guarantees were met by providing deadlines to cells. Then the switch would implement a buffer management policy that processes cells in order of their deadline. (Such a service is called "earliest deadline first.")

A fast switch with these buffer management facilities would be able to provide the full range of QoS for ATM connections.

10.10 NOTES

Stochastic models for packet switching and traffic are developed in [H90]. A collection of papers on high-speed switching is available in [R93]. A general discussion on queuing models in data networks is available in [BG92]. A recent approach to the analysis and control of high-speed packet networks is proposed in [FMM91]. The distributed buffer switch was invented as a means of connecting multiple processors to multiple memories. The unbuffered switch is modeled and analyzed in [P81, KS83], which also estimate the hardware complexity. The buffered case is analyzed in [J83, KL90, YLL90]. The input buffered switch is analyzed in [KHM87, HA87]. The benefit of speeding up an input buffered switch is evaluated in [OMKM89]; look ahead schemes to overcome HOL blocking are studied in [L90, MVW93]; the effect of output expansion is investigated in [LL91]. The knockout switch is presented in [YHA87].

10.11 PROBLEMS

1. A circuit switch has M input ports and N output ports as in Figure 10.1. The switch is controlled by an operator who has one cable that can be

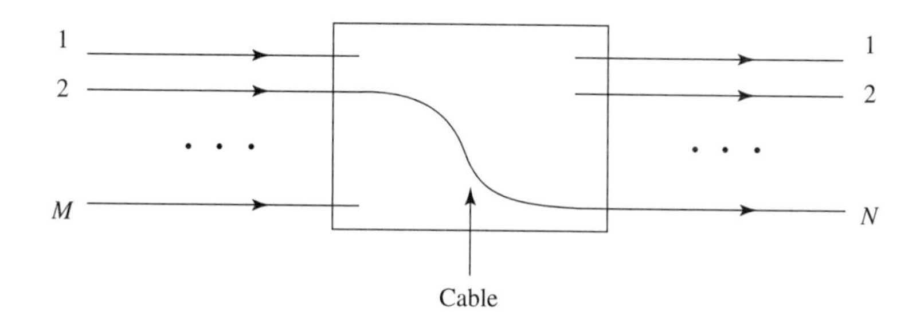

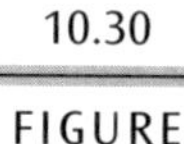
10.30

FIGURE

The cable is used by the operator to connect any input port to any output port.

used to connect any one input port to any one output port as in Figure 10.30. How many switch configurations can the operator reach? Design an arrangement with crosspoints that achieves the same configurations. Repeat the question for the case that the operator has k cables.

2. Determine the tag-to-route encoding for the Baseline and Flip networks of Figure 10.14. Assume as in Figure 10.16 that 1 denotes down and 0 denotes up.
3. Show that if two banyan switches are connected in tandem, the resulting switch is rearrangeably nonblocking. How many paths are there between any pair of input-output ports?
4. In a $2^n \times 2^n$ DBS switch show that there is a unique route through the switch between any pair of input-output ports. How many such routes are there in a Benes network?
5. Show in the omega switch that if cells arriving at the input ports are sorted in increasing order by output port, then the routes of these cells are disjoint. What happens in the other switches?
6. In the Batcher-banyan switch there is no need for buffering in the second stage. However, if cells appear in two or more input lines destined for the same output port, then all except one of these must be buffered. Show that this network behaves like an input buffer switch. In particular, show that the switch is subject to HOL blocking.
7. How would you obtain the steady-state distribution of the HOL queue?
8. How would the analysis of section 10.8 be modified for the case where the cell arrivals are iid but the destination is not uniform over all output ports?

9. A shared buffer switch is built from RAM with an access speed of 1 μs. Suppose the RAM is organized in 16-bit words, and the cells are 48 bytes long. What is the throughput of this switch? If this is a 4×4 local area ATM switch, what is the maximum line rate the switch can support?
10. Analyze the buffers saved in output buffer compared with distributed buffer, and shared buffer compared with output buffer.

11

CHAPTER

Towards a Global Multimedia Network

In Chapters 3, 4, and 5 we studied networks from the viewpoints of the physical layer (bit ways) and the bearer services they provide. Datagram networks provide best-effort transfer of messages, circuit-switched networks dedicate bandwidth along a route for transfer of constant bit rate streams, and ATM networks can dedicate resources for transfer of variable bit rate traffic with differentiated quality of service. In Chapters 6, 7, and 8 we studied the technical and economic means of network control. Finally, in Chapters 9 and 10 we described optical-link and fast switching technologies that make high-speed networking affordable.

In this brief chapter we take a broader perspective starting with two questions. The first question asks what will be the distinctive features of the global, multimedia network of the next century. We address this question by discussing five necessary attributes of a global network. These attributes accommodate service quality variation, heterogeneity, mobility, extensibility, and security.

The second question asks what advances are needed to build a network with these five attributes. We frame these advances as challenges to three technology areas: signal processing, networking, and applications. (This book is primarily concerned with networking technology. However, the two other technology areas are essential to a successfully deployed network: without successful applications there will not be enough users, and without signal-processing advances, it will be prohibitively expensive to use the network to transfer video and images.)

We indicate some recent developments that suggest how these advances might come about.

These three sets of technologies must be used to implement an architecture that can accommodate the five attributes of the global network. We suggest requirements that this architecture must meet, pointing to the shortcomings of some of the current architectures. Since this chapter raises many questions, few of which are answered, the chapter might be regarded as a formulation of a research agenda for a global network.

In section 11.1 we explain the five attributes of the global network, and in section 11.2 we introduce the technological areas and selective recent advances. In section 11.3 we propose requirements of the global network.

11.1 ATTRIBUTES OF THE GLOBAL NETWORK

The phrase *global network* conveys the misleading image of a single unitary network. In reality, the global network will be an interoperable collection of networks that supports multimedia applications incorporating data, audio, graphics, video, images, and animation. This network will offer cost-effective, high-performance service, including entertainment-quality video. The network will be scalable to support millions of users, flexible and extensible to accommodate future applications.

The global network will be heterogeneous in many dimensions. There will be multiple constituent networks, including the public telephone network, the Internet, extensions of the current CATV distribution systems, satellite networks, packet radio networks, and local area networks. These networks individually and collectively incorporate disparate transmission technologies, including fiber optics, wire pair, coaxial cable, microwave, radio, and infrared wireless. There will be a variety of terminals with widely differing capabilities for display, playback, and processing, ranging from battery-operated wireless personal digital assistants or PDAs and PCs to multiprocessor supercomputers.

This heterogeneous infrastructure of networks and terminal equipment will support a wide and dynamic mix of applications, addressing the needs of small, specialized groups of users (e.g., users remotely running supercomputer applications requiring very high communications bandwidth) as well as common applications like telephony, e-mail, videoconferencing, information and entertainment delivery, and electronic commerce with millions of users.

Several major industries (software, semiconductor, computer, telecommunications, and content providers), tens of standards bodies, and hundreds of hardware and software vendors will participate in the design and deployment of this global multimedia network.

Largely unaware of the technologies and organizations involved, users will want applications to operate seamlessly across the network infrastructure, with applications and networks appropriately scaling and configuring to whatever detailed technological components are involved. Users may want their applications to be restricted to a portion of the network (and hence to a subset of the other users) or to equipment from a particular vendor. Most of the parties involved—users, service providers, content providers, and equipment vendors—will want a flexible and dynamic network that can scale and evolve to meet whatever demands are placed on it and to accommodate new, unanticipated applications without major dislocation or investment.

These considerations suggest that the global network must have the following five attributes:

- Heterogeneity—the ability to deal with a large variety of transport and terminal technologies and applications;
- Quality of service—the ability to reserve resources within the network and terminal devices so as to ensure that certain perceptual or objective performance measures are met;
- Mobility—the ability to provide a moving access point to the network;
- Extensibility—the ability to accommodate a variety of new applications and users in the future. There are two aspects to extensibility: first, all architectural features of the network must scale to an arbitrary expansion of users and applications needs; second, the architecture must be able to accommodate new technologies and applications;
- Security and reliability—including the ability to ensure that user communications are not intercepted and that their location is not tracked and also to assure the high availability of network services.

The following example illustrates some of these attributes. Audio and video compression algorithms in the global network must accomplish much more than minimizing the bit rate, as in the past. They must be capable of scaling to a variety of transport scenarios, from circuit to packet switched, from broadband fiber optics to wireless, from high reliability (fiber) to low reliability (mobile wireless), and from low time jitter (constant bit rate or circuit switched) to high jitter transmission. The compression algorithms must be able to work with a concatenation of several of these transmission scenarios, dealing, for example, with broadband backbone connections with and without wireless access.

These algorithms must also be able to scale to differing levels of processing power and resolution in the receiving terminals while providing the best subjective quality possible. They must also accommodate new types of services, such as multicast (several simultaneous receivers, as in a videoconference) and mobile users who have a changing access point to the network. In addition, these functions must be carried out in a way that is compatible with end-to-end encryption for privacy that prevents the network from "looking into" the compression syntax or processing it in any way.

All these objectives must be accommodated in an architecture that manages the inherent complexity possibly by providing the appropriate abstractions and modularity, so that the different underlying technologies can evolve separately and new technologies can be accommodated. This implies that highly configurable generic resources (such as reliable bandwidth, resolution, and delay) can be used in a negotiation between applications and networks at the time of service establishment. All this points to an elaborate set of interactions among the constituent parts of the system and the need for a carefully crafted architecture that structures those interactions in a useful manner.

11.2 TECHNOLOGY AREAS

Three sets of technologies are used to build the global network: networking, signal processing, and applications. These technologies are embodied in elements that must be organized within an appropriate architecture in order to provide the attributes listed in the previous section.

Networking technologies are concerned with providing quality end-to-end transport between service providers and consumers. Networking technologies must be able to transport diverse multimedia data types across heterogeneous link technologies and independent subnetworks while supporting user mobility and protecting user privacy. Networking technologies include transmission, multiplexing, switching and routing, communication protocols, administration (billing and security), and network management.

Signal processing is concerned with the encoding, compression, storage, and playback of video, audio, and image signals. Signal-processing technologies include algorithms for coding and compression, encryption, and error correction and implementations of those algorithms in hardware or software. The algorithms must be well matched to both the transport capabilities of the network and the multimedia demands of the applications.

Applications technologies are concerned with determining and developing generic applications that will ease the development of user applications. The File Transfer Protocol (FTP) is an example of a generic application. A large fraction of future applications will incorporate multimedia (data, graphics, audio, video, and animation), as illustrated by the success of the CD-ROM and the World Wide Web. The development of HTML (Hypertext Markup Language) and associated browsers provides another example of a generic application that has popularized the use of Web pages.

A third example is offered by MBone. MBone is an IP application that creates connections between one user (speaker) and a group of other users (subscribers); new subscribers may join or leave. A naive way to implement MBone would create one separate connection between the speaker and each subscriber. If a connection uses up a bandwidth of B bps and the speaker's link has a capacity of NB bps, then there could be at most N subscribers. As the number of users attempting to subscribe approaches N, the speaker's node would be unable to cope, and service to each subscriber would deteriorate. Thus the naive approach does not scale.

Instead of the naive approach, MBone creates a spanning tree of subscribers rooted at the speaker: a new subscriber is attached to the "nearest" connected subscriber, who forwards a copy of the speaker's packets to the new subscriber. The copy function is now spread over many subscribers, and there is no limit to the number of subscribers that can be connected. The MBone example has many interesting features. First, the approach scales. Second, the idea can be extended to other functions. For instance, frequently accessed information from a particular Web site could be stored at intermediate nodes, reducing the burden on the source Web site. Third, MBone service can be introduced incrementally, being available only from those routers that implement the associated spanning tree and routing protocols.

11.2.1 Architecture

The network architecture provides a framework for organizing the functional elements needed for the global network. The elements must be modular (that is, specified independently of each other) so that different implementations can realize those elements in ways that encourage the use and development of technological innovations. The modularity of the Internet and OSI architectures has permitted the immediate incorporation into networks of higher-speed computers, links, and switches.

However, as we have often stressed, many applications, particularly multimedia applications, demand dedicated resources at the physical layer. Meeting such demands may create a dependence between the application layer and physical layer of the architecture. The architecture must be carefully designed so that such dependence does not compromise modularity. An equally important requirement on the architecture is that it must accommodate existing networks.

We are concerned primarily with technological challenges to global networking, but economic and social challenges must be overcome as well. Access to the global network must be sufficiently cheap if it is going to be global. (In many developing countries today, large numbers of people do not have access to telephone service.) And the global network must be deployed in ways that contribute to social progress.

11.2.2 Networking

The current networking strategies (multiplexing, switching, flow control, and congestion control) of the Internet are remarkably successful for applications that require only best-effort transmissions. These strategies must evolve to provide the stringent QoS required by multimedia applications. The QoS must be estimated for network services, and a signaling procedure must be defined to enable the selection of the QoS across different networks. The strategies must take signal processing into account for several reasons:

- The subjective fidelity at the application layer depends on the compression, encryption, and temporal characteristics of the losses at the buffers. (For example, are losses independent, positively or negatively correlated?) Currently, such temporal characteristics are not known, and methods for controlling them are not available.
- The losses depend on the statistics of the traffic and the buffer and bandwidth allocation strategies (as we have seen in Chapters 6 and 7).
- The statistics of the traffic depend on the compression, encryption, traffic shaping, flow control, buffer, and bandwidth allocation, and on adaptive/predictive routing for mobile applications.
- If the application is carried by a number of streams (e.g., one stream for audio and another for video), these streams may have different traffic statistics and QoS requirements.
- The QoS provided to mobile users may specify that the resolution of the video stream can drop when a user leaves the building.
- The security of an application may prescribe different levels of protection for distinct subsets of users.

For multimedia networking, wireless access techniques must provide high bit rates and low latency, often while accommodating a high density of users. Facing limited transmission bandwidth, available approaches seek to increase multiuser system transmission capacity (the product of bit rate and number of users per unit area or volume) by implementing sophisticated signal-processing techniques, such as joint detection and interference cancellation. However, the high complexity of these techniques will increase latency.

11.2.3 Signal Processing

The main standards in video processing are H.261, MPEG1, and MPEG2. These are well suited for storage and continuous bit rate transmission in the absence of errors, but they are less than ideal for networking, for these reasons:

- They are subject to propagation of errors over many frames, which impacts the QoS in case of losses, delays, or transmission errors.
- They do not permit multiresolution representation of video, which reduces their usefulness in heterogeneous environments.
- They cannot adapt to changing network conditions (congestion, variable throughput), which again limits the QoS.

The encoding of information currently enforces a separation of source and channel coding. This separation limits the performance when applications share wireless channels.

Encryption and compression accumulate their error-propagation effects. Such error propagation could be controlled by an integrated approach that develops compression and coding algorithms that match the network environment.

11.2.4 Applications

Current applications are often designed assuming extreme properties of the service offered by the network. Thus, a client-server application may be designed to work properly even when packets are lost, replicated, or misordered. For multimedia applications that often involve QoS and real-time constraints, it will be impossible to design those applications to provide satisfactory performance despite such worst-case assumptions about network service. Instead, the application must be designed under the assumption that the host's operating system running the application knows (for instance) the delay and loss characteristics of the network and schedules operations accordingly. But adapting applications to the network in this way couples the operating system to the network. Such dependence of the application on the network is a major departure from current practice. Here are examples of such dependence:

- A video connection may still be satisfactory if the picture occasionally switches to black and white.
- A mobile application must be designed to handle variations in the delay and loss rate, possibly through fail-soft solutions.
- The selection of servers should take transmission delays into account.
- The setup of multimedia connections may involve resource reservations that require information about pricing and network characteristics (as opposed to RSVP where the routing is decided beforehand).
- The selection of parameters of connections (e.g., leaky-bucket parameters or other traffic-shaping options) requires estimation of the impact of a given selection on the subjective quality of the application.

11.3 CHALLENGES

Historically, the disciplines of signal processing, networking, and applications have evolved with little interaction. As explained above, we believe that to address successfully the challenges of multimedia networking, researchers must collaborate across these disciplines and focus on the ultimate design goals. This need for a synergy across research disciplines can be organized in six areas: (1) architecture, (2) QoS, (3) mobility, (4) heterogeneity, (5) extensibility, and (6) security.

11.3.1 Architecture

A multimedia network incorporates many different technologies, as well as myriad functional and applications requirements. Unfortunately, these interact in complex ways, creating undesirable dependencies. Here are some examples:

- Joint source/channel coding (coordination of channel impairments with source characteristics and subjective effects) is necessary for high traffic capacity on wireless access links, where high capacity is important.

However, this creates an unfortunate dependence between the design of channel coders (at the physical/link layer) and source coders (audio and video compression) that runs directly counter to the desire for generic networks running generic and flexible applications.

- Encryption (for privacy and security) hides the basic syntactical and semantic components of a media stream, precluding any operations that "look into" or "process" that stream, such as bridging for multicast connections, conversions from one resolution to another, or conversions from one protocol or compression standard to another.
- Delay is lower-bounded by propagation delay, which is already appreciable for interactive applications on a global scale (on the order of several hundred milliseconds). The architecture must be quite disciplined to avoid adding appreciable processing or queuing delays, an issue that has been largely ignored for applications initially deployed in local area and national networks. This affects decisions about conversions from one protocol or compression standard to another, packet sizes, network topologies, etc.

Research on architecture definition attempts to identify functional interdependencies, quantifies the relationship between overall performance parameters and architectural choices, and defines architectural concepts that satisfy user and application needs. One large challenge is to define the appropriate modularity and system abstractions so as to maintain the greatest possible independence in the design of the different parts of the network and at the same time meet performance, cost, and functionality objectives. As mentioned before, the objective of the research is not the definition of a totally new infrastructure; rather, the objective is to incorporate into the future multimedia networking infrastructure existing component networks (such as the Internet, public telephone network, CATV networks, and future extensions) with minimal modification (this is one aspect of the heterogeneity issue discussed later). What are the minimal necessary changes?

11.3.2 Quality of Service

Quality of service refers to network performance measures such as delay, loss, and corruption, as seen by the users and applications. (*Corruption* is the reduction in the quality of the information perceived by the user because of quantization, compression, and loss.) A traditional approach is for the application to place constraints on the offered traffic and the network to provide QoS guarantees. While many data-oriented applications can live with best-effort networking (with no QoS guarantees, other than reliable delivery), audio and video generally require QoS guarantees. Delay must be upper-bounded to ensure real-time delivery and loss and corruption must be bounded to ensure a predictable subjective quality. The wide range of requirements, from those with relaxed to stringent QoS parameters, suggests that it would be highly advantageous (mostly in terms of carried traffic) for applications to have control over the QoS. This is especially critical in wireless access, where providing all services a reliability high enough for the most stringent service (like MPEG video) would be prohibitively expensive. A few of the many research issues raised by QoS are described next.

A major challenge is defining appropriate measures of QoS that strike a balance between (from the application view) capturing the essence of impairments with a detail sufficient to predict subjective quality to the user and (from the network view) simplicity to enable monitoring and control mechanisms for guaranteeing QoS. These measures should (in contrast to current approaches) capture temporal behavior (time correlations), and they have to support features such as aggregation and disaggregation, explained below. Various other dependencies have to be accommodated. The subjective quality of audio and video presentations depends on the compression and encryption algorithms used and their error-propagation effects and the temporal characteristics of the losses and corruption. The temporal characteristics of losses depend on the aggregate source traffic statistics. The traffic statistics depend on the compression, encryption, traffic shaping, flow control, buffer, and bandwidth allocation, as well as adaptive/predictive routing for mobile applications. To date, only oversimplified measures of QoS, security, and reliability have been defined, such as the delay-throughput characteristic, cell delay variation, cell loss rates, security of point-to-point links, and the probability of failure of unreliable links.

The first step in controlling QoS is to be able to predict it (using analytical and numerical tools) on the basis of resource allocations within the network. In addition to the models described in previous chapters, new traffic models are needed in broadband networks, possibly including self-similar models and multiple-time-scale models (see references at the end of this chapter). The next step is to define control algorithms to complement stochastic dynamic programming. As an example we mention a game-theoretic model for the study of resource allocation problems based on the observation that regulated traffic within the network is bounded by an affine function of the time interval, as is the case with traffic regulated by GCRA or leaky bucket (see section 7.4.1). Methods for predicting and controlling QoS may be based on statistics collected in a systematic way from actual network implementations (see section 6.4.3). On-line estimation methods and associated adaptive control strategies will be essential. Related work is required for routing, fault detection, adaptive compression, and other control functions.

One key concern is to understand the *aggregation* of QoS, that is the prediction of QoS based on multiple sources of impairment, such as in tandem transport links. For example, an aggregation approach based on affine bounds turns specified bounds on the traffic into bounds on end-to-end delays. (ATM Forum recommendations based on GCRA follow this approach.) The product-form result for average delays in datagram networks discussed in section 7.3.3 is another example. Based on effective bandwidth theory, decoupling techniques for aggregating the performance measures of fast packet switches provide a third example. It is necessary to extend these approaches or to develop new approaches to aggregation, especially emphasizing the heterogeneous networking context where qualitatively different impairments must be aggregated.

The inverse of aggregation is the problem of *disaggregation* of QoS, that is to determine the characteristics of network elements needed to achieve a given end-to-end QoS for the application. For instance, given the acceptable average end-to-end delay of a connection across several networks, we must allocate acceptable delays to the different networks and then to the various links of each network. This is straightforward for simple measures like worst-case delay, but further research is needed to treat more sophisticated measures that capture temporal behavior and statistical fluctuation. Both aggregation and disaggregation are management plane functions, together with admission control, routing, and network configuration.

There are other QoS-related research issues. Unlike today's networks, which simply react to application needs, multimedia networks require a back-and-forth negotiation between application and network, based on trade-offs between price and quality measures and dependent on current traffic conditions. There are many questions relating to the implementation of this negotiation and the delay it will introduce in session establishment, especially in the face of heterogeneous networks with many service providers. One approach uses software agents that allow the negotiation to be logically distributed but physically centralized, reducing the establishment latency while allowing complex traffic-dependent negotiations.

Limited transmission bandwidth will be a major factor constraining the QoS obtained in wireless access to multimedia networks. Achieving high reliability in wireless links is also expensive in terms of traffic capacity. There is also a strong dependence between delay and reliability on wireless links.

These factors imply that the ability to configure applications and specify desired QoS to the wireless link is especially critical. There is a need to examine technologies that hold the promise of high-bandwidth wireless links, such as wireless infrared and millimeter-wave radio. Experimental research suggests that the infrared wireless links may provide a bandwidth of 100 Mbps with a bit error rate of 10^{-9}. Millimeter-wave radio can provide ample bandwidth, but it is necessary to overcome difficulties like multipath.

Maintaining QoS in the presence of failures is important in some cases. Whereas a datagram network adapts smoothly to link and node failures, such failures are typically fatal for virtual circuit connections. There are promising suggestions for network designs that are robust against failures. One approach is the hierarchical organization of virtual circuits that keep the size of routing information in the nodes under control and that enable automatic rerouting when the network detects a failure. Another approach is redundant transmission of critical information along disjoint paths (this is implemented at the bit way level in SONET rings). In some multimedia applications, the stream can be partitioned into substreams with different levels of importance. This opens the possibility for fail-soft scenarios, in which only less-critical portions of streams are lost or excessively impaired.

Although current proposals for QoS specification are based on static application requirements, this is not necessary. One alternative is adaptive control of the application (such as adaptive video compression), possibly based on QoS monitoring parameters. For video and audio, where subjective impairments predominate, it might be better to vary subjective quality measures like distortion than to have excessive time jitter. (This is in sharp contrast to how data should be treated.) For example, video compression has been based on a circuit (fixed delay and bound on loss probability) model of transport. It may prove necessary to reexamine video (and audio) compression within the context of more complex impairments characteristic of packet networks. One idea is that of asynchronous video that reduces the perceptual delay to the user by assuming an asynchronous model for both transport and reconstruction. Similar opportunities for coordinating compression and transport, through control, configurability, and scalability, need to be explored. QoS can also be achieved by structuring the application; for instance, multicast connections allow resource sharing over different receivers.

Current video coding research is moving beyond the definition of fixed compression algorithms. Instead, one can define a toolbox of algorithms from which various compression schemes can be derived. That is, a compression language or syntax is defined, rather than a specific coder. Efficient software for this compression language then becomes a key requirement. The coding algorithm specification now becomes part of the video data, and the choice of the algorithm becomes part of connection establishment. Such an approach to video coding must resolve issues of implementation complexity on general-purpose processors, and this will influence algorithmic choices in a novel way. The flexibility so achieved should also make adaptive coding (i.e., coding tuned to a particular video sequence) much easier. One key advance that is expected is a software-only encoder/decoder for multiresolution multicast.

Pricing is an effective mechanism for allocating scarce resources, and it will be an important component to the establishment of QoS in future networks. Pricing, and the related billing mechanisms, raise several issues: the heterogeneity, security, mobility, and privacy aspects of these mechanisms must be studied together with the impact on the QoS. The use of incentives and punishments to force desired behavior on the part of agents is an old technique in economic theory. The multimedia network will constitute an enormous information economy, and one of the most promising research avenues to investigate is how pricing can provide the right incentives to ensure that all applications can receive their requisite QoS.

11.3.3 Mobility

Many of the existing algorithms for mobile network management were developed for cellular telephone networks. On entering a new cell, the user must be authenticated, billing must be established, and the connection must be rerouted to the new end-point base station. Cellular telephone systems assume centralized tracking and control and relatively large cells, resulting in infrequent handovers. These assumptions are no longer appropriate within a microcellular in-building network or the emerging personal communications service (PCS) networks, where handovers could occur much more frequently, low handoff latency is needed for the seamless delivery of continuous-media streams, and the base stations are likely to be interconnected by high-speed, low-latency wired networks.

For mobile networks, handoff represents the most significant challenge, primarily because of the latencies it introduces, associated with rerouting of connections and restarting packet streams as users move through the network. While establishing the reroute itself may not entail much latency, continuous-media streams are likely to be buffered at base stations, and restarting these streams back from their source will introduce significant latency. For example, the existing Mobile IP algorithms for mobile handoff invoke heavyweight routing and forwarding machinery characterized by interruptions in service that are measured in seconds. Such large overhead is intolerable for many latency-sensitive, media-intensive applications.

For many applications, (near) real-time audio/video and collaborative support for workers on the move is a requirement. To support users and applications roaming across such a heterogeneous collection of networks, the applications must be able to adapt to the available network performance. It may appear that the concept of multiple overlay networks will be impeded by the mobile host's need for multiple transmitter/receiver systems. But already today it is possible to simultaneously configure a laptop with network adapters for in-room diffuse IR, in-building RF, campus area packet radio (connected to the serial port), and wide area CDPD (replacing the floppy drive). New technological developments are likely to yield multimode radios that integrate such alternatives into a more convenient package in the near future. Simply extending the wireless network across these multiple overlays will not be enough. Wireless network management algorithms must be redesigned to better scale with larger numbers of users. And unlike wired networks, the wireless network must handle unpredictable increases in the densities of mobile hosts and their movement patterns.

The key to success is the development of hierarchical and distributed algorithms that can scale as the size of cells decreases, the number of cells increases, and the density of users and bits per second per user dramatically increase. The ability to track the motion of users through the network, and to exploit information about feasible and probable trajectories of such motions, will enable more localized decision making and more scalable management of network resources.

Tracking users and exploiting information about their location and the physical nature of their environment is crucial for effective network management. For example, algorithms can exploit the location of users and their physical environment to yield lower-latency handoffs as well as better allocation of network resources to high-traffic areas (see predictive routing below).

Handoff across cells can introduce significant latencies reducing the quality of continuous-media streams. This problem may be mitigated through predictive routing. If users can be tracked by the system and their likely next cells determined, one strategy for reducing the impact of handoff is to utilize processing, buffering, and backbone-network bandwidth to selectively multicast the packet streams to the probable next base stations. Selective multicast with duplicate packet-stream buffering gives the network management algorithms more flexibility as to when to execute the handoff, since the new base station will have already been "precharged" with the packet stream. Complexity remains, however, in ensuring that duplicate packets are correctly filtered by the mobile device and that buffered packets, no longer needed to support the handoff, are deleted with minimal overhead.

Conventional mobile handoffs are implemented within a homogeneous wireless subnet. We may call these *horizontal handoffs.* The mobile host or associated base station detects a degraded signal as it reaches the fringe area of its cell. In mobile-assisted handoff, the mobile listens for beacon signals from base stations in adjacent cells, choosing to register with the cell with the strongest signal. *Vertical handoffs* allow mobile hosts to roam between heterogeneous wireless overlays. The mobile host, or higher-level network management, determines when to switch the connections to an alternative overlay network, driven by signal quality, network load, or the costs of using one overlay versus an alternative.

When the characteristics of the transport between users change during a connection, it may be advantageous to employ adaptive connections, i.e., the compression and other aspects of the system may be adapted to maintain a suitable QoS. Special buffering techniques can be used to cope with fading channels. Wireless transmitters must be flexible enough to adapt to changing channel characteristics. Source-coding algorithms have been well studied for traditional transmission media, such as telephone channels. However, the study of source-coding methods and their interaction with transmission mechanisms in time-varying and error-prone environments, such as packet networks or wireless channels, is in its infancy.

At a higher level, it may become important to develop application program interfaces (APIs) that make it possible for applications to discover from the network that their communication characteristics have changed as a result of movement through the network. Much of the existing work assumes that applications can negotiate a level of QoS, and if this cannot be obtained, the service is denied; there is no fall-back strategy. Applications like Netscape have been finely crafted to work effectively in communications environments with limited connectivity. No general bandwidth- and location-sensitive API has been proposed that would provide an application with access to a range of bandwidth-sensitive, end-to-end compression, and synchronization strategies.

Connection-oriented services with performance guarantees provide useful network support for multimedia applications. The guarantees are typically achieved by reserving network resources in advance. (Current ATM Forum proposals are of this type.) The dynamic nature of mobile handoffs introduces complications; it is impractical to reserve all possible future channels. Rather than tearing down connections only to rebuild them, there may be an incremental strategy that modifies existing connections by partially reestablishing them after a horizontal handoff. Such an approach exploits the locality of logically adjacent cells to limit the amount of work involved in reestablishing the connection. Since the established channel from the packet stream source to either cell will largely be along the same route, only the "tail" of the connection needs to be rebuilt.

This illustrates a general strategy worth exploiting in mobile networks—the so-called gateway-centered approach investigated in the early days of packet radio. Connections are routed to a logical gateway for a region of the mobile network. Movement between gateways is a relatively rare event. For example, a gateway might map onto a building or a section of a campus. The gateway hides the complexity due to local movement of mobile hosts from the rest of the network by providing local routes and performance guarantees to the mobile host within its region.

For "black pipe" networks with no performance guarantees, it is still desirable to attempt to characterize end-to-end network performance. One possible mechanism is for the network management layer to inject periodic measurement packets into the subnet to characterize the route's latency and variability. An alternative, available in networks that expose some level of control to higher layers, is to exploit out of band measurements of the QoS characteristics. This makes it possible for network management algorithms to make their decisions based on more accurate and timely characterization than would be possible with periodic end-to-end measurements. These kinds of QoS measurements can best be exploited by source- and receiver-based rate control mechanisms in the mobile networking and application support layers. For example, they can be used to determine how aggressively a mobile application can pursue buffering, prefetching, and compression strategies. Wireless spectrum is not free, so read ahead should be applied only when the application can expect a high hit rate in cached data.

11.3.4 Heterogeneity

Heterogeneity will be an inescapable feature of global networks. It manifests itself in the physical layer media (wire pair, wireless, fiber), terminals (telephones, PDAs, workstations), access techniques (time, code, and frequency-division multiplexing), transport assumptions (fixed and variable bit rates), protocols (circuit and packet switched), terminal operating systems, and applications (data, audio, video). Past efforts in network design and standardization have embraced heterogeneity in certain forms, particularly in the physical media, but have strived for homogeneity at some level (typically transport protocol and establishment protocol). Unfortunately there are several such network designs, including the public telephone network (PTN), the Internet, CATV distribution systems (that are evolving into networks), local area networks, ATM and Frame Relay networks, and a proliferation of wireless networks (DECT, GSM, IS-54 and IS-95, JDC, etc.). Most of these networks are likely to persist for a long time, and there is no credible path toward a single relatively homogeneous network (such as the simpler situation in 1950 when all we had was the PTN). In fact, there is widespread consensus that it is desirable to encourage rapid technological and commercial innovation, a process that rapidly spawns new networking technologies. These different networks can coexist on the same media (such as Internet access on CATV), and can even ride on top of one another (such as IP on ATM or the PTN), all the while remaining logically separate.

Heterogeneity is not benign. It makes difficult the realization of a vision in which users seamlessly share multimedia applications over a heterogeneous networking universe. Thus, for example, users of the Internet should be able to telephone users on the PTN, and users with CATV as their primary network access should be able to videoconference with users on Frame Relay networks, PTN, and the Internet. In short, all general-purpose networks should provide unlimited connectivity to all users, regardless of what networks they access. Of course, applications will be subject to the limitations of the underlying networks, such as bandwidths, error rates, etc., but interconnection of networks should not, by itself, introduce substantial additional limitations.

Embracing heterogeneity is one of the prime goals of the network architecture design. A conceptually simple mechanism for internetworking is to allow each network to use its local protocols, signal compression and encryption techniques, etc., and to perform conversions at gateways interconnecting the networks. This approach solves the interoperability problems, but it has some limitations:

- Privacy by end-to-end encryption is precluded wherever encryption will interfere with the conversion function (as in a video or audio compression transcoding operation). This forces decryption before conversion, which is a serious breach of privacy. In fact, for a link entirely internal to the network, there will not even be a way for the user to confirm that encryption was performed at all.
- Conversion adds processing delay. For example, digital cellular base stations add on the order of 80 ms for conversion from one speech code to another. Delay is harder to mitigate than other impairments introduced in the network.

- For continuous-media services, subjective impairments will accumulate. It is also difficult to characterize the accumulation of such subjective impairments across heterogeneous coding techniques, making it difficult or impossible to predict in advance the end-to-end subjective impairments.
- Conversions require detailed knowledge of the applications to be embedded within the networks. This is a serious breach of modularity and will result in a substantial increase in complexity. It is also a major barrier to the introduction of new services, since they will often require uneconomic global upgrades of equipment.

An alternative is to make networks transparent and force interoperability at the network edge. This approach has been successfully applied to the Internet. It stimulates the rapid introduction and deployment of new services, since they need to be embedded only in those terminals (or access points) desiring the new service. However, end-to-end transparency introduces its own set of difficulties, particularly for nonprogrammable implementations. Joint source/channel coding, which is important for achieving high traffic capacity on wireless access channels, is difficult to achieve unless the source and channel coding are designed in close coordination. The transparent network is also more challenging for multicast networks, for which compatibility with multiple destination terminals must be achieved.

11.3.5 Scalability and Configurability

Scalability and configurability of audio, graphics, and video coding algorithms are essential to maintaining interoperability and efficient use of resources. Such scalability also offers a wealth of novel techniques for congestion control by adjusting data rates to the available bandwidth. A key technique in flexible video compression, for example, is multiresolution or scalable coding. Typically, the video source is successively approximated by multiple layers of coded bit streams (such as the flows mentioned earlier). This allows different terminals to access the same source representation at different bit rates and resolutions. In a multicast context, different representations can be spawned within the network in a generic fashion, without knowledge of the video coding algorithms or compression syntax.

While there has been significant progress in the standardization of video compression (such as H.261 and MPEG), these standards miss several elements that appear essential for the future requirements in multimedia networks. They offer either no scalability to bandwidth (like H.261 and MPEG1), or limited scalability (like MPEG2). Their ability to support layered coding is either missing or very limited, and they therefore do not support multicast transport. Their scalability to network QoS is also missing, generally requiring a very high reliability that will be costly on wireless access channels. (Past compression activity has focused on minimum bit rate, which is not the best criterion on wireless channels, where reliability is also a significant factor.)

Not only is scalability to transport rate, to receiver processing, and display capabilities needed, but so is configuration to the QoS parameters of the transport environment. A given transmit terminal will encounter in a heterogeneous network environment a wide variety of rate and QoS characteristics, differing radically depending on whether there is a wireless access link or not, for example. For a given combination of loss, corruption, and delay parameters, the source coder must configure to achieve the highest subjective quality. Such configurability is not a feature of existing source-coding standards, for they are typically designed with a fixed QoS in mind, such as high reliability (like MPEG) or robustness to high error rates (like digital cellular VCELP voice-coding algorithms).

Related to this issue is joint source/channel coding. The classical Shannon "separation theorem of source and channel coding" is often not applicable in heterogeneous networks, where we cannot assume that the channel is fixed, time invariant, and known. As has long been recognized, the separation theorem and other results of information theory do not adequately address the implications of delay in interactive services. In practice, the highest traffic capacity and subjective quality will be achieved only through the coordination of source coding and channel coding. Such coordination will yield the largest benefit in wireless access links. In a heterogeneous network there is need for coordination in the form of configurability of both source and channel coding to the needs of the other. This configurability requires true negotiation in the connection establishment, making call admission control more complex.

Another way in which scalable video can be incorporated into an existing network with minimal change in the networking elements (e.g., switches, routers, etc.) is to apply unequal error protection (UEP) techniques to protect various layers of the scalable video to varying degrees.

CATV, ATM, IP, digital cellular, packet radio, and other networks implement bearer services with different characteristics. It will be necessary to explore the interoperability of these services and its architectural implications. An example that we discussed in section 5.6.3 is the transmission of TCP connections across an ATM network. The question in that situation is when to open and close ATM virtual circuits.

11.3.6 Extensibility and Complexity Management

The network must be designed so that it can evolve and adapt over a wide range of parameters as the internetwork grows in size, speed, complexity, and technology. The ability to gracefully accommodate increasing levels of usage and applications with increasing bandwidth requirements is a central challenge of multimedia networking. It is not clear that existing solutions can cost-effectively scale to millions of users at entertainment qualities. At the same time, the multimedia network architectures must accommodate new technologies. One challenge is to define architectural concepts that are inherently scalable and do not limit future possibilities.

The extensibility of a heterogeneous multimedia network bears some resemblance to problems encountered in large software systems, where the essential element has been found to be complexity management. A fairly small but effective set of management principles has evolved. Two key concepts are modularity (partitioning of functionality into independent and configurable modules) and abstraction (the hiding of irrelevant implementation details, with explicit visibility of only essential characteristics). These principles serve to separate the system into modules that can evolve independently by avoiding unnecessary dependence among them. Independent evolution is essential to extensibility because global upgrades are not economically feasible.

When applying the principles of complexity management to multimedia networks, legitimate and necessary dependencies must be taken into account, while making them as benign as possible through appropriate configuration flexibility. For example, the joint source/channel coding mentioned earlier creates a necessary but undesirable dependency between the design of the source coder and the channel coder. We must be sure that this dependency is not hard wired into the system (as is the case with many existing standards, such as digital cellular voice), but is introduced in a generic and configurable fashion that can accommodate future unanticipated applications. This may be achieved through the abstraction of the channel to appropriate QoS models, explicitly hiding details such as wired or wireless transport, etc. In addition, there must be configurability through a true negotiation of QoS parameters as mentioned earlier.

A key to extensibility is sufficient flexibility in the establishment phase to allow future evolution of the network without massive changes or upgrades. One approach is to use software agents as part of the establishment protocols. In this approach, the traditional static message structures are replaced by dynamic executable messages that embed procedures and associated dynamic data structures. In essence, one module configures another (or more generally conducts a negotiation) by sending an executable message. This increases flexibility, because static message sets limit the semantics to the description of an operating point, whereas agents can specify an individual control interface for each network component. This approach can also dramatically reduce the delay caused by the establishment negotiations. The problem with conventional approaches is that an establishment negotiation may require numerous back-and-forth messages, incurring considerable latency. In agent-based establishment, the different network elements or subnetworks and the requirements of the network can be represented by distinct agents collected in a common processor for negotiation, eliminating communication latency while maintaining desirable logical modularity. Object-oriented Tcl (Tool Command Language) and Java can facilitate the description of agents.

Extensibility to new services and applications is also a critical feature of future multimedia networks. Extensibility will be facilitated by the ability to prototype and deploy new services with minimal effort. Rapidly deployable applications can be based on a service description language, a platform at each terminal incorporating the networked operating system, an interpreter of the service language, and a collection of resources or primitives for the realization of services. These resources include hardware and processing capability, as well as stored software definitions of useful service elements, such as audio or video compression algorithms. New services can be deployed by transferring a service description agent during call setup. Such descriptions are likely to be large.

This agent approach to service deployment avoids two traditional obstacles. The first is the critical size problem, where a sufficient number of specialized terminals must be deployed before an economically sustainable community of service participants exists. (The critical size problem was described in section 1.2.) In the agent approach, there must exist only a critical mass of platforms, not specialized service terminals. The second obstacle is standardization. While we require standardization of the platform, service description language, and method of transferring service descriptions in call setup, we avoid standardization of specific new services before they are deployed. An additional advantage of the agent approach is that services can be dynamically reconfigured during a call.

11.3.7 Security

The network is a shared resource, providing the advantages of access to a wide community of users and services. This brings with it the disadvantage of potential breaches of security, lack of privacy, and exposure to fraud. Security has many components. These include confidentiality and integrity (inability of unauthorized parties to read or modify information), as well as authentication and nonrepudiation (providing the equivalent of an electronic signature). It may be desirable in some cases even to mask the fact that communication is taking place between a subset of users (masking of traffic patterns). Multimedia networking poses special security questions because of the interaction between compression, encryption, and error propagation.

Encryption hides the basic syntactical and semantic elements of a bit stream and thus obstructs many important processes, such as protocol conversion, standard conversion, and joint source/channel coding. The proper placement of encryption within the network architecture is an important question. Deploying encryption at lower layers such as the network layer or data link layer allows encryption of routing overhead from higher layers, thus masking traffic patterns better. However, at internetwork gateways, such as between OSI and TCP/IP, the user data has to be decrypted and reencrypted, making it vulnerable to eavesdropping. Deploying encryption at higher layers, such as the application layer, has the advantage that user data can be encrypted end to end. But then headers appended by lower layers, being unencrypted, give clues to the traffic patterns. Further, the number of entities that need to be separately encrypted is much larger, as each of the user processes associated with the application now needs to be encrypted. These trade-offs are strongly influenced by the way the applications are structured.

Audio and video can tolerate residual errors at playback or display, and thus error propagation in encryption is important. Further, if multiscale representations of information are used as suggested above, it will be important to structure encryption so that error propagation preserves the hierarchical structure of the information.

In more complicated multiuser, multicast applications (for example videoconferencing) the very definition of privacy becomes problematic. For instance, managers in a conference with labor might wish to briefly have a private conversation that cannot be heard by the labor representatives (or vice versa). This illustrates the importance of considering application requirements, including privacy, in the definition of network service primitives.

Bibliography

[A93] The ATM Forum (1993). *ATM User-Network Interface Specification: Version 3.0*. PTR Prentice Hall, Englewood Cliffs, NJ.

[A94] A. S. Acampora (1994). *An Introduction to Broadband Networks: LANs, MANs, ATM, B-ISDN, and Optical Networks for Integrated Multimedia Telecommunications*. Plenum, New York.

[A95] The ATM Forum (1995). *ATM User-Network Interface (UNI) Specification: Version 3.1*. PTR Prentice Hall, Englewood Cliffs, NJ.

[AA73] R. Artle and C. Averous (1973). The telephone system as a public good. *Bell Journal of Economics and Management Science* 4(1):89–100.

[AL95] A. Alles (1995). ATM Internetworking. Cisco Systems, http://www.cisco.com/, May 1995.

[AMS82] D. Anick, D. Mitra, and M. M. Sondhi (1982). Stochastic theory of a data-handling system with multiple sources. *Bell System Technical Journal* 61(8):1871–1894, October 1982.

[ATM1] The ATM Forum (1995). ATM Forum 94-0471R7: P-NNI draft specification, March 1995.

[ATM2] The ATM Forum (1995). LAN emulation over ATM specification—Version 1, February 1995.

[B90] J. A. Bucklew (1990). *Large Deviation Techniques in Decision, Simulation, and Estimation*. John Wiley & Sons, New York.

[B95] K. Balaji (1995). *Broadband Communications: A Professional's Guide to ATM, Frame Relay, SMDS, SONET, and B-ISDN*. McGraw-Hill, New York.

[BC89] R. Ballart and Y. C. Ching (1989). SONET: Now it's the standard optical network. *IEEE Communications Magazine* 27(3):8–15.

[BD94] D. D. Botvich and N. G. Duffield (1994). Large deviations, the shape of the loss curve, and economies of scale in large multiplexers. Preprint.

[BG92] D. Bertsekas and R. Gallager (1992). *Data Networks*. Prentice Hall, Englewood Cliffs, NJ.

[BR60] R. R. Bahadur and R. R. Rao (1960). On deviations of the sample mean. *Ann. Math. Statis.* 31(1960):1015–1027.

[C88] D. Comer (1988). *Internetworking with TCP/IP: Principles, Protocols, and Architecture*. Prentice Hall, Englewood Cliffs, NJ.

[C89] R. L. Carroll (1989). Optical architecture and interface lightguide unit for fiber-to-the-home feature of the AT&T SLC series 5 carrier system. *Journal of Lightwave Technology* 7(11):1727–1732.

[C90] J. A. Chiddix (1990). Fiber backbone trunking in cable television networks: An evolutionary adoption of new technology. *IEEE LCS Magazine* 1(1):32–37.

[C91] R. L. Cruz (1991). A calculus for network delay, I. Network elements in isolation, II. Network analysis. *IEEE Transactions on Information Theory* 37(1):114–131, 132–141.

[C95] D. Comer (1995). *Internetworking with TCP/IP*. Prentice Hall, Englewood Cliffs, NJ.

[Cl88] D. Clark (1988). The design philosophy of the DARPA internet protocols. *Proceedings of the SIGCOMM '88 Symposium*, 106–114.

[CM91] *IEEE Communications Magazine* (1991). The 21st Century Subscriber Loop, vol. 29, no. 3.

[CM92] *IEEE Communications Magazine* (1992). Intelligent Networks, vol. 31, no. 2.

[CM94] *IEEE Communications Magazine* (1994). Video on Demand, vol. 32, no. 5.

[CN95] CommerceNet and Nielsen Media Research (1995). *The CommerceNet/Nielsen Internet Demographics Survey: Executive Summary*. Available at: http://www.commerce.net.

[CT91] T. M. Cover and J. A. Thomas (1991). *Elements of Information Theory*. John Wiley & Sons, New York.

[CW89] D. R. Cheriton and C. L. Williamson (1989). VMTP as the transport layer for high performance distributed systems. *IEEE Communications Magazine* 27(6):37–44.

[CW94] W. Y. Chen and W. L. Waring (1994). Applicability of ADSL to support video dial tone in the copper loop. *IEEE Communications Magazine* 32(5):102–109.

[CWe94] C. Courcoubetis and R. Weber (1994). Buffer overflow asymptotics for a switch handling many traffic sources. To appear in *Journal of Applied Probability*.

[CWSA95] J. Crowcroft, Z. Wang, A. Smith, and J. Adams (1995). A rough comparison of the IETF and ATM service models. *IEEE Network* 9(6):12–16.

[CWW96] C. Courcoubetis, J. Walrand, and R. Weber (1996). Pricing models for multiclass networks. In preparation.

[D91] R. Durrett (1991). *Probability: Theory and Examples*. Wadsworth and Brooks/Cole, Pacific Grove, CA.

[DV93] G. de Veciana (1993). *Design Issues in ATM Networks: Traffic Shaping and Congestion Control*. Ph.D. thesis, Dept. of EECS, University of California, Berkeley.

[DZ93] A. Dembo and O. Zeitouni (1993). *Large Deviations Techniques and Applications*. Jones and Bartlett, Boston, MA.

[EMV95] R. Edell, N. McKeown, and P. P. Varaiya (1995). Billing users and pricing for TCP. *IEEE J. Selected Areas in Communications* 13(7):1162–1175.

[F95] D. Frankel (1995). ISDN reaches the market. *IEEE Spectrum* 32(6):20–25.

[FMM91] K. W. Fendick, D. Mitra, I. Mitrani, M. A. Rodriguez, J. B. Seery, and A. Weiss (1991). An approach to high-performance, high-speed data networks. *IEEE Communications Magazine*, October 1991, 74–82.

[G93] P. E. Green, Jr. (1993). *Fiber Optic Networks*. Prentice Hall, Englewood Cliffs, NJ.

[GG92] A. Gersho and R. M. Gray (1992). *Vector Quantization and Signal Compression*. Kluwer, Norwell, MA.

[GH91] R. J. Gibbens and P. J. Hunt (1991). Effective bandwidths for the multi-type UAS channel. *Queueing Systems* 9(1):17–28.

[H88] F. Halsall (1988). *Data Communications, Computer Networks and OSI*. Addison-Wesley, Reading, MA.

[H90] J. Hui (1990). *Switching and Traffic Theory for Integrated Broadband Networks*. Kluwer, Norwell, MA.

[H95] I. Hsu (1995). *Admission Control and Resource Management for Multi-Service ATM Networks*. Ph.D. thesis, Dept. of EECS, University of California, Berkeley.

[HA87] J. Y. Hui and E. Arthurs (1987). A broadband packet switch for integrated transport. *IEEE J. Selected Areas in Communications* 5(8):1264–1273.

[Hu88] J. Y. Hui (1988). Resource allocation for broadband networks. *IEEE J. Selected Areas in Communications* 6(9):1598–1608.

[HW94] I. Hsu and J. Walrand (1994). Admission control for ATM networks. *Proceedings of the IMA Workshop on Stochastic Networks*, March 1994. IMA volumes in *Mathematics and Its Applications*, vol. 71, 411–427. Springer-Verlag (1995).

[I120] ITU-T Recommendation I.120, Integrated services digital network (ISDN). 1992 (rev).

[IN95] *IEEE Network* (1995). Digital Interactive Broadband Video Dial Tone Networks, vol. 9, no. 5.

[J83] Y.-C. Jenq (1983). Performance analysis of a packet switch based on single-buffered banyan network. *IEEE J. Selected Areas in Communications* 1(6):1014–1021.

[JSAC95] *IEEE Journal on Selected Areas in Communications* (1995). Copper wire access technologies for high performance networks, vol. 13, no. 9.

[JV91] S. Jordan and P. P. Varaiya (1991). Throughput in multiple service, multiple resource communication networks. *IEEE Transactions on Communications* 39(8):1216–1222.

[JV94] S. Jordan and P. P. Varaiya (1994). Control of multiple service, multiple resource communication networks. *IEEE Transactions on Communications* 42(11):2979–2988.

[K75] L. Kleinrock (1975). *Queueing Systems*. John Wiley & Sons, New York.

[K79] F. P. Kelly (1979). *Reversibility and Stochastic Networks*. John Wiley & Sons, New York.

[K91] F. P. Kelly (1991). Effective bandwidths at multi-class queues. *Queueing Systems*, no. 9:5–16.

[K94] F. P. Kelly (1994). Dynamic routing in stochastic networks. IMA volumes in *Mathematics and Its Applications*, ed. F. P. Kelly and R. J. Williams, vol. 71, 169–186. Springer-Verlag (1995).

[KHM87] M. J. Karol, M. G. Hluchyj, and S. P. Morgan (1987). Input versus output queueing on a space-division packet switch. *IEEE Transactions on Communications* 25(12):1347–1356.

[KL90] H. S. Kim and A. Leon-Garcia (1990). Performance of buffered banyan networks under nonuniform traffic patterns. *IEEE Transactions on Communications* 38(5):648–658.

[Kle94] L. Kleinrock (1994). *Realizing the Future: The Internet and Beyond*. National Academy Press, Washington, D.C.

[KS83] C. P. Kruskal and M. Snir (1983). The performance of multistage interconnection networks for multiprocessors. *IEEE Transactions on Computers* C-32(12):1091–1098.

[KWC93] G. Kesidis, J. Walrand, and C. S. Chang (1993). Effective bandwidths for multiclass Markov fluids and other ATM sources. *IEEE/ACM Transactions on Networking*, 424–428, August 1993.

[L90] T. T. Lee (1990). Modular architecture for very large packet switches. *IEEE Transactions on Communications* 38(7):1097–1106.

[LL91] S. C. Liew and K. W. Lu (1991). Comparison of buffering strategies for asymmetric packet switch modules. *IEEE J. Selected Areas in Communications* 9(3):428–438.

[LLKS95] H. C. Lucas, Jr., H. Levecq, R. Kraut, and L. Streeter (1995). France's grass-roots data net. *IEEE Spectrum* 32(11):71–77.

[LV95] S. Low and P. Varaiya (1995). Burst reducing servers in ATM networks. *Queueing Systems* 20:61–84.

[MV95] J. K. MacKie-Mason and H. R. Varian (1995). Pricing congestible network resources. *IEEE J. Selected Areas in Communications* 13(7):1141–1149.

[MVW93] N. McKeon, P. Varaiya, and J. Walrand (1993). Scheduling cells in an input-queued switch. *Electronics Letters*, 29(25):2174–2175, December.

[OMKM89] Y. Oie, M. Murata, K. Kubota, and H. Miyahara (1989). Effect of speedup in nonblocking packet switch. *Proceedings of IEEE International Conference on Communications '89*, pages 410–414.

[P81] J. H. Patel (1981). Performance of processor-memory interconnections for multiprocessors. *IEEE Transactions on Computers* C-30(10):771–780.

[P93] M. Padovano (1993). *Networking Applications on UNIX System V Release 4*. Prentice Hall, Englewood Cliffs, NJ.

[P94] C. Partridge (1994). *Gigabit Networking*. Addison-Wesley, Reading, MA.

[P95] G. Pettersson (1995). ISDN: From custom to commodity service. *IEEE Spectrum* 32(6):26–31.

[P96] C. Perkins (1996). IP mobility support. IETT Internet Draft, February 9, 1996.

[R91] T. R. Rowbotham (1991). Local loop development in the U.K. *IEEE Communications Magazine* 29(3):50–59.

[R93] T. G. Robertazzi, ed. (1993). *Performance Evaluation of High Speed Switching Fabrics and Networks*. IEEE Press, Piscataway, NJ.

[S92] W. Stallings (1992). *ISDN and Broadband ISDN*. Macmillan, New York.

[S96] K. Sayood (1996). *Introduction to Data Compression*. Morgan Kaufmann, San Francisco.

[SW95] A. Shwartz and A. Weiss (1995). *Large Deviations for Performance Analysis*. Chapman and Hall, New York.

[T88] A. S. Tanenbaum (1988). *Computer Networks*. Prentice Hall, Englewood Cliffs, NJ, second edition.

[TG87] P. Temin and L. Galambos (1987). *The Fall of the Bell System: A Study in Prices and Politics*. Cambridge University Press, New York.

[V93] H. R. Varian (1993). *Intermediate Microeconomics: A Modern Approach*. W. W. Norton, New York.

[W86] A. Weiss (1986). A new technique for analyzing large traffic systems. *Adv. Appl. Prob.*, 506–532.

[W88] J. Walrand (1988). *An Introduction to Queuing Networks*. Prentice Hall, Englewood Cliffs, NJ.

[W91] J. Walrand (1991). *Communication Networks: A First Course*. R. Irwin, Homewood, IL.

[W95] A. Weiss (1995). An introduction to large deviations for communication networks. *IEEE J. Selected Areas in Communications* 13(6):938–952.

[WL89] S. S. Wagner and H. L. Lemberg (1989). Technology and system issues for the WDM-based fiber loop architecture. *Journal of Lightwave Technology* 7(11):1759–1768.

[WL92] T.-H. Wu and R. C. Lau (1992). A class of self-healing ring architectures for SONET network applications. *IEEE Transactions on Communications* 40(11):1746–1756.

[Wu95] T.-H. Wu (1995). Emerging technologies for fiber network survivability. *IEEE Communications Magazine* 33(2):60–74.

[X92] XTP Forum (1992). *Xpress Transfer Protocol Version 4.0*. Technical Report. Ptp://dancer.ca.sandia.gov/pub/xtp4.0/xtp4.0-specification-25.ps

[YHA87] Y.-S. Yeh, M. G. Hluchyj, and A. S. Acampora (1987). The knockout switch: A simple modular architecture for high performance packet switch. *IEEE J. Selected Areas in Communications* 5(8):1274–1283.

[YLL90] H. Yoon, K. Y. Lee, and M. T. Liu (1990). Performance analysis of multibuffered packet-switching networks in multiprocessor systems. *IEEE Transactions on Computers* 39(3):319–327.

[ZDESZ93] L. Zhang, S. Deering, D. Estrin, S. Shenker and D. Zappala (1993). RSVP: A new resource ReSerVation protocol. *IEEE Network* 7(5):8–18.

Index

A

AAL. *See* adaptation layer (AAL) of ATM
ABP (Alternating Bit protocol), 69–72
ABR. *See* available bit rate (ABR)
access line, 54
adaptation layer (AAL) of ATM
  header structure, 210–214
  multiprotocol encapsulation over AAL5, 227
  overview, 214–217
add/drop multiplexer (ADM), 163
addresses
  care-of address, 140
  home IP address, 139
  network address, 132–133
  telephone network conventions, 158
Address Resolution Protocol (ARP), 13
ADM (add/drop multiplexer), 163
admission control
  ATM networks, 270–274
  circuit-switched networks, 252
  overview, 239–240
ADSL (Asymmetric Digital Subscriber Line), 19
Advanced Research Projects Agency (ARPA), 12
agent advertisement messages, 140
agent solicitation messages, 140
aggregate traffic demand, 382–383
aggregation of QoS, 489
allocation control. *See* resource allocation
Alternating Bit protocol (ABP), 69–72
amplifiers, optical, 408, 433
aperiodic transition probability matrix, 307–308
application layer
  ODN model, 84–85
  OSI model, 100
applications, 39–45
  bit error rate (BER), 41, 43, 44
  constant bit rate (CBR), 41, 42
  defined, 39
  demands on bearer services, 52–53
  examples, 42–43
  global multimedia network technologies, 483, 485–486
  messages, 41–42
  network dependence, 485–486

applications *(cont.)*
- overview, 39–40
- rates, 43, 44
- reliability, 45
- security, 45
- variable bit rate (VBR), 41, 42

architecture. *See also* Intelligent Networks (IN); layered architecture
- defined, 77
- global multimedia network, 483–484, 486–487

ARPA (Advanced Research Projects Agency), 12

ARPANET, 12, 13

Asymmetric Digital Subscriber Line (ADSL), 19

Asynchronous Transfer Mode (ATM)
- adaptation layer, 214–217
- BISDN, 225–226
- connection-oriented service, 199–203
- CRC sequence, 203–204
- delay, 204–208, 394
- development of, 15–16, 197–198
- fixed cell size, 203–208
- Frame Relay over, 232
- future of, 28–29, 30–31
- header structure, 210–214
- and Internet development, 27–28
- internetworking, 226–232
- IP over ATM, 228–231
- LAN emulation layer, 227–228
- layer service characteristics, 200–201
- main features, 198–199
- management and control, 217–224, 264–296
  - admission control, 270–274
  - ATM forum recommendations, 266–270
  - averaging rate fluctuations, 278–280
  - Bahadur-Rao theorem, 355–356
  - buffering, 284–287
  - burstiness, 338–339, 393, 394
  - congestion control, 223
  - control plane, 218, 219
  - control problems, 264–266
  - deterministic approaches, 266–276, 335–339
  - deterministic vs. statistical, 293–296
  - fault management, 219–222
  - GCRA selection, 291–293
  - large deviations of iid random variables, 339–343
  - large deviations of multiclass queues, 349–355
  - large deviations of queues, 344–349
  - layer management plane, 218, 219
  - leaky bucket, 336–338
  - linear bounds, 335–336
  - Markov-modulated fluid (MMF), 277–278
  - multiclass case, 290–291
  - multiplexing, 280–281
  - network status monitoring and configuration, 223–224
  - on-line estimation, 282–284
  - plane management, 218–219
  - pricing calls, 274–276
  - pricing services, 389–396
  - signaling functions, 224
  - statistical multiplexing and buffering, 287–290
  - statistical procedures, 276–293
  - straight-line large deviations, 343
  - traffic models, 276–280
  - traffic shaping, 223, 293
- maximizing revenue, 395–396
- messages, 392–393

model of resources and services, 390–394
multiprotocol encapsulation, 227
multiprotocol over ATM (MPOA), 231–232
OAM cells, 220–222
queuing analysis (average delay), 207–208
reserving bandwidth for users, 390, 391, 394, 396
resource allocation, 209–210
routing procedure, 201–203
SMDS over, 232
statistical multiplexing, 208
user plane, 218, 219
asynchronous transmission over FDDI networks, 115–116
AT&T Subscriber Loop Carrier (SLC) system, 172–173
ATM. *See* Asynchronous Transfer Mode (ATM)
ATM Forum user-network interface recommendations, 266–270
attenuation of fiber, 404–409
couplers and power loss, 407
as exponential in the fiber length, 404–405
low-loss windows, 406
maximum usable length, 406–407
physical causes, 405–407
power increases vs. attenuation reduction, 408
Rayleigh scattering, 405–406
attenuators in optical links, 432
audio compression, 481–482
available bit rate (ABR)
ATM service characteristics, 200, 201
GCRA for, 269
averaging rate fluctuations, 278–280

B

backbone FDDI network configuration, 113
back-end FDDI network configuration, 113
Bahadur-Rao theorem, 355–356
bandgap energy, 416, 417
bandwidth
bit rate vs., 60
and buffering, 285–287
decoupling bandwidth, 287, 347–349, 351–355
effective bandwidth of arrival stream, 345–346
of fiber, 406
and fiber dispersion, 409–411
GCRA effective bandwidth, 273–274
on-line estimation, 282–284
reserving for ATM users, 390, 391, 394, 396
statistical multiplexing and buffering, 287–290
banyan network
Batcher network before, 464–466
distributed buffer switches, 454, 456, 457
randomization stage for, 463–464
baseline network, 454–455
Batcher networks, 464–466
BCH (Bose-Chaudhury-Hocquenghem) codes, 67
bearer service layer, 83, 91
bearer services. *See also* Asynchronous Transfer Mode (ATM); network services
and application demands, 52–53
defined, 39
and internetworking, 85
ISDN, 178
Bell, Alexander Graham, 5

Bellman-Ford algorithm, 134–136, 258–259
Benes network, 447–449
  distributed buffer switches (DBS) vs., 453–454
BER. *See* bit error rate (BER)
best-effort service, 244
  Internet as, 367–368
BGP (Border Gateway Protocol), 137–138
bibliography, 501–506
billing system for Internet connections, 368–377. *See also* charges
  functions of Internet charges, 374–377
  required features, 368–369
  stage 1: single site, 369–372
  stage 2: multiple cooperating sites, 372–373
  stage 3: noncooperative sites, 373–374
BISDN. *See* Broadband ISDN (BISDN)
bit error rate (BER)
  acceptable rates of applications, 43, 44
  for CBR and VBR applications, 41
  defined, 41
  of microwave receivers, 407
  of optical link receivers, 404, 422
  for optical links, 402–403
bit link, and physical layer, 92
bit rate, bandwidth vs., 60
bit times, 56
bit ways layer, 83
"black box," switch as, 437, 438, 439
"black pipe" networks, 494
blocked connection requests, 158
blocking probability, 158, 159–161, 246–248
  complexity of calculations, 318–319
  Erlang fixed-point approximation, 319
  Erlang fixed-point equations, 319
  Erlang loss formula, 315–316
  insensitivity of, 316, 331–335
block mode (FTP), 144
Border Gateway Protocol (BGP), 137–138
Bose-Chaudhury-Hocquenghem (BCH) codes, 67
bridges
  internetworking with SMDS, 127–129
  in OSI model, 96
  spanning tree routing, 109
  transparent routing, 108–109
British Telecom TPON system, 174–176
(B, R)-leaky bucket, 336–338
Broadband ISDN (BISDN)
  ATM bearer service, 225–226
  evolution of, 8–9
  future of, 16
(B, R)-traffic theory, 336
BT. *See* burst tolerance (BT)
buffers/buffering
  ATM admission control, 270–274
  averaging rate fluctuations, 278–280
  distributed buffer switches (DBS), 453–466
  input buffer switches, 470–475
  leaky-bucket scheme, 126, 267
  loss rate analysis, 284–287
  MMF traffic model, 277–278
  occupancy for MMF source, 328–331
  output buffer switches, 468–470
  shared buffer switches (SBS), 466–468
  and statistical multiplexing, 287–290
burstiness, 338–339, 393, 394

burst tolerance (BT)
  ATM service characteristics, 200, 201
  defined, 200
  GCRA algorithm for VBR traffic, 268–269

C

cable, coaxial. *See* coaxial cable
cable television networks. *See also* video dial tone
  future of, 29, 30, 31
  key innovations, 16–20
  overview, 157
  penetration, 16
  video dial tone, 29, 186–192
call, defined, 61, 159
call admission, 158
call forwarding service, 182–183
care-of address, 140
Carrier Sense Multiple Access with Collision Detection (CSMA/CD) protocol, 105–107
CATV networks. *See* cable television networks
CBR. *See* constant bit rate (CBR)
CCS (common channel signaling), 6
CD-ROM specifications, 23
CDV (cell delay variation), 201
CDVT. *See* cell delay variation tolerance (CDVT)
cell delay variation (CDV), 201
cell delay variation tolerance (CDVT)
  ATM service characteristics, 200, 201
  defined, 200
  GCRA algorithms, 267–269
cell loss priority (CLP), 210, 213
cell loss ratio (CLR), 201
cells in ATM, 203–208
  conformant and nonconformant, 267
  CRC sequence, 203–204
  delay, 204–208
  header structure, 210–214
centralization trends, 2
central office, 54
channels
  defined, 56
  joint source/channel coding, 486–487, 497
  utilization, 56
charges. *See also* costs; economics; revenue
  congestion charges, 364–365, 369, 376, 379–380, 385–387
  congestion cost, 386–387
  economic principles, 363–366
  fixed charges, 364, 366, 375–376
  Internet billing system, 368–374
  Internet charge functions, 374–377
  prices vs. charges, 363
  pricing a single resource, 380–389
  pricing ATM calls, 274–276
  pricing ATM services, 389–396
  pricing schemes, 366–367
  quality charges, 365–366, 367, 376–377
  reservation price, 387
  social welfare optimum, 384–385
  two-part (fixed and usage), 375–376
  usage charges, 364, 365, 367, 375–376, 381–385
checksum code (CKS), 65
circuits, defined, 61
circuit-switched networks. *See also* *specific types of networks*
  control, 246–252, 313–319
    admission control, 252
    blocking, 246–248
    complexity of calculations, 318–319

circuit-switched networks *(cont.)*
control *(cont.)*
network mathematics, 316–319
routing optimization, 248–252
single switch mathematics, 313–316
fiber to the home, 156, 172–177
Intelligent Network Architecture (INA), 181–186
ISDN, 178–180
overview, 155–157
performance of, 157–161
setup time, 158
SONET, 161–171
video dial tone, 186–192
circuit switching. *See also* virtual circuit switching
Benes network, 447–449
Clos networks, 444–450
development of, 6
distributed buffer switches (DBS), 453–466
Markov chain models, 313–314, 316–317
modular design, 446–450
overview, 61–62, 76
packet switching compared to, 63–64, 65, 450–451
and resource allocation, 75
space-division, 443
statistical multiplexing compared to, 11
time-division, 442–443
CKS (checksum code), 65
classical IP model, 229–230
Clos networks, 444–450
Benes network, 447–449
built from RNB modules, 446
built from SNB modules, 445
modular switch design, 446–450
CLP (cell loss priority), 210, 213
CLR (cell loss ratio), 201
coaxial cable
for 10BASE2 networks, 102
for 10BASE5 networks, 102, 103
fiber vs., 407–408
maximum usable length, 407
coherent detection, 425–426
heterodyne detection, 426
homodyne detection, 426, 430
commands, FTP, 144
common channel signaling (CCS), 6
communications industry revenue, 4
complexity
global multimedia network management, 498–499
switch measures, 439–441
compression
FTP compressed mode, 144
global multimedia network requirements, 481–482, 485, 490–491, 496–497
MPEG video compression standards, 191–192
scalability and configurability, 496–497
computer communications industry revenue, 4–5
computer networks
key innovations, 10–16, 20
speed increases in, 16–17
conduction band, 416
configurability, 496–497
conformant cells, 267
congestion charges, 364–365. *See also* charges
billing system, 376
congestion cost, 386–387
Internet pricing, 379–380
and on-line reporting, 369
pricing a single resource, 385–387
reservation price, 387
usage charges vs., 365
congestion control
ATM networks, 223

defined, 240
congestion cost, 386–387
connectionless service
OSI model, 98
overview, 46
connection-oriented service. *See also* Asynchronous Transfer Mode (ATM); Frame Relay networks
advantages, 199–200
ATM, 199–203
disadvantages, 200
mobile networks, 491–494
OSI model, 98
overview, 45–46
connection request, 185
connection setup, 61
connection statistics, Internet, 378
connection teardown, 61
constant bit rate (CBR)
ATM service characteristics, 200, 201
examples, 42
GCRA for, 267–268
overview, 41
continuous-time Markov chains, 308–312
definition, 309–310
exponential distribution, 308–309
invariant distribution, 311–312
irreducibility, 311
Markov property, 310–311
rate matrix, 309
reversing time, 312
control functions in INA, 184, 185–186
control of networks. *See also* admission control; flow control; resource allocation; routing
ATM networks, 217–224, 264–296
admission control, 270–274
ATM forum recommendations, 266–270
averaging rate fluctuations, 278–280
Bahadur-Rao theorem, 355–356
buffering, 284–287
burstiness, 338–339, 393, 394
congestion control, 223
control plane, 218, 219
control problems, 264–266
deterministic approaches, 266–276, 335–339
deterministic vs. statistical, 293–296
fault management, 219–222
GCRA selection, 291–293
large deviations of iid random variables, 339–343
large deviations of multiclass queues, 349–355
large deviations of queues, 344–349
layer management plane, 218, 219
leaky bucket, 336–338
linear bounds, 335–336
Markov-modulated fluid (MMF), 277–278
multiclass case, 290–291
multiplexing, 280–281
network status monitoring and configuration, 223–224
on-line estimation, 282–284
overview, 240, 243
plane management, 218–219
pricing calls, 274–276
signaling functions, 224
statistical multiplexing and buffering, 287–290
statistical procedures, 276–293
straight-line large deviations, 343
traffic models, 276–280

control of networks *(cont.)*
ATM networks *(cont.)*
traffic shaping, 223, 293
circuit-switched networks, 246–252, 313–319
admission control, 252
blocking, 246–248
network mathematics, 316–319
overview, 240, 242
routing optimization, 248–252
single switch mathematics, 313–316
datagram networks, 252–263, 320–335
buffer occupancy for an MMF source, 328–331
discrete-time queue, 322–326
flow control, 261–263
insensitivity of blocking probability, 331–335
Jackson network, 326–328
key queuing result, 254–255
M/M/1 queue, 320–322
overview, 240, 242–243, 252–253
queuing model, 253–254
routing optimization, 255–261
examples, 242–243
INA control functions, 184, 185–186
Markov chains, 301–312
methods, 239–241
overview, 239
quality of service (QoS), 244–246
time scales, 241–242
control of processing, 185
control plane of ATM, 218, 219
convergence sublayer (CS)
overview, 214–215
and Type 1 traffic, 215–216
and Type 2 traffic, 216
and Type 3 traffic, 216
and Type 4 traffic, 217
and Type 5 traffic, 217
cooperative sites' billing system, 372–373
corruption of QoS, 487
costs
congestion cost, 386–387
cost recovery and optimum link capacity, 387–389
telecommuting and ISDN costs, 51–52
couplers in optical links, 407, 432
Cramer's theorem, 339–340
comments and sharpening, 340–341
proof of, 341–343
crankback, 202
CRC. *See* cyclic redundancy check (CRC)
crossbar switches, 439–440
as space-division switches, 443
crosstalk
in FDM, 59
in serial transmission, 10
CS. *See* circuit switching; convergence sublayer (CS)
CSMA/CD (Carrier Sense Multiple Access with Collision Detection) protocol, 105–107
cyclic redundancy check (CRC)
in ATM cells, 203–204
overview, 11, 65–67, 68

D

dark current in optical link receivers, 422, 424, 430, 431–432
datagram network control, 252–263, 320–335
buffer occupancy for MMF source, 328–331
discrete-time queue, 322–326
flow control, 261–263
insensitivity of blocking probability, 331–335

Jackson network, 326–328
key queuing result, 254–255
M/M/1 queue, 320–322
overview, 252–253
queuing model, 253–254
routing optimization, 255–261
datagram packet switching, 62–64
and resource allocation, 75
data link connection identifiers (DLCIs), 122–123
data link layer, 92–96. *See also specific sublayers*
DQDB networks, 118–119
Ethernet networks, 104–109
FDDI networks, 114–117
ISDN networks, 180
logical link control (LLC) sublayer, 94–96
media access control (MAC) sublayer, 94, 95, 96
overview, 92–94
token ring networks, 107–109, 110–112
DBS. *See* distributed buffer switches (DBS)
DCS (digital cross-connect systems), 164–165
DD (depacketization delay) over ATM, 206–207
decoupling bandwidth, 287, 347–349
multiclass queues, 351–355
DE (discard eligibility) bit, 123
delay. *See also* congestion charges
of ATM cells, 204–208
of ATM messages, 394
average packet delay, 24–25
calculating for packets, 50–51
congestion cost, 386–387
delay-insensitive traffic, 381
in distributed buffer switches (DBS), 460, 462, 463–466
as function of transmission rate, 53
global multimedia network requirements, 487
packet-switched networks summary, 129
demand
aggregate traffic demand, 382–383
application demands on bearer services, 52–53
demand factors, 362
derived, 362
demultiplexing. *See also* multiplexing
defined, 56
in optical domains, 401
depacketization delay (DD) over ATM, 206–207
derived demand, 362
detectors for optical links. *See* receivers for optical links
deterministic ATM control
admission control, 270–274
ATM Forum recommendations, 266–270
burstiness, 338–339, 393, 394
leaky bucket, 336–338
linear bounds, 335–336
pricing calls, 274–275
statistical approach vs., 293–296
deviations, large. *See* large deviations
digital carrier systems, 7
digital cross-connect systems (DCS), 164–165
digital signal (DS) hierarchy, 7, 8, 163
digitization, 20–23
quantization, 21–23
sampling, 21, 22
Dijkstra's algorithm, 136–137
disaggregation of QoS, 489
discard eligibility (DE) bit, 123
discrete-time Markov chains, 302–308
aperiodicity, 307–308
irreducibility, 305–306
discrete-time queue, 322–326
Markov chain model, 322–323

discrete-time queue *(cont.)*
M/GI/∞ queue, 324–326
dispersion of fiber, 409–415
and bandwidth-distance product limits, 409–411
defined, 409
and fiber length, 411
graded-index (GRIN) fiber, 412–413
material dispersion, 413–415
maximum repeaterless distance, 415
modal dispersion, 412–414
multimode vs. single-mode fiber, 413–414
propagation modes, 412–414
step-index fiber, 411–412
distributed buffer switches (DBS), 453–466
Benes network vs., 453–454
blocking in, 456–457
combating hot spots, 462–466
crosspoint complexity, 456
hot spots and throughput reduction, 457–458
input buffers, 458–462
randomization stage for, 463–464
self-routing in, 455–456, 457
sorting for, 464–466
distributed-gradient algorithm, 259–261
Distributed Queue Dual Bus (DQDB) networks, 117–120
frame format, 118
MAC sublayer, 118–119
overview, 117–118, 129
topology, 117
DLCIs (data link connection identifiers), 122–123
DQDB networks. *See* Distributed Queue Dual Bus (DQDB) networks
DS-1 and DS-3 networks, 123–124. *See also* Switched Multimegabit Data Service (SMDS) networks
DS (digital signal) hierarchy, 7, 8, 163
dual ring SONET systems, 164–165
dynamic routing
circuit-switched networks, 249
datagram networks, 257–261

E

economics. *See also* charges; costs; revenue
demand factors, 362
Internet traffic measurements, 377–380
principles, 363–366
supply factors, 362–363
economies of scale, 23–24
economies of scope, 25–26
effective bandwidth
of arrival stream, 345–346
GCRA, 273–274
efficiency. *See also* performance
ABP protocol, 71–72
CSMA/CD protocol, 106–107
DQDB MAC protocol, 119
Ethernet networks, 106–107
Go Back *N* protocol, 72, 74, 121
release after reception MAC protocol, 111–112
spectral, 60
token ring networks, 111–112
electron-hole pairs, 417–418
elements. *See* network elements
encapsulation of PDUs, 227
encryption for global multimedia network, 485, 487, 495, 500
Erlang fixed-point approximation, 319
Erlang fixed-point equations, 319
Erlang loss formula, 315–316
error control, 65–74
cyclic redundancy check (CRC), 11, 65–67, 68

error correction, 67
error detection, 65–69
Frame Relay vs. X.25, 120–123
overview, 55, 76
protocols, 67–69
retransmission, 69–74
Ethernet networks
development of, 13
frame format, 105
LLC sublayer, 107–109
MAC sublayer, 104–107
overview, 101–102, 129
physical layer, 102–104
standards, 101
exponential distribution of continuous-time Markov chains, 308–309
extensibility, 481, 498–499
externalities, 24–25

F

fast packet switches (FPS), 450–453
distributed buffer switches (DBS), 453–466
input buffer switches, 470–475
output buffer switches, 468–470
shared buffer switches (SBS), 466–468
fault management over ATM, 219–222
FD (fixed processing delay) over ATM, 206
FDDI networks. *See* Fiber Distributed Data Interface (FDDI) networks
FDM. *See* frequency-division multiplexing (FDM)
FECN (forward explicit congestion notification) bit, 122–123
fiber, 404–415. *See also* attenuation of fiber; dispersion of fiber
attenuation, 404–409
bandwidth, 406
coaxial cable vs., 407–408
dispersion, 409–415
graded-index (GRIN) fiber, 412–413
maximum repeaterless distance, 415
maximum usable length, 406–407, 411
multimode vs. single-mode fiber, 413–414
optical amplifiers, 408
power increases vs. attenuation reduction, 408
propagation modes, 412–414
step-index fiber, 411–412
Fiber Distributed Data Interface (FDDI) networks, 112–117
advantages, 117
asynchronous transmission, 115–116
configurations, 113
development of, 15
frame format, 114
internetworking with SMDS, 126–129
MAC sublayer, 114–117
maximum length of fibers, 113
overview, 112–113, 129
physical layer, 113–114
synchronous transmission, 116–117
fiber-to-the-curb (FTTC) networks, 18, 190, 427
fiber to the home, 172–177
AT&T Subscriber Loop Carrier (SLC) system, 172–173
British Telecom TPON system, 174–176
hybrid schemes, 177–178
overview, 156
passive photonic loop (PPL), 176–177
File Transfer Protocol (FTP), 144, 145, 483

fixed charges, 364. *See also* charges
data network charges, 366
disadvantages, 375
by online services, 366
with usage charges, 375–376
fixed processing delay (FD) over ATM, 206
flip network, 454–455
flow control
ATM networks, 210, 213
datagram networks, 261–263
defined, 240, 261
generic flow control (GFC), 210, 213
open-loop flow control, 75
overview, 55, 74–75, 76
rate flow control, 75, 76, 263
usefulness of, 261–262
window flow control, 74–75, 76, 243, 262–263
fluid continuity, 349–350
decoupling fluid-continuous queues, 351–355
of FCFS queues, 350–351
foreign agents, 140
forward explicit congestion notification (FECN) bit, 122–123
FPS. *See* fast packet switches (FPS)
Frame Relay networks, 120–123
ATM packet transport, 232
error control in, 120–123
frame format, 122
overview, 120, 129
usage charges, 367
X.25 protocol modifications, 120, 123
frames
BISDN, 225–226
DQDB networks, 118
Ethernet networks, 105
FDDI networks, 114
Frame Relay networks, 122
SMDS networks, 124–125
SONET networks, 165–171
in time-division multiplexing, 56–57
token ring networks, 110
TPON system, 174–176
frequency-division multiplexing (FDM)
and CATV services, 29
disadvantages, 59
overview, 16–18, 59–60, 76
front-end FDDI network configuration, 113
FTP (File Transfer Protocol), 144, 145, 483
FTTC (fiber-to-the-curb) networks, 18, 190, 427
future networks. *See also* global multimedia network
global multimedia network, 479–500
pure ATM, 28–29, 30–31
the Internet, 26–28, 30
video dial tone, 29

G

gain, multiplexing, 58, 280, 355–356
gateway-centered mobile networks, 494
gateway-to-gateway protocol (GGP), 134
GCRA. *See* generalized cell rate algorithm (GCRA)
generalized cell rate algorithm (GCRA)
for ABR traffic, 269
and admission control, 270–274
ATM Forum recommendations, 266–270
for CBR traffic, 267–268
effective bandwidth, 273–274
as leaky bucket, 267
overview, 266–267

selecting a GCRA, 291–293
for UBR traffic, 270
for VBR traffic, 268–269
generic flow control (GFC), 210, 213
GFC (generic flow control), 210, 213
GGP (gateway-to-gateway protocol), 134
global multimedia network, 479–500
applications technologies, 483, 485–486
architecture, 483–484, 486–487
attributes, 480–482
audio and video compression algorithms, 481–482
challenges, 486–500
complexity management, 498–499
compression algorithm requirements, 481–482, 485, 490–491, 496–497
encryption, 485, 487, 495, 500
extensibility, 481, 498–499
heterogeneity, 480–481, 494–496
joint source/channel coding, 486–487, 497
mobility, 481, 491–494
networking technologies, 482, 484–485
quality of service, 481, 484, 487–491
scalability and configurability, 496–497
security and reliability, 481, 499–500
signal processing technologies, 482, 485
globalization trends, 2
Go Back *N* protocol, 72–74
disadvantages of, 120, 121
in X.25 networks, 120
graded-index (GRIN) fiber, 412–413

H

handoffs in mobile services, 492–493
HDSL (High Bit-Rate Digital Subscriber Line), 19
header error control (HEC), 210, 213–214
headers
ATM header structure, 210–214
IPv4 header, 131–132
IPv6 header, 141–142
of packets, 11
Transmission Control Protocol (TCP), 143
HEC (header error control), 210, 213–214
heterodyne detection, 426
heterogeneity, 480–481, 494–496
High Bit-Rate Digital Subscriber Line (HDSL), 19
high-performance networks (HPN), 47–48, 478
history of communication networks, 4–20
cable television networks, 16–20
computer networks, 10–16
telephone networks, 5–9
HOL blocking, 471–473
overcoming, 473–475
holding time, 61, 160
home IP address, 139
homodyne detection, 426, 430
horizontal handoffs, 493
hot spots
combating, 462–466
and throughput reduction, 457–458
HPN (high-performance networks), 47–48, 478
HyperText Markup Language (HTML), 483

I

ICMP (Internet Control Message Protocol), 134
IEEE 802.3. *See* Ethernet networks
IEEE 802.5. *See* token ring networks

IEEE 802.6. *See* Distributed Queue Dual Bus (DQDB) networks
ILMI (Intermediate Local Management Interface) protocol, 223–224
IN. *See* Intelligent Networks (IN)
INA. *See* Intelligent Network Architecture (INA)
information revolution, 1–3
input buffer switches, 470–475
  HOL blocking, 471–473
  look ahead, 473–474
  output expansion, 474–475
  overview, 470–471
insensitivity of blocking probability
  circuit-switched networks, 316
  datagram networks, 331–335
Institute of Electrical and Electronic Engineers
  IEEE 802.3. *See* Ethernet networks
  IEEE 802.5. *See* token ring networks
  IEEE 802.6. *See* Distributed Queue Dual Bus (DQDB) networks
Integrated Services Digital Network (ISDN), 178–180
  bearer services, 178
  Broadband ISDN (BISDN), 8–9
  evolution of, 8–9
  OSI layers, 180
  overview, 156–157
  telecommuting and ISDN costs, 51–52
Intelligent Network Architecture (INA), 183–185
  network elements, 184–185
  separation of control functions and network resources, 184
Intelligent Networks (IN), 181–186
  functional components, 185–186
  Intelligent Network Architecture (INA), 183–185
  overview, 157
  service examples, 181–183
intelligent peripherals (IPs), 184–185
interarrival time, 160
Intermediate Local Management Interface (ILMI) protocol, 223–224
Internet, 130–149
  addressing, 132–133
  billing systems, 368–374
  charge functions, 374–377
  charges, 366–367, 379–380
  development of, 12, 27, 130
  faster transfer protocols, 146
  File Transfer Protocol (FTP), 144, 145
  future of, 27–28, 30
  gateway-to-gateway protocol (GGP), 134
  growth of, 12, 130
  Internet Control Message Protocol (ICMP), 134
  Internet Protocol (IP), 12–13, 131–143
  limitations, 148–149
  ODN implementation, 147
  overview, 26–27, 130–132
  protocol hierarchy, 130
  real-time protocols, 145–146
  remote login service (rlogin), 144–145
  routing, 133–136
  Simple Mail Transfer Protocol (SMTP), 144
  success of, 146–148
  TCP/IP shortcomings, 145–146
  traffic measurements, 377–380
  traffic shaping, 379–380
  Transmission Control Protocol (TCP), 143–144
  Trivial File Transfer Protocol (TFTP), 145
  User Datagram Protocol (UDP), 143

vulnerability to abuse, 367–368
Internet Control Message Protocol (ICMP), 134
Internet Protocol (IP), 131–143
addressing, 132–133
Address Resolution Protocol (ARP), 13
advantages of, 12–13
Bellman-Ford algorithm, 134–136
Border Gateway Protocol (BGP), 137–138
classical model, 229–230
Dijkstra's algorithm, 136–137
gateway-to-gateway protocol (GGP), 134
integrated model, 231
Internet Control Message Protocol (ICMP), 134
IPv4, 131–138
IPv4 header, 131–132
IPv6, 140–142
IPv6 header, 141–142
Mobile IP, 139–140
multicast IP, 138–139
multicast IP over ATM, 231
over ATM, 228–231
Resource Reservation Protocol (RSVP), 142–143
segmentation/reassembly, 133, 134
shortcut models, 230–231
tunneling, 140–141
internetworking
with ATM, 226–232
and bearer service complexity, 85
with SMDS, 126–129
intrinsic semiconductors, 416
invariant distribution
in circuit-switched networks, 314–315, 317–318
of continuous-time Markov chains, 311–312
in Jackson networks, 327–328
in M/M/1 queue, 320
of on-off MMF source, 331, 332
invariant probability distribution, 304
IP. *See* Internet Protocol (IP)
IPs (intelligent peripherals), 184–185
irreducibility
continuous-time Markov chains, 311
discrete-time Markov chains, 305–308
ISDN. *See* Integrated Services Digital Network (ISDN)
isolators in optical links, 432

J

Jackson network, 326–328
jitter, 206–207
joint source/channel coding, 486–487, 497

K

knockout switch, 469–470

L

LAN emulation with ATM, 227–228
large deviations
of iid random variables, 339–343
of multiclass queues, 349–355
of queues, 344–349
straight-line, 343
laser diodes (LDs), 403, 419–420
bandgap energy, 416
mirrors in, 419
mode partition noise (MPN), 419–420
modulation, 419–420
semiconductor materials, 416
stimulated emission, 419
wavelengths of emitted photons, 416–417

layered architecture, 77–82. *See also* Open Data Network (ODN) model; Open Systems Interconnection (OSI) model
advantages, 78
implementation, 78–82
layers, 7–8
ODN model, 82–85
OSI model, 13, 91–101
SONET, 166–167
layer management plane of ATM, 218, 219
LDs. *See* laser diodes (LDs)
leaky-bucket scheme, 126
ATM control, 336–339
GCRA as, 267
LEDs. *See* light-emitting diodes (LEDs)
light-emitting diodes (LEDs), 403, 416–419
bandgap energy, 416, 417
and electron-hole pairs, 417–418
modulation, 418–419
semiconductor materials, 416
spontaneous radiative emission, 416
wavelengths of emitted photons, 416–417
light sources for optical links. *See* transmitters for optical links
linear bounds, 335–336
line layer, SONET, 166–167
line overhead (LOH), 167, 168, 169, 225
lines, 166
links. *See also* optical links
defined, 92
overview, 48–49
Little's result, 321–322
LLC sublayer. *See* logical link control (LLC) sublayer
local area network (LAN) emulation with ATM, 227–228
local loop, 54, 155
local syntax, 100
logical link control (LLC) sublayer
Ethernet networks, 107–109
OSI model, 94–96
token ring networks, 107–109, 112
LOH (line overhead), 167, 168, 169, 225
loss
cell loss priority (CLP), 210, 213
cell loss ratio (CLR), 201
Erlang loss formula, 315–316
fiber couplers and power loss, 407
loss rate analysis of buffering, 284–287
low-loss windows in fiber, 406
network model of loss system, 331, 333

M

MAC sublayer. *See* media access control (MAC) sublayer
management information base (MIB), 223–224
Markov chains, 301–312
circuit switch model, 313–314
continuous time, 308–312
defined, 301
discrete time, 302–308
discrete-time queue, 322–323
Jackson network, 326–328
and metastability, 250–251
M/M/1 queue, 320–321
overview, 301–302
single switch circuit-switched networks, 313–314
Markov-modulated fluids (MMF), 277–278
and buffering, 287–290
buffering occupancy for MMF source, 328–331

invariant distribution of on-off source, 331, 332
Markov property, 310–311
MARS (Multicast Address Resolution Server), 231
material dispersion of fiber, 413–415
material transformation, 1
maximum cell transfer delay (Max CTD), 201
MBone, 483
MCR (minimum cell rate), 200, 201
mean cell transfer delay (Mean CTD), 201
media access control (MAC) sublayer
DQDB networks, 118–119
Ethernet networks, 104–107
FDDI networks, 114–117
OSI model, 94, 95, 96
release after reception MAC protocol, 111
release after transmission MAC protocol, 111
messages
agent advertisement messages, 140
agent solicitation messages, 140
applications, 41–42
ATM messages, 392–394
defined, 41
message passing, 79–80
OSI transport layer functions, 98–99
path messages, 142
reservation messages, 142
metastability, 250–251
M/GI/∞ queue, 324–326
quasi reversibility of, 326
MIB (management information base), 223–224
microwave receivers' BER, 407
microwave satellite transmission, 19
middleware layer, 83–84
minimum cell rate (MCR), 200, 201
Minitel, 30
M/M/1 queue
Jackson network, 326–328
Markov chain model, 320–321
quasi reversibility of, 321–322
MMF. *See* Markov-modulated fluids (MMF)
mobile handoffs, 492–493
Mobile IP, 139–140
mobility, 481, 491–494
modal dispersion of fiber, 412–414
modems, 10
mode partition noise (MPN), 419–420
modes, FTP, 144
modes of propagation in fiber, 412–414
modulation in optical links
coherent detection, 425–426
heterodyne detection, 426
homodyne detection, 426, 430
of LDs, 419–420
of LEDs, 418–419
nondirect, 425–429
subcarrier multiplexing, 426–427
wave-division multiplexing (WDM), 427–429
monitoring ATM network status, 223–224
Moving Pictures Expert Group (MPEG)
MPEG1, 41
MPEG2, 41
TV codec, 18
video compression standards, 191–192
MPN (mode partition noise), 419–420
MPOA (multiprotocol over ATM), 231–232
Multicast Address Resolution Server (MARS), 231
multicast IP, 138–139
over ATM, 231
multicast routers, 138–139
multicast switching, 441, 468

multiclass queues, 290–291
large deviations of, 349–355
multimedia network. *See* global multimedia network
multimode fiber vs. single-mode fiber, 413–414
multiplexing, 56–60. *See also* Asynchronous Transfer Mode (ATM)
add/drop (ADM), 163
in AT&T SLC system, 172, 173
ATM control, 280–281, 287–290
averaging rate fluctuations, 278–280
defined, 56
demultiplexing, 56, 401
frequency-division (FDM), 16–18, 59–60, 76
gain, 58, 280, 355–356
in optical domains, 401
overview, 54–55, 76
statistical (SM), 11, 57–59, 76, 208, 280–281, 287–289
subcarrier, 426–427
time-division (TDM), 56–57, 76
wave-division (WDM), 176–177, 427–429
multipoint switches, 439, 441
modular construction, 449–450
multiprotocol over ATM (MPOA), 231–232

N

NBMA (nonbroadcast multiaccess) link layer, 230–231
network access providers (NAPs), 27
network address, 132–133
network elements, 48–53
examples, 51–53
principal elements, 48–50
queuing network model, 49–50
and service characteristics, 50–51
network externalities, 24–25
network information revision request, 186
networking principles
digitization, 20–23
economies of scale, 23–24
network externalities, 24–25
service integration, 25–26
networking technologies, 482, 484–485
network layer, 96–97
and bearer service layer, 91
Internet Protocol (IP), 12–13, 131–143
ISDN networks, 180
X.25 networks, 120
network resource status request, 186
network resources in INA, 184
network services
application demands, 52–53
applications, 39–45
ATM services model, 390–394
bearer services, 39, 52–53, 85, 178
charges, 366–367
connectionless, 46
connection-oriented, 45–46
delay, 50–53
network elements and service characteristics, 50–51
overview, 37–39
noise
in optical link receivers, 422–424
quantization, 22
signal-to-noise ratio (SNR), 22–23, 422
nonblocking switches, 444–450
Benes network construction, 447–449
Clos network theory, 445–446
modular construction of, 446–449
multipoint switches, 449–450
nonbroadcast multiaccess (NBMA) link layer, 230–231
nonconformant cells, 267

noncooperative sites' billing system, 373–374
nondirect modulation, 425–429
n-type semiconductors, 416
Nyquist's theorem, 22

O

OAM cells, 220–222
ODN model. *See* Open Data Network (ODN) model
omega network, 454, 455–456
Open Data Network (ODN) model, 82–85
  application layer, 84–85
  bearer layer, 83
  bit way layer, 83
  cooperation among layers, 85
  Internet implementation, 147
  middleware layer, 83–84
open-loop flow control, 75
Open Systems Interconnection (OSI) model, 91–101
  development of, 13, 91
  DQDB networks, 118–119
  Ethernet networks, 102–109
  FDDI networks, 113–117
  ISDN networks, 180
  layer 1: physical, 92, 93
  layer 2: data link, 92–96
  layer 3: network, 91, 96–97
  layer 4: transport, 98–99
  layer 5: session, 99
  layer 6: presentation, 99–100
  layer 7: application, 100
  overview, 99, 100–101
  token ring networks, 107–109, 110–112
optical amplifiers, 408, 433
optical links, 401–435. *See also* fiber; receivers for optical links; transmitters for optical links
  attenuation of fiber, 404–409
  attenuators, 432
  bandwidth, 406
  bit error rate (BER), 402–403
  coaxial cable vs. fiber, 407–408
  couplers, 407, 432
  dispersion of fiber, 409–415
  isolators, 432
  maximum repeaterless distance, 415
  maximum usable fiber length, 406–407, 411
  modeling as communication systems, 402–403
  nondirect modulation, 425–429
  optical amplifiers, 408, 433
  overview, 401–404
  power budget, 407
  power budget calculation, 432–433
  power increases vs. attenuation reduction, 408
  propagation modes, 412–414
  receivers (detectors), 404, 406, 421–424
  regeneration, 433
  required receiver power, 408–409
  splices, 432
  transmitters (light sources), 403, 406, 415–420
optical local loop. *See* fiber to the home
optical networks. *See* Fiber Distributed Data Interface (FDDI); fiber to the home; optical links; Synchronous Optical Network (SONET)
OSI model. *See* Open Systems Interconnection (OSI) model
output buffer switches, 468–470
  knockout switch, 469–470
  overview, 468–469

P

packetization delay (PD) over ATM, 204, 205

packets
ATM packet stream types, 214–217
delay calculation, 50–51
Ethernet packet format, 105
overview, 10–11
packet-switched networks. *See also* Open Systems Interconnection (OSI) model; *specific types of networks*
datagram network control, 252–263, 320–325
buffer occupancy for an MMF source, 328–331
discrete-time queue, 322–326
flow control, 261–263
insensitivity of blocking probability, 331–335
Jackson network, 326–328
key queuing result, 254–255
M/M/1 queue, 320–322
overview, 252–253
queuing model, 253–254
routing optimization, 255–261
DQDB networks, 117–120
Ethernet networks, 101–109
FDDI networks, 112–117
Frame Relay networks, 120–123
Internet, 130–149
OSI model, 91–101
overview, 129
SMDS networks, 123–129
token ring networks, 109–112
packet switching. *See also* virtual circuit switching
circuit switching compared to, 63–64, 65, 450–451
datagram, 62–64
distributed buffer switches (DBS), 453–466
fast packet switching, 450–453
input buffer switches, 470–475
output buffer switches, 468–470
overview, 62–63, 76
shared buffer switches (SBS), 466–468
store-and-forward, 11–12, 62
passive photonic loop (PPL), 176–177
path layer, SONET, 166–167
path messages, 142
path overhead (POH), 167, 170, 225
paths
defined, 166
in state transition diagrams of Markov chains, 304
payload type (PT), 210, 213
PCR. *See* peak cell rate (PCR)
PD (packetization delay) over ATM, 204, 205
PDUs (protocol data units), 124–126, 227
peak cell rate (PCR)
ATM service characteristics, 200, 201
defined, 200
GCRA algorithms, 267–269
performance. *See also* efficiency; rates
of circuit-switched networks, 157–161
and queuing network model, 49–50
switching performance measures, 438–441
Permanent Virtual Circuits (PVCs), 122–123
photonic layer, SONET, 166–167
physical layer
ATM fault management, 219–222
BISDN, 225–226
Ethernet networks, 102–104
FDDI networks, 113–114
ISDN networks, 180
OSI model, 92, 93
token ring networks, 110
physical medium dependent (PMD) sublayer, 113–114
plain old telephone service (POTS), 181–182

plane management over ATM, 218–219
PMD (physical medium dependent) sublayer, 113–114
POH (path overhead), 167, 170, 225
ports, 98
postal system, 37–38
POTS (plain old telephone service), 181–182
power budget calculation in optical links, 432–433
PPL (passive photonic loop), 176–177
preamble, 92
predicting quality of service, 488–489
presentation layer, 99–100
pricing. *See* charges; costs
print news industry revenue, 4
propagation delay over ATM, 205
propagation modes of fiber, 412–414
protocol data units (PDUs), 124–126, 227
protocols. *See also* efficiency
  Address Resolution Protocol (ARP), 13
  Alternating Bit protocol (ABP), 69–72
  Border Gateway Protocol (BGP), 137–138
  CSMA/CD, 105–107
  DQDB MAC protocol, 118–119
  error detection, 67–69
  faster Internet transfer protocols, 146
  File Transfer Protocol (FTP), 144, 145
  gateway-to-gateway protocol (GGP), 134
  Go Back *N*, 72–74
  Intermediate Local Management Interface (ILMI), 223–224
  and Internet billing systems, 368–369
  Internet Control Message Protocol (ICMP), 134
  Internet protocol hierarchy, 130
  Internet Protocol (IP), 12–13, 131–143, 228–231
  and layers, 77–82
  multiprotocol over ATM (MPOA), 231–232
  real-time Internet protocols, 145–146
  Resource Reservation Protocol (RSVP), 142–143
  Selective Repeat Protocol (SRP), 74
  Simple Mail Transfer Protocol (SMTP), 144
  SMT, 113, 114
  timed-token, 114–116
  Transmission Control Protocol (TCP), 143–144
  Trivial File Transfer Protocol (TFTP), 145
  User Datagram Protocol (UDP), 143
PS. *See* packet switching
PT (payload type), 210, 213
p-type semiconductors, 416
pulse spread. *See* dispersion of fiber
PVCs (Permanent Virtual Circuits), 122–123

Q

QAM (quadrature amplitude modulation), 190
QD. *See* queuing delay (QD)
QoS. *See* quality of service (QoS)
QPSK (quadrature phase-shift keying), 190
quadrature amplitude modulation (QAM), 190
quadrature phase-shift keying (QPSK), 190
quality charges, 365–366. *See also* charges
  Internet billing system, 376–377
  in practice, 367

quality of service (QoS). *See also* delay; resource allocation
- aggregation of, 489
- ATM attributes, 201
- ATM management, 217–218
- best-effort service, 244
- for connectionless service, 46
- for connection-oriented service, 46
- control problems, 264–266
- corruption, 487
- disaggregation of, 489
- and global multimedia network, 481, 484, 487–491
- overview, 244–246
- predicting, 488–489
- and signal processing technologies, 485
- SMDS networks, 125–126

quantization, 21–23
quantization intervals, 22
quantization noise, 22
quantum limits on detection, 429–432
quasi reversibility
- of M/GI/∞ queue, 326
- of M/M/1 queue, 321–322

queues
- decoupling fluid-continuous queues, 351–355
- fluid continuity of FCFS queues, 350–351
- large deviations of, 344–355
- for message passing, 80
- multiclass, 290–291, 349–355

queuing delay (QD). *See also* delay
- over ATM, 205–206
- overview, 50–51

queuing model, 253–254
queuing network model, 49–50

R

rate flow control
- datagram networks, 263
- overview, 75, 76

rate matrix of continuous-time Markov chains, 309
rates. *See also* performance
- 10BASE2 transmission, 102
- 10BASE5 transmission, 102, 103
- 10BASE-T transmission, 103, 104
- ADSL, 19
- applications, 43, 44
- ATM transmission, 15
- averaging rate fluctuations, 278–280
- BISDN, 9
- bit rate vs. bandwidth, 60
- CATV transmission, 18
- of data networks (1970–1995), 16–17
- DS hierarchy, 7
- Ethernet transmission, 13, 102, 103, 104, 129
- FDDI transmission, 15, 112, 117
- Frame Relay transmission, 120, 129
- HDSL, 19
- HPNs, 47–48
- Internet traffic, 26, 27–28
- ISDN, 178–179, 180
- packet-switched networks, 129
- RS-232-C standard, 10
- SLC systems, 173
- SMDS transmission, 123–124, 129
- SONET, 156
- STS hierarchy, 7–8, 161–163
- subscriber loop, 155
- token ring transmission, 14, 109, 110, 129
- TPON systems, 174, 175

Rayleigh scattering, 405–406
rearrangeably nonblocking (RNB) switches
- Benes network construction, 447–449
- Clos networks built from, 446

defined, 445
modular construction of, 447, 448
reassembly, IP, 133, 134
receivers for optical links, 404
additive effect of noise, 422
bit error rate (BER), 404, 422
coherent detection, 425–426
components, 421
dark current, 422, 424, 430, 431–432
heterodyne detection, 426
homodyne detection, 426, 430
maximum fiber length, 406–407
nondirect modulation, 425–429
quantum limits on detection, 429–432
required power, 408–409
sensitivity, 406–407, 422
shot noise, 422–424, 430–432
signal power, 422
signal to noise ratio (SNR), 422
subcarrier multiplexing, 426–427
thermal noise, 422, 424
wave-division multiplexing (WDM), 427–429
Reed-Solomon (RS) codes, 67
regenerators, 165–166, 433
release after reception MAC protocol, 111
release after transmission MAC protocol, 111
reliability
of applications, 45
global multimedia network, 481
remote login service (rlogin), 144–145
Request for Comment (RFC), 150
reservation messages, 142
reservation price, 387
reserving bandwidth for ATM users, 390, 391, 394, 396
resource allocation. *See also* quality of service (QoS)
ATM, 209–210
overview, 55–56, 75, 76, 241
Resource Reservation Protocol (RSVP), 142–143
retransmission
Alternating Bit protocol (ABP), 69–72
Go Back *N* protocol, 72–74
Selective Repeat Protocol (SRP), 74
revenue. *See also* charges; costs; economics
communications industry, by sector, 4
maximizing ATM revenue, 395–396
and static routing, 249
reversing the arrow, 327–328
reversing time of continuous-time Markov chains, 312
RFC (Request for Comment), 150
rlogin (remote login service), 144–145
RNB switches. *See* rearrangeably nonblocking (RNB) switches
route, 60
routers. *See also* switches
internetworking with SMDS, 127–129
multicast, 138–139
in OSI model, 97
output buffer switch, 468–470
routing
ATM procedure, 201–203
Bellman-Ford algorithm, 134–136, 258–259
circuit-switched networks, 248–252
datagram networks, 255–261
Dijkstra's algorithm, 136–137
distributed-gradient algorithm, 259–261
dynamic, 249, 257–261
Internet, 133–136
metastability and trunk reservations, 250–251
overview, 96–97, 240
separable, 251–252

routing *(cont.)*
spanning tree, 109
static, 248–249, 255–257
telephone network algorithms, 158
transparent, 108–109
RS-232-C standard, 10
RS (Reed-Solomon) codes, 67
RSVP. *See* Resource Reservation Protocol (RSVP)

S

sampling, 21, 22
SAR. *See* segmentation and reassembly sublayer (SAR)
satellite transmission, 19
SBS. *See* shared buffer switches (SBS)
scalability, 496–497
scale economies, 23–24
scope economies, 25–26
SCPs (service control points), 184–185
SCR. *See* sustained cell rate (SCR)
SDH. *See* Synchronous Digital Hierarchy (SDH)
SDLC (Synchronous Data Link Control), 10–11
SDUs (service data units), 214–215
section layer, SONET, 166–167
section overhead (SOH), 167, 168–169, 225
sections, 166
security
of applications, 45
global multimedia network, 481, 499–500
segmentation and reassembly sublayer (SAR)
overview, 214, 215
and Type 1 traffic, 216
and Type 2 traffic, 216
and Type 3 traffic, 216–217
and Type 4 traffic, 217
and Type 5 traffic, 217
segmentation, IP, 133, 134
Selective Repeat Protocol (SRP), 74, 243
semiconductors for fiber light sources, 416
separable routing, 251–252
serial transmission, 10
service control points (SCPs), 184–185
service data units (SDUs), 214–215
service integration, 25–26
service switching points (SSPs), 184–185
services. *See* bearer services; network services
session layer, 99
setup time, 158
shared buffer switches (SBS), 466–468
advantages and disadvantages, 466–467
overview, 466
queuing analysis, 467–468
shared memory for message passing, 79–80
shortcut IP models, 230–231
shortest path algorithms
Bellman-Ford, 134–136, 258–259
Dijkstra's (shortest-path first), 136–137
shot noise in optical link receivers, 422–424, 430–432
SIEs (small information enterprises), 373–374
signaling functions of ATM, 224
signal processing technologies, 482, 485
signal-to-noise ratio (SNR), 22–23, 422
Simple Mail Transfer Protocol (SMTP), 144
single-mode fiber vs. multimode fiber, 413–414
single resource pricing, 380–389
congestion prices, 385–387

cost recovery and optimum link capacity, 387–389
overview, 380–381
usage-based prices, 381–385
single-site billing system, 369–372
SLC system. *See* AT&T Subscriber Loop Carrier (SLC) system
slots in TDM, 56–57
SM. *See* statistical multiplexing (SM)
small information enterprises (SIEs), 373–374
SMDS networks. *See* Switched Multimegabit Data Service (SMDS) networks
SMT (station management) protocols, 113, 114
SMTP (Simple Mail Transfer Protocol), 144
SNB switches. *See* strictly nonblocking (SNB) switches
SNR (signal-to-noise ratio), 22–23, 422
social welfare optimum, 384–385
SOH (section overhead), 167, 168–169, 225
SONET. *See* Synchronous Optical Network (SONET)
space-division switching, 443
spanning tree routing, 109
SPE (synchronous payload envelope), 167–168
spectral efficiency, 60
speed. *See* performance; rates
splices in optical links, 432
spontaneous radiative emission, 416
SRP (Selective Repeat Protocol), 74, 243
SSPs (service switching points), 184–185
state space, 302
state transition diagrams of Markov chains, 303–304
static routing
circuit-switched networks, 248–249
datagram networks, 255–257
station management (SMT) protocols, 113, 114
statistical ATM control
averaging rate fluctuations, 278–280
buffering, 284–287
deterministic approach vs., 293–296
GCRA selection, 291–293
general traffic model, 278
large deviation of a multiclass queue, 349–355
large deviation of a queue, 344–349
large deviations of iid random variables, 339–343
Markov-modulated fluids, 277–278
multiclass case, 290–291
multiplexing, 280–281
on-line estimation, 282–284
statistical multiplexing and buffering, 287–290
straight-line large deviations, 343
traffic models, 277–280
traffic shaping, 293
statistical multiplexing (SM)
ATM, 208
ATM control, 280–281, 287–290
and buffering, 287–290
gain, 355–356
multiplexing gain, 58
overview, 11, 57–59, 76
step-index fiber, 411–412
stimulated emission, 419
store-and-forward packet switching, 11–12, 62
store-and-forward virtual circuit switching, 243
straight-line large deviations, 343
stream mode (FTP), 144
strictly nonblocking (SNB) switches
Clos networks built from, 445
defined, 445

strictly nonblocking (SNB) switches *(cont.)*
modular construction of, 446–447
STS. *See* Synchronous Transfer Signal (STS)
subcarrier multiplexing, 426–427
subscriber loop, 54, 155
Subscriber Loop Carrier (SLC) system, 172–173
supply factors, 362
sustained cell rate (SCR)
ATM service characteristics, 200, 201
defined, 200
GCRA algorithm for VBR traffic, 268–269
switch configuration, 440
Switched Multimegabit Data Service (SMDS) networks, 123–129
frame structure, 124–125
internetworking with, 126–129
leaky-bucket scheme, 126
over ATM, 232
overview, 123–124, 129
protocol data units (PDUs), 124–126
protocol stack, 125
service-quality levels, 125–126
usage charges, 367
switches. *See also* routers
ATM switches, 199–200
Benes network construction, 447–449
as "black box," 437, 438, 439
Clos network theory, 445–446
configurations possible, 440
crossbar switches, 439–440, 443
distributed buffer switches (DBS), 453–466
input buffer switches, 470–475
modular construction of, 446–450
multipoint switches, 439, 441, 449–450
nonblocking, 444–450
output buffer switches, 468–470
overview, 49, 60–61
shared buffer switches (SBS), 466–468
space-division, 443
time-division, 442–443
switching, 60–65, 437–478. *See also* circuit switching; packet switching; virtual circuit switching
Benes network, 447–449
"black box" view, 437, 438, 439
Clos networks, 444–450
complexity measures, 439–441
function of, 437, 438
in optical domains, 401
overview, 54–55, 76, 437
performance measures, 438–441
space-division, 443
throughput, 441
time-division, 442–443
synchronization bits, 92, 93
synchronization points, 99
Synchronous Data Link Control (SDLC), 10–11
Synchronous Digital Hierarchy (SDH), 7–8, 161–163
DS hierarchy vs., 8, 163
Synchronous Optical Network (SONET), 161–171. *See also* Asynchronous Transfer Mode (ATM)
add/drop multiplexer (ADM), 163
dual ring systems, 164–165
frame structure, 165–171
layered overhead structure, 166–167
line overhead (LOH), 167, 168, 169
overview, 156, 161–165
path overhead (POH), 167, 170
section overhead (SOH), 167, 168–169

STS hierarchy, 7–8, 161–163
synchronous payload envelope (SPE), 167–168
time-division multiplexing, 161, 162
synchronous payload envelope (SPE), 167–168
Synchronous Transfer Signal (STS)
DS hierarchy compared to, 8, 163
hierarchy, 7–8, 161–163
synchronous transmission
over FDDI networks, 116–117
standards, 10–11
syntax, 100
System Network Architecture (SNA), 243

T

target token rotation time (TTRT), 115–116
tat (theoretical arrival time), 267
TCP (Transmission Control Protocol), 143–144, 243
TCP/IP. *See* Internet; protocols
TD (transmission delay) over ATM, 205
TDM. *See* time-division multiplexing (TDM)
telecommuting, 51–52
telephone industry revenue, 4
telephone networks. *See also* specific types of networks
key innovations, 5–9, 20
need for redundancy, 163–164
pricing schemes, 366
Telephony on a Passive Optical Network (TPON) system, 174–176
10BASE2 (thin Ethernet) networks, 102
10BASE5 (thick Ethernet) networks, 102, 103
10BASE-T (twisted pair Ethernet) networks, 102–103, 104
terminal points, 166
TFTP (Trivial File Transfer Protocol), 145
theoretical arrival time (tat), 267
thermal noise in optical link receivers, 422, 424
thick Ethernet (10BASE5) networks, 102, 103
thin Ethernet (10BASE2) networks, 102
throughput
and DBS hot spots, 457–458
and HOL blocking, 473
of switches, 441
THT. *See* token holding time (THT)
time-division multiplexing (TDM)
overview, 56–57, 76
in SONET, 161, 162
in time-division switching, 442–443
time-division switching, 442–443
timed-token protocol, 114–116
time scales and control actions, 241–242
token holding time (THT)
FDDI networks, 115–116
token ring networks, 111
token ring networks
advantages of, 109–110
development of, 14
disadvantages of, 14–15
frame format, 110
LLC sublayer, 107–109, 112
MAC sublayer, 110–112
overview, 109–110, 129
physical layer, 110
token rotation time (TRT), 115–116
topology
10BASE5 networks, 103
10BASE-T networks, 104
DQDB networks, 117
FDDI networks, 113

topology *(cont.)*
SLC system, 173
token ring networks, 110
TPON system, 174
TPON system, 174–176
traffic measurements on the Internet, 377–380
congestion pricing and traffic shaping, 379–380
connection statistics, 378
diversity of usage, 378–379
traffic shaping
ATM networks, 223, 293
on the Internet, 379–380
overview, 240–241
transfer syntax, 100
transition probability matrix, 303–308
aperiodicity, 307–308
balance equations, 305
invariant probability distribution, 304
irreducible, 305–306
Transmission Control Protocol (TCP), 143–144, 243
transmission delay (TD) over ATM, 205
transmission rates. *See* rates
transmitters for optical links, 403, 415–420. *See also* laser diodes (LDs); light-emitting diodes (LEDs)
bandgap energy, 416, 417
conduction band, 416
factors affecting usefulness of, 416
laser diodes (LDs), 403, 419–420
light-emitting diodes (LEDs), 403, 416–419
semiconductor materials, 416
and spectrum of light source, 418
transmitted power and maximum fiber length, 406–407
valence band, 416
wavelengths of emitted photons, 416–417
transparent routing, 108–109
transport layer
of Internet, 143–144
OSI model, 98–99
transport service access points (TSAPs), 98
trigger table, 182
Trivial File Transfer Protocol (TFTP), 145
TRT (token rotation time), 115–116
trunk, 54
trunk reservation, 251
TSAPs (transport service access points), 98
TTRT (target token rotation time), 115–116
TV codec, 18
twisted pair Ethernet (10BASE-T) networks, 102–103, 104
Type 1 ATM traffic, 215–216
Type 2 ATM traffic, 216
Type 3 ATM traffic, 216–217
Type 4 ATM traffic, 217
Type 5 ATM traffic, 217

U

UBR. *See* unspecified bit rate (UBR)
UDP (User Datagram Protocol), 143
unspecified bit rate (UBR)
ATM service characteristics, 200, 201
and quality of service, 270
usage charges, 364. *See also* charges
congestion charges vs., 365
with fixed charges, 375–376
for network services, 367
pricing a single resource, 381–385
social welfare optimum, 384–385
utility function, 381–382
User Datagram Protocol (UDP), 143

user interaction request, 186
user plane of ATM, 218, 219
utility function, 381–382
utilization, 56, 159

V

valence band, 416
variable bit rate-non-real time (VBR-NRT), 200, 201
variable bit rate-real time (VBR-RT), 200, 201
variable bit rate (VBR)
  examples, 42
  GCRA for, 268–269
  overview, 41
VC. *See* virtual circuit switching
VCI. *See* virtual channel identifier (VCI)
vertical handoffs, 493
vestigial side band (VSB) modulation, 190
video compression
  global multimedia network requirements, 481–482, 485, 490–491, 496–497
  scalability and configurability, 496–497
  standards, 191–192
video dial tone, 186–192. *See also* cable television networks
  control and video networks, 188–191
  future of, 29
  MPEG standards, 191–192
  overview, 186–187
  physical layout, 187–188
video on demand, 18–19
virtual channel identifier (VCI)
  ATM header structure, 210, 211–212
  fast packet switching, 451–453
  overview, 199
  reserved combinations, 214
  and resource allocation, 209–210
  and routing, 208
virtual channels, 199, 208
virtual circuit switching. *See also* Asynchronous Transfer Mode (ATM)
  fast packet switching, 451–453
  overview, 63–65, 76, 198, 199
  and resource allocation, 75
  store-and-forward networks, 243
virtual path identifier (VPI)
  ATM header structure, 210, 212–213
  reserved combinations, 214
VMTP, 146
VPI. *See* virtual path identifier (VPI)
VSB (vestigial side band) modulation, 190

W

wave-division multiplexing (WDM), 176–177, 427–429
window flow control
  datagram networks, 262–263
  link-level, 243
  overview, 74–75, 76
  TCP, 243
wireless Ethernet networks, 103–104
wireless transmission, 19
World Wide Web (WWW), 27

X

X.25 protocols, 120, 123
XTP, 146

Related Titles from Morgan Kaufmann

Computer Networks: A Systems Approach
Larry L. Peterson and Bruce S. Davie
1996; 552 pages; cloth; ISBN 1-55860-368-9

A systems-oriented view of computer network design that goes beyond current technology to help readers grasp the underlying concepts and build a foundation for making sound network design decisions. By providing an understanding of the components of a network and a feel for how these components fit together, this book empowers readers to design real networks that are both efficient and elegant. It emphasizes network software that transforms raw hardware into richly functional, high-performance network systems.

Introduction to Data Compression
Khalid Sayood
1995; 475 pages; cloth; ISBN 1-55860-346-8

The fundamental theories and techniques of data compression, with the most complete coverage available of both lossy and lossless methods. Sayood explains the theoretical underpinnings of the algorithms so that readers learn how to model structures in data and design compression packages of their own. Practitioners, researchers, and students will benefit from the balanced presentation of theoretical material and implementations.

Web Server Technology: The Advanced Guide for World Wide Web Information Providers
Nancy J. Yeager and Robert E. McGrath
1996; 407 pages; paper; ISBN 1-55860-376-X

The success of the Web depends not only on the creation of stimulating and valuable information, but also on the speed, efficiency and convenient delivery of this information to the Web consumer. This authoritative presentation of Web server technology takes you beyond the basics to provide the underlying principles and technical details of how WWW servers really work. It explains current technology and suggests enhanced and expanded methods for disseminating information via the Web.